Basic Complex Analysis

Basic Complex Analysis

Jerrold E. Marsden
University of California, Berkeley

with the assistance of

Michael Buchner   Michael Hoffman   Clifford Risk

W. H. Freeman and Company San Francisco

cover:

From a photograph by Gary Koopmann
at the U.S. Naval Research Laboratory, Washington, D.C.

AMS 1970 subject classifications:
30-01; 81A48, 30A84, 44A10.

Library of Congress Cataloging in Publication Data

Marsden, Jerrold E.
Basic complex analysis.
    1. Functions of complex variables.   I.   Title.
QA331.M378   515′.9   72–89894
ISBN 0–7167–0451–X

8 9 MP 1 0 8 9 8 7 6 5 4 3

to Glynis and Chris

Contents

Preface

This book is intended for undergraduates in mathematics, the physical sciences, or engineering who are taking a course in complex analysis or complex variables for the first time. Two years of calculus are adequate preparation for the course. The text contains some references to linear algebra and basic ϵ-δ analysis, but such subjects are easily avoided if that background is lacking.

In this text the student will find abundant motivation, examples, problems, and applications. Each section corresponds to between one and two lectures and each includes examples and problems to help the student master the material as it is presented. Adequate motivation has been supplied to help the student gain an intuitive understanding of the subject. For example, Section 2.2 presents a version of Cauchy's Theorem that is based on Green's Theorem, so that the main content of Cauchy's Theorem quickly becomes apparent. Section 2.3 then supplies a more careful proof of Cauchy's Theorem using the "modern" bisection method and homotopy of curves (both treated in an elementary way). Such a method enables the applications-oriented student to skip Section 2.3 and spend time on applications of complex variable theory. These applications are varied. They include Laplace's Equation in the context of electric potentials, heat conduction, and hydrodynamics; the Laplace transform; and scattering theory.

Whenever technical proofs interfere with the explanation of basic ideas, the proofs have been placed at the end of the section. Experience gained from teaching mathematics majors has shown that Chapters 1–4, Chapter 5 (Sections 1 and 2 only), and Chapter 6 can be covered in

a half-year course. When the course was taught to engineers and physicists, Chapters 1–5 (omitting Section 2.3), Chapter 7, and parts of Chapter 8 could be covered in the same period. If time permits, presentation
of the applications concurrently with the theoretical material enables
the mathematics major to see the historical origins of his subject.

The symbols used in this text are standard except possibly for the
following: $\mathbb{R}$ denotes the real number line, $\mathbb{C}$ denotes the complex plane
(see Section 1.1), "iff" stands for "if and only if," and ■ denotes the
end of a proof. The notation $]a, b[$ is used to indicate the open interval
consisting of all real numbers x satisfying $a < x < b$. This European
convention avoids confusion with the ordered pair (a, b). The notation
$z \mapsto f(z)$ indicates that z is mapped to $f(z)$ by f. The notation $f: A \subset \mathbb{C}$
$\to \mathbb{C}$ means that f maps the domain A into $\mathbb{C}$. Occasionally $\Rightarrow$ is used
to denote "implies." The symbol $A\backslash B$ denotes the members of the set
A that are not members of B, and $z \in A$ means that z is a member of A.

Sections, theorems, and definitions are numbered consecutively
within each chapter. A reference such as "Theorem 24" or "exercise 3"
applies to material within the present chapter or section; otherwise, the
chapter or section number is cited.

Thanks are due to Nora Lee of the Department of Mathematics at
Berkeley for supplying typing and secretarial help during the preparation of the course notes and for allowing the use of old exams for additional problems, and to Arthur Fischer, who used the notes in his course
and supplied several corrections. Special thanks go to Michael Buchner,
who, during the summer of 1969, helped to turn rough course notes and
an outline of the Committee for Undergraduate Programs in Mathematics of the Mathematical Association of America into a preliminary
manuscript, to Clifford Risk, who contributed the bulk of Section 8.4
and added numerous suggestions, and to Michael Hoffman, who checked
the problems, prepared the answers, and made numerous helpful suggestions. Finally, thanks are extended to Ralph Abraham, Kenneth
McAloon, Anthony Tromba, and Michael O'Nan, officers of Eagle
Mathematics, Incorporated (an organization of mathematics authors),
for their suggestion that this book be written and for their subsequent encouragement.

I thank the many kind readers who supplied corrections for this
printing. These include, in chronological order, P. Roeder, W. Barker,
G. Hill, J. Seitz, J. Brudowski, H. O. Cordes, M. Choi, W. T. Stallings,
E. Green, R. Iltis, N. Starr, D. Fowler, L. L. Campbell, D. Goldschmidt,
T. Kato, J. Mesizov, P. Kenschaft, and K. L. Teo.

Jerrold E. Marsden

August, 1975

Basic Complex Analysis

Chapter 1

Analytic Functions

In this chapter the basic ideas about complex numbers and analytic functions are introduced. The organization of the text is analogous to that of an elementary calculus textbook, in which the student begins with the real line $\mathbb{R}$ and a function $f(x)$ of a real variable x, then studies differentiation of f. Similarly, in complex analysis he begins with complex numbers z and studies functions $f(z)$, which are differentiable. (These are called analytic functions.) The analogy is, however, deceptive, because complex analysis is a much richer theory; a lot more can be said about an analytic function than about a differentiable function of a real variable. The properties of analytic functions will be fully developed in subsequent chapters.

In addition to becoming familiar with the theory, the student should gain some facility with the standard (or "elementary") functions, such as polynomials, e^z, $\log z$, $\sin z$, used in calculus. These functions are studied in Sections 3 and 5 and appear frequently throughout the text.

1.1 Introduction to Complex Numbers

Historical sketch

The following discussion will assume some familiarity with the main properties of real numbers. The real number system resulted from the search for a system (an abstract set together with certain rules) that included the rationals but that also provided solutions to such polynomial equations as $x^2 - 2 = 0$.

Historically, a similar consideration gave rise to an extension of the real numbers. As early as the sixteenth century Geronimo Cardano considered quadratic (and cubic) equations such as $x^2 + 2x + 2 = 0$, which is satisfied by no real number x. The quadratic formula $(-b \pm \sqrt{b^2 - 4ac})/2a$ yields "formal" expressions for the two solutions of the equation $ax^2 + bx + c = 0$. But this formula may involve square roots of negative numbers; for example, $-1 \pm \sqrt{-1}$ for the equation $x^2 + 2x + 2 = 0$. Cardano noticed that if these "complex numbers" were treated as ordinary numbers with the added rule that $\sqrt{-1} \cdot \sqrt{-1} = -1$, they did indeed solve the equations.

The important expression $\sqrt{-1}$ is now given the widely accepted designation $i = \sqrt{-1}$. (This convention is not followed by the electrical engineers, who prefer the symbol $j = \sqrt{-1}$ since they wish to reserve the symbol i for electric current.) However, in the past it was felt that no meaning could actually be assigned to such expressions, which were therefore termed "imaginary." Gradually, especially as a result of the work of Leonhard Euler in the eighteenth century, these imaginary quantities came to play an important role. For example, Euler's formula $e^{i\theta} = \cos \theta + i\sin \theta$ revealed the existence of a profound relationship between complex numbers and the trigonometric functions. The rule $e^{i(\theta_1 + \theta_2)} = e^{i\theta_1}e^{i\theta_2}$ was found to summarize the rules for expanding sine and cosine of a sum in a very neat way, and this result alone indicated that some meaning should be attached to these "imaginary" numbers.

However, not until the work of Casper Wessel (ca. 1797), Jean Robert Argand (1806), Karl Friedrich Gauss (1831), Sir William R. Hamilton (1837), and others, was the meaning of complex numbers clarified, and it was realized that there is nothing "imaginary" about them at all, although this term is still used.

The complex analysis that is the subject of this book was developed in the nineteenth century, mainly by Augustin Cauchy (1789–1857). Later his theory was made more rigorous and extended by such mathematicians as Peter Dirichlet (1805 – 1859), Karl Weierstrass (1815 – 1897), and George Friedrich Bernhard Riemann (1826 – 1866).

The search for a method to describe heat conduction influenced the development of the theory, which has found many uses outside mathematics. Subsequent chapters will discuss some of these applications to problems in physics and engineering, like hydrodynamics and electrostatics. The theory also has mathematical applications to problems that at first do not seem to involve complex numbers. For example, the proof that

$$\int_0^\infty \frac{\sin^2 x}{x^2}\, dx = \frac{\pi}{2}$$

or that

$$\int_0^\infty \frac{x^{\alpha-1}}{1+x}\, dx = \frac{\pi}{\sin(\alpha\pi)} \qquad \text{for } 0 < \alpha < 1$$

or that

$$\int_0^{2\pi} \frac{d\theta}{a+\sin\theta} = \frac{2\pi}{\sqrt{a^2-1}}$$

may be difficult or impossible using elementary calculus, but these equations can be readily handled using the techniques of complex variables.

Complex analysis has become an indispensable and standard tool of the working mathematician, physicist, and engineer. Neglect of it can prove to be a severe handicap in most areas of research and application involving mathematical ideas and techniques.

Definition of the complex numbers

The first task in this section will be to define complex numbers and to show that they possess properties that are adequate for the usual algebraic manipulations to hold. The basic idea of complex numbers is credited to Jean Robert Argand, who suggested using points in the plane to represent complex numbers. The student will recall that the plane, denoted by $\mathbb{R}^2$, consists of all ordered pairs (x,y) of real numbers.

Definition 1. *The system of complex numbers, denoted $\mathbb{C}$, is the set $\mathbb{R}^2$ together with the usual rules of vector addition and scalar multiplication by a real number a:*

$$(x_1,y_1) + (x_2,y_2) = (x_1 + x_2, y_1 + y_2)$$

$$a(x,y) = (ax,ay)$$

and the operation of complex multiplication defined by

$$(x_1,y_1)(x_2,y_2) = (x_1 x_2 - y_1 y_2, x_1 y_2 + y_1 x_2).$$

The geometric meaning of vector addition is reviewed in Figure 1.1. The geometric meaning of multiplication will be discussed in Section 1.2.

Rather than using (x,y) to represent a complex number, we will find it more convenient to return to more standard notation as follows. Let us identify *real* numbers x with points on the x-axis; thus x and $(x,0)$ stand for the same point $(x,0)$ in $\mathbb{R}^2$. The y-axis will be called the *imaginary axis* and the unit point $(0,1)$ will be denoted i. Thus, by definition, $i = (0,1)$. Then

$$(x,y) = x + yi$$

because the right side of the equation stands for $(x,0) + y(0,1) = (x,0) + (0,y) = (x,y)$. We also have, using $y = (y,0)$ and complex multiplication, $iy = (0,1)(y,0) = (0 \cdot y - 1 \cdot 0, y \cdot 1 + 0 \cdot 0) = (0,y) = y(0,1) = yi$, so we can also write $(x,y) = x + iy$. A single symbol such as $z = a + ib$ is generally used to indicate a complex number. The notation $z \in \mathbb{C}$ means that z belongs to the set of complex numbers.

Note that $i^2 = i \cdot i = (0,1) \cdot (0,1) = (0 \cdot 0 - 1 \cdot 1, (1 \cdot 0 + 0 \cdot 1)) = (-1,0) = -1$, so we do have the property we want:

$$i^2 = -1.$$

If we remember this equation, then the rule for multiplication of complex numbers is also easy to remember:

$$(a + ib)(c + id) = ac + iad + ibc + i^2 bd$$
$$= (ac - bd) + i(ad + bc).$$

Thus, for example, $2 + 3i$ is the complex number $(2,3)$, and $(2 + 3i)(1 - 4i) = 2 - 12i^2 + 3i - 8i = 14 - 5i$ is another way of saying that $(2,3)(1,-4) = (2 \cdot 1 - 3 (-4), 3 \cdot 1 + 2 (-4)) = (14,-5)$. The reason for using the expression $a + bi$ is twofold. First, it is conventional. Second, the rule $i^2 = -1$ is easier to use than the rule $(a,b)(c,d) = (ac - bd, bc + ad)$ although both rules produce the same result.

Notice that because multiplication of real numbers is associative, commutative, and distributive, multiplication of complex numbers is also; that is, for all complex numbers z,w,s we have $(zw)s = z(ws)$, $zw = wz$, and $z(w + s) = zw + zs$. Let us verify the first of these properties; the others can be similarly verified.

Let $z = a + ib$, $w = c + id$, and $s = e + if$. Then $zw = (ac - bd) + i(bc + ad)$, and so $(zw)s = e(ac - bd) - f(bc + ad) + i\{e(bc + ad) + f(ac - bd)\}$. Similarly, $z(ws) = (a + bi)\{(ce - df) + i(cf + de)\} = a(ce - df) - b(cf + de) + i\{a(cf + de) + b(ce - df)\}$. Comparing these expressions and accepting the usual properties of real numbers, we conclude that $(zw)s = z(ws)$. Thus we can write, without ambiguity, an expression like $z^n = z \cdot \cdot \cdot z$ (n times).

Note that $a + ib = c + id$ means $a = c$ and $b = d$ (since this is what equality means in $\mathbb{R}^2$) and that 0 stands for $0 + i0 = (0,0)$. Thus $a + ib = 0$ means that *both* $a = 0$ and $b = 0$.

In what sense are these complex numbers an extension of the reals? We have already said that if a is real we also write a to stand for $a + 0i = (a,0)$. In other words, the reals $\mathbb{R}$ are identified with the x-axis in $\mathbb{C} = \mathbb{R}^2$; we are thus regarding the real numbers as those complex

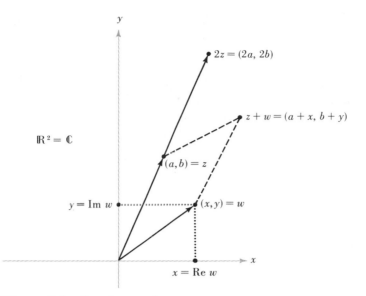

Figure 1.1 Complex numbers.

numbers $a + bi$, where $b = 0$. If, in the expression $a + bi$ the term $a = 0$, we call $bi = 0 + bi$ a *pure imaginary number*. In the expression $a + bi$ we say that a is the *real part* and b is the *imaginary part*. This is sometimes written Re $z = a$, Im $z = b$ where $z = a + bi$. Note that Re z and Im z are always real numbers (see Figure 1.1).

Actually, $\mathbb{C}$ obeys all the algebraic rules that ordinary real numbers do. For example, it will be shown in the following discussion that multiplicative inverses exist for nonzero elements. This means that if $z \neq 0$, then there is a (complex) number z' such that $zz' = 1$, and we write $z' = z^{-1}$. We can write this expression unambiguously (in other words, z' is uniquely determined), because if $zz'' = 1$ as well, then $z' = z' \cdot 1 = z'(zz'') = (z'z)z'' = 1 \cdot z'' = z''$, so $z'' = z'$. To show that z' exists, suppose that $z = a + bi \neq 0$. Then at least one of $a \neq 0$, $b \neq 0$ holds, and so $a^2 + b^2 \neq 0$. To find z', we set $z' = a' + b'i$. The condition $zz' = 1$ imposes conditions that will enable us to compute a' and b'. Computing the product gives $zz' = (aa' - bb') + (ab' + a'b)i$. The equations $aa' - bb' = 1$ and $ab' + a'b = 0$ can be solved for a' and b' by letting $a' = a/(a^2 + b^2)$ and $b' = -b/(a^2 + b^2)$, since $a^2 + b^2 \neq 0$. Thus for $z = a + ib \neq 0$, we may set

$$z^{-1} = \frac{a}{a^2 + b^2} - \frac{ib}{a^2 + b^2}.$$

If z and w are complex numbers with $w \neq 0$, then the symbol z/w means zw^{-1}; we call z/w the *quotient* of z by w. Thus $z^{-1} = 1/z$. To compute z^{-1}, the following series of equations is common and is a useful way to remember the preceding formula for z^{-1}:

$$\frac{1}{a+ib} = \frac{a-ib}{(a+ib)(a-ib)} = \frac{a-ib}{a^2+b^2} = \frac{a}{a^2+b^2} - \frac{b}{a^2+b^2} \, i.$$

In short, all the usual algebraic rules for manipulating real numbers, fractions, polynomials, and so on hold in complex analysis.

Formally, the system of complex numbers is an example of a *field*. The crucial rules for a field, stated here for reference only, are:

Additivity Rules:
i. $z + w = w + z$;
ii. $z + (w + s) = (z + w) + s$;
iii. $z + 0 = z$;
iv. $z + (-z) = 0$.

Multiplication Rules:
i. $zw = wz$;
ii. $(zw)s = z(ws)$;
iii. $1z = z$;
iv. $z(z^{-1}) = 1$ for $z \neq 0$.

Distributive Law: $z(w + s) = zw + zs$.

In summary, we have:

Theorem 1. *The complex numbers* $\mathbb{C}$ *form a field.*

The student is cautioned that we generally *cannot* say that $z \leq w$ for complex z and w. If the usual ordering properties for reals are to hold, then *such an ordering is impossible.* This statement can be proved as follows.

Suppose that such an ordering exists. Then either $i \geq 0$ or $i \leq 0$. Suppose that $i \geq 0$. Then $i \cdot i \geq 0$ and so $-1 \geq 0$, which is absurd. Alternatively, suppose that $i \leq 0$. Then $-i \geq 0$, so $(-i)(-i) \geq 0$, or $-1 \geq 0$, again absurd.

If $z = a + ib$ and $w = c + id$, we could say that $z \leq w$ iff $a \leq c$ and $b \leq d$. This is an ordering of sorts, but it does not satisfy all the rules that might be required, such as those obeyed by real numbers. Thus in this text the notation $z \leq w$ will be avoided unless z and w happen to be real.

Roots of quadratic equations

As mentioned previously, one of the reasons for using complex numbers is to enable us to take square roots of negative real numbers. That this can, in fact, be done for all complex numbers is verified in Theorem 2.

Theorem 2.   *Let $z \in \mathbb{C}$. Then there exists $w \in \mathbb{C}$ such that $w^2 = z$. (Notice that $-w$ also satifies this equation.)*

Proof of this theorem follows more easily after Section 2, where the machinery of polar coordinates is introduced, but we can attempt a purely algebraic proof here.

Proof.   Let $z = a + bi$. We are trying to find $w = x + iy$ such that $(x + iy)^2 = a + bi$, or, equivalently, $(x^2 - y^2) + (2xy)i = a + bi$. To do so we are required to solve $x^2 - y^2 = a$ and $2xy = b$. We know that $(x^2 + y^2)^2 = (x^2 - y^2)^2 + 4x^2y^2 = a^2 + b^2$. Hence $x^2 + y^2 = \sqrt{a^2 + b^2}$, so $x^2 = (a + \sqrt{a^2 + b^2})/2$ and $y^2 = (-a + \sqrt{a^2 + b^2})/2$.

If we let

$$\alpha = \sqrt{\left(\frac{a + \sqrt{a^2 + b^2}}{2}\right)} \qquad \text{and} \qquad \beta = \sqrt{\left(\frac{-a + \sqrt{a^2 + b^2}}{2}\right)}$$

(the usual square root of positive real numbers), then, in the event that b is positive, we have $x = \alpha$, $y = \beta$, or $x = -\alpha$, $y = -\beta$; in the event that b is negative, we have $x = \alpha$, $y = -\beta$, or $x = -\alpha$, $y = \beta$. We conclude that the equation $w^2 = z$ has solutions $\pm(\alpha + \mu\beta i)$, where $\mu = 1$ if $b \geq 0$ and $\mu = -1$ if $b < 0$. ∎

From the expressions for α and β we can conclude that (1) the square roots of a complex number are real if and only if the complex number is real and positive; (2) the square roots of a complex number are pure imaginary if and only if the complex number is real and negative; and (3) the two square roots of a complex number coincide if and only if the complex number is zero. (The student should check these conclusions.)

We can easily check that the quadratic equation $az^2 + bz + c = 0$ for complex numbers a,b,c has solutions $z = (-b \pm \sqrt{b^2 - 4ac})/2a$, where the square root is as previously constructed.

Uniqueness of the complex numbers

The following discussion will not be fully rigorous; rather it will, at times, be somewhat informal. An absolutely precise explanation would be tantamount to a short course in abstract algebra. However, the student should nevertheless be able to grasp the important points of the presentation.

We have constructed the field $\mathbb{C}$, which contains the reals and in which every quadratic equation has a solution. It is only natural to ask whether there are any other fields containing the reals and in which every quadratic equation has a solution.

The answer is that any other such field, say F, must already contain the complex numbers; therefore, the complex numbers are the *smallest* field containing $\mathbb{R}$ in which all quadratic equations are solvable. In that sense it is unique. The reason is quite simple. Let j be any solution in F to the equation $z^2 + 1 = 0$. Consider, in F, all numbers of the form $a + jb$ for real numbers a and b. This set is, algebraically, the same as $\mathbb{C}$ because of the simple fact that since $j^2 = -1$, j plays the role of and may be identified with i. We must also check that $a + jb = c + jd$ implies that $a = c$ and $b = d$ to be sure that equality in this set coincides with that in $\mathbb{C}$. Indeed, $(a - c) + j(b - d) = 0$, so we must prove that $e + jf = 0$ implies that $e = 0$ and $f = 0$ (where $e = a - c$ and $f = b - d$). If $f = 0$, then clearly $e = 0$ as well. But if $f \neq 0$, then $j = -e/f$, which is real, and no *real* number satisfies $j^2 = -1$ because the square of any real number is nonnegative, so f must be zero. This proves our claim. We can rephrase our result by saying that $\mathbb{C}$ *is the smallest field extension of* $\mathbb{R}$ *in which quadratic equations are solvable.*

Another question arises at this point. We made $\mathbb{R}^2$ into a field. For what other n can $\mathbb{R}^n$ be made into a field? Let us demand at the outset that the algebraic operations agree with those on $\mathbb{R}$, assuming that $\mathbb{R}$ is the x_1-axis. The answer is: only in the case $n = 2$. A field-like structure, called the *quaternions*, can be obtained for $n = 4$, except that the rule $zw = wz$ fails. Such a structure is called a *noncommutative field*. The proof of these facts can be found in an advanced abstract algebra text.

Worked Examples

1.   Prove that $\dfrac{1}{i} = -i$ and that $\dfrac{1}{i + 1} = \dfrac{1 - i}{2}$.

Solution.   First,

$$\frac{1}{i} = \frac{1}{i} \cdot \frac{-i}{-i} = -i$$

because $i \cdot -i = -(i^2) = -(-1) = 1$. Also,

$$\frac{1}{i + 1} = \left(\frac{1}{i + 1}\right)\left(\frac{1 - i}{1 - i}\right) = \frac{1 - i}{2},$$

since $(1 + i)(1 - i) = 1 + 1 = 2$.

2.   Find the real and imaginary parts of $\dfrac{z + 2}{z - 1}$ where $z = x + iy$.

Solution.

$$\frac{z + 2}{z - 1} = \frac{(x + 2) + iy}{(x - 1) + iy} = \frac{(x + 2) + iy}{(x - 1) + iy} \cdot \frac{(x - 1) - iy}{(x - 1) - iy}$$

$$= \frac{(x + 2)(x - 1) + y^2 + i[y(x - 1) - y(x + 2)]}{(x - 1)^2 + y^2}.$$

Hence,

$$\mathrm{Re}\left(\frac{z+2}{z-1}\right) = \frac{x^2 + x - 2 + y^2}{(x-1)^2 + y^2}$$

and

$$\mathrm{Im}\left(\frac{z+2}{z-1}\right) = \frac{-3y}{(x-1)^2 + y^2}.$$

3. Solve $x^4 + i = 0$ for x.

Solution.   Let $x^2 = y$. Then $y^2 + i = 0$. Substituting in the formula we developed for taking square roots, letting $a = 0$ and $b = -1$, we get

$$y = \pm\left(\frac{1}{\sqrt{2}} - \frac{1}{\sqrt{2}}i\right).$$

Consider the equation $x^2 = (1 - i)/\sqrt{2}$. Again substituting in the formula, letting $a = 1/\sqrt{2}$ and $b = -1/\sqrt{2}$, we obtain the two solutions

$$x = \pm\left\{\frac{\sqrt{1 + \sqrt{2}}}{2^{3/4}} - \left(\frac{\sqrt{-1 + \sqrt{2}}}{2^{3/4}}\right)i\right\}.$$

From the other value for y we obtain two further solutions:

$$x = \pm\left\{\frac{\sqrt{-1 + \sqrt{2}}}{2^{3/4}} + \left(\frac{\sqrt{1 + \sqrt{2}}}{2^{3/4}}\right)i\right\}.$$

Remark.   In the next section de Moivre's Formula will be developed, which will enable us to find, quite simply, the nth root of any complex number.

4. Prove that, for complex numbers z and w,

$$\mathrm{Re}(z + w) = \mathrm{Re}\ z + \mathrm{Re}\ w$$

and

$$\mathrm{Im}(z + w) = \mathrm{Im}\ z + \mathrm{Im}\ w.$$

Solution.   Let $z = x + iy$ and $w = a + ib$. Then $z + w = (x + a) + i(y + b)$, and so $\mathrm{Re}(z + w) = x + a = \mathrm{Re}\ z + \mathrm{Re}\ w$. Similarly, $\mathrm{Im}(z + w) = y + b = \mathrm{Im}\ z + \mathrm{Im}\ w$.

Exercises

1. Express the following complex numbers in the form $a + ib$:
  a. $(2 + 3i) + (4 + i)$.
  b. $(2 + 3i)(4 + i)$.
  c. $\dfrac{2 + 3i}{4 + i}$.
  d. $(8 + 6i)^2$.
  e. $\dfrac{1}{i} + \dfrac{3}{1 + i}$.

f. $\left(1 + \dfrac{3}{1+i}\right)^2$.

2. Find the solutions to:

a. $x^2 = 3 - 4i$.

b. $x^4 - i = 0$.

3. Find the real and imaginary parts of the following, where $z = x + iy$:

a. $\dfrac{1}{z^2}$.

b. $\dfrac{1}{3z + 2}$.

c. $\dfrac{z + 1}{2z - 5}$.

d. z^3.

4. Show that $\mathrm{Re}(iz) = -\mathrm{Im}(z)$ and that $\mathrm{Im}(iz) = \mathrm{Re}(z)$ for any complex number z.

5. If a is real and z is complex, prove that $\mathrm{Re}(az) = a\,\mathrm{Re}\,z$ and that $\mathrm{Im}(az) = a\,\mathrm{Im}\,z$. Generally, show that $\mathrm{Re}\colon \mathbb{C} \to \mathbb{R}$ is a linear map; that is, that $\mathrm{Re}(az + bw) = a\,\mathrm{Re}\,z + b\,\mathrm{Re}\,w$ for a,b real, z,w complex.

6. Is it true that $\mathrm{Re}(zw) = (\mathrm{Re}\,z)(\mathrm{Re}\,w)$?

7. a. Fix a complex number $z = x + iy$ and consider the linear mapping $\varphi_z\colon \mathbb{R}^2 \to \mathbb{R}^2$ (that is, of $\mathbb{C} \to \mathbb{C}$) defined by $\varphi_z(w) = z \cdot w$ (that is, multiplication by z). Prove that the matrix of φ_z in the standard basis $(1,0)$, $(0,1)$ of $\mathbb{R}^2$ is given by

$$\begin{pmatrix} x & -y \\ y & x \end{pmatrix}.$$

b. Show that $\varphi_{z_1 z_2} = \varphi_{z_1} \circ \varphi_{z_2}$.

8. Using the axioms for a field, give a formal proof (including all details) for the following:

a. $\dfrac{1}{z_1 z_2} = \dfrac{1}{z_1} \cdot \dfrac{1}{z_2}$.

b. $\dfrac{1}{z_1} + \dfrac{1}{z_2} = \dfrac{z_1 + z_2}{z_1 z_2}$.

9. Show that z is real if and only if $\mathrm{Re}\,z = z$.

10. Let $\dfrac{x - iy}{x + iy} = a + ib$. Prove that $a^2 + b^2 = 1$.

11. Prove the binomial theorem for complex numbers; that is, letting z,w be complex numbers and n be a positive integer,

$$(z + w)^n = z^n + \binom{n}{1} z^{n-1} w + \binom{n}{2} z^{n-2} w^2 + \cdots + \binom{n}{n} w^n$$

where

$$\binom{n}{r} = \frac{n!}{r!(n-r)!}.$$

Use induction on n.

12. Prove that, for any integer k,

$$i^{4k} = 1, \quad i^{4k+1} = i, \quad i^{4k+2} = -1, \quad i^{4k+3} = -i.$$

Argue that this result gives a formula for i^n for any n by setting $n = 4k + j$, $0 \leq j \leq 3$.

13. Show that the following rules uniquely determine complex multiplication:
   a. $(z_1 + z_2)w = z_1 w + z_2 w$.
   b. $z_1 z_2 = z_2 z_1$.
   c. $i \cdot i = -1$.
   d. $z_1(z_2 z_3) = (z_1 z_2)z_3$.
   e. If z_1 and z_2 are real, $z_1 \cdot z_2$ is the usual product of real numbers.

14. Simplify the following:
   a. $(1 + i)^4$.
   b. $(-i)^{-1}$.
   c. $(1 - i)^{-1}$.
   d. $(1 + i)/(1 - i)$.

15. Simplify the following:
   a. $\sqrt{1 + \sqrt{i}}$.
   b. $\sqrt{1 + i}$.
   c. $\sqrt{\sqrt{-i}}$.

1.2   Properties of Complex Numbers

In mathematics it is very important to be able to picture the concepts being studied and to develop what is called geometric intuition. This ability is particularly valuable in dealing with complex numbers. In studying them we shall also encounter the important concepts of the absolute value, argument, polar representation, and complex conjugate of a complex number. These concepts have simple geometric interpretations that are important to understand.

In the preceding section a complex number was defined to be a point in the plane $\mathbb{R}^2$. In addition to addition and multiplication by a real scalar, the operation of complex multiplication was defined. A complex number may be thought of geometrically as a (two-dimensional) vector and pictured as an arrow from the origin to the point in $\mathbb{R}^2$ given by the complex number (see Figure 1.2).

Because the points $(x,0) \in \mathbb{R}^2$ correspond to real numbers, the horizontal or x-axis is called the *real axis*. Similarly, the vertical axis (the y-axis) is called the *imaginary axis*, because points on it have the form $iy = (0,y)$ for y real. Addition of complex numbers can thus be pictured as addition of vectors (see Figures 1.1 and 1.3).

It would be helpful if we could picture complex multiplication in the same manner. To do this we must first write complex numbers in what is called polar coordinate form.

Recall that the *length* of the vector $(a,b) = a + ib$ is defined as $r = \sqrt{a^2 + b^2}$ and suppose that the vector makes an angle θ with the positive direction of the real axis, where $0 \leq \theta < 2\pi$ (see Figure 1.4).

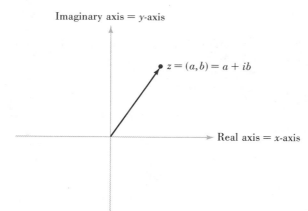

Figure 1.2   Vector representation of complex numbers.

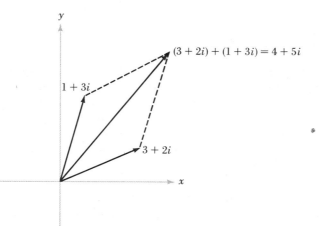

Figure 1.3   Addition of complex numbers.

Thus $\tan \theta = b/a$. Since $a = r \cos \theta$ and $b = r \sin \theta$, we thus have $a + bi = r \cos \theta + (r \sin \theta)i = r(\cos \theta + i \sin \theta)$. This way of writing the complex number is called the *polar coordinate representation*. The length of the vector $z = (a,b) = a + ib$ is denoted $|z|$ and is called the *norm*, or *modulus*, or *absolute value* of z. The angle θ is called the *argument* or *amplitude* of the complex numbers and is denoted $\theta = \arg z$.

If we restrict our θ, as we actually did, to the interval $[0,2\pi[$—that is, $0 \le \theta < 2\pi$—then a given nonzero complex number has an unambiguously defined argument. (We accept this as known from trigonometry.)

However, it is clear that we, can add integral multiples of 2π to θ and still obtain the same complex number.

We shall find it convenient to be flexible in our requirements for the values that θ is to assume. For example, we could equally well allow the range of θ to be $-\pi < \theta \le \pi$. Such an interval will always be specified in subsequent chapters. The student should remember that once this interval is specified for each $z \ne 0$, a unique θ is determined that lies within that specified interval. It is clear that any $\theta \in \mathbb{R}$ can be brought into our specified interval ($[0,2\pi[$ or $]-\pi,\pi]$, for example) by addition of some (positive or negative) integral multiple of 2π.

Polar representation of complex numbers simplifies the task of describing geometrically the product of two complex numbers. Let $z_1 = r_1(\cos \theta_1 + i\sin \theta_1)$ and $z_2 = r_2(\cos \theta_2 + i\sin \theta_2)$. Then $z_1 z_2 = r_1 r_2([\cos \theta_1 \cdot \cos \theta_2 - \sin \theta_1 \cdot \sin \theta_2] + i[\cos \theta_1 \cdot \sin \theta_2 + \cos \theta_2 \cdot \sin \theta_1]) = r_1 r_2(\cos (\theta_1 + \theta_2) + i\sin(\theta_1 + \theta_2))$, by the addition formula for the sine and cosine functions used in trigonometry. Thus we have proven

Theorem 3.

$$|z_1 z_2| = |z_1| \cdot |z_2|$$

and

$$\arg(z_1 z_2) = \arg z_1 + \arg z_2.$$

Translated into words, the product of two complex numbers is the complex number that has a length equal to the product of the lengths

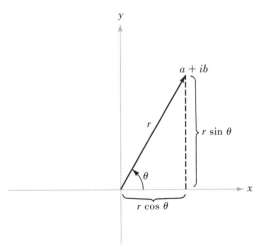

Figure 1.4   Polar coordinate representation of complex numbers.

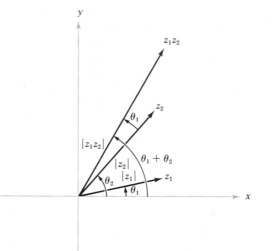

Figure 1.5 Multiplication of complex numbers.

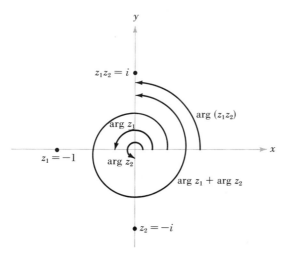

Figure 1.6 Multiplication of the complex
numbers -1 and $-i$.

of the two complex numbers and an argument equal to the sum of the
arguments of those numbers. This is the basic geometric representation
of complex multiplication (see Figure 1.5).

Let us agree that in the equality for $\arg(z_1 z_2)$ in Theorem 3, if $\arg z_1 +$
$\arg z_2$ lies outside the interval that we specify, we should adjust it by a
multiple of 2π to bring it within that interval. For example, if our interval

is $[0,2\pi[$ and $z_1 = -1$ and $z_2 = -i$, then $\arg z_1 = \pi$ and $\arg z_2 = 3\pi/2$ (see Figure 1.6), but $z_1 z_2 = i$, so $\arg z_1 z_2 = \pi/2$, and $\arg z_1 + \arg z_2 = \pi + 3\pi/2 = 2\pi + \pi/2$. We can obtain the correct answer by subtracting 2π to bring it within the interval $[0,2\pi[$.

As a result of the preceding discussion, the second equality in Theorem 3 should be written as $\arg z_1 z_2 = \arg z_1 + \arg z_2 \pmod{2\pi}$, "mod 2π" meaning that the left and right sides of the equation agree after addition of a multiple of 2π to the right side.

There is also a geometric construction using similar triangles that enables us to construct the product of two complex numbers. The two shaded triangles in Figure 1.7 are similar. The construction of the point $z_1 z_2$ is effected by drawing the two angles θ_1 and the angle σ. That the constructed point is the correct one follows at once from Theorem 3 and standard properties of similar triangles.

Multiplication of complex numbers can be analyzed in another useful way. Let $z \in \mathbb{C}$ and define $\psi_z : \mathbb{C} \to \mathbb{C}$ by $\psi_z(w) = wz$; that is, ψ_z is the map "multiplication by z." By Theorem 3, it is clear that *the effect of this map is to rotate a complex number through an angle equal to* $\arg z$ *in the counterclockwise direction and to stretch its length by the factor* $|z|$. For example, ψ_i (multiplication by i) simply rotates complex numbers by $\pi/2$ in the counterclockwise direction (see Figure 1.8).

The map ψ_z is a linear transformation on the plane, in the sense that $\psi_z(\lambda w_1 + \mu w_2) = \lambda \psi_z(w_1) + \mu \psi_z(w_2)$ where λ, μ are real numbers and w_1, w_2 are complex numbers.

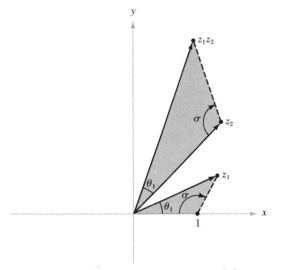

Figure 1.7 Geometric construction of the product of two complex numbers.

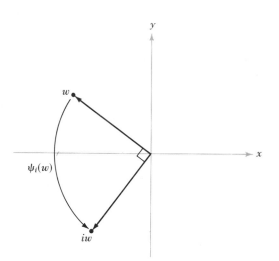

Figure 1.8 Multiplication by i.

Remark.   If $z = a + ib = (a,b)$, then the matrix of ψ_z is

$$\begin{pmatrix} a & -b \\ b & a \end{pmatrix}.$$

since

$$\begin{pmatrix} a & -b \\ b & a \end{pmatrix}\begin{pmatrix} x \\ y \end{pmatrix} = \begin{pmatrix} ax - by \\ ay + bx \end{pmatrix}$$

(compare exercise 7, Section 1.1).

The formula we derived for multiplication, using the polar coordinate representation, provides more than geometric intuition. We can obtain a formula that enables us to find nth roots of any complex number.

Theorem 4 (de Moivre's Formula).   If $z = r(\cos \theta + i\sin \theta)$ and n is a positive integer, then $z^n = r^n(\cos n\theta + i\sin n\theta)$.

Proof.   By Theorem 3, $z^2 = r^2(\cos(\theta + \theta) + i\sin(\theta + \theta)) = r^2(\cos 2\theta + i\sin 2\theta)$. Multiplying again by z gives $z^3 = z \cdot z^2 = r \cdot r^2[\cos(2\theta + \theta) + i\sin(2\theta + \theta)] = r^3(\cos 3\theta + i\sin 3\theta)$. This procedure may be continued by induction to obtain the desired result for any integer n. ∎

Let w be a complex number; that is, let $w \in \mathbb{C}$. Using de Moivre's Formula we will solve the equation $z^n = w$ for z when w is given. Suppose that $w = r(\cos \theta + i\sin \theta)$ and $z = \rho(\cos \psi + i\sin \psi)$. Then, by Theorem 4, $z^n = \rho^n(\cos n\psi + i\sin n\psi)$. It follows that $\rho^n = r = |w|$ by uniqueness of the polar representation and $n\psi = \theta + k(2\pi)$ where k is some integer.

Thus

$$z = \sqrt[n]{r}\left(\cos\left(\frac{\theta}{n} + \frac{k}{n}2\pi\right) + i\sin\left(\frac{\theta}{n} + \frac{k}{n}2\pi\right)\right).$$

Each value of $k = 0,1, \cdots , n-1$ gives a different value of z. Any other value of k merely repeats one of the values of z corresponding to $k = 0,1,2, \cdots , n-1$. Thus there are exactly n nth roots of any complex number.

For example, the preceding formula gives the three solutions to the equation $z^3 = 1 = 1(\cos 0 + i\sin 0)$ as

$$z = \cos\frac{k2\pi}{3} + i\sin\frac{k2\pi}{3}, \; k = 0,1,2;$$

that is,

$$z = 1, \frac{-1}{2} + \frac{i\sqrt{3}}{2}, -\frac{1}{2} - \frac{i\sqrt{3}}{2}.$$

This result is stated formally as:

Theorem 5. *Let w be a given (nonzero) complex number with polar representation $w = r(\cos\theta + i\sin\theta)$. Then the nth roots of w are given by the n complex numbers*

$$z_k = \sqrt[n]{r}\left[\cos\left(\frac{\theta}{n} + \frac{2\pi k}{n}\right) + i\sin\left(\frac{\theta}{n} + \frac{2\pi k}{n}\right)\right], \; k = 0,1, \cdots , n-1.$$

As a special case of this formula we note that the n roots of 1 (that is, the nth roots of unity) are 1 and $n-1$ other points equally spaced around the unit circle, as illustrated in Figure 1.9 for the case $n = 8$.

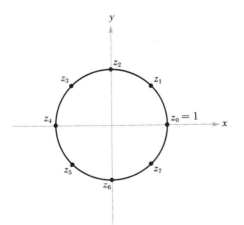

Figure 1.9   The nth roots of unity $(n = 8)$.

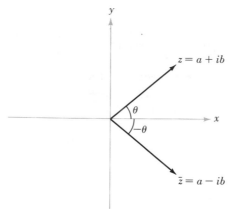

Figure 1.10 Complex conjugation.

Subsequent chapters will include many references to the simple idea of conjugation, which is defined as follows: If $z = a + ib$, then $\bar{z}$, the *complex conjugate* of z, is defined by $\bar{z} = a - ib$. Complex conjugation can be pictured geometrically as reflection in the real axis (see Figure 1.10).

Theorem 6 summarizes the main properties of complex conjugation.

Theorem 6.
  i. $\overline{z + z'} = \bar{z} + \overline{z'}$.
 ii. $\overline{zz'} = \bar{z}\overline{z'}$.
iii. $\overline{z/z'} = \bar{z}/\overline{z'}$ for $z' \neq 0$.
 iv. $z\bar{z} = |z|^2$ *and hence if* $z \neq 0$, *we have* $z^{-1} = \bar{z}/|z|^2$.
  v. $z = \bar{z}$ *if and only if* z *is real*.
 vi. $\operatorname{Re} z = \dfrac{z + \bar{z}}{2}$ *and* $\operatorname{Im} z = \dfrac{z - \bar{z}}{2i}$.
vii. $\bar{\bar{z}} = z$.

Proof.
  i. Let $z = a + ib$ and let $z' = a' + ib'$. Then $z + z' = a + a' + i(b + b')$, so $\overline{z + z'} = (a + a') - i(b + b') = a - ib + a' - ib' = \bar{z} + \overline{z'}$.
 ii. Let $z = a + ib$ and let $z' = a' + ib'$. Then

$$\overline{zz'} = \overline{(aa' - bb') + i(ab' + a'b)} = (aa' - bb') - i(ab' + a'b).$$

On the other hand, $\bar{z}\,\overline{z'} = (a - ib)(a' - ib') = (aa' - bb') - i(ab' + a'b)$.
iii. By (ii) we have $\overline{z'}\,\overline{z/z'} = \overline{z'z/z'} = \bar{z}$. Hence $\overline{z/z'} = \bar{z}/\overline{z'}$.
 iv. $z\bar{z} = (a + ib)(a - ib) = a^2 + b^2 = |z|^2$.

v. If $a + ib = a - ib$, then $ib = -ib$, so $b = 0$.

vi. This assertion is clear by the definition of $\bar{z}$.

vii. This assertion is also clear by the definition. ∎

The absolute value of a complex number $|z| = |a + ib| = \sqrt{a^2 + b^2}$, which is merely the usual Euclidean length of the vector representing the complex number, has already been defined. From Theorem 6(iv) we note that $|z|$ is also given by $|z|^2 = z\bar{z}$. The absolute value of a complex number is encountered throughout complex analysis; the following properties of the absolute value are quite basic.

Theorem 7.
i. $|zz'| = |z| \cdot |z'|$.

ii. If $z' \neq 0$, then $|z/z'| = |z|/|z'|$.

iii. $-|z| \le \operatorname{Re} z \le |z|$ and $-|z| \le \operatorname{Im} z \le |z|$; that is, $|\operatorname{Re} z| \le |z|$ and $|\operatorname{Im} z| \le |z|$.

iv. $|\bar{z}| = |z|$.

v. $|z + z'| \le |z| + |z'|$.

vi. $|z - z'| \ge ||z| - |z'||$.

vii. $|z_1 w_1 + \cdots + z_n w_n| \le \sqrt{|z_1|^2 + \cdots + |z_n|^2}\, \sqrt{|w_1|^2 + \cdots + |w_n|^2}$.

Note.   Result (iv) is clear geometrically from Figure 1.10; (v) is the usual triangle inequality for vectors in $\mathbb{R}^2$ (see Figure 1.11), and (vii) is referred to as Cauchy's inequality. By repeated application of (v) we get the general statement $|z_1 + \cdots + z_n| \le |z_1| + \cdots + |z_n|$.

Proof.
i. This equality was shown in Theorem 3.

ii. By (i), $|z'||z/z'| = |z' \cdot (z/z')| = |z|$, so $|z/z'| = |z|/|z'|$.

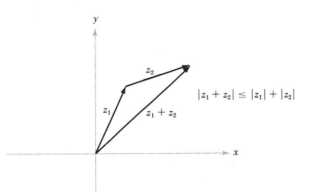

$|z_1 + z_2| \le |z_1| + |z_2|$

Figure 1.11   Triangle inequality.

iii. If $z = a + ib$, then $-\sqrt{a^2 + b^2} \le a \le \sqrt{a^2 + b^2}$ since $b^2 \ge 0$. The other inequality asserted in (iii) is similarly proved.

iv. If $z = a + ib$, then $\bar{z} = a - ib$, and we clearly have $|z| = \sqrt{a^2 + b^2}$ $= \sqrt{a^2 + (-b)^2} = |\bar{z}|$.

v. By Theorem 6(iv),

$$|z + z'|^2 = (z + z')\overline{(z + z')}$$
$$= (z + z')(\bar{z} + \bar{z}')$$
$$= z\bar{z} + z'\bar{z}' + z'\bar{z} + z\bar{z}'.$$

But $\overline{zz'}$ is the conjugate of $z'\bar{z}$ (Why?), so by Theorem 6(vi) and (iii) in this proof, $|z|^2 + |z'|^2 + 2 \operatorname{Re} z'\bar{z} \le |z|^2 + |z'|^2 + 2|z'\bar{z}| = |z|^2$ $+ |z'|^2 + 2|z| |z'|$. But this equals $(|z| + |z'|)^2$, so we get our result.

vi. By applying (v) to z' and $z - z'$ we get $|z| = |z' + (z - z')| \le |z'| +$ $|z - z'|$, so $|z - z'| \ge |z| - |z'|$. By interchanging the roles of z and z', we similarly get $|z - z'| \ge |z'| - |z| = -(|z| - |z'|)$, which is what we originally claimed.

vii. This inequality is less evident, and the proof of it requires a slight mathematical trick. Let us suppose that not all the $w_k = 0$ (or else the result is clear). Let $c = (z_1 w_1 + \cdots + z_n w_n)/(|w_1|^2$ $+ \cdots + |w_n|^2)$ and let

$$v = \sum_{k=1}^{n} |z_k|^2, \qquad t = \sum_{k=1}^{n} |w_k|^2, \qquad \text{and} \qquad s = \sum_{k=1}^{n} z_k w_k.$$

Therefore, $c = s/t$. Now let us consider

$$\sum_{k=1}^{n} |z_k - c\bar{w}_k|^2,$$

which is ≥ 0 and equals

$$v + |c|^2 t - c \sum_{k=1}^{n} \bar{z}_k \bar{w}_k - \bar{c} \sum_{k=1}^{n} z_k w_k = v + |c|^2 t - 2 \operatorname{Re} \bar{c}s$$

$$= v + \frac{|s|^2}{t} - 2 \operatorname{Re}\left(\frac{\bar{s}s}{t}\right).$$

Since t is real and $s\bar{s} = |s|^2$ is real, $v + (|s|^2/t) - 2(|s|^2/t)$ $= v - |s|^2/t \ge 0$. Hence $|s|^2 \le vt$, which is the desired result. ∎

Worked Examples

1. Solve $z^8 = 1$ for z.

   Solution.   Since $1 = \cos k2\pi + i\sin k2\pi$ when k equals any integer, we have, by Theorem 5.

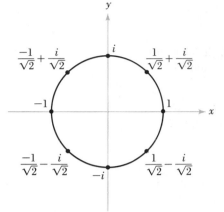

Figure 1.12 The eight 8th roots
of unity.

$$z = \cos \frac{k2\pi}{8} + i\sin \frac{k2\pi}{8}, \ k = 0, 1, 2, \cdots, 7$$

$$= 1, \frac{1}{\sqrt{2}} + \frac{i}{\sqrt{2}}, i, \frac{-1}{\sqrt{2}} + \frac{i}{\sqrt{2}}, -1, \frac{-1}{\sqrt{2}} - \frac{i}{\sqrt{2}}, -i, \frac{1}{\sqrt{2}} - \frac{i}{\sqrt{2}}.$$

These may be pictured as points evenly spaced on the circle in the complex
plane (see Figure 1.12).

2. Show that $\overline{\left[\dfrac{(3+7i)^2}{(8+6i)}\right]} = \dfrac{(3-7i)^2}{(8-6i)}.$

Solution. The point is that it is not necessary first to work out $(3 + 7i)^2/$
$(8 + 6i)$ if we simply use the properties developed in the text, namely, $\overline{z^2}$
$= (\bar{z})^2$ and $\overline{z/z'} = \bar{z}/\bar{z}'$. Thus we obtain:

$$\overline{\left[\frac{(3+7i)^2}{(8+6i)}\right]} = \frac{\overline{(3+7i)^2}}{\overline{(8+6i)}} = \frac{\overline{(3+7i)}^2}{8-6i} = \frac{(3-7i)^2}{8-6i}.$$

3. If $|z| = 1$, prove that

$$\left|\frac{az+b}{\bar{b}z+\bar{a}}\right| = 1$$

for any complex numbers a and b.

Solution. Since $|z| = 1$, we have $z = \bar{z}^{-1}$. Thus

$$\frac{az+b}{\bar{b}z+\bar{a}} = \frac{az+b}{\bar{b}+\bar{a}\bar{z}} \cdot \frac{1}{z}.$$

Because of the properties of the absolute value and because $|z| = 1$,

$$\left|\frac{az+b}{\bar{b}z+\bar{a}}\right| = \left|\frac{az+b}{\bar{a}z+\bar{b}}\right| \cdot \frac{1}{|z|} = 1,$$

Since $|az + b| = |\overline{az+b}| = |\bar{a}\bar{z} + \bar{b}|.$

4. Show that the maximum absolute value of $z^2 + 1$ on the unit disc $\{z \mid |z| \leq 1\}$ is 2.

Solution. By the triangle inequality, $|z^2 + 1| \leq |z^2| + 1 = |z|^2 + 1 \leq 1^2 + 1 = 2$, since $|z| \leq 1$. The maximum is achieved at $z = 1$.

5. The aim of this example is to show how to express some equations from analytic geometry in the language of complex numbers. Write, in complex form, the equation of a straight line; of a circle; of an ellipse.

Solution.
The straight line is most conveniently expressed in parametric form: $z = a + bt$, $a, b \in \mathbb{C}$, $t \in \mathbb{R}$, which represents a line in the direction of b and passing through the point a.
The circle can be expressed as $|z - a| = r$ (radius r, center a).
The ellipse can be expressed as $|z - d| + |z + d| = 2a$; the foci are located at $\pm d$ and the semi-major axis equals a.
These equations, in which $|\cdot|$ is interpreted as length, are merely the usual geometric definitions of these loci.

6. Express $\cos 3\theta$ in terms of $\cos \theta$ and $\sin \theta$ using de Moivre's Theorem.

Solution. We have the identity

$$(\cos \theta + i\sin \theta)^3 = \cos 3\theta + i\sin 3\theta.$$

The left side of this equation, when expanded (see exercise 11 at the end of Section 1.1), becomes

$$\cos^3 \theta + i3\cos^2 \theta \sin \theta - 3\cos \theta \sin^2 \theta - i\sin^3 \theta.$$

By equating real and imaginary parts, we get

$$\cos 3\theta = \cos^3 \theta - 3\cos \theta \sin^2 \theta$$

and the additional formula

$$\sin 3\theta = -\sin^3 \theta + 3\cos^2 \theta \sin \theta.$$

Exercises

1. Solve the following equations:
   a. $z^5 - 2 = 0$.
   b. $z^4 + i = 0$.
2. Prove that $|z| = 0$ iff $z = 0$.
3. What is the complex conjugate of $(3 + 8i)^4/(1 + i)^{10}$?
4. Let w be an nth root of unity, $w \neq 1$. Show that $1 + w + w^2 + \cdots w^{n-1} = 0$.
5. Express $\cos 5x$ and $\sin 5x$ in terms of $\cos x$ and $\sin x$.

6. Show that the roots of a polynomial with real coefficients occur in conjugate pairs.

7. If $a,b \in \mathbb{C}$, show that $|a - b|^2 + |a + b|^2 = 2(|a|^2 + |b|^2)$.

8. Find the absolute value of $[i(2 + 3i)(5 - 2i)]/(-2 - i)$.

9. When does equality hold in the triangle inequality $|z_1 + z_2 + \cdots + z_n| \leq |z_1| + |z_2| + \cdots + |z_n|$? Interpret your result geometrically.

10. If either $|z| = 1$ or $|w| = 1$, then prove that

$$\left|\frac{z - w}{1 - \bar{z}w}\right| = 1.$$

(Assume that $\bar{z}w \neq 1$.)

11. Does $z^2 = |z|^2$? If so, prove this equality. If not, for what z is it true?

12. Letting $z = x + iy$, prove that $|x| + |y| \leq \sqrt{2}|z|$.

13. Let $z = a + ib$ and $z' = a' + ib'$. Prove that $|zz'| = |z| \, |z'|$ by evaluating each side.

14. Prove the following:
   a. Arg $\bar{z} = -$Arg z.
   b. Arg $(z/w) = $ Arg $z - $ Arg w (mod 2π).

15. What is the equation of the circle with radius 3 and center $8 + 5i$ in complex notation?

16. Describe the set of all z such that Im$(z + 5) = 0$.

17. Using the formula $z^{-1} = \bar{z}/|z|^2$, show how to construct z^{-1} geometrically.

18. Prove Lagrange's identity:

$$\left|\sum_{k=1}^{n} z_k w_k\right|^2 = \left(\sum_{k=1}^{n} |z_k|^2\right)\left(\sum_{k=1}^{n} |w_k|^2\right) - \sum_{k<j} |z_k\bar{w}_j - z_j\bar{w}_k|^2.$$

Deduce the Cauchy inequality from your proof.

19. Find the maximum of $|z^n + a|$ for those z with $|z| \leq 1$.

20. Compute the least upper bound (that is, supremum) of the following set of real numbers: $\{\text{Re}(iz^3 + 1) \mid |z| < 2\}$.

21. Prove Lagrange's trigonometric identity:

$$1 + \cos\theta + \cos 2\theta + \cdots + \cos n\theta = \frac{1}{2} + \frac{\sin\left[\left(n + \frac{1}{2}\right)\theta\right]}{2\sin\left(\frac{\theta}{2}\right)}.$$

(Assume that $\sin(\theta/2) \neq 0$.)

22. Suppose that the complex numbers z_1, z_2, z_3 satisfy the equation

$$\frac{z_2 - z_1}{z_3 - z_1} = \frac{z_1 - z_3}{z_2 - z_3}.$$

Prove that $|z_2 - z_1| = |z_3 - z_1| = |z_2 - z_3|$.

Hint.   Argue geometrically, interpreting the meaning of each statement.

23. Give a necessary and sufficient condition for
   a. z_1, z_2, z_3 to lie on a straight line.
   b. z_1, z_2, z_3, z_4 to lie on a straight line or a circle.

24. Prove the identity

$$\sin\left(\frac{\pi}{n}\right) \sin\left(\frac{2\pi}{n}\right) \cdots \sin\left(\frac{(n-1)\pi}{n}\right) = \frac{n}{2^{n-1}}.$$

Hint. Try to show that the given product can be written as $1/2^{n-1}$ times the product of the nonzero roots of the polynomial $(1-z)^n - 1$.

25. What is the locus of points satisfying $|z - z_1|/|z - z_2| = $ constant, where z_1 and z_2 are given?

1.3 Some Elementary Functions

The trigonometric functions sine and cosine, as well as the exponential function and the logarithmic function, are covered in elementary calculus. Let us recall that the trigonometric functions are defined in terms of the ratios of sides of a right-angled triangle. The definition of "angle" may be extended to include any real value, and thus $\cos \theta$ and $\sin \theta$ become real-valued functions of the real variable θ. It is a basic mathematical fact that $\cos \theta$ and $\sin \theta$ are differentiable with derivatives given by $d\cos \theta/d\theta = -\sin \theta$ and $d\sin \theta/d\theta = \cos \theta$. Alternatively, $\cos \theta$ and $\sin \theta$ can be defined by their power series:

$$\sin x = x - \frac{x^3}{3!} + \frac{x^5}{5!} - \cdots, \quad \cos x = 1 - \frac{x^2}{2!} + \frac{x^4}{4!} - \cdots$$

Convergence must also be proved; such a proof can be found in Chapter 3 and in any calculus text. Alternatively, $\sin x$ can be defined as the unique solution $f(x)$ to the differential equation $f''(x) + f(x) = 0$, and satisfying $f(0) = 0, f'(0) = 1$; and $\cos x$ can be defined as the unique solution to $f''(x) + f(x) = 0, f(0) = 1, f'(0) = 0$ (again see a calculus text for proofs).

The exponential function, denoted e^x, may be defined as the unique solution to the differential equation $f'(x) = f(x)$, subject to the initial condition that $f(0) = 1$. (It can be shown that a unique solution exists.) The exponential function can also be defined by its power series:

$$e^x = 1 + x + \frac{x^2}{2!} + \frac{x^3}{3!} + \cdots$$

We accept from calculus the fact that e^x is a positive, strictly increasing function of x. Therefore, for $y > 0$, $\log y$ can be defined as the inverse function of e^x; that is, $e^{\log y} = y$. Another approach that is often used in calculus books is to begin by defining

$$\log y = \int_1^y \frac{1}{t}\, dt$$

for $y > 0$ and then to define e^x, as the inverse function of $\log y$.

In this section these functions will be extended to the complex plane. In other words, the functions $\sin z$, $\cos z$, e^z, and $\log z$ will be defined on $\mathbb{C}$, and their restrictions to the real line will be the usual $\sin x$, $\cos x$, e^x, and $\log x$. The extension should be natural in that the familiar properties of sin, cos, exp, and log are retained. These functions, as well as the power function z^n, will be studied geometrically later in this section.

The first step is to extend the exponential function. We know from elementary calculus that for real x, e^x can be represented by its Maclaurin series:

$$e^x = 1 + \frac{x}{1!} + \frac{x^2}{2!} + \frac{x^3}{3!} + \cdots$$

Thus it would be most natural to define e^{iy} by

$$1 + \frac{(iy)}{1!} + \frac{(iy)^2}{2!} + \cdots$$

for $y \in \mathbb{R}$. Of course, this definition is not quite legitimate as convergence of series in $\mathbb{C}$ has not yet been discussed. Chapter 3 will show that this series does indeed represent a well-defined complex number for each given y, but for the moment the series is used informally as the basis for the definition that follows, which will be precise. A slight rearrangement of the series (using exercise 12, Section 1.1) shows that

$$e^{iy} = \left(1 - \frac{y^2}{2!} + \frac{y^4}{4!} - \cdots \right) + i\left(y - \frac{y^3}{3!} + \frac{y^5}{5!} - \cdots \right).$$

But we recognize this as being simply $\cos y + i\sin y$. So we define

$$e^{iy} = \cos y + i\sin y.$$

Thus we have defined e^z for z along both the real and imaginary axes. How do we define $e^z = e^{x+iy}$? We desire our extension of the exponential to retain the familiar properties, and among these is the law of exponents: $e^{a+b} = e^a \cdot e^b$. This requirement forces us to define $e^{x+iy} = e^x\, e^{iy}$. (Actually, there are more significant reasons for stating the definition in this manner—see p. 308.) This can be stated in a formal definition:

Definition 2. *If $z = x + iy$, then e^z is defined by $e^x(\cos y + i\sin y)$.*

Note that if z is real (that is, if $y = 0$), this definition agrees with the usual exponential function e^x.

The student is cautioned that he should not, at this stage, think of e^z as "e raised to the 'power' of z," since the concept of complex exponents has not yet been defined. The notation e^z is merely shorthand for the function defined by $e^x(\cos y + i\sin y)$.

There is another, again purely formal, reason for defining $e^{iy} = \cos y + i\sin y$. It should be emphasized that these manipulations, as well as the preceding power series operations, are not yet precise because the terms and operations used have not been carefully justified. Nevertheless, they help the student understand why the definitions are formulated the way they are. If we write $e^{iy} = f(y) + ig(y)$, we note that since we want $e^0 = 1$, we should have $f(0) = 1$, and $g(0) = 0$. Differentiating with respect to y gives us $ie^{iy} = f'(y) + ig'(y)$, so when $y = 0$ we get $f'(0) = 0$, $g'(0) = 1$. Differentiating again gives us $-e^{iy} = f''(y) + ig''(y)$. Comparing this equation with $e^{iy} = f(y) + ig(y)$, we get $f''(y) + f(y) = 0$, $f(0) = 1, f'(0) = 0$, so $f(y) = \cos y$; and $g''(y) + g(y) = 0$, $g(0) = 0$, $g'(0) = 1$, so $g(y) = \sin y$, by the definition of $\cos y$ and $\sin y$ in terms of differential equations. Thus we would obtain $e^{iy} = \cos y + i\sin y$ as in Definition 2.

Some of the important properties of e^z are summarized in the following theorem. For this theorem, we need to recall the definition of a periodic function; namely, a function $f\colon \mathbb{C} \rightarrow \mathbb{C}$ is called *periodic* if there exists a $w \in \mathbb{C}$ such that $f(z + w) = f(z)$ for all $z \in \mathbb{C}$ and w is called a *period* for f.

Theorem 8.
i. $e^{z+w} = e^z e^w$ *for all* $z, w \in \mathbb{C}$.
ii. e^z *is never zero*.
iii. *If x is real,* $e^x > 1$ *if $x > 0$;* $e^x < 1$ *if $x < 0$.*
iv. $|e^{x+iy}| = e^x$.
v. $e^{\pi i/2} = i$, $e^{\pi i} = -1$, $e^{3\pi i/2} = -i$, $e^{2\pi i} = 1$.
vi. e^z *is periodic; any period for e^z must have the form $2\pi n i$, n an integer*.
vii. $e^z = 1$ *iff $z = 2n\pi i$ for some integer n (positive, negative, or zero)*.

Proof.
i. Let $z = x + iy$, and $w = s + it$. Then, by our definition of e^z,

$$e^{z+w} = e^{(x+s)+i(y+t)}$$

$$= e^{x+s}(\cos(y + t) + i\sin(y + t))$$

$$= \{e^x(\cos y + i\sin y)\}\{e^s(\cos t + i\sin t)\}$$

by the addition formulas for sine and cosine and the property, well known from elementary calculus, that $e^{x+s} = e^x \cdot e^s$ for real numbers x and s. Thus $e^{z+w} = e^z \cdot e^w$ for all complex numbers z and w.

ii. For any z, we have $e^z \cdot e^{-z} = e^0 = 1$ since we know that the usual exponential satisfies $e^0 = 1$. Thus e^z can never be zero, because if it were, then $e^z \cdot e^{-z}$ would be zero, which is not true.

iii. This follows from elementary calculus. (For example, obviously

$$1 + \frac{x}{1!} + \frac{x^2}{2!} + \frac{x^3}{3!} + \cdots > 1 \qquad \text{when } x > 0.)$$

[Another proof utilizing the definition of e^x in terms of differential equations is as follows. Recall that e^x is the unique solution to $f'(x) = f(x)$ with $e^0 = 1$ (x real). Since e^x is continuous and is never zero, it must be strictly positive. Hence $(e^x)' = e^x$ is always positive and consequently e^x is strictly increasing. Thus for $x > 0$, $e^x > 1$, and for $x < 0$, $e^x < 1$].

iv. Using $|zz'| = |z| \, |z'|$ (see Theorem 7), we get

$$|e^{x+iy}| = |e^x e^{iy}| = |e^x| \, |e^{iy}|$$
$$= e^x \, |\cos y + i \sin y|, \text{ since } e^x > 0,$$
$$= e^x, \text{ since } \cos^2 y + \sin^2 y = 1.$$

v. From Definition 2, $e^{\pi i/2} = \cos \dfrac{\pi}{2} + i \sin \dfrac{\pi}{2} = i$. The proofs of the other formulas are similar.

vi. Suppose that $e^{z+w} = e^z$ for all $z \in \mathbb{C}$. Setting $z = 0$ we get $e^w = 1$. If $w = s + ti$, then, using (iv), $e^w = 1$ implies that $e^s = 1$ and so $s = 0$. Hence any period is of the form ti, $t \in \mathbb{R}$. Suppose that $e^{ti} = 1$, that is, $\cos t + i \sin t = 1$. Then $\cos t = 1$, $\sin t = 0$, and so $t = 2\pi n$ for some integer n.

vii. $e^0 = 1$ as we have seen, and $e^{2\pi n i} = 1$ because e^z is periodic, by (vi). Conversely, $e^z = 1$ implies that $e^{z+z'} = e^{z'}$ for all z'; so by (vi), $z = 2\pi n i$ for some integer n. ∎

How can we picture e^{iy}? As y goes from 0 to 2π, e^{iy} moves along the unit circle in a counterclockwise direction, reaching i at $y = \pi/2$, -1 at π, $-i$ at $3\pi/2$, and 1 again at 2π. Thus e^{iy} is the point on the unit circle with argument y (see Figure 1.13).

Note that in exponential form, the polar representation of a complex number becomes

$$z = |z| \, e^{i(\arg z)},$$

which is sometimes abbreviated to $z = re^{i\theta}$.

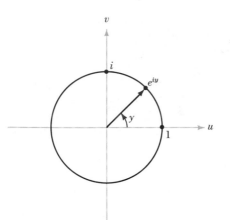

Figure 1.13 Points on the unit circle.

The trigonometric functions

Next we wish to extend the definitions of cosine and sine to the complex plane. The extension of the exponential to the complex plane suggests a way to extend the definitions of sine and cosine. We have $e^{iy} = \cos y + i\sin y$, which implies that

$$\sin y = \frac{e^{iy} - e^{-iy}}{2i} \text{ and } \cos y = \frac{e^{iy} + e^{-iy}}{2}.$$

But since e^{iz} is now defined for any $z \in \mathbb{C}$, we are led to formulate the following definition:

Definition 3.

$$\sin z = \frac{e^{iz} - e^{-iz}}{2i} \text{ and } \cos z = \frac{e^{iz} + e^{-iz}}{2}$$

for any complex number z.

Again if z is real, these definitions agree with the usual definitions of sine and cosine learned in elementary calculus.

Theorem 9 lists some of the properties of the sine and cosine functions that have now been defined over the whole of $\mathbb{C}$ and not merely on $\mathbb{R}$.

Theorem 9.
i. $\sin^2 z + \cos^2 z = 1$.
ii. $\sin(z + w) = \sin z \cdot \cos w + \cos z \cdot \sin w$ *and*
$\cos(z + w) = \cos z \cdot \cos w - \sin z \cdot \sin w$.

Again the student is cautioned that these formulas, although plausible, must be proven, since at this stage we know their validity only when w and z are real.

Proof.

i. $\sin^2 z + \cos^2 z = \left[\dfrac{e^{iz} - e^{-iz}}{2i} \right]^2 + \left[\dfrac{e^{iz} + e^{-iz}}{2} \right]^2$

$$= \left(\frac{e^{2iz} - 2 + e^{-2iz}}{-4} \right) + \left(\frac{e^{2iz} + 2 + e^{-2iz}}{4} \right) = 1.$$

ii. $\sin z \cdot \cos w + \cos z \cdot \sin w$

$$= \left(\frac{e^{iz} - e^{-iz}}{2i} \right) \cdot \left(\frac{e^{iw} + e^{-iw}}{2} \right) + \left(\frac{e^{iz} + e^{-iz}}{2} \right) \cdot \left(\frac{e^{iw} - e^{-iw}}{2i} \right),$$

which, using $e^{iz} e^{iw} = e^{i(z+w)}$, simplifies to:

$$\left(\frac{e^{i(z+w)} - e^{-i(z+w)}}{4i} \right) + \left(\frac{e^{i(z+w)} - e^{-i(z+w)}}{4i} \right)$$

$$= \frac{e^{i(z+w)} - e^{-i(z+w)}}{2i} = \sin(z+w).$$

The student can similarly check the addition formula for $\cos(z + w)$. ∎

In addition to $\cos z$, $\sin z$, we can define $\tan z = \sin z / \cos z$ when $\cos z \neq 0$, and similarly obtain the other trigonometric functions.

The logarithmic function

We would now like to be able to define the logarithm in such a way that our definition will agree with the usual definition of $\log x$ when x is located on the positive part of the real axis. In the real case we can define the logarithm as the inverse of the exponential (that is, $\log x = y$ is the solution of $e^y = x$). When we allow z to range over $\mathbb{C}$ we must be more careful because, as has been shown, the exponential is periodic and thus does not have an inverse. Furthermore, the exponential is never zero and so we cannot expect to be able to define the logarithm at zero. Thus we must be more careful in our choice of the domain in $\mathbb{C}$ on which we can define the logarithm. Theorem 10 indicates how this may be done.

Theorem 10. *Let A_{y_0} denote the set of complex numbers $x + iy$ such that $y_0 \leq y < y_0 + 2\pi$; symbolically,*

$$A_{y_0} = \{x + iy \,|\, x \in \mathbb{R} \text{ and } y_0 \leq y < y_0 + 2\pi\}.$$

Then e^z maps A_{y_0} in a one-to-one manner onto the set $\mathbb{C}\backslash\{0\}$.

Note.   Recall that a map is *one-to-one* when the map takes every two distinct points to two distinct points; in other words, two distinct points never get mapped to the same point. The statement that a map is *onto* a set B means that every point of B is the image of some point under the mapping. The notation $\mathbb{C}\backslash\{0\}$ means the whole plane $\mathbb{C}$ minus the point 0; that is, the plane with the origin removed.

Proof.   If $e^{z_1} = e^{z_2}$, then $e^{z_1 - z_2} = 1$, and so $z_1 - z_2 = 2\pi in$ for some integer n, by Theorem 8. But because z_1 and z_2 both lie in A_{y_0}, where the difference between the imaginary parts of any points is less than 2π, we must have $z_1 = z_2$. This argument shows that e^z is one-to-one. Let $w \in \mathbb{C}$ with $w \neq 0$. We claim the equation $e^z = w$ has a solution z in A_{y_0}. The equation $e^{x+iy} = w$ is then equivalent to the two equations $e^x = |w|$ and $e^{iy} = w/|w|$. (Why?) The solution of the first equation is $x = \log |w|$ where "log" is the ordinary logarithm defined on the positive part of the real axis. The second equation has infinitely many solutions y, each differing by integral multiples of 2π, but exactly one of these is in the interval $[y_0, y_0 + 2\pi[$. This y is merely arg w where the specified range for the arg function is $[y_0, y_0 + 2\pi[$. Thus e^z is onto $\mathbb{C}\backslash\{0\}$. ∎

The sets denoted in Theorem 10 may be depicted as in Figure 1.14. Here e^z maps the horizontal strip between $y_0 i$ and $(y_0 + 2\pi)i$ one-to-one onto $\mathbb{C}\backslash\{0\}$.

Note.   The notation $z \mapsto f(z)$ is used to indicate that z is sent to $f(z)$ under the mapping f.

In Theorem 10 an explicit expression was also derived for the inverse of e^z restricted to the strip $y_0 \leq \operatorname{Im} z < y_0 + 2\pi$ and this expression is stated formally in the following definition.

Definition 4.   *The function* log: $\mathbb{C}\backslash\{0\} \rightarrow \mathbb{C}$, *with range* $y_0 \leq$ Im log $z < y_0 + 2\pi$, *is defined by* log $z = \log |z| + i$ arg z, *where* arg z *takes values in* $[y_0, y_0 + 2\pi[$. *Here* $\log |z|$ *is the usual logarithm of the positive real number* $|z|$.

This function is sometimes referred to as the "branch of the logarithm function lying in $\{x + iy | y_0 \leq y < y_0 + 2\pi\}$." But we must remember that the function log z is only well defined when we specify an interval of length 2π in which arg z takes its values, that is, when a

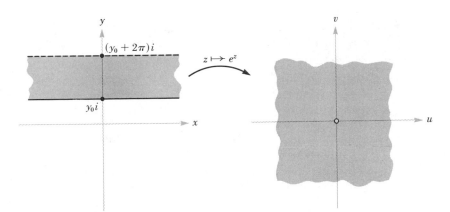

Figure 1.14 e^z as a one-to-one function onto $\mathbb{C}\setminus\{0\}$.

specific branch is chosen. For example, suppose that the specified interval is $[0, 2\pi[$. Then $\log (1 + i) = \log \sqrt{2} + i\,\pi/4$. However, if the specified interval is $[\pi, 3\pi[$, then $\log (1 + i) = \log \sqrt{2} + i\,9\pi/4$.

Theorem 11 summarizes our construction of $\log z$.

Theorem 11. $\log z$ *is the inverse of* e^z *in the following sense: For any branch of* $\log z$, *we have* $e^{\log z} = z$, *and if we choose the branch lying in* $y_0 \le y < y_0 + 2\pi$, *then* $\log(e^z) = z$ *for* $z = x + iy$ *and* $y_0 \le y < y_0 + 2\pi$.

Note. In Section 1.5 the domain of $\log$ will be restricted further to $y_0 < y < y_0 + 2\pi$ so that $\log z$ will be a continuous function.

Proof. Since $\log z = \log |z| + i \arg z$,

$$e^{\log z} = e^{\log|z|} e^{i\arg z}$$

$$= |z|\, e^{i\arg z} = z.$$

Conversely, suppose that $z = x + iy$ and $y_0 \le y < y_0 + 2\pi$. By definition, $\log e^z = \log |e^z| + i\arg e^z$. But $|e^z| = e^x$ and $\arg e^z = y$ by our choice of branch. Thus $\log e^z = \log e^x + iy = x + iy = z.$ ∎

The logarithm defined on $\mathbb{C}\setminus\{0\}$ behaves the same way with respect to products as the logarithm restricted to the positive part of the real axis. This is proved in Theorem 12.

Theorem 12. *If* $z_1, z_2 \in \mathbb{C}\setminus\{0\}$, *then* $\log (z_1 z_2) = \log z_1 + \log z_2$ *(up to the addition of integral multiples of* $2\pi i$).

Proof. $\log z_1 z_2 = \log |z_1 z_2| + i \arg(z_1 z_2)$ where an interval $[y_0, y_0 + 2\pi[$ has been chosen for the values of the arg function. We know that $\log |z_1 z_2| = \log |z_1| \, |z_2| = \log |z_1| + \log |z_2|$ and $\arg(z_1 z_2) = \arg z_1 + \arg z_2$ (up to integral multiples of 2π). Thus $\log z_1 z_2 = (\log |z_1| + i \arg z_1) + (\log |z_2| + i \arg z_2) = \log z_1 + \log z_2$ (up to integral multiples of $2\pi i$). ∎

To illustrate: Let us find $\log[(-1-i)(1-i)]$ where the range for the arg function is chosen as, say, $[0, 2\pi[$. Thus $\log[(-1-i)(1-i)] = \log(-2) = \log 2 + \pi i$. On the other hand, $\log(-1-i) = \log \sqrt{2} + i5\pi/4$ and $\log(1-i) = \log \sqrt{2} + i7\pi/4$. Thus $\log(-1-i) + \log(1-i) = \log 2 + i3\pi = (\log 2 + \pi i) + 2\pi i$; so in this case, when $z_1 = -1-i$ and $z_2 = 1-i$, $\log z_1 z_2$ differs from $\log z_1 + \log z_2$ by $2\pi i$.

Complex powers

We are now in a position to define the term a^b where $a,b \in \mathbb{C}$ and $a \neq 0$ (read "a raised to the power b"). Of course, however we define a^b, the definition should reduce to the usual one in which a and b are real numbers. The trick is to notice that a can also be written $e^{\log a}$ by Theorem 11. If b is an integer, we have $a^b = (e^{\log a})^b = e^{b \log a}$. This last equality holds since if n is an integer and z is any complex number, $(e^z)^n = e^z \cdots e^z = e^{nz}$ by Theorem 8(i). Thus we are led to formulate the following definition.

Definition 5. a^b *(where $a,b \in \mathbb{C}$ and $a \neq 0$) is defined to be $e^{b \log a}$; it is understood that some interval $[y_0, y_0 + 2\pi[$ (that is, some branch of $\log$) has been chosen within which the arg function takes its values.*

It is of the utmost importance to understand precisely what this definition involves. Note especially that in general $\log z$ is "multiple-valued"; that is, $\log z$ can be assigned many different values because different intervals $[y_0, y_0 + 2\pi[$ can be chosen. This is not surprising, for if $b = 1/q$, where q is an integer, then our previous work with de Moivre's Theorem would lead us to expect that a^b is one of the qth roots of a and thus should have q distinct values. The following theorem elucidates this point.

Theorem 13.   *Let $a,b \in \mathbb{C}$, $a \neq 0$. Then a^b is single-valued (that is, the value of a^b does not depend on the choice of branch for $\log$) if and only if b is an integer. If b is a real rational number, and $b = p/q$ is in its lowest terms (in other words, if p and q have no common factor), then a^b has exactly q distinct values; namely, the q roots of a^p. If b is real and irrational or if b has a nonzero imaginary part, then a^b has* *infinitely many values. Where a^b has distinct values, these values differ by factors of the form $e^{2\pi nbi}$.*

Proof.   Choose some interval, say, $[0, 2\pi[$, for the values of the arg function. Let $\log z$ be the corresponding branch of the logarithm. If we were to choose any other branch of the log function, we would obtain $\log a + 2\pi ni$ rather than $\log a$, for some integer n. Thus $a^b = e^{b\log a + 2\pi nbi} = e^{b\log a} \cdot e^{2\pi nbi}$, where the value of n depends on the branch of the logarithm (that is, on the interval chosen for the values of the arg function). But by Theorem 8, $e^{2\pi nbi}$ remains the same for different values of n if and only if b is an integer. Similarly, it can be shown that $e^{2\pi nip/q}$ has q distinct values if p and q have no common factor. If b is irrational, and if $e^{2\pi nbi} = e^{2\pi mbi}$, it follows that $e^{(2\pi bi)(n-m)} = 1$ and hence $b(n-m)$ is an integer; since b is irrational, this implies that $n - m = 0$. Thus if b is irrational, $e^{2\pi nbi}$ has infinitely many distinct values. If b is of the form $x + iy$, $y \neq 0$, then $e^{2\pi nbi} = e^{-2\pi ny} \cdot e^{2\pi nix}$, which also has infinitely many distinct values. ∎

To repeat: When we write $e^{b\log a}$, it is understood that some branch of log has been chosen, and accordingly $e^{b\log a}$ has a single well-defined value. But as we change the branch of log, we get values for $e^{b\log a}$ that differ by factors of $e^{2\pi inb}$. This is what we mean when we say that $a^b = e^{b\log a}$ is "multiple-valued."

An example should make this clear. Let $a = 1 + i$ and let b be some real irrational number. Then the infinitely many different possible values of a^b are given by $e^{b[\log(1+i)+2\pi ni]} = e^{b(\log \sqrt{2} + i\pi/4 + 2\pi ni)} = (e^{b\log \sqrt{2} + ib\pi/4}) e^{b2\pi ni}$ as n takes on all integral values (corresponding to different choices of the branch). For instance, if we used the branch corresponding to $[-\pi,\pi[$ or $[0,2\pi[$, we would set $n = 0$.

Some general properties of a^b are found in the exercises at the end of this section, but we are now interested in the special case when b is of the form $1/n$, because this gives the nth root.

The nth root function

We know that $\sqrt[n]{z}$ has exactly n values for $z \neq 0$. To make it a specific function we must again single out a branch of log as described in the preceding paragraphs. The function $\sqrt[n]{z}$ is defined formally in Definition 6.

Definition 6.   *The nth root function is defined by*

$$\sqrt[n]{z} = z^{1/n} = e^{(\log z)/n}$$

for a specific choice of branch of $\log z$; with this choice, $\sqrt[n]{z} = e^{(\log z)/n}$ is called a branch of the nth root function.

The following theorem should not surprise us.

Theorem 14. $\sqrt[n]{z}$ *so defined is an nth root of z. It is obtained as follows. If $z = re^{i\theta}$, then*

$$\sqrt[n]{z} = \sqrt[n]{r}\, e^{i\theta/n}$$

where θ is chosen so that it lies within a particular interval corresponding to the branch choice. As we add multiples of 2π to θ, we run through the n nth roots of z.

Proof.   By our definition,

$$\sqrt[n]{z} = e^{(\log z)/n}.$$

But $\log z = \log r + i\theta$; so

$$e^{(\log z)/n} = e^{(\log r)/n} \cdot e^{i\theta/n} = \sqrt[n]{r}e^{i\theta/n}.$$

The assertion is then clear.∎

The student should convince himself that this way of describing the n nth roots of z is the same as that described in Theorem 5.

Geometry of the elementary functions

To further understand the functions z^n, $\sqrt[n]{z}$, e^z, and $\log z$, we shall consider the geometric interpretation of each in the remainder of this section.

Let us begin with the power function z^n and let $n = 2$. We know that z^2 has length $|z|^2$ and argument $2\arg z$. Thus the map $z \mapsto z^2$ squares lengths and doubles arguments (see Figure 1.15).

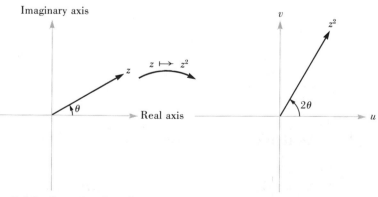

Figure 1.15   Squaring function.

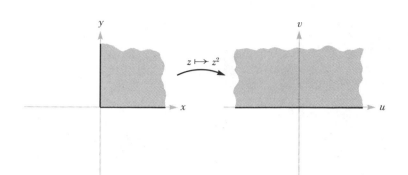

Figure 1.16 Effect of the squaring function on the first quadrant.

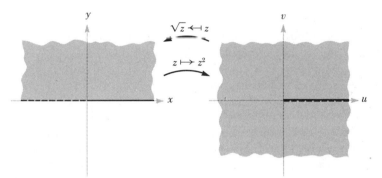

Figure 1.17 The squaring function and its inverse.

It should be clear from this doubling of angles that the power function z^2 maps the first quadrant to the whole upper half plane (see Figure 1.16). Similarly, the upper half plane is mapped to the whole plane.

Now compare the square root function $\sqrt{z} = \sqrt{r}e^{i\theta/2}$. Suppose that we choose a branch by using the interval $0 \le \theta < 2\pi$. Then $0 \le \theta/2 < \pi$, so $\sqrt{z}$ will always lie in the upper half plane, and the angles thus are cut in half. The situation is similar to that involving the exponential function in that $z \mapsto \sqrt{z}$ is the inverse for $z \mapsto z^2$ when the latter is restricted to a region on which it is one-to-one. In like manner, if we choose the branch $-\pi \le \theta < \pi$, we have $-\pi/2 \le \theta/2 < \pi/2$, so $\sqrt{z}$ takes its values in the right half plane instead of the upper half plane. (Generally, any "half plane" could be used—see Figure 1.17.) If we choose a specific branch of $\sqrt{z}$, we also choose which of the two possible square roots we shall obtain.

Various geometric statements can be made concerning the map $z \mapsto z^2$ that also give information about the inverse, $z \mapsto \sqrt{z}$. For

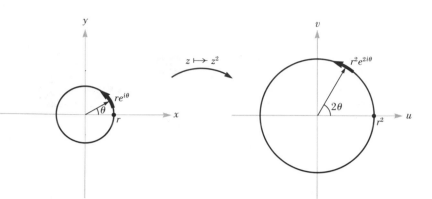

Figure 1.18   Effect of the squaring function on a circle of radius r.

example, a circle of radius r, described by the set of points $re^{i\theta}$, $0 \le \theta < 2\pi$, is mapped to $r^2 e^{i2\theta}$, a circle of radius r^2; as $re^{i\theta}$ moves once around the first circle, the image point moves twice around (see Figure 1.18). The inverse map does the opposite; as z moves along the circle $re^{i\theta}$ of radius r, $\sqrt{z}$ moves half as fast along the circle $\sqrt{r}e^{i\theta/2}$ of radius $\sqrt{r}$.

The correct domains on which $z \mapsto e^z$ and $z \mapsto \log z$ are inverses have already been discussed (see Figure 1.14). Let us note that a line $y = \text{constant}$; that is, points $x + iy$ as x varies are mapped to points $e^x e^{iy}$, which is a ray with argument y. As x ranges from $-\infty$ to $+\infty$, the image point on the ray goes from 0 out to infinity (see Figure 1.19). Similarly, the vertical line $x = \text{constant}$ is mapped to a circle of radius e^x. If we restrict y to an interval of length 2π, the image circle is described once, but if y is unrestricted, the image circle is described infinitely many times as y ranges from $-\infty$ to $+\infty$. The logarithm, being the inverse of e^z, maps points in the opposite direction to e^z, as shown in Figure 1.19. Because of the special nature of the strip-like regions in Figures 1.14 and 1.19 (on them e^z is one-to-one) and because of the periodicity of e^z, these regions deserve a name. They are usually called *period strips* of e^z.

Worked Examples

1.   Find the real and imaginary parts of $\exp(e^z)$. ($\exp w$ is another way of writing e^w.)

Solution.   Let $z = x + iy$; then $e^z = e^x \cos y + i e^x \sin y$. Thus $\exp(e^z) = e^{e^x \cos y} \{\cos(e^x \sin y) + i\sin(e^x \sin y)\}$. Hence $\text{Re}(\exp(e^z)) = (e^{e^x \cos y}) \cos(e^x \sin y)$ and $\text{Im}(\exp(e^z)) = (e^{e^x \cos y}) \sin(e^x \sin y)$.

2.   Show that $\sin(iy) = i\sinh y$, where

$$\sinh y = \frac{e^y - e^{-y}}{2}$$

by definition.

Solution.

$$\sin(iy) = \frac{e^{i(iy)} - e^{-i(iy)}}{2i} = i\left\{\frac{e^y - e^{-y}}{2}\right\}$$

$$= i\sinh y.$$

3.   Find all the values of i^i.

Solution.

$$i^i = e^{i\log i}$$

$$= e^{i(\log 1 + (i\pi/2) + (2\pi n)i)}$$

$$= (e^{-2\pi n})\, e^{-\pi/2}.$$

All the values of i^i are given by the last expression as n takes integral values, $n = 0, \pm 1, \pm 2, \cdots$.

4.   Solve $\cos z = 1/2$. (Of course, we know that $z_n = \pm[(\pi/3) + 2\pi n]$, n an integer, solves $\cos z = 1/2$, but we shall show that z_n, $n = 0, \pm 1, \cdots$, are the only solutions; that is, there are no solutions off the real axis.)

Solution.   We are given

$$\cos z = \frac{e^{iz} + e^{-iz}}{2} = \frac{1}{2}.$$

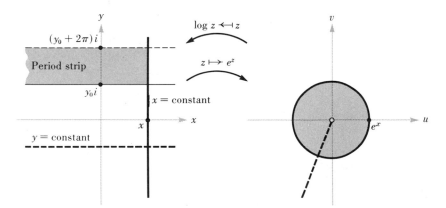

Figure 1.19   Geometry of e^z and $\log z$.

Hence $e^{2iz} - e^{iz} + 1 = 0$ or $e^{iz} = (1 \pm \sqrt{-3})/2 = (1/2) \pm \sqrt{3}i/2$. So $iz = \log [(1/2) \pm (\sqrt{3}i/2)] = \pm \log [(1/2) + (\sqrt{3}i/2)]$, since $(1/2) \pm (\sqrt{3}i/2)$ are reciprocals of one another. We thus obtain

$$z = \pm i \log\left(\frac{1}{2} + \frac{\sqrt{3}}{2}i\right)$$

$$= \pm i\left(\log 1 + \frac{\pi}{3}i + 2\pi ni\right)$$

$$= \pm \left(\frac{\pi}{3} + 2\pi n\right).$$

5. Consider the mapping $z \mapsto \sin z$. Show that lines parallel to the real axis are mapped to ellipses and that lines parallel to the imaginary axis are mapped to hyperbolas.

Solution. Using Theorem 9 (see also example 2), we get

$$\sin z = \sin(x + iy)$$
$$= \sin x \cos(iy) + \sin(iy) \cos x$$
$$= \sin x \cosh y + i \sinh y \cos x,$$

where

$$\cosh y = \frac{e^y + e^{-y}}{2} \text{ and } \sinh y = \frac{e^y - e^{-y}}{2}.$$

Suppose that $y = y_0$ is constant; then if we write

$$\sin z = u + iv$$

we have

$$\frac{u^2}{\cosh^2 y_0} + \frac{v^2}{\sinh^2 y_0} = 1,$$

since $\sin^2 x + \cos^2 x = 1$. This is an ellipse.

Similarly, if $x = x_0$ is constant, from $\cosh^2 y - \sinh^2 y = 1$ we obtain

$$\frac{u^2}{\sin^2 x_0} - \frac{v^2}{\cos^2 x_0} = 1,$$

which is a hyperbola.

Exercises

1. Compute e^{2+i}; $\sin(1 + i)$; $\cos(2 + 3i)$.

2. Solve $\cos z = (3/4) + (i/4)$ and $\cos z = 4$.

3. Let $\sqrt{}$ denote the particular square root defined by $\sqrt{r(\cos \theta + i \sin \theta)}$ $= r^{1/2}[\cos(\theta/2) + i \sin(\theta/2)], \; 0 \le \theta < 2\pi$; the other square root is

$r^{1/2}\{\cos[(\theta + 2\pi)/2] + i\sin[(\theta + 2\pi)/2]\}$. For what values of z does the equation $\sqrt{z^2} = z$ hold?

4. Compute all values of $\log 1$; $\log i$; $\log -i$; $\log (1 + i)$.

5. Find all the values of $(-i)^i$, $(-1)^i$, 2^i, $(1 + i)^{1+i}$.

6. Along which rays through the origin (a ray is determined by $\arg z = $ constant does $\lim_{z \to \infty} |e^z|$ exist?

7. Prove that

$$z = \tan\left\{\frac{1}{i} \log\left(\frac{1 + iz}{1 - iz}\right)^{1/2}\right\}.$$

8. Simplify e^{z^2}; e^{iz}; $e^{1/z}$; where $z = x + iy$. For $e^{1/z}$ we specify that $z \neq 0$.

9. For what values of z do we have $\overline{(e^{iz})} = e^{i\bar{z}}$?

10. Examine the behavior of e^{x+iy} as $x \to \pm\infty$ and the behavior of e^{x+iy} as $y \to \pm \infty$.

11. Prove that $\sin(-z) = -\sin z$; that $\cos(-z) = \cos z$; and that $\sin[(\pi/2) - z] = \cos z$.

12. Define sinh and cosh on all of $\mathbb{C}$ by $\sinh z = (e^z - e^{-z})/2$, $\cosh z = (e^z + e^{-z})/2$.
   Prove that:
   a. $\cosh^2 z - \sinh^2 z = 1$.
   b. $\sinh(z_1 + z_2) = \sinh z_1 \cosh z_2 + \cosh z_1 \sinh z_2$.
   c. $\cosh(z_1 + z_2) = \cosh z_1 \cosh z_2 + \sinh z_1 \sinh z_2$.
   d. $\sinh(x + iy) = \sinh x \cos y + i \cosh x \sin y$.
   e. $\cosh(x + iy) = \cosh x \cos y + i \sinh x \sin y$.

13. Use $\sin z = \sin x \cosh y + i \sinh y \cos x$ where $z = x + iy$ to prove that $|\sinh y| \leq |\sin z| \leq |\cosh y|$.

14. If b is real, prove that $|a^b| = |a|^b$.

15. a. The map $z \mapsto z^3$ maps the first quadrant onto what?
   b. Discuss the geometry of $z \mapsto \sqrt[3]{z}$ as was done in the text for $\sqrt{z}$.

16. The map $z \mapsto 1/z$ takes the exterior of the unit circle to the interior (excluding zero) and vice versa. What are lines $\arg z = $ constant mapped to?

17. Using polar coordinates, show that $z \mapsto z + 1/z$ maps the circle $|z| = 1$ to the interval $[-2,2]$ on the x-axis.

18. a. For complex numbers a,b,c, prove that $a^b a^c = a^{b+c}$ using a fixed branch of log.
   b. Show that $(ab)^c = a^c b^c$ if we choose branches so that $\log(ab) = \log a + \log b$ (with no extra $2\pi ni$).

19. What is the image of vertical and horizontal lines under $z \mapsto \cos z$?

20. a. Show that, under $z \mapsto z^2$, lines parallel to the real axis are mapped to parabolas.
   b. Show that, under (a branch of) $z \mapsto \sqrt{z}$, lines parallel to the real axis are mapped to hyperbolas.

21. Under what conditions does $\log(a^b) = b \log a$ for complex numbers a,b? (Use the branch of log with $-\pi \leq \theta < \pi$.)

22. Is it true that $|a^b| = |a|^{|b|}$ for all $a,b \in \mathbb{C}$?

23. Show that the trigonometric identities can formally be deduced if $e^{i(x_1 + x_2)} = e^{ix_1} \cdot e^{ix_2}$ is assumed.

24. Show that n nth roots of unity are $1, w, w^2, w^3, \cdots, w^{n-1}$ where $w = e^{2\pi i/n}$.

25. Show that $\sin z = 0$ iff $z = k\pi$, $k = 0, \pm 1, \pm 2, \cdots$.

26. Show that $\log z = 0$ iff $z = 1$, using the branch with $-\pi < \arg z \le \pi$.

27. Compute the following numerically to two significant figures: $e^{3.2 + 6.1i}$; $\log(1.2 - 3.0i)$; $\sin(8.1i - 3.2)$.

28. Find the maximum of $|\cos z|$ on the square $0 \le \operatorname{Re} z \le 2\pi, 0 \le \operatorname{Im} z \le 2\pi$.

29. Show that $\sin z$ maps the strip $-\pi/2 < \operatorname{Re} z < \pi/2$ onto $\mathbb{C} \backslash \{ z \mid \operatorname{Im} z = 0$ and $|\operatorname{Re} z| \ge 1 \}$.

30. Discuss the inverse functions $\sin^{-1} z$, $\cos^{-1} z$. For example, is $\sin z$ one-to-one on $0 \le \operatorname{Re} z < 2\pi$?

1.4　Analytic Functions

In this section the fundamental notion of complex differentiability and its main properties will be analyzed. The rules for differentiation of a function of a complex variable are very similar to those for differentiation of a function of a real variable and the student should have no trouble adapting his knowledge of the latter to the former. This section will be concerned mostly with the underlying theory; the elementary functions will be considered in Section 5.

Because this theory is very similar to that used in calculus, most of the technical proofs will appear at the end of this section, on the

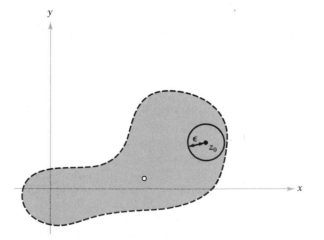

Figure 1.20　An open set.

Figure 1.21 Neighborhood and deleted neighborhood.

assumption that the student has a knowledge of calculus for functions of $\mathbb{R}^2$ to $\mathbb{R}^2$.

Since $\mathbb{C}$ is merely $\mathbb{R}^2$ with the extra structure of complex multiplication, we can immediately translate many concepts from $\mathbb{R}^2$ into complex notation. In fact we have already done so in analyzing the absolute value $|z|$, which is the same as the norm, or length, of z regarded as an element of $\mathbb{R}^2$.

Open sets

It is necessary first to recall the notion of an open set. A set $A \subset \mathbb{C} = \mathbb{R}^2$ is *open* iff, for each point $z_0 \in A$, there is a real number ϵ, $\epsilon > 0$, such that if $|z - z_0| < \epsilon$, then $z \in A$. This means that some small disk around z_0 lies in A (see Figure 1.20).

Note that the value of ϵ depends on z_0; as z_0 gets close to the "edge" of A, ϵ gets smaller. Intuitively, a set is open if it does not contain any of its "boundary," or "edge" points.

Given $r > 0$, the *r-neighborhood* or *r-disk* around a point $z_0 \in \mathbb{C}$ is defined to be the set $D(z_0, r) = \{z \in \mathbb{C} \mid |z - z_0| < r\}$.

For review the student should prove that for each $w_0 \in \mathbb{C}$ and $r > 0$, the disk $A = \{z \in \mathbb{C} \mid |z - w_0| < r\}$ itself is open.

A *deleted r-neighborhood* is an *r*-neighborhood whose center point has been removed. Thus a deleted neighborhood has the form $D(z_0, r) \backslash \{z_0\}$, which stands for the set $D(z_0, r)$ minus the set $\{z_0\}$ (see Figure 1.21). A *neighborhood* of a point z_0 is merely an open set containing z_0.

Thus we can rephrase the definition of "open" as follows. A set A is open iff, for each $z_0 \in A$, there is an *r*-neighborhood of z_0 wholly contained in A.

Mappings and continuity

Let $A \subset \mathbb{C}$ be a set. We recall that a mapping $f \colon A \to \mathbb{C}$ is merely an assignment of a specific point $f(z) \in \mathbb{C}$ to each $z \in A$, A being the domain of f. When the domain is a set in $\mathbb{C}$ and when the range (the set of values f assumes) consists of complex numbers, we speak of f as a *complex function of a complex variable.*

Alternatively, we can think of f as a map $f: A \subset \mathbb{R}^2 \to \mathbb{R}^2$; therefore, f becomes a vector-valued function of two real variables.

For $f: A \subset \mathbb{C} \to \mathbb{C}$, we can let $z = x + iy = (x,y)$ and define $u(x,y) = \text{Re } f(z)$ and $v(x,y) = \text{Im } f(z)$. Thus u and v are merely the components of f thought of as a vector function. Hence we may write uniquely $f(x + iy) = u(x,y) + iv(x,y)$, where u and v are *real*-valued functions defined on A.

Let us next transcribe the notion of limit to the notation of complex numbers.

Definition 7.   *Let f be defined on a deleted neighborhood of z_0. The expression*

$$\underset{z \to z_0}{\text{limit}} f(z) = a$$

means that for every $\epsilon > 0$, there is a $\delta > 0$ such that $z \in D(z_0, r)$, $z \neq z_0$, and $|z - z_0| < \delta$ imply that $|f(z) - a| < \epsilon$.

The expression in this definition has the same intuitive meaning as it has in calculus; namely, $f(z)$ is close to a when z is close to z_0.

The limit a is unique. (See the end of this section for proofs of this and the following assertions regarding limits and derivatives. Extensive proofs will not be given because the student should already be familiar with many of the ideas included in them.)

It is not necessary to define f on a whole deleted neighborhood to have a valid theory of limits, but deleted neighborhoods are used here for the sake of simplicity and also because such usage will be appropriate later in the text.

The following properties of limits hold:

If $\underset{z \to z_0}{\text{limit}} f(z) = a$ and $\underset{z \to z_0}{\text{limit}} g(z) = b$, then

  i. $\underset{z \to z_0}{\text{limit}} [f(z) + g(z)] = a + b$.

 ii. $\underset{z \to z_0}{\text{limit}} [f(z)g(z)] = ab$.

iii. $\underset{z \to z_0}{\text{limit}} [f(z)/g(z)] = a/b$ if $b \neq 0$.

Also, if h is defined at the points $f(z)$ and $\underset{w \to a}{\text{limit}} h(w) = c$, then

 iv. $\underset{z \to z_0}{\text{limit}} h(f(z)) = c$.

Definition 8.   *Let $A \subset \mathbb{C}$ be an open set and let $f: A \to \mathbb{C}$ be a given function. We say f is continuous at $z_0 \in A$ iff*

$$\underset{z \to z_0}{\text{limit}} f(z) = f(z_0),$$

and f is continuous on A if f is continuous at each $z_0 \in A$.

This definition has the same intuitive meaning as it has in elementary calculus: If z is close to z_0, then $f(z)$ is close to $f(z_0)$. However, complex functions are more difficult to picture than are real functions because the graph of $f(z)$ requires four-dimensional space for its full visualization. The geometric interpretations included in the specific examples of Section 1.3 are more instructive than graphic visualization.

From (i), (ii), and (iii) we can immediately deduce that if f and g are continuous on A, then so are the sum $f + g$ and the product fg, and so is f/g if $g(z_0) \neq 0$ for all $z_0 \in A$. Also, if h is defined and continuous on the range of f, then the composition $h \circ f$, defined by $h \circ f(z) = h(f(z))$, is continuous by (iv) (see exercise 3 at the end of this section for another proof of (iv)).

Sequences

The concept of a convergent sequence is analogous to that of a real sequence studied in calculus. A sequence z_n of points of $\mathbb{C}$ *converges* to z_0 iff, for every $\epsilon > 0$, there is an N such that $n \geq N$ implies that $|z_n - z_0| < \epsilon$. In notation, the limit of a sequence is expressed as

$$\underset{n \to \infty}{\text{limit}}\ z_n = z_0, \quad \text{or} \quad z_n \to z_0.$$

Sequences have the same properties and are obtained by the same proofs as limits of functions. For example, the limit is unique if it exists; and if $z_n \to z_0$ and $w_n \to w_0$, then

i. $z_n + w_n \to z_0 + w_0$,
ii. $z_n w_n \to z_0 w_0$, and
iii. $z_n/w_n \to z_0/w_0$ (if w_0, $w_n \neq 0$).

Also, $z_n \to z_0$ iff $\operatorname{Re} z_n \to \operatorname{Re} z_0$ and $\operatorname{Im} z_n \to \operatorname{Im} z_0$. Proof of this statement is found in exercise 1 at the end of this section.

A sequence z_n is called *Cauchy* if, for any $\epsilon > 0$, there is an N such that $n, m \geq N$ implies that $|z_n - z_m| < \epsilon$. We know from calculus that any Cauchy sequence in $\mathbb{R}$ converges. From the fact that $z_n \to z_0$ iff $\operatorname{Re} z_n \to \operatorname{Re} z_0$ and $\operatorname{Im} z_n \to \operatorname{Im} z_0$, we can conclude that any Cauchy sequence in $\mathbb{C}$ converges. This is a technical point that is often useful in convergence proofs as we shall see in Chapter 3.

It should be noted that a link exists between sequences and continuity; namely, $f: A \subset \mathbb{C} \to \mathbb{C}$ is continuous iff, for any convergent sequence $z_n \to z_0$ of points in A (that is, $z_n \in A$ and $z_0 \in A$), we have $f(z_n) \to f(z_0)$. The student is requested to prove this in exercise 20 at the end of this section.

Although continuity is an important concept, its importance in complex analysis is completely overshadowed by that of the complex derivative, which is defined next.

Differentiability

Definition 9.   *Let $f: A \rightarrow \mathbb{C}$ where $A \subset \mathbb{C}$ is an open set. Then f is said to be differentiable in the complex sense at $z_0 \in A$ if*

$$\underset{z \rightarrow z_0}{\text{limit}} \frac{f(z) - f(z_0)}{z - z_0}$$

exists. This limit is denoted by $f'(z_0)$, or sometimes by df/dz (z_0). Thus $f'(z_0)$ is a complex number. f is said to be analytic on A if f is complex differentiable at each $z_0 \in A$. The word "holomorphic," which is sometimes used, is synonymous with the word "analytic." The phrase "analytic at z_0" means analytic on a neighborhood of z_0.

Note that

$$\frac{f(z) - f(z_0)}{z - z_0}$$

is undefined at $z = z_0$, and this is the reason why deleted neighborhoods were used in the definition of limit.

The student is cautioned that although the definition of the derivative $f'(z_0)$ is similar to that of the usual derivative of a function of a real variable and although many of their properties are similar, the complex case is much richer. Note also that in the definition of $f'(z_0)$, we are dividing by the complex number $z - z_0$ and the special nature of division by complex numbers is also a consideration. The limit as $z \rightarrow z_0$ is taken for an arbitrary z approaching z_0 but not along any particular direction.

The existence of f' implies a great deal about f. It will be proven in Section 2.4 that if f' exists, then all the derivatives of f exist (that is, f'', the (complex) derivative of f', exists, and so on). This is in marked contrast to the case of a function $g(x)$ of the real variable x, in which $g'(x)$ can exist without the existence of $g''(x)$.

The analysis of what are called the Cauchy-Riemann equations in Theorem 19 will show how the complex derivative of f is related to the usual partial derivatives of f as a function of the real variables (x, y) and will supply a useful criterion for determining the existence of $f'(z_0)$.

Theorem 15.   *If $f'(z_0)$ exists, then f is continuous at z_0.*

As in elementary calculus, continuity of f does not imply differentiability; for example, $f(z) = |z|$ is continuous but is not differentiable (see exercises 1 and 7 at the end of this section).

Theorem 16 demonstrates that the usual rules of calculus can be used when differentiating analytic functions.

Theorem 16. *Suppose that f and g are analytic on A, where $A \subset \mathbb{C}$ is an open set. Then*
  i. *$af + bg$ is analytic on A and $(af + bg)'(z) = af'(z) + bg'(z)$ for any complex numbers a and b.*
 ii. *fg is analytic on A and $(fg)'(z) = f'(z)g(z) + f(z)g'(z)$.*
iii. *If $g(z) \neq 0$ for all $z \in A$, then f/g is analytic on A and*

$$\left(\frac{f}{g}\right)'(z) = \frac{f'(z)g(z) - g'(z)f(z)}{[g(z)]^2}.$$

 iv. *Any polynomial $a_0 + a_1 z + \cdots + a_n z^n$ is analytic on all of $\mathbb{C}$ with derivative $a_1 + 2a_2 z + \cdots + n a_n z^{n-1}$.*
  v. *Any rational function*

$$\frac{a_0 + a_1 z + \cdots + a_n z^n}{b_0 + b_1 z + \cdots + b_m z^m}$$

is analytic on the set A, consisting of all z except those (at most, m) points where the denominator is zero.

The student will also recall that one of the most important rules for differentiation is the Chain Rule, or "function of a function" rule. To illustrate:

$$\frac{d}{dz}(z^3 + 1)^{10} = 10(z^3 + 1)^9 \cdot 3z^2$$

$$= 30 z^2 (z^3 + 1)^9.$$

This procedure for differentiating should be familiar; it is justified by Theorem 17.

Theorem 17 (Chain Rule). *Let $f: A \to \mathbb{C}$ and $g: B \to \mathbb{C}$ be analytic (A,B are open sets) and let $f(A) \subset B$. Then $g \circ f: A \to \mathbb{C}$ defined by $g \circ f(z) = g(f(z))$ is analytic and*

$$\frac{d}{dz} g \circ f(z) = g'(f(z)) \cdot f'(z).$$

The basic idea of the proof of Theorem 17 is that if $w = f(z)$, and $w_0 = f(z_0)$, then

$$\frac{g(f(z)) - g(f(z_0))}{z - z_0} = \frac{g(w) - g(w_0)}{w - w_0} \cdot \frac{f(z) - f(z_0)}{z - z_0},$$

and if we let $z \to z_0$, we also have $w \to w_0$, and the right side of the preceding equation thus becomes $g'(w_0)f'(z_0)$. The trouble is that even if $z \neq z_0$, we could have $w = w_0$. Therefore, this proof is specious. A more careful proof is given at the end of this section. An argument similar to the one given there shows that if $c: \,]a,b[\,\rightarrow \mathbb{C}$ is differentiable, we can differentiate the curve $d(t) = f(c(t))$ and obtain $d'(t) = f'(c(t)) \cdot c'(t)$. Here $c'(t)$ is the derivative of c as a function $]a,b[\,\rightarrow \mathbb{R}^2$; that is, if $c(t) = (x(t), y(t))$, then $c'(t) = (x'(t), y'(t)) = x'(t) + iy'(t)$.

Conformal maps

The existence of the complex derivative f' places severe, but very useful, restrictions on f. The first of these restrictions will be briefly discussed here. Another restriction will be mentioned when the Cauchy-Riemann equations are analyzed in Theorem 19.

It will be shown that "infinitesimally" near a point z_0 at which $f'(z_0) \neq 0$, f is a rotation by $\arg f'(z_0)$ and a magnification by $|f'(z_0)|$. The term "infinitesimally" is defined more precisely below, but intuitively it means that locally f is approximately a rotation together with a magnification (see Figure 1.22). If $f'(z_0) = 0$, the structure of f is more complicated. (This point will be studied further in Chapter 6.)

Definition 10. *A map $f: A \rightarrow \mathbb{C}$ is called conformal at z_0 if there is a $\theta \in [0, 2\pi[$ and $r > 0$ such that for any curve $c(t)$, which is differentiable at $t = 0$, $c(t) \in A$, $c(0) = z_0$ and $v = c'(0) \neq 0$ (v is the tangent vector to c at $c(0)$), the curve $d(t) = f(c(t))$ is differentiable at $t = 0$, and, setting $u = d'(0)$, we have $|u| = r\,|v|$ and $\arg u = \arg v + \theta \pmod{2\pi}$.*

In this text a map will be called conformal when it is conformal at every point.

Thus a conformal map merely rotates and stretches tangent vectors to curves. This is the precise meaning of "infinitesimal" as previously

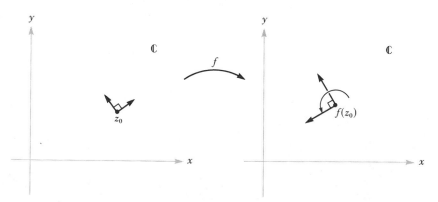

Figure 1.22   Conformal map at z_0.

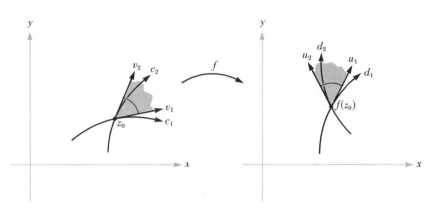

Figure 1.23 Preservation of angles by a conformal map.

used. It should be noted that a conformal map *preserves angles* between intersecting curves. (By definition, the angle between two curves is the angle between their tangent vectors — see Figure 1.23.)

Theorem 18. *If* $f\colon A \to \mathbb{C}$ *is analytic and if* $f'(z_0) \neq 0$, *then* f *is conformal at* z_0 *with* $\theta = \arg f'(z_0)$ *and* $r = |f'(z_0)|$ *fulfilling Definition 10.*

The proof of this theorem is remarkably simple.

Proof. Using the preceding notation and the Chain Rule, we get $u = d'(0) = f'(z_0) \cdot c'(0) = f'(z_0) \cdot v$. Thus $\arg(u) = \arg f'(z_0) + \arg v$ (mod 2π) and $|u| = |f'(z_0)| \cdot |v|$, as required ∎

The point of this proof is that we get a fixed complex number, namely, $f'(z_0)$, times v, no matter in which direction v is pointing. This is because, in the definition of $f'(z_0)$, $\displaystyle\lim_{z \to z_0}$ is "taken through all possible directions" as $z \to z_0$.

Cauchy-Riemann equations

Let us recall that if $f\colon A \subset \mathbb{C} = \mathbb{R}^2 \to \mathbb{R}^2$ and if $f(x,y) = (u(x,y),\, v(x,y)) = u(x,y) + iv(x,y)$, then the *Jacobian matrix* of f is defined as the matrix of partial derivatives given by

$$Df(x,y) = \begin{pmatrix} \dfrac{\partial u}{\partial x} & \dfrac{\partial u}{\partial y} \\[2mm] \dfrac{\partial v}{\partial x} & \dfrac{\partial v}{\partial y} \end{pmatrix}$$

at each point (x,y). We want to relate these partial derivatives to the complex derivative.

More precisely, f is called *differentiable* (in the sense of real variables) with derivative the matrix $Df(x_0,y_0)$ at (x_0,y_0) iff, for any $\epsilon > 0$, there is a $\delta > 0$ such that $|(x_0,y_0) - (x,y)| < \delta$ implies that $|f(x,y) - f(x_0,y_0) - Df(x_0,y_0)[(x,y) - (x_0,y_0)]| \leq \epsilon |(x,y) - (x_0,y_0)|$, where $Df(x_0,y_0) \cdot [(x,y) - (x_0,y_0)]$ means the matrix $Df(x_0,y_0)$ applied to the (column) vector

$$\begin{pmatrix} x - x_0 \\ y - y_0 \end{pmatrix},$$

and $|w|$ stands for the length of a vector w.

If f is differentiable, then the usual partials $\partial u/\partial x$, $\partial u/\partial y$, $\partial v/\partial x$, and $\partial v/\partial y$ exist and $Df(x_0,y_0)$ is given by the Jacobian matrix. The expression $Df(x_0,y_0)[w]$ represents the derivative of f in the direction w. If the partials exist and are continuous, then f is differentiable. Generally then, differentiability is a bit stronger than existence of the individual partials. (Proofs of the preceding statements are not included here but can be found in any advanced calculus text, such as J. Marsden, *Elementary Classical Analysis* (San Francisco: W. H. Freeman and Co., 1974), Chapter 6.)

The main result connecting the partial derivatives and analyticity is stated in Theorem 19.

Theorem 19. *Let $f: A \subset \mathbb{C} \to \mathbb{C}$ be a given function, with A an open set. Then $f'(z_0)$ exists if and only if f is differentiable in the sense of real variables and, at $(x_0, y_0) = z_0$, u,v satisfy*

$$\frac{\partial u}{\partial x} = \frac{\partial v}{\partial y} \quad and \quad \frac{\partial u}{\partial y} = -\frac{\partial v}{\partial x}$$

(called the Cauchy-Riemann equations).

Thus, if $\partial u/\partial x$, $\partial u/\partial y$, $\partial v/\partial x$, and $\partial v/\partial y$ exist, are continuous on A, and satisfy the Cauchy-Riemann equations, then f is analytic on A.

If $f'(z_0)$ does exist, then

$$f'(z_0) = \frac{\partial u}{\partial x} + i \frac{\partial v}{\partial x}$$

$$= \frac{\partial v}{\partial y} - i \frac{\partial u}{\partial y}.$$

Because the proof of this theorem that appears at the end of the section is somewhat devious, a separate and more direct proof is given here to show that if $f'(z_0)$ exists, then u and v satisfy the Cauchy-Riemann equations.

Proof.   In the limit

$$f'(z_0) = \lim_{z \to z_0} \frac{f(z) - f(z_0)}{z - z_0},$$

let us take the special case that $z = x + iy_0$. Then

$$\frac{f(z) - f(z_0)}{z - z_0} = \frac{u(x,y_0) + iv(x,y_0) - u(x_0,y_0) - iv(x_0,y_0)}{x - x_0}$$

$$= \frac{u(x,y_0) - u(x_0,y_0)}{x - x_0} + i \frac{v(x,y_0) - v(x_0,y_0)}{x - x_0}.$$

As $x \to x_0$, the left side of the equation converges to the limit $f'(z_0)$. Thus both real and imaginary parts of the right side must converge to a limit (see exercise 1 at the end of this section). From the definition of the partial derivatives, this limit is $\partial u/\partial x(x_0,y_0) + i \, \partial v/\partial x(x_0,y_0)$. Thus $f'(z_0) = \partial u/\partial x + i \, \partial v/\partial x$.

Next let $z = x_0 + iy$. Then we similarly have

$$\frac{f(z) - f(z_0)}{z - z_0} = \frac{u(x_0,y) + iv(x_0,y) - u(x_0,y_0) - iv(x_0,y_0)}{i(y - y_0)}$$

$$= \frac{u(x_0,y) - u(x_0,y_0)}{i(y - y_0)} + \frac{v(x_0,y) - v(x_0,y_0)}{y - y_0}.$$

As $y \to y_0$, we get

$$\frac{1}{i} \frac{\partial u}{\partial y} + \frac{\partial v}{\partial y} = \frac{\partial v}{\partial y} - i \frac{\partial u}{\partial y}.$$

Thus, since $f'(z_0)$ exists and has the same value regardless of how z approaches z_0, we get

$$f'(z_0) = \frac{\partial u}{\partial x} + i \frac{\partial v}{\partial x} = \frac{\partial v}{\partial y} - i \frac{\partial u}{\partial y}.$$

By comparing real and imaginary parts of these equations, we derive the Cauchy-Riemann equations as well as the two formulas for $f'(z_0)$. ∎

We can also express the Cauchy-Riemann equations in terms of polar coordinates, but care must be exercised because the change of coordinates defined by $r = \sqrt{x^2 + y^2}$ and $\theta = \arg(x + iy)$ is a differentiable change only if θ is restricted to the *open* interval $]0,2\pi[$ or any

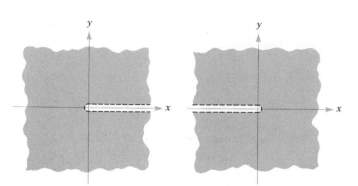

Figure 1.24 Two regions of validity of polar coordinates.

other open interval of length 2π and if the origin $(r = 0)$ is omitted. Without such a restriction θ is discontinuous because it jumps by 2π on crossing the x-axis. Using $\partial x/\partial r = \cos\theta$, $\partial y/\partial r = \sin\theta$, we easily see that the Cauchy-Riemann equations are equivalent to saying that

$$\frac{\partial u}{\partial r} = \frac{1}{r}\frac{\partial v}{\partial \theta}, \quad \frac{\partial v}{\partial r} = \frac{-1}{r}\frac{\partial u}{\partial \theta}$$

on a region contained in a region such as those shown in Figure 1.24. Here we are employing standard abuse of notation by writing $u(r,\theta) = u(r\cos\theta, r\sin\theta)$. (For a more precise statement, see exercise 11 at the end of this section.

Inverse Function Theorem

Another basic theorem of calculus is the Inverse Function Theorem. On p. 57 the relevant theorem for functions of real variables is stated and is used to prove the following version for analytic functions.

Theorem 20. *Let $f : A \to \mathbb{C}$ be analytic (with f' continuous) and assume that $f'(z_0) \neq 0$. Then there is a neighborhood U of z_0 and V of $f(z_0)$ such that $f : U \to V$ is a bijection (that is, is one-to-one and onto) and $f^{-1}(z)$ is analytic with*

$$\frac{d}{dw}f^{-1}(w) = \frac{1}{f'(z)}, \quad where\ w = f(z).$$

Note. In the proof we shall assume that the partials of u,v are continuous (that is, that f' is continuous), even though this follows automatically, as will be proved in Section 2.4. This proof may be omitted without loss of continuity if the student has not encountered the Inverse Function Theorem in advanced calculus. Applications of Theorem 20 are given in Section 5. The student is cautioned that application of Theorem 20 allows him only to conclude the existence of a *local* inverse for f. For example, let us consider $f(z) = z^2$ defined on $A = \mathbb{C}\backslash\{0\}$. Then $f'(z) = 2z \neq 0$ at each point of A. Theorem 20 says that f has a unique local analytic inverse, which is, in fact, merely some branch of the square root function. But f is not one-to-one on all of A since, for example, $f(1) = f(-1)$. Thus f will be one-to-one only within sufficiently small neighborhoods surrounding each point.

Connected sets

Let $A \subset \mathbb{C}$ be an open set. We want to give precise meaning to the assertion that "A consists of one piece." Actually, there are several possible interpretations of this statement, but for our purposes the following definition can be used (see also the appendix to Section 2.5).

Definition 11. *The open set A is called connected iff, for every $z_1, z_2 \in A$, there is a differentiable path $c:[a,b] \rightarrow A$ with $c(a) = z_1$ and $c(b) = z_2$. We say that c is a path joining z_1 to z_2.*

In any particular situation we usually can tell at a glance if a given set A is connected, as illustrated in Figure 1.25.

A main consequence of connectedness is stated in Theorem 21.

Theorem 21. *Let $A \subset \mathbb{C}$ be open and connected and let $f:A \rightarrow \mathbb{C}$ be analytic. If $f'(z) = 0$ on A, then f is constant on A.*

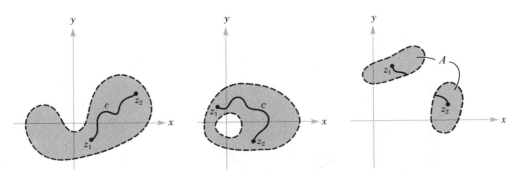

Figure 1.25   Connectedness. *Left and center*, regions connected: *right*, region not connected.

Clearly, connectedness is needed because if A consisted of two pieces, we could let $f = 1$ on one piece and $f = 0$ on the other. Then $f'(z) = 0$ but f would not be constant on A.

Proof.   Let $z_1, z_2 \in A$. We want to show that $f(z_1) = f(z_2)$. Let $c(t)$ be a path joining z_1 to z_2. By the Chain Rule, $df(c(t))/dt = f'(c(t)) \cdot c'(t) = 0$, since $f' = 0$. Thus if $f = u + iv$, we have $du(c(t))/dt = 0$ and $dv(c(t))/dt = 0$. From calculus, we know this implies that $u(c(t)), v(c(t))$ are constant functions of t. Comparing the values at $t = a$ and $t = b$ gives us $f(z_1) = f(z_2)$. ∎

Because of the importance of open connected sets, they are called *regions*, or *domains*.

Technical proofs of some theorems in Section 1.4

Theorem.   *Limits are unique if they exist* (see p. 42).

Proof.   Suppose that $\lim\limits_{z \to z_0} f(z) = a$ and $\lim\limits_{z \to z_0} f(z) = b$ with $a \neq b$. Let $2\epsilon = |a - b|$; then $\epsilon > 0$. There is a $\delta > 0$ such that $|z - z_0| < \delta$ implies that $|f(z) - a| < \epsilon$ and $|f(z) - b| < \epsilon$. Choose such a point $z \neq z_0$ (because f is defined in a deleted neighborhood of z_0). Then, by the triangle inequality, $|a - b| \leq |a - f(z)| + |f(z) - b| < 2\epsilon$, a contradiction. Thus $a = b$. ∎

Next let us consider the "limit theorems" (see pp. 42–43).

Theorem.   *If* $\lim\limits_{z \to z_0} f(z) = a$ *and* $\lim\limits_{z \to z_0} g(z) = b$, *then*

i. $\lim\limits_{z \to z_0} [f(z) + g(z)] = a + b$.

ii. $\lim\limits_{z \to z_0} [f(z) \cdot g(z)] = ab$.

iii. $\lim\limits_{z \to z_0} f(z)/g(z) = a/b$ *if* $b \neq 0$.

iv. $\lim\limits_{z \to z_0} h(f(z)) = c$ *if* $\lim\limits_{w \to a} h(w) = c$.

Proof.   Only assertion (ii) will be proved here. The proof of assertion (i) is easy and proofs of assertions (iii) and (iv) are slightly more challenging, but the student can easily get the necessary clues from the real variable case. To prove assertion (ii), we can choose $\delta_1 > 0$ so that $|z - z_0| < \delta_1$ implies that $|f(z) - a| < 1$, and thus $|f(z)| < |a| + 1$ since $|f(z) - a| \geq |f(z)| - |a|$ by Theorem 7(vi). Given $\epsilon > 0$, we can choose $\delta_2 > 0$ such that $|z - z_0| < \delta_2$ implies that

$$|f(z) - a| < \begin{cases} \dfrac{\epsilon}{2|b|} & \text{if } b \neq 0 \\ 1 & \text{if } b = 0 \end{cases}$$

and $\delta_3 > 0$ such that if $|z - z_0| < \delta_3$, then

$$|g(z) - b| < \frac{\epsilon}{2(|a| + 1)}.$$

Let δ be the minimum of $\delta_1, \delta_2, \delta_3$; then, by the triangle inequality, if $|z - z_0| < \delta$, then

$$|f(z)g(z) - ab| \leq |f(z)g(z) - f(z)b| + |f(z)b - ab|$$
$$= |f(z)| \, |g(z) - b| + |f(z) - a| \, |b|$$
$$< \frac{\epsilon}{2(|a| + 1)}|f(z)| + \frac{\epsilon}{2|b|}|b| \text{ (if } b \neq 0; \text{ if } b = 0, \text{ then the second term is 0)}$$
$$\leq \frac{\epsilon}{2} + \frac{\epsilon}{2} = \epsilon. \blacksquare$$

Theorem 15. *If $f'(z_0)$ exists, then f is continuous at z_0.*

Proof.   By assertion (i) for limits, we need only show that

$$\lim_{z \to z_0} (f(z) - f(z_0)) = 0.$$

But

$$\lim_{z \to z_0} (f(z) - f(z_0)) = \lim_{z \to z_0} \left[\frac{f(z) - f(z_0)}{z - z_0} \right] [z - z_0],$$

which, by assertion (ii) for limits, equals $f'(z_0) \cdot 0 = 0$. $\blacksquare$

Theorem 16.   *Suppose that f and g are analytic on A, where $A \subset \mathbb{C}$ is an open set. Then*
  i. *$af + bg$ is analytic on A and $(af + bg)'(z) = af'(z) + bg'(z)$ for any complex numbers a and b.*
 ii. *fg is analytic on A and $(fg)'(z) = f'(z)g(z) + f(z)g'(z)$.*
iii. *If $g(z) \neq 0$ for all $z \in A$, then f/g is analytic on A and*

$$\left[\frac{f}{g} \right]'(z) = \frac{f'(z)g(z) - g'(z)f(z)}{[g(z)]^2}.$$

 iv. *Any polynomial $a_0 + a_1 z + \cdots + a_n z^n$ is analytic on all of $\mathbb{C}$ with derivative $a_1 + 2a_2 z + \cdots + na_n z^{n-1}$.*

v. *Any rational function*

$$\frac{a_0 + a_1 z + \cdot\cdot\cdot + a_n z^n}{b_0 + b_1 z + \cdot\cdot\cdot + b_m z^m}$$

is analytic on the set A consisting of all z except those (at most, m) points where the denominator is zero.

Proof.   The proofs of (i), (ii), and (iii) are all similar to the proofs found in calculus. The procedure can be illustrated with a proof of (ii). Applying the limit theorems and the fact that $\underset{z \to z_0}{\text{limit}} f(z) = f(z_0)$ (Theorem 15), we get

$$\underset{z \to z_0}{\text{limit}} \left[\frac{f(z)g(z) - f(z_0)g(z_0)}{z - z_0} \right]$$

$$= \underset{z \to z_0}{\text{limit}} \left[\frac{f(z)g(z) - f(z)g(z_0)}{z - z_0} + \frac{f(z)g(z_0) - f(z_0)g(z_0)}{z - z_0} \right]$$

$$= \underset{z \to z_0}{\text{limit}} \left[f(z) \right] \left[\frac{g(z) - g(z_0)}{z - z_0} \right] + \underset{z \to z_0}{\text{limit}} \left[\frac{f(z) - f(z_0)}{z - z_0} \right] \left[g(z_0) \right]$$

$$= f(z_0)g'(z_0) + f'(z_0)g(z_0).$$

To prove (iv) we must first show that $f' = 0$ if f is constant. This is immediate from the definition of derivative because $f(z) - f(z_0) = 0$. It is equally easy to prove that $dz/dz = 1$. Then, using (ii), we can prove that

$$\frac{d}{dz} z^2 = 1 \cdot z + z \cdot 1 = 2z, \text{ and}$$

$$\frac{d}{dz} z^3 = \frac{d}{dz} z \cdot z^2 = 1 \cdot z^2 + z \cdot 2z = 3z^2.$$

In general, we see by induction that $dz^n/dz = nz^{n-1}$. Then (iv) follows easily, and (v) follows from (iv) and (iii) (see exercise 2 at the end of this section for the proof that A is open).∎

Theorem 17 (Chain Rule).   *Let $f: A \to \mathbb{C}$ and $g: B \to \mathbb{C}$ be analytic (A,B are open sets) and let $f(A) \subset B$. Then $g \circ f: A \to \mathbb{C}$ defined by $g \circ f(z) = g(f(z))$ is analytic and*

$$\frac{d}{dz} g \circ f(z) = g'(f(z)) \cdot f'(z).$$

Although this proof can easily be deduced from the Chain Rule for the usual derivative of functions of several variables (see the proof of ' Theorem 19 that follows), a separate proof is given here.

Proof.   Let $w_0 = f(z_0)$ and define, for $w \in B$,

$$h(w) = \begin{cases} \dfrac{g(w) - g(w_0)}{w - w_0} - g'(w_0); & w \neq w_0 \\ 0; & w = w_0 \end{cases}$$

Since $g'(w_0)$ exists, h is continuous. Since the composite of continuous functions is continuous (see exercise 3 at the end of this section or (iv) of the preceding theorem),

$$\operatorname*{limit}_{z \to z_0} h(f(z)) = h(w_0) = 0.$$

From the definition of h, we get, letting $w = f(z)$, $g \circ f(z) - g(w_0) = [h(f(z)) + g'(w_0)][f(z) - w_0]$. Note that this still holds if $f(z) = w_0$. For $z \neq z_0$, we get

$$\frac{g \circ f(z) - g \circ f(z_0)}{z - z_0} = [h(f(z)) + g'(w_0)]\frac{f(z) - f(z_0)}{z - z_0}.$$

As $z \to z_0$, the right side of the equation converges to $[0 + g'(w_0)] \cdot [f'(z_0)]$, so the theorem is proved. ■

Theorem 19.   *Let $f: A \subset \mathbb{C} \to \mathbb{C}$ be a given function, with A an open set. Then $f'(z_0)$ exists if and only if f is differentiable in the sense of real variables and at $(x_0, y_0) = z_0$, u, v satisfy*

$$\frac{\partial u}{\partial x} = \frac{\partial v}{\partial y} \quad and \quad \frac{\partial u}{\partial y} = -\frac{\partial v}{\partial x}$$

(called the Cauchy-Riemann equations).

   Thus, if $\partial u/\partial x$, $\partial u/\partial y$, $\partial v/\partial x$, and $\partial v/\partial y$ exist, are continuous on A, and satisfy the Cauchy-Riemann equations, then f is analytic on A. If $f'(z_0)$ does exist, then

$$f'(z_0) = \frac{\partial u}{\partial x} + i\frac{\partial v}{\partial x}$$

$$= \frac{\partial v}{\partial y} - i\frac{\partial u}{\partial y}.$$

To prove this theorem, we first need a lemma (compare exercise 7, Section 1.1).

Lemma.   *A matrix*

$$\begin{pmatrix} a & b \\ c & d \end{pmatrix}$$

represents, under matrix multiplication, multiplication by a complex

number iff $a = d$ *and* $b = -c$. *The complex number in question is*
$a + ic = d - ib$.

Proof.   First, let is consider multiplication by the complex number
$a + ic$. It sends $x + iy$ to $(a + ic)(x + iy) = ax - cy + i(ay + cx)$,
which is the same as

$$\begin{pmatrix} a & -c \\ c & a \end{pmatrix} \begin{pmatrix} x \\ y \end{pmatrix} = \begin{pmatrix} ax - cy \\ cx + ay \end{pmatrix}.$$

Conversely, let us suppose that

$$\begin{pmatrix} a & b \\ c & d \end{pmatrix} \begin{pmatrix} x \\ y \end{pmatrix} = z \cdot (x + iy)$$

for a complex number $z = \alpha + i\beta$. Then we get

$$ax + by = \alpha x - \beta y$$

and

$$cx + dy = \alpha y + \beta x$$

for all x, y. This implies (setting $x = 1$, $y = 0$, then $x = 0$, $y = 1$) that
$a = \alpha$, $b = -\beta$, $c = \beta$, and $d = \alpha$, so the proof is complete.∎

We can now complete the proof of Theorem 19.

From the definition of f', the statement that $f'(z_0)$ exists is equiva-
lent to: For any $\epsilon > 0$, there is a $\delta > 0$ such that $|z - z_0| < \delta$ implies
that

$$|f(z) - f(z_0) - f'(z_0)(z - z_0)| < \epsilon |z - z_0|.$$

First let us suppose that $f'(z_0)$ exists. We recall that by definition,
$Df(x_0, y_0)$ is the unique matrix with the property that for any $\epsilon > 0$
there is a $\delta > 0$ such that, setting $z = (x, y)$ and $z_0 = (x_0, y_0)$, $|z - z_0| < \delta$
implies that

$$|f(z) - f(z_0) - Df(z_0)(z - z_0)| < \epsilon |z - z_0|.$$

If we compare this equation with the preceding one and recall
that multiplication by a complex number is a linear map, we conclude
that f is differentiable in the sense of real variables and that the
matrix $Df(z_0)$ exactly represents multiplication by the complex
number $f'(z_0)$. Thus, applying the lemma to the matrix

$$Df(z_0) = \begin{pmatrix} \dfrac{\partial u}{\partial x} & \dfrac{\partial u}{\partial y} \\ \dfrac{\partial v}{\partial x} & \dfrac{\partial v}{\partial y} \end{pmatrix}$$

with $a = \partial u/\partial x$, $b = \partial u/\partial y$, $c = \partial v/\partial x$, $d = \partial v/\partial y$, we have $a = d$, $b = -c$, which are the Cauchy-Riemann equations.

Conversely, if the Cauchy-Riemann equations hold, $Df(z_0)$ represents multiplication by a complex number (by the lemma) and then as above, the definition of differentiability in the sense of real variables reduces to that for the complex derivative.

The formula for $f'(z_0)$ follows from the last statement of the lemma. ■

Theorem 20.   *Let $f: A \rightarrow \mathbb{C}$ be analytic (with f' continuous) and assume that $f'(z_0) \neq 0$. Then there is a neighborhood U of z_0 and V of $f(z_0)$ such that $f: U \rightarrow V$ is a bijection (that is, one-to-one and onto) and $f^{-1}(z)$ is analytic with*

$$\frac{d}{dw} f^{-1}(w) = \frac{1}{f'(z)}, \text{ where } w = f(z).$$

To prove this theorem, let us recall the statement for real variables.

Theorem.   *If $f: A \subset \mathbb{R}^2 \rightarrow \mathbb{R}^2$ is continuously differentiable and $Df(x_0,y_0)$ has a nonzero determinant, then there are neighborhoods U of (x_0,y_0) and V of $f(x_0,y_0)$ such that $f: U \rightarrow V$ is a bijection, $f^{-1}: V \rightarrow U$ is differentiable, and*

$$Df^{-1}(f(x,y)) = [Df(x,y)]^{-1}$$

(this is the inverse of the matrix of partials).

The proof of this theorem may be found in advanced calculus texts. See, for instance, J. Marsden, *Elementary Classical Analysis* (San Francisco: W. H. Freeman and Co., 1974), Chapter 7.

Accepting this statement and assuming that f' in Theorem 20 is continuous, we can prove Theorem 20.

Proof.   For analytic functions such as $f(z)$, we have seen that the matrix of partial derivatives is

$$Df = \begin{pmatrix} \dfrac{\partial u}{\partial x} & \dfrac{\partial u}{\partial y} \\[2mm] \dfrac{\partial v}{\partial x} & \dfrac{\partial v}{\partial y} \end{pmatrix} = \begin{pmatrix} \dfrac{\partial u}{\partial x} & -\dfrac{\partial v}{\partial x} \\[2mm] \dfrac{\partial v}{\partial x} & \dfrac{\partial u}{\partial x} \end{pmatrix},$$

which has determinant

$$\left[\frac{\partial u}{\partial x}\right]^2 + \left[\frac{\partial v}{\partial x}\right]^2 = |f'(z)|^2,$$

since $f'(z) = \partial u/\partial x + i \partial v/\partial x$. All these functions are to be evaluated at the point $(x_0,y_0) = z_0$. Now $f'(z_0) \neq 0$, so the determinant is also not equal to 0; that is: det $Df(x_0,y_0) = |f'(z_0)| \neq 0$. Thus the theorem just quoted applies. By Theorem 19, we need only verify that the entries of $[Df(x,y)]^{-1}$ satisfy the Cauchy-Riemann equations and give $(f^{-1})'$ as stated.

As we have just seen,

$$Df = \begin{pmatrix} \dfrac{\partial u}{\partial x} & \dfrac{\partial u}{\partial y} \\ \dfrac{\partial v}{\partial x} & \dfrac{\partial v}{\partial y} \end{pmatrix},$$

and the inverse of this matrix is

$$\frac{1}{\text{Det } Df} \begin{pmatrix} \dfrac{\partial v}{\partial y} & \dfrac{-\partial u}{\partial y} \\ -\dfrac{\partial v}{\partial x} & \dfrac{\partial u}{\partial x} \end{pmatrix}.$$

Thus if we write $f^{-1}(x,y) = t(x,y) + is(x,y)$, then, comparing

$$D(f^{-1}) = \begin{pmatrix} \dfrac{\partial t}{\partial x} & \dfrac{\partial t}{\partial y} \\ \dfrac{\partial s}{\partial x} & \dfrac{\partial s}{\partial y} \end{pmatrix}$$

with the inverse matrix for Df, we get, with the partials of t and s evaluated at $f(x_0,y_0)$,

$$\frac{\partial t}{\partial x} = \frac{1}{\text{Det}(Df)} \frac{\partial v}{\partial y} = \frac{1}{\text{Det}(Df)} \frac{\partial u}{\partial x}$$

and

$$\frac{\partial s}{\partial x} = \frac{1}{\text{Det}(Df)} \frac{-\partial v}{\partial x} = \frac{1}{\text{Det}(Df)} \frac{\partial u}{\partial y}.$$

Similarly,

$$\frac{\partial t}{\partial y} = \frac{1}{\text{Det}(Df)} \frac{\partial v}{\partial x}$$

and

$$\frac{\partial s}{\partial y} = \frac{1}{\text{Det}(Df)} \frac{\partial v}{\partial y}.$$

Thus the Cauchy-Riemann equations hold for t and s since they hold for u and v. Therefore, f^{-1} is complex differentiable. From Theorem 19 we see that at the point $f(z_0)$,

$$(f^{-1})' = \frac{\partial t}{\partial x} + i\frac{\partial s}{\partial x}$$

$$= \frac{1}{\mathrm{Det}\,(Df)}\left[\frac{\partial u}{\partial x} - i\frac{\partial v}{\partial x}\right]$$

$$= \frac{\overline{f'(z_0)}}{|f'(z_0)|^2} = \frac{1}{f'(z_0)} \cdot \blacksquare$$

An alternate way to show that $(f^{-1})'(f(z)) = 1/f'(z)$ is outlined in exercise 19.

Remark.   After we have proven Cauchy's Theorem in Chapter 2, we will be able to prove Theorem 20 in another way that does not depend on the real variable theorem stated on p. 57 (see the first theorem in Chapter 6).

Worked Examples

1. Where is

$$f(z) = \frac{z^3 + 2z + 1}{z^3 + 1}$$

analytic? Compute the derivative.

Solution.   By Theorem 16, f is analytic on the set $A = \{z \in \mathbb{C} \mid z^3 + 1 \neq 0\}$; that is, f is analytic on the whole plane except the cube roots of $-1 = e^{\pi i}$; namely, the points $e^{\pi i/3}$, $e^{\pi i}$, and $e^{5\pi i/3}$. By the formula for differentiating a quotient, the derivative is:

$$f'(z) = \frac{(z^3 + 1)(3z^2 + 2) - (z^3 + 2z + 1)(3z^2)}{(z^3 + 1)^2}$$

$$= \frac{(2 - 4z^3)}{(z^3 + 1)^2}.$$

2. Consider $f(z) = z^3 + 1$. Study the infinitesimal behavior of f at $z_0 = i$.

Solution.   We use Theorem 18. In this case $f'(z_0) = 3i^2 = -3$. Thus f rotates locally by $\pi = \arg(-3)$ and multiplies lengths by $3 = |f'(z_0)|$. More precisely, if c is any curve through $z_0 = i$, the image curve will, at $f(z_0)$, have its tangent vector rotated by π and stretched by a factor 3.

3. Show that $f(z) = \bar{z}$ is not analytic.

Solution.   Let $f(z) = u(x,y) + iv(x,y) = x - iy$ where $z = (x,y) = x + iy$. Thus $u(x,y) = x$, $v(x,y) = -y$. But $\partial u/\partial x = 1$ and $\partial v/\partial y = -1$ and hence $\partial u/\partial x \neq \partial v/\partial y$, so the Cauchy-Riemann equations do not hold. Therefore, $f(z) = \bar{z}$ cannot be analytic by Theorem 19.

4.   We know by Theorem 16 that $f(z) = z^3 + 1$ is analytic. Verify the Cauchy-Riemann equations for this function.

Solution.   If $f(z) = u(x,y) + iv(x,y)$ when $z = (x,y) = x + iy$, then in this case $u(x,y) = x^3 - 3xy^2 + 1$ and $v(x,y) = 3x^2y - y^3$. Therefore, $\partial u/\partial x = 3x^2 - 3y^2$, $\partial u/\partial y = -6xy$, $\partial v/\partial x = 6xy$, and $\partial v/\partial y = 3x^2 - 3y^2$, from which we see that $\partial u/\partial x = \partial v/\partial y$ and $\partial u/\partial y = -\partial v/\partial x$.

5.   If f is analytic, $f = u + iv$, and u,v are twice continuously differentiable, show that they must be *harmonic*; that is, that

$$\frac{\partial^2 u}{\partial x^2} + \frac{\partial^2 u}{\partial y^2} = 0 \text{ and } \frac{\partial^2 v}{\partial x^2} + \frac{\partial^2 v}{\partial y^2} = 0.$$

Remark.   In Section 2.4 we shall see that u and v are automatically infinitely differentiable.

Solution.   We apply the Cauchy-Riemann equations, $\partial u/\partial x = \partial v/\partial y$ and $\partial u/\partial y = -\partial v/\partial x$. Differentiating the first equation with respect to x and the second equation with respect to y, we get

$$\frac{\partial^2 u}{\partial x^2} = \frac{\partial^2 v}{\partial x \partial y} \text{ and } \frac{\partial^2 u}{\partial y^2} = -\frac{\partial^2 v}{\partial y \partial x}.$$

We know from calculus that the second partials are symmetric:

$$\frac{\partial^2 v}{\partial x \partial y} = \frac{\partial^2 v}{\partial y \partial x}.$$

Adding the equations gives us

$$\frac{\partial^2 u}{\partial x^2} + \frac{\partial^2 u}{\partial y^2} = \frac{\partial^2 v}{\partial x \partial y} - \frac{\partial^2 v}{\partial y \partial x} = 0.$$

The equation is proved for v in the same way.

Exercises

1.   a. Show that $|\operatorname{Re} z_1 - \operatorname{Re} z_2| \leq |z_1 - z_2| \leq |\operatorname{Re} z_1 - \operatorname{Re} z_2| + |\operatorname{Im} z_1 - \operatorname{Im} z_2|$ for any two complex numbers z_1, z_2.
   b. If $f(z) = u(x,y) + i\,v(x,y)$, show that

$$\lim_{z \to z_0} f(z) = \lim_{\substack{x \to x_0 \\ y \to y_0}} u(x,y) + i \lim_{\substack{x \to x_0 \\ y \to y_0}} v(x,y)$$

exists if both limits on the right of the equation exist. Conversely, if the limit on the left exists, show that both limits on the right exist as well and equality holds. Show that $f(z)$ is continuous iff u and v are. Apply this result to $f(z) = \bar{z}$ and $f(z) = |z|$ to establish their continuity.

2.   Prove: The complement of a finite number of points is an open set.

3.   Prove: A map $f: A \to \mathbb{C}$, where A is an open set, is continuous iff, for every open set $U \subset \mathbb{C}$, $f^{-1}(U)$ is open. Use this statement to prove that,

if $f: A \to \mathbb{C}$ and $g: B \to \mathbb{C}$ are continuous and $f(A) \subset B$, then $g \circ f$ is continuous.

Note. $f^{-1}(U) = \{z \in \mathbb{C} \mid f(z) \in U\}$ and $f(A) = \{f(z) \mid z \in A\}$.

4. Use the Inverse Function Theorem to show that if $f: A \to \mathbb{C}$ is analytic and $f'(z) \neq 0$ for all $z \in A$, then f maps open sets to open sets.

5. Prove: If f is continuous and $f(z_0) \neq 0$, there is a neighborhood of z_0 on which f is $\neq 0$.

6. Study the infinitesimal behavior of the following functions at the indicated points:
a. $f(z) = z + 3$, $z_0 = 3 + 4i$.
b. $f(z) = z^6 + 3z$, $z_0 = 0$.
c. $f(z) = \dfrac{z^2 + z + 1}{z - 1}$, $z_0 = 0$.

7. Prove that $f(z) = |z|$ is not analytic.

8. Determine the sets on which the following functions are analytic and compute their derivatives:
a. $(z + 1)^3$.
b. $z + \dfrac{1}{z}$.
c. $\left(\dfrac{1}{z - 1}\right)^{10}$.
d. $\dfrac{1}{(z^3 - 1)(z^2 + 2)}$.

9. For $c: \,]a,b[\to \mathbb{C}$ differentiable and $f: A \to \mathbb{C}$ analytic with $c(\,]a,b[) \subset A$, prove that $d = f \circ c$ is differentiable with $d'(t) = f'(c(t)) \cdot c'(t)$ by imitating the proof of Theorem 17.

10. Without justifying your computations, show, by changing variables, that the Cauchy-Riemann equations in terms of polar coordinates become

$$\frac{\partial u}{\partial r} = \frac{1}{r}\frac{\partial v}{\partial \theta}, \quad \frac{\partial v}{\partial r} = -\frac{1}{r}\frac{\partial u}{\partial \theta}.$$

11. Carefully perform the computations in exercise 10 by the following procedure. Let f be defined on the open set $A \subset \mathbb{C}$ (that is, $f: A \subset \mathbb{C} \to \mathbb{C}$) and suppose that $f(z) = u(z) + iv(z)$. Let $T: \,]0,2\pi[\times \mathbb{R}^+ \to \mathbb{R}^2$, where $\mathbb{R}^+ = \{x \in \mathbb{R} \mid x > 0\}$, be given by $T(\theta,r) = (r\cos \theta, r\sin \theta)$. Thus T is one-to-one and onto the set $\mathbb{R}^2 \backslash \{(x,0) \mid x \geq 0\}$. Define $\tilde{u}(\theta,r) = u(r\cos \theta, r\sin \theta)$ and $\tilde{v}(\theta,r) = v(r\cos \theta, r\sin \theta)$. Show that:
a. T is continuously differentiable and has a continuously differentiable inverse.
b. f is analytic on $A \backslash \{x + iy \mid y = 0, x \geq 0\}$ if and only if $(\tilde{u},\tilde{v}): T^{-1}(A) \to \mathbb{R}^2$ is differentiable and

$$\frac{\partial \tilde{u}}{\partial r} = \frac{1}{r}\frac{\partial \tilde{v}}{\partial \theta}, \quad \frac{\partial \tilde{v}}{\partial r} = -\frac{1}{r}\frac{\partial \tilde{u}}{\partial \theta}$$

on $T^{-1}(A)$.

12. Define $f: \mathbb{C} \to \mathbb{C}$ by setting $f(0) = 0$ and by setting $f(r[\cos\theta + i\sin\theta]) = \sin\theta$ if $r > 0$. Show that f is discontinuous at 0 but is continuous everywhere else.

13. Define the symbol

$$\frac{\partial f}{\partial \bar{z}} = \frac{1}{2}\left(\frac{\partial f}{\partial x} - \frac{\partial f}{i\partial y}\right).$$

Show that the Cauchy-Riemann equations are equivalent to $\partial f/\partial \bar{z} = 0$. (It is sometimes said, because of this result, that analytic functions are not functions of $\bar{z}$ but of z alone. This statement should be taken only as a rough guide, since $\partial f/\partial \bar{z}$ is not really a derivative of f with respect to $\bar{z}$ but merely a shorthand notation for $(\partial f/\partial x - \partial f/i\partial y)/2$.)

14. Let f be an analytic function on an open connected set A and suppose that $f^{(n+1)}(z)$ (the $n + 1$st derivative) exists and is zero on A. Then show that f is a polynomial of degree $\leq n$.

15. On what sets are the following functions analytic? Compute the derivative for each.

a. z^n, n being an integer (positive or negative).

b. $\dfrac{1}{\left(z + \dfrac{1}{z}\right)^2}$.

c. $\dfrac{z}{z^n - 2}$, n being a positive integer.

16. Verify directly that the real and imaginary parts of $f(z) = z^4$ are harmonic (see example 5, p. 60).

17. On what set is

$$u(x,y) = \operatorname{Re}\left(\frac{z}{z^3 - 1}\right)$$

harmonic?

18. Assume that f is analytic on $A = \{z \mid \operatorname{Re} z > 1\}$ and that $\partial u/\partial x + \partial v/\partial y = 0$ on A. Show that f' is constant on A.

19. Prove that $df^{-1}/dw = 1/f'(z)$ where $w = f(z)$ by differentiating $f^{-1}(f(z)) = z$, using the Chain Rule. Assume that f^{-1} is defined and is analytic.

20. Prove:

a. $f: A \subset \mathbb{C} \to \mathbb{C}$ is continuous iff $z_n \to z_0$ in A implies that $f(z_n) \to f(z_0)$.

b. $f: \mathbb{C} \to \mathbb{C}$ is continuous iff the inverse image of every open set is open.

21. For what z does the sequence $z_n = nz^n$ converge?

22. a. Let $f(z) = u(x,y) + iv(x,y)$ be an analytic function in a connected open set A. If $au(x,y) + bv(x,y) = c$ in A, where a,b,c are real constants not all 0, prove that $f(z)$ is constant in A.

b. Is the result obtained in (a) still valid if a,b,c are complex constants?

23. Let $f(z) = z^5/|z|^4$ if $z \neq 0$ and $f(0) = 0$.

a. Show that $f(z)/z$ does not have a limit as $z \to 0$.

b. If $u = \operatorname{Re} f$ and $v = \operatorname{Im} f$, show that $u(x,0) = x$, $v(0,y) = y$, $u(0,y) = v(x,0) = 0$.

c. Conclude that the partials of u,v exist, that the Cauchy-Riemann equations hold, but that $f'(0)$ does not exist. Does this conclusion contradict Theorem 19?

 d. Repeat exercise (c) letting $f = 1$ on the x- and y-axes and 0 elsewhere.

 e. Repeat exercise (c) letting $f(z) = \sqrt{|xy|}$.

24. Let $f(z) = (z+1)/(z-1)$.

 a. Where is f analytic?

 b. Is f conformal at $z = 0$?

 c. What are the images of the x- and y-axes under f?

 d. At what angle do these images intersect?

25. Let u have continuous second partials on an open set A and let $\partial^2 u/\partial x^2 + \partial^2 u/\partial y^2 = 0$ (see example 5, p. 60). Let $f = \partial u/\partial x - i\,\partial u/\partial y$. Show that f is analytic.

26. Write an expression for the m roots of $z^m + i - 0$.

1.5 Differentiation of the Elementary Functions

This section will discuss the properties of differentiability of the elementary functions discussed in Section 1.3.

Exponential function and logarithm

> ***Theorem 22.*** *The map $f: \mathbb{C} \to \mathbb{C}$, $z \mapsto e^z$ is analytic on $\mathbb{C}$ and*
>
> $$\frac{de^z}{dz} = e^z.$$

A function that is defined and analytic on the whole plane $\mathbb{C}$ is called *entire*. Note that at this point in the text complex numbers are not allowed to assume the value ∞.

Proof. By definition, $f(z) = e^x(\cos y + i\sin y)$, so $u(x,y) = e^x \cos y$ and $v(x,y) = e^x \sin y$. These are C^∞ (infinitely often differentiable) functions, so f is differentiable in the sense of real variables. To show that f is analytic, we must verify the Cauchy-Riemann equations. But

$$\frac{\partial u}{\partial x} = e^x \cos y, \quad \frac{\partial u}{\partial y} = -e^x \sin y;$$

$$\frac{\partial v}{\partial x} = e^x \sin y, \quad \frac{\partial v}{\partial y} = e^x \cos y.$$

Thus $\partial u/\partial x = \partial v/\partial y$ and $\partial u/\partial y = -\partial v/\partial x$, so the Cauchy-Riemann equations hold, and hence f is analytic. Also, since

$$\frac{df}{dz} = \frac{\partial u}{\partial x} + i\frac{\partial v}{\partial x} = e^x(\cos y + i\sin y) = e^z,$$

Theorem 22 is proved. ∎

Using the rules in Theorem 16 as they are used in elementary calculus, we can differentiate e^z in combination with various other functions. For instance, $e^{z^2} + 1$ is entire because $z \mapsto z^2$ and $w \mapsto e^w$ are analytic, so by the Chain Rule, $z \mapsto e^{z^2}$ is analytic; the constant 1 is analytic and the sum of analytic functions is analytic. By the Chain Rule and Theorem 16, $d(e^{z^2} + 1)/dz = 2ze^{z^2}$.

We recall that $\log z \colon \mathbb{C}\backslash\{0\} \to \mathbb{C}$ is an inverse for e^z when e^z is restricted to a period strip $\{x + iy | y_0 \leq y < y_0 + 2\pi\}$. However, for differentiability of $\log z$ we must restrict $\log z$ to a set that is smaller than $\mathbb{C}\backslash\{0\}$. The reason is simple: $\log z = \log |z| + i \arg z$, for, say, $0 \leq \arg z < 2\pi$. But the arg function is discontinuous; it jumps by 2π as we cross the real axis. If we remove the real axis, then we are excluding the usual positive reals on which we want $\log z$ defined. Therefore, it is convenient to use the branch $-\pi < \arg z < \pi$. Then an appropriate set on which $\log z$ is analytic is given as follows.

Theorem 23.   *Let A be the open set that is $\mathbb{C}$ minus the negative real axis including zero (that is, $\mathbb{C}\backslash\{x + iy | x \leq 0, \text{ and } y = 0\}$). Define the branch of $\log$ on A by*

$$\log z = \log |z| + i \arg z, \ -\pi < \arg z < \pi,$$

called the principal branch of the logarithm. Then $\log z$ is analytic on A with

$$\frac{d}{dz} \log z = \frac{1}{z}.$$

Two proofs are given here.

First Proof (using the Inverse Function Theorem).   From Section 1.3 we know that $\log z$ is the unique inverse of the function $f(z) = e^z$ restricted to the set $\{z \mid z = x + iy, -\pi < y < \pi\}$. Since $de^z/dz = e^z \neq 0$, Theorem 20 implies that locally, e^z has an analytic inverse. Since the inverse is unique, it must be $\log z$. The derivative of $f^{-1}(w)$ is $1/f'(z)$. In this case $f'(z) = f(z) = w$, so $df^{-1}/dw = 1/w$ at each point $w \in A$, and the theorem is proved. ∎

Second Proof (using the Cauchy-Riemann equations in polar form). In polar form, $\log z = \log r + i\theta$, so $u(r,\theta) = \log r$, $v(r,\theta) = \theta$, which are C^∞ functions of r, θ. Also, the Cauchy-Riemann equations expressed in polar coordinates are a valid tool on the region A, as explained in Section 1.4 and exercises 10 and 11 at the end of that section. But

$$\frac{\partial u}{\partial r} = \frac{1}{r} = \frac{1}{r}\frac{\partial v}{\partial \theta}$$

and

$$-\frac{1}{r}\frac{\partial u}{\partial \theta} = 0 = \frac{\partial v}{\partial r},$$

so the Cauchy-Riemann equations hold. We also have

$$\frac{d}{dz}\log z = \frac{\partial}{\partial x}\log r + i\,\frac{\partial \theta}{\partial x}$$

$$= \frac{1}{r}\frac{\partial r}{\partial x} + i\,\frac{\partial \theta}{\partial x}.$$

It is obvious that on A, $\partial r/\partial x = x/r$ and $\partial \theta/\partial x = -y/r^2$, using, for example, $\theta = \tan^{-1}(y/x)$, so

$$\frac{d}{dz}\log z = \frac{x}{r^2} - \frac{iy}{r^2} = \frac{\bar{z}}{|z|^2} = \frac{1}{z}.\;\blacksquare$$

The domain on which the principal branch of $\log z$ is analytic is given in Figure 1.26. A similar theorem exists for other branches; for instance, $\log z = \log |z| + i\arg z,\; 0 < \arg z < 2\pi$ is analytic on $\mathbb{C}\backslash\{x + iy\,|\,x \geq 0,\; y = 0\}$.

When using $\log z$ in compositions, we must be careful to stay in the domain of $\log$. For example, let us consider $f(z) = \log(z^2)$ using the principal branch of $\log$.

This is analytic on $A = \{z\,|\,z \neq 0 \text{ and } \arg z \neq \pm\,\pi/2\}$ by the following reasoning. Theorem 16 showed that z^2 is analytic on all of $\mathbb{C}$. The image of A under the map $z \mapsto z^2$ is precisely $\mathbb{C}\backslash\{x + iy\,|\,x \leq 0,\; y = 0\}$ (Why?), which is the set on which the principal branch of $\log$ is defined and

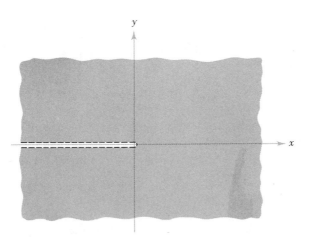

Figure 1.26 Domain of $\log z$.

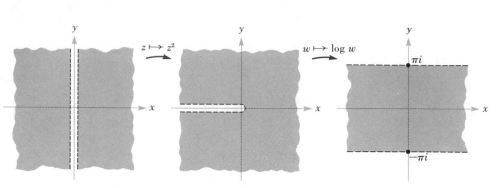

Figure 1.27   Region of analyticity of $\log(z^2)$.

analytic. By the Chain Rule, $z \mapsto \log(z^2)$ is analytic on A (see Figure 1.27).

The trigonometric functions

Once we have e^z and the Chain Rule, the differentiation of sine and cosine is less difficult.

Theorem 24. $\sin z$ *and* $\cos z$ *are entire functions with derivatives given by*

$$\frac{d}{dz}\sin z = \cos z \qquad and \qquad \frac{d}{dz}\cos z = -\sin z.$$

Proof. $\sin z = (e^{iz} - e^{-iz})/2i$; using the sum rule and Chain Rule and the fact that the exponential function is entire, we conclude that $\sin z$ is entire and that $d\sin z/dz = \{ie^{iz} - (-ie^{-iz})\}/2i = (e^{iz} + e^{-iz})/2 = \cos z$.
Similarly,

$$\frac{d}{dz}\cos z = \frac{d}{dz}\frac{1}{2}(e^{iz} + e^{-iz}) = \frac{i}{2}(e^{iz} - e^{-iz})$$

$$= -\frac{1}{2i}(e^{iz} - e^{-iz}) = -\sin z.\blacksquare$$

We can also discuss $\sin^{-1} z$ and $\cos^{-1} z$ in somewhat the same way that we discussed $\log z$ (which is $\exp^{-1}(z)$ with appropriate domains and ranges). Since these functions are not as important, they will be analyzed in exercise 4 at the end of this section.

The power function

Let a and b be complex numbers. We recall that $a^b = e^{b\log a}$, which is, in general, multivalued according to the different branches for log that we choose. We now wish to consider the functions $z \mapsto z^b$ and $z \mapsto a^z$. Although these functions appear to be similar, their properties of analyticity are quite different. The case in which b is an integer was covered in Theorem 16. The situation for general b is slightly more complicated.

Theorem 25.
i. *For any choice of a branch for the* log *function, the function* $z \mapsto a^z$ *is entire and has derivative* $z \mapsto (\log a)a^z$.
ii. *Fix a branch of the* log *function—say, the principal branch; then the function* $z \mapsto z^b$ *is analytic on the domain of the branch of* log *chosen and the derivative is* $z \mapsto bz^{b-1}$.

Proof.
i. $a^z = e^{z\log a}$. By the Chain Rule ($\log a$ is merely a constant), this function is analytic on $\mathbb{C}$ with derivative $(\log a)e^{z\log a} = (\log a)a^z$.
ii. $z^b = e^{b\log z}$. This function is analytic on the domain of $\log z$ since $w \mapsto e^{bw}$ is entire. Also, by the Chain Rule,

$$\frac{d}{dz}z^b = \frac{b}{z}e^{b\log z} = \frac{b}{z}z^b .$$

(That this equals bz^{b-1} follows from exercise 18 at the end of Section 1.3.)∎

Of course in (ii), if b is an integer ≥ 0 we know that z^b is entire (with derivative bz^{b-1}). But in general, z^b is analytic only on the domain of $\log z$.

Let us note what is stated in (ii). If we choose the principal branch of $\log z$, which has domain $\mathbb{C}\backslash\{x + iy \mid y = 0 \text{ and } x \leq 0\}$ and range $\mathbb{R} \times]-\pi,\pi[$ (Why?), then $z \mapsto z^a$ is analytic on $\mathbb{C}\backslash\{x + iy \mid y=0, x \leq 0\}$ (see Figure 1.28). We could also choose the branch of log that has domain

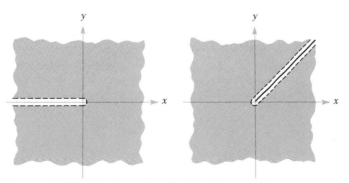

Figure 1.28 Regions of analyticity for $z \rightarrow z^a$.

$\mathbb{C}\backslash\{x + ix \mid x \geq 0\}$ and range $\mathbb{R} \times \,]-7\pi/4, \pi/4[$; then $z \mapsto z^a$ would be analytic on $\mathbb{C}\backslash\{x + ix \mid x \geq 0\}$.

The nth root function

One of the nth roots of z is given by $z^{1/n}$, for a choice of branch of $\log z$. The other roots are given by the other choices of branches as in Section 1.3. The principal branch is the one that is usually used. Thus from Theorem 25(ii) we get, as a special case,

Theorem 26. *The function $z \mapsto z^{1/n} = \sqrt[n]{z}$ is analytic on the domain of $\log z$ (say, the principal branch) and has derivative*

$$z \mapsto \frac{1}{n} z^{\frac{1}{n} - 1}.$$

As with $\log z$, we must exercise care with the functions $z \mapsto z^b$, $z \mapsto \sqrt[n]{z}$ when composing with other functions to be sure we stay in the domain of analyticity. The procedure is illustrated for the square root function in example 3.

Worked Examples

1. Differentiate the following functions, giving the appropriate region on which the functions are analytic:
    a. e^{e^z}.
    b. $\sin(e^z)$.
    c. $e^z/(z^2 + 3)$.
    d. $\sqrt{e^z + 1}$.
    e. $\cos \bar{z}$.
    f. $1/(e^z - 1)$.
    g. $\log(e^z + 1)$.

    Solution.
    a. $z \mapsto e^{e^z}$.
       e^z is entire; by the Chain Rule $z \mapsto e^{e^z}$ is entire. Also, by the Chain Rule the derivative at z is $e^z e^{e^z}$.
    b. $z \mapsto \sin e^z$.
       Both $z \mapsto e^z$ and $w \mapsto \sin w$ are entire, so by the Chain Rule $z \mapsto \sin e^z$ is entire and the derivative at z is $(\cos e^z)e^z$.
    c. $z \mapsto e^z/(z^2 + 3)$.
       The map $z \mapsto e^z$ is entire and the map $z \mapsto 1/(z^2 + 3)$ is analytic on $\mathbb{C}\backslash\{\pm\sqrt{3}i\}$ by Theorem 16. Hence $z \mapsto e^z/(z^2 + 3)$ is analytic on $\mathbb{C}\backslash\{\sqrt{3}i, -\sqrt{3}i\}$ and has derivative at $z \neq \pm \sqrt{3}i$ given by

$$\frac{e^z}{z^2 + 3} - \frac{e^z \cdot 2z}{(z^2 + 3)^2} = \frac{e^z(z^2 - 2z + 3)}{(z^2 + 3)^2}.$$

d. $z \mapsto \sqrt{e^z + 1}$.

Choose the branch of the function $w \mapsto \sqrt{w}$ that is analytic on $\mathbb{C}\backslash\{x + iy \mid y = 0, x \leq 0\}$. Then we must choose the region A, such that if $z \in A$, then $e^z + 1$ is not both real and ≤ 0. Notice that e^z is real iff $y = \operatorname{Im} z = n\pi$ for some integer n (Why?). When n is even, e^z is positive; when n is odd, e^z is negative. Here $|e^z| = e^x$, where $x = \operatorname{Re} z$ and $e^x \geq 1$ iff $x \geq 0$. So if we define $A = \mathbb{C}\backslash\{x + iy \mid x \geq 0, y = (2n+1)\pi, n$ running through positive and negative integers and zero$\}$ (as in Figure 1.29), then $e^z + 1$ is real and ≤ 0 iff $z \in A$. Since $e^z + 1$ is entire, it is certainly analytic on A. By the Chain Rule, $\sqrt{e^z + 1}$ is analytic on A with derivative at z given by $(e^z + 1)^{-1/2}(e^z)/2$.

e. $z \mapsto \cos \bar{z}$.

Since $z = x + iy$, by Theorem 9 we have $\cos \bar{z} = \cos(x - iy) = \cos x \cos(-iy) - \sin x \sin(-iy) = \cos x \cosh y + i\sin x \sinh y$, so $u(x,y) = \cos x \cosh y$ and $v(x,y) = \sin x \sinh y$. Thus $\partial u/\partial x = -\sin x \cosh y$, $\partial v/\partial y = \sin x \cosh y$. If $\cos \bar{z}$ were analytic, $\partial u/\partial x = \partial v/\partial y$, which would occur iff $\sin x = 0$ (that is, if $x = 0$, or if $x = \pi n$, $n = \pm 1, \pm 2, \cdots$). Thus there is not an (nonempty) open set A on which $z \mapsto \cos \bar{z}$ is analytic.

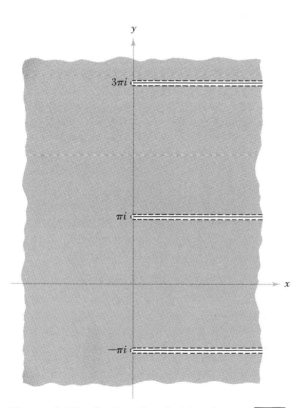

Figure 1.29 Region of analyticity of $z \to \sqrt{e^z + 1}$.

f. $z \mapsto 1/(e^z - 1)$.

By Theorem 16(iii) and the fact that $z \mapsto e^z - 1$ is entire, we conclude that $z \mapsto 1/(e^z - 1)$ is analytic on the set on which $e^z - 1 \neq 0$; namely, the set $A = \mathbb{C}\backslash\{z = 2\pi ni \,|\, n = 0, \pm 1, \pm 2, \cdots\}$. The derivative at z is $-e^z/(e^z - 1)^2$.

g. $z \mapsto \log(e^z + 1)$. Since (the principal branch of) the log is defined and analytic on the same region as the square root, namely, $A = \mathbb{C}\backslash\{x + iy \,|\, y = 0, x \leq 0\}$, we can use the results of (d). By the Chain Rule and the results of (d), $z \mapsto \log(e^z + 1)$ is analytic on the same region as that depicted in Figure 1.29.

2. Verify directly that the function e^z preserves the angles between lines parallel to the coordinate axes (see Theorem 18).

Solution. The line determined by $y = y_0$ is mapped to the ray $\{x + ix\tan y_0 \,|\, x > 0\}$ and the line determined by $x = x_0$ is mapped to the circle $\{z \,|\, |z| = e^{x_0}\}$ — see Figure 1.19. The angle between any such ray and the tangent vector to the circle at the point of contact of the ray and the circle is $\pi/2$. Thus angles are preserved.

3. Show that a branch of the function $w \mapsto \sqrt{w}$ can be defined such that $z \mapsto \sqrt{z^2 - 1}$ is analytic in the region diagrammed in Figure 1.30, and using the notations of that figure, show that $\sqrt{z^2 - 1} = \sqrt{r_1 r_2} \, e^{i(\theta_1 + \theta_2)/2}$, where $0 < \theta_1 < 2\pi$, $-\pi < \theta_2 < \pi$.

Solution. If $\sqrt{z - 1}$ is a square root of $z - 1$ and $\sqrt{z + 1}$ is a square root of $z + 1$, then $\sqrt{z - 1} \cdot \sqrt{z + 1}$ is a square root of $z^2 - 1$ (Why?). For $z \mapsto$

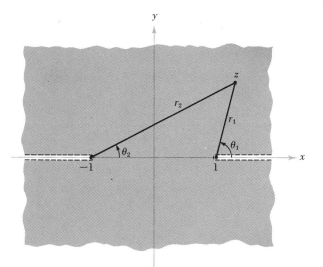

Figure 1.30 Region of analyticity of $\sqrt{z^2 - 1}$.

$\sqrt{z-1}$ we may choose $\sqrt{}$ defined and analytic on $\mathbb{C}\backslash\{x + iy|y = 0$ and $x \geq 0\}$; thus $z \mapsto \sqrt{z-1}$ is analytic on $\mathbb{C}\backslash\{x + iy|y = 0, x \geq 1\}$. For $z \mapsto \sqrt{z+1}$ we may choose $\sqrt{}$ defined and analytic on $\mathbb{C}\backslash\{x + iy|y = 0$ and $x \leq 0\}$; therefore, $z \mapsto \sqrt{z+1}$ is analytic on $\mathbb{C}\backslash\{x + iy|y = 0, x \leq -1\}$. Thus $z \mapsto \sqrt{z-1} \sqrt{z+1}$ is analytic on $\mathbb{C}\backslash\{x + iy|y = 0, |x| \geq 1\}$ with the appropriate branches of $\sqrt{}$ as indicated. With these branches we have $\sqrt{z-1} = \sqrt{r_1}\, e^{i\theta_1/2}$ and $\sqrt{z+1} = \sqrt{r_2}\, e^{i\theta_2/2}$, so $\sqrt{z-1}\sqrt{z+1} = \sqrt{r_1 r_2}\, e^{i(\theta_1 + \theta_2)/2}$. Since $z \mapsto \sqrt{z-1}\sqrt{z+1}$ is analytic on $\mathbb{C}\backslash\{x + iy|y=0, |x| \geq 1\}$, it should correspond to some branch of the square root function $\sqrt{}$ in the map $z \mapsto \sqrt{z^2 - 1}$. To see which one, note that $z \in \mathbb{C}\backslash\{x + iy \mid y = 0, |x| \geq 1\}$ iff $z^2 - 1 \in \mathbb{C}\backslash\{x + iy \mid y = 0, x \geq 0\}$ (Why?), so if we take the $\sqrt{}$ defined and analytic on $\mathbb{C}\backslash\{x + iy \mid y = 0, x \geq 0\}$, then $z \mapsto \sqrt{z^2 - 1}$ is analytic on $\mathbb{C}\backslash\{x + iy \mid y = 0, |x| \geq 1\}$.

Note.   The symbol $\sqrt{}$ has been used here to mean different branches of the square root function, the particular branch being clear from the context. The student should get in the habit of thinking of $\sqrt{}$ together with a choice of branch.

Exercises

1. Differentiate and give the appropriate region of analyticity for each of the following:
    a. $z^2 + z$.
    b. $1/z$.
    c. $\sin z/\cos z$.
    d. $\exp\left(\dfrac{z^3 + 1}{z - 1}\right)$.
    e. 3^z.
    f. $\log(z + 1)$.
    g. $z^{(1 + i)}$.
    h. $\sqrt{z}$.
    i. $\sqrt[3]{z}$.

2. Determine whether the following complex limits exist and find their value if they do:
    a. $\displaystyle\lim_{z \to 0} \left(\dfrac{e^z - 1}{z}\right)$.
    b. $\displaystyle\lim_{z \to 0} \left(\dfrac{\sin |z|}{z}\right)$.

3. Is it true that $|\sin z| \leq 1$ for all $z \in \mathbb{C}$?

4. Solve $\sin z = w$ and show how to choose a domain and thus how to pick a particular branch of $\sin^{-1}(z)$ so that it is analytic on the domain. Give the derivative of this branch of $\sin^{-1}(z)$ — see exercise 30 at the end of Section 1.3.

5. Prove Theorem 26 using the method of the first proof of Theorem 23.

6. Let $f: A \subset \mathbb{C} \to \mathbb{C}$ be analytic and suppose that A is an open, connected set. Show that if $f = u + iv$ and u is constant, then f is constant.

7. Let $f(z) = 1/(1 - z)$; is it continuous on the interior of the unit circle?

8. Let $u(x,y)$ and $v(x,y)$ be real-valued functions defined on an open set $A \subset \mathbb{R}^2 = \mathbb{C}$ and suppose that they satisfy the Cauchy-Riemann equations on A. Show that $u_1(x,y) = [u(x,y)]^2 - [v(x,y)]^2$ and $v_1(x,y) = 2u(x,y)v(x,y)$ satisfy the Cauchy-Riemann equations on A and that $u_2(x,y) = e^{u(x,y)}$ $\cos v(x,y)$ and $v_2(x,y) = e^{u(x,y)} \sin v(x,y)$ also satisfy the Cauchy-Riemann equations on A. Can you do this without performing any computations?

9. Suppose that $f: \mathbb{C} \to \mathbb{C}$ is entire and that $f(z) = f(2z)$ for all $z \in \mathbb{C}$. Show that f is constant on $\mathbb{C}$. (You may assume that f' is continuous.)

10. Find the region of analyticity and the derivative of each of the following functions.
    a. $z/(z^2 - 1)$.
    b. $e^{z + (1/z)}$.
    c. $\sqrt{z^3 - 1}$.
    d. $\sin \sqrt{z}$.

11. Suppose that $f: \mathbb{C} \to \mathbb{C}$ is entire and that $f = f \circ f$. What can you say about f? (You may assume that f' is continuous.)

12. Solve $\cos z = \sqrt{3}$ for z.

13. Where is $z \mapsto 2^{z^2}$ analytic? $z \mapsto z^{2z}$?

14. Find the minimum of $|e^{z^2}|$ for those z with $|z| \leq 1$.

15. Define a branch of $\sqrt{1 + \sqrt{z}}$ and show that it is analytic.

Review Exercises for Chapter 1

1. Compute e^i; $\log(1 + i)$; $\sin i$; 3^i; $e^{2 \log(-1)}$.

2. Find the 8th roots of i.

3. Let $f: A \subset \mathbb{C} \to \mathbb{C}$ be analytic on an open set A. Let $\bar{A} = \{\bar{z} | z \in A\}$.
   a. Describe $\bar{A}$ geometrically.
   b. Define $g: \bar{A} \to \mathbb{C}$ by $g(z) = \overline{f(\bar{z})}$. Show that g is analytic.

4. Differentiate the following expressions on appropriate regions:
   a. $z^3 + 8$.
   b. $\dfrac{1}{z^3 + 1}$.
   c. $\exp(z^4 - 1)$.
   d. $\sin(\log z^2)$.

5. Describe geometrically the set of points $z \in \mathbb{C}$ satisfying:
   a. $|z + i| = |z - i|$.
   b. $|z - 1| = 3|z - 2|$.

6. Find the real and imaginary parts of $f(z) = z^3$ and verify directly that they satisfy the Cauchy-Riemann equations.

7. On what set is $\sqrt{z^2 - 2}$ analytic? Compute the derivative.

8. Let f be analytic on A. Define $g: A \to \mathbb{C}$ by $g(z) = \overline{f(\bar{z})}$. When is g analytic?

9. Describe the sets on which the following functions are analytic and compute their derivatives:

   a. $e^{1/z}$.

   b. $\dfrac{1}{(1 - \sin z)^2}$.

   c. $\dfrac{e^{az}}{a^2 + z^2}$; a real.

   d. $\exp\left(\dfrac{1}{1 - az}\right)$, $a \in \mathbb{C}$.

   e. $\dfrac{\sin z}{z}$.

10. Can a single-valued (analytic) branch of $\log z$ be defined on the following sets?

   a. $\{z \mid 1 < |z| < 2\}$.

   b. $\{z \mid \operatorname{Re} z > 0\}$.

   c. $\{z \mid \operatorname{Re} z > \operatorname{Im} z\}$.

11. Show that the map $z \mapsto z + 1/z$ maps the circle $\{z \mid |z| = c\}$ onto the ellipse described by $\{w = u + iv \mid u = (c + 1/c) \cos\theta, v = (c - 1/c) \sin\theta, 0 \le \theta \le 2\pi\}$. Can we allow $c = 1$?

12. Prove the Cauchy-Riemann equations as follows. Let $f: A \subset \mathbb{C} \to \mathbb{C}$ be differentiable at $z_0 = x_0 + iy_0$. Let $g_1(t) = t + iy_0$ and $g_2(t) = x_0 + it$. Apply the Chain Rule to $f \circ g_1$ and $f \circ g_2$ to prove the result.

13. Let $f(z)$ be analytic in the disk $|z - 1| < 1$. Suppose that $f'(z) = 1/z, f(1) = 0$. Prove that $f(z) = \log z$.

14. Use the Inverse Function Theorem to prove the following result. Let $f: A \subset \mathbb{C} \to \mathbb{C}$ be analytic (where A is open and connected) and suppose that $f(A) \subset \{z \mid |z| = 3\}$. Then f is constant.

15. Prove that

$$\lim_{h \to 0} \frac{(z_0 + h)^n - z_0^n}{h} = n z_0^{n-1}$$

   for any $z_0 \in \mathbb{C}$.

16. a. If a polynomial $p(z) = a_0 + a_1 z + \cdots + a_n z^n$ has a root c, then show that we can write $p(z) = (z - c)h(z)$ where $h(z)$ is a polynomial of degree $n - 1$.

   Hint.   Review the division of polynomials to show that $z - c$ divides $p(z)$.

   b. Use 16(a) to show that p can have no more than n roots.

   c. When is c a root of both $p(z)$ and $p'(z)$?

17. Let $f(x + iy) = (x^2 + 2y) + i(x^2 + y^2)$. At what points does $f'(z_0)$ exist?

18. Suppose that $f: A \subset \mathbb{C} \to \mathbb{C}$ is analytic on the open connected set A and that $f(z)$ is real for all $z \in A$. Then show that f is constant.

19. Let $f: A \subset \mathbb{C} \to \mathbb{C}$ be analytic on the open set A and let $f'(z_0) \neq 0$ for $z_0 \in A$. Show that $\{\operatorname{Re} f(z) \mid z \in A\} \subset \mathbb{R}$ is open.

20.   On what set is the function $z \longmapsto z^z$ analytic? Compute its derivative.

21.   Let g be analytic on the open set A. Let $B = \{z \in A \,|\, g(z) \neq 0\}$. Show that B is open and that $1/g$ is analytic on B.

22.   Find and plot all solutions of $z^3 = -8i$.

23.   Show that $u(x,y) = x^3 - 3xy^2$, $v(x,y) = 3x^2y - y^3$ satisfy the Cauchy-Riemann equations. Comment on the result.

24.   Prove that the following functions are continuous at $z = 0$:

    a. $f(z) = \begin{cases} [\mathrm{Re}(z^2)]^2/|z|^2, & z \neq 0; \\ 0, & z = 0. \end{cases}$

    b. $f(z) = |z|$.

25.   Use de Moivre's Theorem to find the sum of $\sin x + \sin 2x + \cdots + \sin nx$.

26.   Suppose that $f(z)$ is analytic and satisfies the condition $|f(z)^2 - 1| < 1$ in a region Ω. Show that either $\mathrm{Re}\, f(z) > 0$ or that $\mathrm{Re}\, f(z) < 0$ throughout Ω.

27.   At what points z are the following functions differentiable?
    a. $f(z) = |z|^2$.
    b. $f(z) = y - ix$.

28.   For the function $u(x,y) = y^3 - 3x^2y$,
    a. Show that u is harmonic (see example 5, Section 1.4).
    b. Determine a function (called the *conjugate function*) $v(x,y)$ such that $u + iv$ is analytic.

29.   Consider the function $w(z) = 1/z$. Draw the level curves $u = $ constant. Discuss.

30.   Determine the four different values of z that are mapped to unity by the function $w(z) = z^4$.

Chapter 2

Cauchy's Theorem

A convenient feature of complex analysis is that it is based on a few simple, yet powerful theorems from which most of the results of the subject follow. Foremost among these theorems is Cauchy's Theorem, which enables us to prove, for example, that if f is analytic, then all the derivatives of f exist. Cauchy's Theorem is the key to the development of the rest of the subject and its applications.

2.1   Contour Integrals

To be able to study Cauchy's Theorem we first need to define contour integrals and to study their basic properties.

Let $h: [a,b] \subset \mathbb{R} \to \mathbb{C}$ be a function and set $h(t) = u(t) + iv(t)$. Suppose, for the sake of simplicity, that u and v are continuous. Define

$$\int_a^b h(t) \, dt = \int_a^b u(t) \, dt + i \int_a^b v(t) \, dt \in \mathbb{C}$$

giving the integrals of u and v their usual meaning.

We want to extend this definition to integrals of functions along curves in $\mathbb{C}$. A continuous *curve* or *contour* in $\mathbb{C}$ is, by definition, a continuous map $\gamma: [a,b] \to \mathbb{C}$. The curve is called *piecewise C^1* if we can divide up the interval $[a,b]$ into finitely many subintervals $a = a_0 < a_1 < \cdots < a_n = b$ such that $\gamma'(t)$ exists on each open subinterval $]a_i, a_{i+1}[$ and is continuous on $[a_i, a_{i+1}]$; that is, the limits $\lim_{t \to a_i^+} \gamma'(t)$ and $\lim_{t \to a_{i+1}^-} \gamma'(t)$ exist (see, for example, Figure 2.1). Unless otherwise specified, curves will always be assumed to be continuous and piecewise C^1.

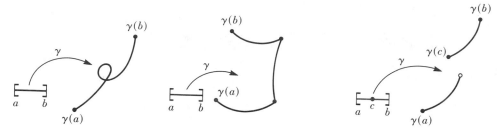

Figure 2.1   Curves in $\mathbb{C}$. *Left*, smooth curve; *center*, piecewise C^1 curve; *right*, discontinuous curve.

Definition 1.   *Suppose that f is continuous and defined on an open set $A \subset \mathbb{C}$ and that $\gamma: [a,b] \to \mathbb{C}$ is a piecewise smooth curve, with $\gamma([a,b]) \subset A$. The expression*

$$\int_\gamma f = \int_\gamma f(z) \, dz = \sum_{i=0}^{n-1} \int_{a_i}^{a_{i+1}} f(\gamma(t))\gamma'(t) \, dt$$

is called the integral of f along γ.

This definition is analogous to the following definition of a line integral that we learned in calculus: Let $P(x,y)$ and $Q(x,y)$ be given real-valued functions of x and y and let γ be a given curve. Then define

$$\int_\gamma P(x,y) \, dx + Q(x,y) \, dy = \sum_{i=0}^{n-1} \int_{a_i}^{a_{i+1}} \left\{ P(x(t),y(t)) \frac{dx}{dt} + Q(x(t),y(t)) \frac{dy}{dt} \right\} dt,$$

in which $\gamma(t) = (x(t),y(t))$. The two definitions are related as follows.
If $f(z) = u(x,y) + iv(x,y)$, then

$$\int_\gamma f = \int_\gamma (u(x,y) \, dx - v(x,y) \, dy) + i \int_\gamma (u(x,y) \, dy + v(x,y) \, dx).$$

Proof.

$$\begin{aligned}
f(\gamma(t)) \cdot \gamma'(t) &= [u(x(t), y(t)) + iv(x(t), y(t))] \cdot [x'(t) + iy'(t)] \\
&= (u(x(t), y(t))x'(t) - v(x(t), y(t))y'(t)) \\
&\quad + i(v(x(t), y(t))x'(t) + u(x(t), y(t))y'(t)).
\end{aligned}$$

Integrating both sides over $[a_i, a_{i+1}]$ with respect to t and using Definition 1 gives the desired result. ∎

This result can easily be remembered by formally writing $f(z)dz = (u + iv)(dx + idy) = udx - vdy + i(vdx + udy)$.

For a curve γ: $[a,b] \to \mathbb{C}$, we define the *opposite curve* $-\gamma$ by $-\gamma$: $[a,b] \to \mathbb{C}$, $(-\gamma)(t) = \gamma(a + b - t)$. This is merely γ traversed in the opposite sense (see Figure 2.2).

We also want to define the *join* or *sum* or *union* $\gamma_1 + \gamma_2$ of two curves γ_1 and γ_2. Intuitively, we want to join them at their end points to make a single curve (see Figure 2.3).

Precisely, let us suppose that γ_1: $[a,b] \to \mathbb{C}$ and that γ_2: $[b,c] \to \mathbb{C}$, with $\gamma_1(b) = \gamma_2(b)$. Let us define $\gamma_1 + \gamma_2$: $[a,c] \to \mathbb{C}$ by

$$(\gamma_1 + \gamma_2)(t) = \begin{cases} \gamma_1(t) \text{ if } t \in [a,b] \\ \gamma_2(t) \text{ if } t \in [b,c]. \end{cases}$$

Clearly, if γ_1 and γ_2 are piecewise smooth, then so is $\gamma_1 + \gamma_2$. If the intervals $[a,b]$ and $[b,c]$ for γ_1 and γ_2 are not of this special form (in that the first interval ends where the second begins), then the formula is more complicated, but the special form will suffice for this text. The general sum $\gamma_1 + \cdots + \gamma_n$ is defined similarly.

Theorem 1 gives some properties of the integral that follow at once from the definitions given in this section. The student is asked to prove them in exercise 1 at the end of this section.

Figure 2.2   Opposite curve.

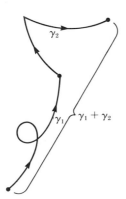

Figure 2.3   Join of two curves.

Theorem 1. *For (continuous) functions f,g and piecewise C^1 curves γ, γ_1, γ_2, we have*

i.
$$\int_\gamma (c_1 f + c_2 g) = c_1 \int_\gamma f + c_2 \int_\gamma g$$

for complex numbers c_1, c_2.

ii.
$$\int_{-\gamma} f = -\int_\gamma f.$$

iii.
$$\int_{\gamma_1 + \gamma_2} f = \int_{\gamma_1} f + \int_{\gamma_2} f.$$

Of course, more general statements (that actually follow from the preceding) could be made, namely:

i.'
$$\int_\gamma \left(\sum_{i=1}^n c_i f_i \right) = \sum_{i=1}^n c_i \int_\gamma f_i.$$

iii.'
$$\int_{\gamma_1 + \cdots + \gamma_n} f = \sum_{i=1}^n \int_{\gamma_i} f.$$

To compute specific examples of the integrals $\int_\gamma f$ requires that we be given f and γ. In such computations it is sometimes convenient to use the formula

$$\int_\gamma f = \int_\gamma (u\,dx - v\,dy) + i \int_\gamma (u\,dy + v\,dx).$$

However, we are usually not given γ as a map but are told only that it is, say, "the straight line joining 0 to $i + 1$" or "the unit circle traversed counterclockwise." Thus we need to choose some explicit map $\gamma(t)$ that describes this geometrically given curve. Obviously, the same geometric curve can be described in different ways, so the question naturally arises whether the integral $\int_\gamma f$ is independent of that description.

To answer this question, we need the following definition.

Definition 2. *A (piecewise smooth) curve $\tilde{\gamma}: [\tilde{a},\tilde{b}] \rightarrow \mathbb{C}$ is called a reparametrization of γ iff there is a C^1 function $\alpha: [a,b] \rightarrow [\tilde{a},\tilde{b}]$ with $\alpha'(t) > 0$, $\alpha(a) = \tilde{a}$, $\alpha(b) = \tilde{b}$ such that $\gamma(t) = \tilde{\gamma}(\alpha(t))$.*

The conditions $\alpha'(t) > 0$ (hence α is increasing) and $\alpha(a) = \tilde{a}$, $\alpha(b) = \tilde{b}$, mean that $\tilde{\gamma}$ traverses the curve in the same sense as γ does. This is the precise meaning of the statement that γ and $\tilde{\gamma}$ represent the

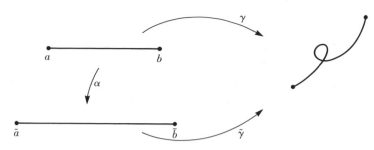

Figure 2.4 Reparametrization.

same (oriented) geometric curve (see Figure 2.4). Also, the points in $[\tilde{a},\tilde{b}]$ at which $\tilde{\gamma}'$ does not exist correspond under α to the points of $[a,b]$ at which γ' does not exist. (This is clear because α has a strictly increasing C^1 inverse.)

Theorem 2. *Let $\tilde{\gamma}$ be a reparametrization of γ as defined in Definition 2. Then*

$$\int_{\gamma} f = \int_{\tilde{\gamma}} f$$

for any continuous f defined on an open set containing the image of $\gamma =$ image of $\tilde{\gamma}$.

Proof. We can, by breaking up $[a,b]$, assume that γ is C^1. By definition,

$$\int_{\gamma} f = \int_{a}^{b} f(\gamma(t)) \cdot \gamma'(t)\,dt.$$

By the Chain Rule, $\gamma'(t) = d\gamma(t)/dt = d\tilde{\gamma}(\alpha(t))/dt = \tilde{\gamma}'(\alpha(t)) \cdot \alpha'(t)$. Let $s = \alpha(t)$ be a new variable, so that $s = \tilde{a}$ when $t = a$ and $s = \tilde{b}$ when $t = b$. Then

$$\int_{a}^{b} f(\gamma(t))\gamma'(t)\,dt = \int_{a}^{b} f(\tilde{\gamma}(\alpha(t)))\tilde{\gamma}'(\alpha(t))\frac{d\alpha}{dt}\,dt$$

$$= \int_{\tilde{a}}^{\tilde{b}} f(\tilde{\gamma}(s))\tilde{\gamma}'(s)\,ds.$$

The changing of variables in a complex integral is justified by applying the usual rule to its real and imaginary parts. ∎

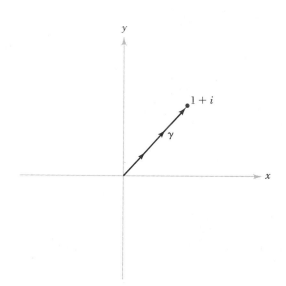

Figure 2.5

Thus Theorem 2 "justifies" the use of any curve γ that describes a given oriented geometric curve.* This statement can be illustrated with an example. Let us evaluate $\int_\gamma x\,dz$ where γ is the straight line from $z = 0$ to $z = 1 + i$ (see Figure 2.5).

We choose the curve $\gamma \colon [0,1] \longrightarrow \mathbb{C}$, defined by $\gamma(t) = t + it$. Of course when we write x in $\int_\gamma x\,dz$, we mean the function that gives the real parts of any complex number (that is, $f(z) = x = \operatorname{Re} z$).

$$\int_\gamma x\,dz = \int_0^1 x \circ \gamma(t)\gamma'(t)\,dt.$$

Hence

$$\int_\gamma x\,dz = \int_0^1 t(1 + i)\,dt = \frac{1 + i}{2}.$$

Actually, the orientation can be described by saying that "γ goes from z_1 to z_2." However, if γ is a closed curve, on which $z_1 = z_2$, we need a different prescription. When solving examples, where the curves are always easy to visualize, the student should assume that γ is traversed

---

*Strictly speaking, this statement is not correct since two maps with the same image need not be reparametrizations of one another. However, they are reparametrizations if we ignore points where $\gamma'(t) = 0$. Theorem 2 can be generalized to cover this case as well, but the complications that result from generalizing Theorem 2 to cover this case have been omitted to make the exposition less encumbered.

in the counterclockwise direction unless he is advised to the contrary.

We recall that the *arc length* of a given curve $\gamma: [a,b] \rightarrow \mathbb{C}$ is defined by

$$l(\gamma) = \int_a^b |\gamma'(t)| \, dt = \int_a^b \sqrt{x'(t)^2 + y'(t)^2} \, dt.$$

Arc length, too, is independent of the parametrization, by a similar proof to that of Theorem 2. The student is familiar with the fact that the arc length of the unit circle is 2π, the perimeter of the unit square is 4, and so on.

Theorem 3 explains an important property of integrals.

Theorem 3.   *Let f be continuous on an open set A and let γ be a piecewise C^1 curve in A. If there is a constant $M \geq 0$ such that $|f(z)| \leq M$ for all points z on γ (that is, $z = \gamma(t)$ for some t), then*

$$\left| \int_\gamma f \right| \leq Ml(\gamma).$$

More generally, we have

$$\left| \int_\gamma f \right| \leq \int_\gamma |f| \, |dz|$$

where the latter integral is defined by

$$\int_\gamma |f| \, d|z| = \int_a^b |f(\gamma(t))| |\gamma'(t)| dt.$$

Proof.   For a complex-valued function $g(t)$ on $[a,b]$, we have

$$\text{Re} \int_a^b g(t) dt = \int_a^b \text{Re } g(t) dt,$$

since $\int_a^b g(t) dt = \int_a^b u(t) dt + i \int_a^b v(t) dt$ if $g(t) = u(t) + iv(t)$. Let us use this fact to prove that

$$\left| \int_a^b g(t) dt \right| \leq \int_a^b |g(t)| dt.$$

(We would know, from calculus, how to prove this if g were real-valued, but here it is complex-valued.) For our proof, we let $\int_a^b g(t)\,dt = re^{i\theta}$ for fixed r,θ, so $r = e^{-i\theta}\int_a^b g(t)\,dt = \int_a^b e^{-i\theta}g(t)\,dt$. Thus

$$r = \mathrm{Re}\ r = \mathrm{Re}\int_a^b e^{-i\theta}g(t)\,dt$$

$$= \int_a^b \mathrm{Re}\ (e^{-i\theta}g(t))\,dt.$$

But by Theorem 7(iii) in Chapter 1, $\mathrm{Re}\ e^{-i\theta}g(t) \le |e^{-i\theta}g(t)| = |g(t)|$, since $|e^{-i\theta}| = 1$. *Thus* $\int_a^b \mathrm{Re}\ (e^{-i\theta}g(t))\,dt \le \int_a^b |g(t)|\,dt$, and so

$$\left|\int_a^b g(t)\,dt\right| = r \le \int_a^b |g(t)|\,dt. \tag{1}$$

Therefore,

$$\left|\int_\gamma f\right| = \left|\int_a^b f(\gamma(t))\gamma'(t)\,dt\right|$$

$$\le \int_a^b |f(\gamma(t))\gamma'(t)|\,dt$$

$$= \int_a^b |f(\gamma(t))||\gamma'(t)|\,dt \tag{2}$$

by inequality (1) and the fact that $|zz'| = |z|\ |z'|$. Expression (2) is merely an ordinary real integral, and therefore since $|f(\gamma(t))| \le M$, expression (2) is bounded by $M\int_a^b |\gamma'(t)|\,dt = Ml(\gamma)$. ∎

This result is a basic tool that we shall use in subsequent proofs to estimate the size of integrals.

The student might try to prove this result directly in terms of the expression

$$\int_\gamma f = \int_\gamma u\,dx - v\,dy + i\int_\gamma (u\,dy + v\,dx)$$

to convince himself that the result is not altogether trivial.

This section is concluded with a useful method for evaluating integrals. It is the analogue of the Fundamental Theorem of Calculus, whereby we can evaluate integrals if we have found an antiderivative.

Theorem 4. *Let* $f: A \subset \mathbb{C} \longrightarrow \mathbb{C}$ *be continuous and such that* $f = F'$ *for some analytic function* $F: A \longrightarrow \mathbb{C}$. *Let* γ *be a piecewise* C^1 *curve lying in* A *and joining* z_1 *to* z_2. *Then*

$$\int_\gamma f = F(z_2) - F(z_1).$$

In particular, if $z_1 = z_2$, *then*

$$\int_\gamma f = 0.$$

A curve γ whose end points z_1 and z_2 are equal is called a *closed curve*. Theorem 4 states that if f is the derivative of some analytic function, then the integral of f around any closed curve is 0. This statement is related to but should not be confused with Cauchy's Theorem, which is studied in Sections 2.2 and 2.3.

Proof.   We have, using the Chain Rule,

$$\int_\gamma f = \int_a^b f(\gamma(t))\gamma'(t)dt$$

$$= \int_a^b F'(\gamma(t)) \cdot \gamma'(t)dt$$

$$= \int_a^b \frac{d}{dt}F(\gamma(t))\,dt$$

$$= F(\gamma(b)) - F(\gamma(a)). \blacksquare$$

Application of Theorem 4 can save a lot of effort in working examples. For instance, consider $\int_\gamma z^3\, dz$, where γ is the portion of the ellipse $x^2 + 4y^2 = 1$ that joins $z = 1$ to $z = i/2$. To evaluate the integral we merely note that $z^3 = (1/4)(dz^4/dz)$ and so, by Theorem 4,

$$\int_\gamma z^3 dz = \frac{z^4}{4}\Big|_1^{i/2} = \frac{1}{4}\left(\frac{i}{2}\right)^4 - \frac{1^4}{4} = -\frac{15}{64}.$$

Notice that we did not even need to parametrize the curve. In fact, by applying Theorem 4 we would have obtained the same answer for any curve joining these two points.

Worked Examples

1. Evaluate the following integrals:
    a. $\int_\gamma x\,dz$ (γ is the circumference of the unit square).
    b. $\int_\gamma e^z\,dz$ (γ is the part of the unit circle joining 1 to i in a counterclockwise direction).

Solution.

a. Define $\gamma: [0,4] \rightarrow \mathbb{C}$ as follows. $\gamma = \gamma_1 + \gamma_2 + \gamma_3 + \gamma_4$ where the four sides of the unit square are:

$$\gamma_1(t) = t + 0i; \; 0 \le t \le 1.$$
$$\gamma_2(t) = 1 + (t-1)i; \; 1 \le t \le 2.$$
$$\gamma_3(t) = (3-t) + i; \; 2 \le t \le 3.$$
$$\gamma_4(t) = 0 + (4-t)i; \; 3 \le t \le 4.$$

We compute as follows:

$$\int_{\gamma_1} x\,dz = \int_0^1 [\text{Re } (\gamma_1(t))]\gamma_1'(t)\,dt = \int_0^1 t\,dt = \frac{1}{2};$$

$$\int_{\gamma_2} x\,dz = \int_1^2 [\text{Re } (\gamma_2(t))]\gamma_2'(t)\,dt = \int_1^2 i\,dt = i;$$

$$\int_{\gamma_3} x\,dz = \int_2^3 [\text{Re } (\gamma_3(t))]\gamma_3'(t)\,dt = \int_2^3 -(3-t)\,dt = -\frac{1}{2};$$

$$\int_{\gamma_4} x\,dz = \int_3^4 [\text{Re } (\gamma_4(t))]\gamma_4'(t)\,dt = \int_3^4 0\,dt = 0.$$

Hence

$$\int_\gamma x\,dz = \int_{\gamma_1} x\,dz + \int_{\gamma_2} x\,dz + \int_{\gamma_3} x\,dz + \int_{\gamma_4} x\,dz$$

$$= \frac{1}{2} + i - \frac{1}{2} + 0 = i.$$

b. e^z is the derivative of the function e^z, and e^z is analytic on all of $\mathbb{C}$. Thus whatever parametrization we use for the part of the unit circle joining 1 to i in a counterclockwise direction, we have $\int_\gamma e^z\,dz = e^i - e^1$ by Theorem 4.

We can return to the original definition to evaluate the integral directly. Define $\gamma(t) = \cos t + i\sin t$, $0 \le t \le \pi/2$. Hence

$$\int_\gamma e^z\,dz = \int_0^{\pi/2} (e^{\cos t + i\sin t})(-\sin t + i\cos t)\,dt$$

$$= \int_0^{\pi/2} (-e^{\cos t}\cos\,(\sin\,t)\cdot\sin\,t - e^{\cos t}\sin\,(\sin\,t)\cos\,t)dt$$

$$+ i\int_0^{\pi/2} (-e^{\cos t}\sin\,(\sin\,t)\cdot\sin\,t + e^{\cos t}\cos\,(\sin\,t)\cdot\cos\,t)dt$$

$$= e^{\cos t}\cos\,(\sin\,t)\Big|_0^{\pi/2} + ie^{\cos t}\sin\,(\sin\,t)\Big|_0^{\pi/2}$$

$$= e^{\cos t + i\sin t}\Big|_0^{\pi/2}$$

$$= e^i - e^1.$$

Clearly, the first method is easier and should be used whenever possible.

2. Let γ be the upper half of the unit circle described counterclockwise. Show that

$$\left| \int_\gamma \frac{e^z}{z}\,dz \right| \le \pi e.$$

Solution.   We use Theorem 3. The arc length of γ is

$$l(\gamma) = \int_0^\pi |\gamma'(t)|\,dt = \pi,$$

since we can take $\gamma(t) = e^{it}$, $0 \le t \le \pi$, and $\gamma'(t) = ie^{it}$, so $|\gamma'(t)| = 1$. Of course, this is what we would expect. The absolute value of e^z/z is, for $z = e^{it} = \cos t + i\sin t$,

$$\left|\frac{e^z}{z}\right| = \frac{e^{\cos t}}{1} \le e,$$

since $\cos t \le 1$. Thus $e = M$ is a bound for $|e^z/z|$ along γ, and therefore,

$$\left| \int_\gamma \frac{e^z}{z}dz \right| \le Ml(\gamma) = e\pi.$$

3. Let γ be the circle of radius r around $a \in \mathbb{C}$. Evaluate $\int_\gamma (z - a)^n\,dz$ for all integers $n = 0, \pm1, \pm2, \cdots$.

Solution.   First, let $n \ge 0$. Then

$$(z - a)^n = \frac{d}{dz}\frac{1}{n + 1}(z - a)^{n + 1},$$

which is the derivative of an analytic function, so by Theorem 4, $\int_\gamma (z - a)^n \cdot dz = 0$.

Second, let $n \le -2$. Then again

$$(z-a)^n = \frac{d}{dz}\frac{1}{n+1}(z-a)^{n+1},$$

which is analytic on $A = \mathbb{C}\backslash\{a\}$. (Note that this formula fails if $n = -1$.) Since γ lies in A, Theorem 4 again shows that $\int_\gamma (z-a)^n \, dz = 0$.

Finally, let $n = -1$. It is easiest to proceed directly. We parametrize γ by $\gamma(\theta) = re^{i\theta} + a$, $0 \le \theta \le 2\pi$ (see Figure 2.6). Then, by the Chain Rule, $\gamma'(\theta) = rie^{i\theta}$, and so

$$\int_\gamma \frac{1}{z-a}\, dz = \int_0^{2\pi} \frac{1}{(re^{i\theta}+a)-a}\, ire^{i\theta}\, d\theta = \int_0^{2\pi} i\, d\theta = 2\pi i.$$

In summary then,

$$\int_\gamma (z-a)^n \, dz = \begin{cases} 0, & n \ne -1; \\ 2\pi i, & n = -1. \end{cases}$$

Remark.   This result is worth noting since it is a useful formula and we shall have occasion to use it later.

4. Prove that there does not exist an analytic function f defined on $\mathbb{C}\backslash\{0\}$ such that $f'(z) = 1/z$.

Solution.   If such an f existed, then, using Theorem 4, we would conclude that $\int_\gamma (1/z)dz = 0$, where γ is the unit circle. But by example 3, $\int_\gamma (1/z)dz = 2\pi i$, so no such f can exist.

Note.   Although $d\log z/dz = 1/z$, it does not contradict this example because $\log z$ is not analytic on $\mathbb{C}\backslash\{0\}$; it is analytic only on $\mathbb{C}$ minus the negative x-axis including zero.

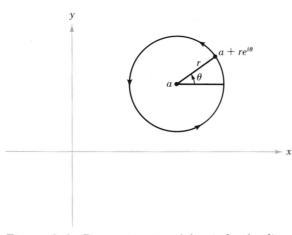

Figure 2.6 Parametrization of the circle of radius r and center a.

Exercises

1. Prove Theorem 1.

2. Does $\text{Re}\{\int_\gamma f dz\} = \int_\gamma \text{Re } f dz$?

3. Evaluate the following:
    a. $\int_\gamma y dz$; γ is the union of the line segments joining 0 to i and then to $i + 2$.
    b. $\int_\gamma \sin 2z \, dz$; γ is the line segment joining $i + 1$ to $-i$.
    c. $\int_\gamma z e^{z^2} \, dz$; γ is the unit circle.

4. Give some conditions on a closed curve γ that will guarantee that $\int_\gamma (1/z) dz = 0$.

5. Let γ be the unit circle. Prove that

$$\left| \int_\gamma \frac{\sin z}{z^2} \, dz \right| \leq 2\pi e.$$

6. Show that the arc length $l(\gamma)$ of a curve γ is unchanged if γ is reparametrized.

7. Evaluate the following:
    a. $\int_\gamma \bar{z} \, dz$; γ is the unit circle traversed in a counterclockwise direction.
    b. $\int_\gamma (x^2 - y^2) \, dz$; γ is the straight line from 0 to i.

8. Evaluate $\int_\gamma \bar{z}^2 \, dz$ along two paths joining $(0,0)$ to $(1,1)$ as follows:
    a. γ is the straight line joining $(0,0)$ to $(1,1)$.
    b. γ is the broken line joining $(0,0)$ to $(1,0)$, then joining $(1,0)$ to $(1,1)$.

    In view of your answers to 8(a) and (b) and Theorem 4, could $\bar{z}^2$ be the derivative of any analytic function $F(z)$?

9. Estimate the absolute value of

$$\int_\gamma \frac{dz}{2 + z^2}$$

    where γ is the upper half of the unit circle.

10. Let C be the arc of the circle $|z| = 2$ that lies in the first quadrant. Show that

$$\left| \int_C \frac{dz}{z^2 + 1} \right| \leq \frac{\pi}{3}.$$

11. Evaluate the following:

    a.
$$\int_{|z|=1} \frac{dz}{z}, \int_{|z|=1} \frac{dz}{|z|}, \int_{|z|=1} \frac{|dz|}{z}, \int_{|z|=1} \left| \frac{dz}{z} \right|.$$

    b. $\int_\gamma z^2 \, dz$ where γ is the curve given by $\gamma(t) = e^{it}\sin^3 t$, $0 \leq t \leq \pi/2$.

12. Let γ be a closed curve lying entirely in $\mathbb{C}\backslash\{z \mid \text{Re } z \leq 0\}$. Show that $\int_\gamma (1/z) dz = 0$.

13. Evaluate $\int_\gamma z \sin z^2 \, dz$ where γ is the unit circle.

2.2   Cauchy's Theorem: Intuitive Version

In mathematics it is important to have an intuitive understanding and to be able to express that intuition precisely. In this section Cauchy's Theorem will be discussed rather informally; the proofs will be simple and the definitions somewhat casual (although entirely adequate for most applications). Section 2.3 provides more precise definitions and more complete proofs. That section can be omitted, but such omission is recommended only if there is a desire to proceed quickly to subsequent sections on applications of the theorem.

In its most simple form, Cauchy's Theorem states that if γ is a simple closed curve (the word "simple" meaning that γ intersects itself only at its end points)* and f is analytic *on and inside* γ, then

$$\int_{\gamma} f = 0$$

(see Figure 2.7).

If this theorem is to be valid, f must be analytic on the whole region inside γ. For example, let γ be the unit circle and let $f(z) = 1/z$. Then f is analytic at all points except $z = 0$, and indeed the integral is not zero. In fact,

$$\int_{\gamma} f = 2\pi i$$

by example 3, Section 2.1. Theorem 4 of the preceding section indicated a special case of Cauchy's Theorem; namely, if f happens to be the derivative of another analytic function F. (A method of constructing such an F, using Cauchy's Theorem, will be described in Theorem 9.)

The simplest proof of Cauchy's Theorem utilizes a theorem from advanced calculus called *Green's Theorem*. (The proof of Cauchy's

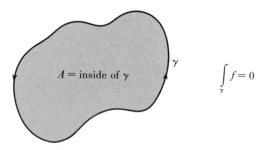

Figure 2.7   Cauchy's Theorem.

*In appropriate situations γ does not actually have to be simple. This more general case is discussed in Section 2.3.

Theorem given in Section 2.3 is not based on this theorem.) Green's Theorem states that, given functions $P(x,y)$ and $Q(x,y)$,

$$\int_{\gamma} P(x,y)dx + Q(x,y)dy = \int\int_{A} \left\{ \frac{\partial Q}{\partial x}(x,y) - \frac{\partial P}{\partial y}(x,y) \right\} dxdy. \qquad (1)$$

Recall that if $\gamma: [a,b] \longrightarrow \mathbb{C}$, $\gamma(t) = (x(t), y(t))$, then we define

$$\int_{\gamma} P(x,y)dx = \int_{a}^{b} P(x(t), y(t))x'(t)dt$$

and

$$\int_{\gamma} Q(x,y)dy = \int_{a}^{b} Q(x(t), y(t))y'(t)dt.$$

In equation (1), A represents the "inside" of γ, γ is traversed in a counterclockwise direction, and P and Q are sufficiently smooth — say, C^1 (see a calculus text such as K. McAloon and A. Tromba, *Calculus* (New York: Harcourt Brace Jovanovich, 1972), Chapter 17, for a proof of equation (1)).

   At this point the student may well ask, "What, precisely, is the inside of γ?" Intuitively, the meaning of "inside" should be clear. The precise expression of the concept is based on the difficult Jordan Curve Theorem (*Any simple closed curve has an "inside" and an "outside"*), which will be stated formally in Section 2.3.

Theorem 5 (Cauchy's Theorem).   *Suppose that f is analytic, with f' continuous, on and inside the simple closed curve γ. Then*

$$\int_{\gamma} f = 0. \qquad (2)$$

Proof.   Setting $f = u + iv$, we have

$$\int_{\gamma} f = \int_{\gamma} f(z)\, dz = \int_{\gamma} (u + iv)(dx + idy)$$

$$= \int_{\gamma} (udx - vdy) + i \int_{\gamma} (udy + vdx).$$

By applying Green's Theorem (formula (1)) to each integral, we get

$$\int_{\gamma} f = \int\int_{A} \left\{ -\frac{\partial v}{\partial x} - \frac{\partial u}{\partial y} \right\} dxdy + i \int\int_{A} \left\{ \frac{\partial u}{\partial x} - \frac{\partial v}{\partial y} \right\} dxdy.$$

Both terms are zero by the Cauchy-Riemann equations. ∎

Actually, the more precise version of Cauchy's Theorem (due to Edouard Goursat) given in Section 2.3 does not require f' to be continuous; actually, continuity of f' follows automatically.

As an example of the use of formula (2), let γ be the unit square and let $f(z) = \sin(e^{z^2})$. Then f is analytic on and inside γ (in fact, f is entire) so $\int_\gamma f = 0$.

Deformation Theorem

It is important to be able to study functions that are not analytic on the entire inside of γ and whose integral therefore might not be zero. For example, $f(z) = 1/z$ fails to be analytic at $z = 0$, and $\int_\gamma f = 2\pi i$ where γ is the unit circle. (The point $z = 0$ is called a *singularity* of f.) To study such functions it is important to be able to replace $\int_\gamma f$ by $\int_{\tilde\gamma} f$ where $\tilde\gamma$ is a simpler curve (say, a circle). Then $\int_{\tilde\gamma} f$ can often be evaluated. The procedure that allows us to pass from γ to $\tilde\gamma$ is based on Cauchy's Theorem and is as follows.

Theorem 6 (Deformation Theorem). *Let f be analytic on a region A and let γ be a simple closed curve in A. Suppose that γ can be continuously deformed to another simple closed curve $\tilde\gamma$ without passing outside of the region A. (We say that γ is homotopic to $\tilde\gamma$ in A.) Then*

$$\int_\gamma f = \int_{\tilde\gamma} f. \tag{3}$$

Note.   The precise definition of "homotopic" is given in Section 2.3.

The Deformation Theorem is illustrated in Figure 2.8. Note that f need not be analytic inside γ, so Cauchy's Theorem does *not* imply

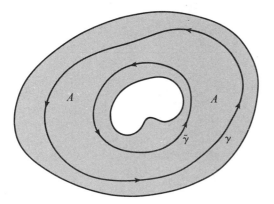

Figure 2.8   Deformation Theorem.

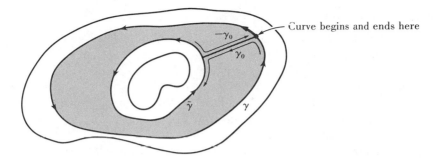

Figure 2.9   Curve used to prove the Deformation Theorem.

that the integrals in formula (3) are zero. It is implicit in the statement of Theorem 6 that both γ and $\tilde{\gamma}$ are traversed in a counterclockwise direction.

Proof.   Consider Figure 2.9, in which a curve γ_0 is drawn joining γ to $\tilde{\gamma}$; such a curve can be drawn in all practical examples. We get a new curve consisting of γ, then γ_0, then $-\tilde{\gamma}$, and then $-\gamma_0$, in that order.

The inside of this curve is the shaded region in Figure 2.9. In this region, f is analytic, so Cauchy's Theorem (formula (2)) gives

$$\int_{\gamma+\gamma_0-\tilde{\gamma}-\gamma_0} f = 0.$$

Strictly speaking, this new curve is not a simple closed curve, but such an objection can be taken care of by drawing two parallel copies of γ_0 and taking the limit as these copies converge together. From

$$\int_{\gamma+\gamma_0-\tilde{\gamma}-\gamma_0} f = 0$$

we get

$$\int_{\gamma} f + \int_{\gamma_0} f - \int_{\tilde{\gamma}} f - \int_{\gamma_0} f = 0;$$

that is,

$$\int_{\gamma} f = \int_{\tilde{\gamma}} f,$$

as required. ∎

The basic idea of this proof is also the basis of the more technical proof of the precise version of the Deformation Theorem considered in Section 2.3, but in such proofs the key ideas sometimes get lost.

Simply connected regions

A region $A \subset \mathbb{C}$ is called *simply connected* if A is connected and every closed curve γ in A can be deformed in A to some constant curve $\tilde{\gamma}(t) = z_0 \in A$; we also say that γ is *homotopic to a point*.

Intuitively, a region is simply connected when it has no holes, because a curve that loops around a hole cannot be shrunk down to a point in A without leaving A (see Figure 2.10).

Therefore, the domain on which a function like $f(z) = 1/z$, which has a singularity, is analytic, is *not* simply connected. Such regions are important because we shall want to study singularities in detail in Chapter 4.

By applying the Jordan Curve Theorem (see p. 110), we can prove that a region is simply connected iff, for every simple closed curve γ in A, the inside of γ also lies in A. This conclusion is, intuitively, quite obvious and the student should try to convince himself that it is. We can also apply the theorem to prove that the inside of a simple closed curve is simply connected.

Thus we can rewrite Cauchy's Theorem as follows.

Theorem 7. *Let f be analytic on a simply connected region A and let γ be a simple closed (piecewise C^1) curve in A. Then*

$$\int_\gamma f = 0.$$

Independence of path and antiderivatives

We now wish to reformulate Theorem 7 in a slightly different way.

Theorem 8. *Suppose that f is analytic on a simply connected region A. Then, for any two curves γ_1 and γ_2 joining two points z_0 and z_1 (as in*

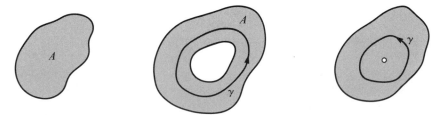

Figure 2.10 Simply connected region (*left*) and regions that are not simply connected (*center and right*).

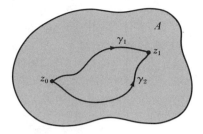

Figure 2.11 Independence of path.

Figure 2.11), *we have*

$$\int_{\gamma_1} f = \int_{\gamma_2} f.$$

Proof.   Let us consider $\gamma = \gamma_2 - \gamma_1$. Then γ is a closed curve. In this proof we assume for the sake of simplicity that γ is a simple closed curve, although this restriction is not necessary. By Theorems 1 and 7,

$$\int_{\gamma} f = 0 = \int_{\gamma_1} f - \int_{\gamma_2} f. \blacksquare$$

Using this result we get the following useful theorem, which is related to Theorem 4 and which is very similar to the Fundamental Theorem of Calculus.

Theorem 9.   *Let f be analytic and defined on a simply connected region A. Then there exists an analytic function F on A unique up to an additive constant such that $F' = f$.*

Proof. F must be unique up to additive constants because if $F' = f = \tilde{F}'$, then $(F - \tilde{F})' = 0$, so $F - \tilde{F} = $ constant, by Theorem 21 of Chapter 1.

To show that F exists, we fix $z_{00} \in A$ and let

$$F(z) = \int_{z_{00}}^{z} f,$$

where the integral means to use any path joining z_{00} to z, which we can do by Theorem 8. Then it follows that for any two points z, $z_0 \in A$, we have the identity

$$\frac{F(z) - F(z_0)}{z - z_0} - f(z_0) = \frac{1}{z - z_0} \int_{z_0}^{z} [f(w) - f(z_0)]dw.$$

Since f is continuous, given $\epsilon > 0$ there is a $\delta > 0$ such that $|w - z_0| < \delta$ implies that $|f(w) - f(z_0)| < \epsilon$. Thus if $|z - z_0| < \delta$ and we integrate along the straight line from z_0 to z, we get

$$\left| \frac{F(z) - F(z_0)}{z - z_0} - f(z_0) \right| = \frac{1}{|z - z_0|} \left| \int_{z_0}^{z} [f(w) - f(z_0)]dw \right|$$

$$\leq \frac{1}{|z - z_0|} \epsilon \cdot |z - z_0| = \epsilon$$

by Theorem 3. The length of the straight line joining z to z_0 is $|z - z_0|$; we can use a straight line if z is close to z_0, since A is open. Thus

$$\lim_{z \to z_0} \frac{F(z) - F(z_0)}{z - z_0} = f(z_0),$$

and so F is analytic on A with $F'(z_0) = f(z_0)$. ■

If A is not simply connected, this proof does not hold. For example, if $A = \mathbb{C} \setminus \{0\}$ and $f(z) = 1/z$, there is no F with $F' = f$ on A (see example 4, Section 2.1). However, on any simply connected region not containing 0 we can find such an F. This is the basis of the following discussion.

The logarithm again

Theorem 10.   *Let A be a simply connected region and let $0 \notin A$. Then there is an analytic function $F(z)$ unique up to the addition of multiples of $2\pi i$ such that $e^{F(z)} = z$.*

We write $F(z) = \log z$ and call such a choice of F a branch of log on A. Clearly, this procedure generalizes the procedure described in Section 1.5 and we get the usual log as defined in that section if A is $\mathbb{C}$ minus 0 and the negative real axis. Note that this A is simply connected. However, the A in Theorem 10 can be more complicated, as depicted in Figure 2.12.

Proof.   By Theorem 9, there is an analytic function F with $F'(z) = 1/z$ on A. Let us fix a point $z_0 \in A$. Then z_0 lies in the domain of some branch of the log function defined in Section 1.5. If we adjust F by adding a constant so that $F(z_0) = \log z_0$, then at z_0, $e^{F(z_0)} = z_0$. We now want to show that $e^{F(z)} = z$ is true on all of A. To do this, we let $g(z) = e^{F(z)}/z$. Then, since $0 \notin A$, g is analytic on A, and since $F'(z) = 1/z$,

$$g'(z) = \frac{z \cdot \dfrac{1}{z} \cdot e^{F(z)} - 1 \cdot e^{F(z)}}{z^2} = 0.$$

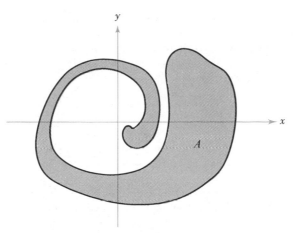

Figure 2.12   A domain for the log function.

Thus g is constant on A. But $g=1$ at z_0, so g is 1 on all of A. Therefore, $e^{F(z)} = z$ on all of A.

For uniqueness, let F and G be functions analytic on A and let $e^{F(z)} = z$ and $e^{G(z)} = z$. Then $e^{[F(z) - G(z)]} = 1$, so at a fixed z_0, $F(z_0) - G(z_0) = 2\pi n i$ for some integer n. But $F'(z) = 1/z = G'(z)$, so we have $d(F - G)/dz = 0$, from which we conclude that $F - G = 2\pi n i$ on all of A. ∎

Worked Examples

1. Evaluate the following integrals:
   a. $\int_{\gamma} e^z\, dz$; γ is the perimeter of the unit square.
   b. $\int_{\gamma} 1/z^2\, dz$; γ is the unit circle.
   c. $\int_{\gamma} 1/z\, dz$; γ is the circle $3 + e^{i\theta}$, $0 \le \theta \le 2\pi$.
   d. $\int_{\gamma} z^2\, dz$; γ is the segment joining $1 + i$ to 2.

   Solution.
   a. e^z is entire; thus by Cauchy's Theorem, $\int_{\gamma} e^z\, dz = 0$, since γ is a simple closed curve. Alternatively, e^z is the derivative of e^z, and since γ is closed we can apply Theorem 4.
   b. $1/z^2$ is defined and analytic on $\mathbb{C}\backslash\{0\}$ and is the derivative of $-1/z$, which is defined and analytic on $\mathbb{C}\backslash\{0\}$. By Theorem 4 and the fact that the unit circle lies in $\mathbb{C}\backslash\{0\}$, we have $\int_{\gamma} 1/z^2\, dz = 0$. Alternatively, we can use example 3, p. 85, for our solution.
   c. The circle $\gamma = 3 + e^{i\theta}$, $0 \le \theta \le 2\pi$, does not pass through 0 or include 0 in its interior. Hence $1/z$ is analytic on γ and the interior of γ, so by Cauchy's Theorem, $\int_{\gamma} 1/z\, dz = 0$. An alternative but less direct solution is the following. The region $\{x + iy \mid x > 0\}$ is simply connected and

$1/z$ is analytic on it. So by Theorem 10, $1/z$ is the derivative of some analytic function $F(z)$ (one of the branches of log z) and thus, since γ is closed, we have by Theorem 4 that $\int_\gamma 1/z \, dz = 0$.

d. z^2 is entire and is the derivative of $z^3/3$, which is also entire. By Theorem 4,

$$\int_\gamma z^2 dz = \frac{1}{3}z^3\Big|_{1+i}^{2} = \frac{(2)^3}{3} - \frac{(1+i)^3}{3} = \frac{10}{3} - \frac{2i}{3}.$$

Remark.   In (a) and (c) the first method uses Cauchy's Theorem; the alternate method is based on the more elementary fact that if $f(z)$ is the derivative of another analytic function, then the integral of f around a closed contour is zero (see Theorem 4). On the other hand, we have Theorem 9, stating that an analytic function defined on a simply connected region is the derivative of some other analytic function. The student is reminded that we had to prove Cauchy's Theorem to be able to prove Theorem 9.

2.   Use the Deformation Theorem to argue informally that if γ is a simple closed curve (not necessarily a circle) containing 0, then

$$\int_\gamma \frac{1}{z} dz = 2\pi i.$$

Solution.   The inside of γ contains 0. Thus we can find an $r > 0$ such that the circle $\tilde{\gamma}$ of radius r and centered at 0 lies entirely on the inside of γ. Our intuition should tell us that we can deform γ to $\tilde{\gamma}$ without passing

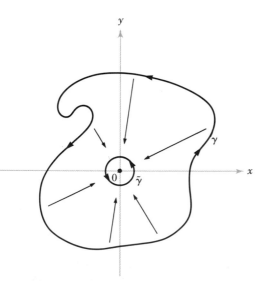

Figure 2.13   Deformation of γ to the circle $\tilde{\gamma}$.

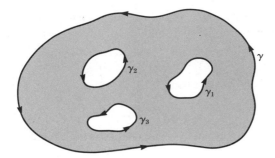

Figure 2.14 Generalized Deformation
Theorem.

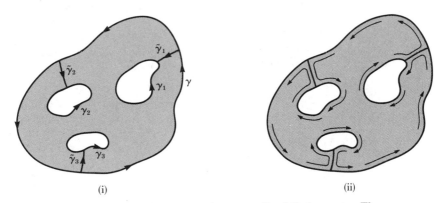

(i)   (ii)

Figure 2.15 Path used to prove the generalized Deformation Theorem.

through 0 (that is, by staying in the region $\varLambda - \mathbb{C}\backslash\{0\}$ of analyticity of $1/z$ – see Figure 2.13). Therefore, by Theorem 6,

$$\int_{\gamma}\frac{1}{z}dz = \int_{\tilde{\gamma}}\frac{1}{z}dz = 2\pi i$$

by our calculation in example 3, Section 2.1.

3. Outline an extension of the Deformation Theorem as follows. Suppose that $\gamma_1, \cdots, \gamma_n$ are simple closed curves and that γ is a simple closed curve with f analytic on the region between γ and $\gamma_1, \cdots, \gamma_n$ (see Figure 2.14). Then

$$\int_{\gamma}f = \sum_{k=1}^{n}\int_{\gamma_k}f.$$

Solution.   Draw curves $\tilde{\gamma}_1, \tilde{\gamma}_2, \cdots, \tilde{\gamma}_n$ joining γ to $\gamma_1, \cdots, \gamma_n$, respectively, as shown in Figure 2.15(i). Let ρ denote the curve drawn in Figure 2.15(ii). The inside of ρ is a region of analyticity of f, so $\int_{\rho}f = 0$.

But ρ consists of $\gamma, -\gamma_1, -\gamma_2, \cdots, -\gamma_n$, and each of $\tilde{\gamma}_i$ traversed twice in opposite directions, so the contributions from these last portions cancel. Thus

$$0 = \int_\gamma f + \int_{-\gamma_1} f + \cdots + \int_{-\gamma_n} f$$

$$= \int_\gamma f - \sum_{i=1}^n \int_{\gamma_i} f,$$

as required.

4. Let $f(z)$ be analytic on a simply connected region A, except possibly not analytic at $z_0 \in A$. Suppose, however, that f is bounded in absolute value near z_0. Show that, for any simple closed curve γ containing z_0, $\int_\gamma f = 0$.

Solution.   Let $\epsilon > 0$ and let γ_ϵ be the circle of radius ϵ and with center z_0. By the Deformation Theorem, $\int_\gamma f = \int_{\gamma_\epsilon} f$. Let $|f(z)| \leq M$ near z_0. Thus

$$\left| \int_{\gamma_\epsilon} f \right| \leq M \cdot 2\pi\epsilon.$$

Hence for any $\epsilon > 0$,

$$\left| \int_\gamma f \right| \leq M \cdot 2\pi\epsilon.$$

Therefore, $\int_\gamma f = 0$.

Exercises

1. Evaluate the following integrals:
   a. $\int_\gamma (z^3 + 3)\,dz$; γ is the upper half of the unit circle.
   b. $\int_\gamma (z^3 + 3)\,dz$; γ is the unit circle.
   c. $\int_\gamma e^{1/z}dz$; γ is a circle of radius 3 centered at $5i + 1$.
   d. $\int_\gamma \cos{(3 + 1/(z - 3))}\ dz$; γ is a unit square with corners at $0, 1, 1 + i$, and i.

2. Let f be entire. Evaluate

$$\int_0^{2\pi} f(z_0 + re^{i\theta})\,e^{ki\theta}\,d\theta$$

for k an integer, $k \geq 1$.

3. Let γ be a simple closed curve containing 0. Argue informally that

$$\int_\gamma \frac{1}{z^2}dz = 0.$$

4. Discuss the formula $\log z = \log r + i\theta$ for log on the region A shown in Figure 2.12.

5. For what simple closed curves γ is

$$\int_\gamma \frac{dz}{z^2 + z + 1} = 0?$$

6. Does Cauchy's Theorem hold separately for the real and imaginary parts of f? If so, prove that it does; if not, give a counterexample.

7. Let γ_1 be the circle of radius 1 and let γ_2 be the circle of radius 2 (traversed counterclockwise and centered at the origin). Show that

$$\int_{\gamma_1} \frac{dz}{z^3(z^2 + 10)} = \int_{\gamma_2} \frac{dz}{z^3(z^2 + 10)}.$$

8. Evaluate $\int_\gamma \sqrt{z}\, dz$ where γ is the upper half of the unit circle: first, directly; then, using Theorem 4.

9. Evaluate $\int_\gamma \sqrt{z^2 - 1}\, dz$ where γ is a circle of radius 1/2 centered at 0.

10. Evaluate

$$\int_\gamma \frac{2z^2 - 15z + 30}{z^3 - 10z^2 + 32z - 32} dz$$

where γ is the circle $|z| = 3$.

Hint. Use partial fractions; one root of the denominator is $z = 2$.

2.3   Cauchy's Theorem: Precise Version

Now that the student is somewhat familiar with Cauchy's Theorem, it can be explained in more precise terms, but in a way that is still based on the intuitive version developed in Section 2.2. In the course of proving the results, Green's Theorem will not be used, and instead a completely different approach will be utilized. The student will notice that in this section, references are made not to "simple closed curves" (except at the end of the section when the link between Cauchy's Theorem and the Jordan Curve Theorem is made) but rather merely to "closed curves." This is one technical advantage of the approach used here.

Local version of Cauchy's Theorem

First we shall prove a version of Cauchy's Theorem "locally" — that is, in a small rectangular neighborhood of each point. The method is the elegant and classical bisection procedure introduced by Edouard Goursat in 1883.

Again, in this section, "curve" means "piecewise C^1 curve." Theorem 11 considers curves that are perimeters of rectangles. Although called a theorem, it is actually the main technical lemma needed to prove Theorem 12 and the subsequent general formulations of Cauchy's Theorem that culminate in Theorem 17.

Theorem 11.   *Let $A \subset \mathbb{C}$ be open and let $f: A \to \mathbb{C}$ be analytic. Let $R = [a,b] \times [c,d]$ be a rectangle contained in A and let γ be its perimeter. Then*

$$\int_{\gamma} f = 0.$$

Note.   In the following method, we do not need to assume that f' is continuous.

Proof.   For any rectangle P contained in A, let γ_P be its boundary and let

$$I(P) = \int_{\gamma_P} f.$$

We divide the given rectangle R into four congruent rectangles, as indicated in Figure 2.16, and call them R_1, R_2, R_3, and R_4.

We know, from Section 2.1, that

$$I(R) = \sum_{i=1}^{4} I(R_i).$$

By the triangle inequality,

$$|I(R)| \le \sum_{i=1}^{4} |I(R_i)|.$$

Figure 2.16   Bisection procedure.

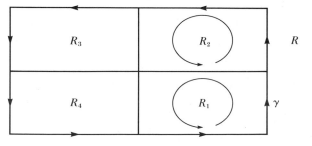

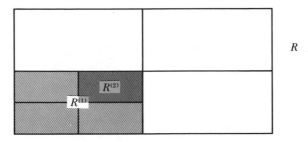

Figure 2.17 Second bisection.

Thus there is at least one R_i such that

$$|I(R_i)| \geq \frac{1}{4}|I(R)|.$$

We call this rectangle $R^{(1)}$.

This is the first step in the bisection procedure. In the second step, we apply the same procedure to $R^{(1)}$ and obtain another rectangle, $R^{(2)}$ (see Figure 2.17). Thus

$$|I(R^{(2)})| \geq \frac{1}{4}|I(R^{(1)})| \geq \frac{1}{4^2}|I(R)|.$$

Inductively we find a sequence of rectangles $R^{(1)}, R^{(2)}, \cdots$, such that:

a.
$$R \supset R^{(1)} \supset R^{(2)} \supset \cdots ;$$

b.
$$|I(R^{(k)})| \geq \frac{1}{4^k}|I(R)|;$$

c. $R^{(k)}$ has diameter $D/2^k$, where D is the diameter of R so that if $z_1, z_2 \in R^{(k)}$, $|z_1 - z_2| \leq D/2^k$; and the perimeter of $R^{(k)}$ is $p/2^k$, where p is the perimeter of R.

We claim that $\bigcap_{k=1}^{\infty} R^{(k)}$ consists of a single point, say z_0. This claim is actually a theorem from advanced calculus, so the proof may be familiar. Since the diameter of $R^{(k)} \to 0$, the intersection cannot contain more than one point. Also, if $z_k \in R^{(k)}$, $|z_k - z_l| \leq D/2^k$ if $l \geq k$, so z_k is a Cauchy sequence and thus converges to some point, say, z_0. This z_0 lies in all the $R^{(k)}$ since they are all closed. This proves our claim.

Since f is analytic at z_0, for any $\epsilon > 0$ there is a $\delta > 0$ such that

$$\left| \frac{f(z) - f(z_0)}{z - z_0} - f'(z_0) \right| < \epsilon;$$

that is, $|f(z) - f(z_0) - f'(z_0)(z - z_0)| < \epsilon|z - z_0|$ if $|z - z_0| < \delta$.

By Theorem 4,

$$\int_{\gamma_{R^{(k)}}} dz = 0 \qquad \text{and} \qquad \int_{\gamma_{R^{(k)}}} z\,dz = 0.$$

Thus

$$I(R^{(k)}) = \int_{\gamma_{R^{(k)}}} \{f(z) - f(z_0) - (z - z_0)f'(z_0)\}\,dz.$$

Therefore, if k is large enough so that $\delta > D/2^k$, then

$$|I(R^{(k)})| \leq \epsilon \cdot \frac{D}{2^k} \cdot \frac{p}{2^k}$$

by Theorem 3.

Thus we have, for k large enough,

$$\frac{1}{4^k}|I(R)| \leq |I(R^{(k)})| \leq \frac{\epsilon D p}{4^k}.$$

Thus $|I(R)| \leq \epsilon D p$. Since ϵ was arbitrarily small, we conclude that $|I(R)| = 0$, and hence that $I(R) = 0.$ ∎

For technical reasons that will be apparent in Section 2.4, it will be necessary also to have available the following variant of Theorem 11:

Theorem 11′.   *Suppose that, in Theorem 11, we assume only that f is analytic on $A\backslash\{z_1\}$ for some fixed $z_1 \in R$, z_1 not lying on the perimeter of R, and that $\lim_{z \to z_1} (z - z_1)f(z) = 0$. Then, as in Theorem 11,*

$$\int_\gamma f = 0.$$

Note.
   i. The conditions of Theorem 11′ hold if f is bounded in a deleted neighborhood of z_1, if f is continuous on A, or if $\lim_{z \to z_1} f(z)$ exists.
   ii. A similar result holds if z_1 is replaced by n points in R.

Proof.   Given $\epsilon > 0$, let us choose a square R_ϵ having side δ and centered around z_1 such that if $z \in R_\epsilon$, then $|f(z)| < \epsilon/|z - z_1|$. We then divide R, by lines extending the sides of R_ϵ, into nine sub-rectangles as shown in Figure 2.18. Let γ_0 be the perimeter of R_ϵ. Then $\int_\gamma f = \int_{\gamma_0} f$ because, by Theorem 11, the integrals around the other subrectangles in Figure 2.18 are zero. But on γ_0,

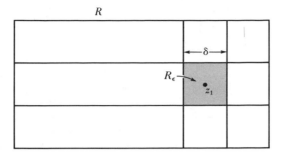

Figure 2.18 Construction of R_ϵ for the proof of Theorem 11'.

$$|f(z)| < \frac{\epsilon}{|z - z_1|} \leq \frac{\epsilon}{\delta/2},$$

since $|z - z_1| \geq \delta/2$ for z on γ_0. Thus

$$\left| \int_{\gamma_0} f \right| \leq \frac{\epsilon}{\delta/2} l(\gamma_0) = \frac{\epsilon}{\delta/2} \cdot 4\delta = 8\epsilon.$$

Therefore, $|\int_\gamma f| < 8\epsilon$ for any $\epsilon > 0$, and hence $\int_\gamma f = 0$. ∎

We shall use Theorem 11 to prove the more general result of Theorem 12, which is an important special case of Cauchy's Theorem.

Theorem 12. *Let $A \subset \mathbb{C}$ be open and let $f \colon A \longrightarrow \mathbb{C}$ be analytic. Let $R = {]}a,b{[} \times {]}c,d{[}$ be an open rectangle contained in A. Then there is a function F analytic on R such that $F' = f$ holds on R. Also, for any closed curve γ in R,*

$$\int_\gamma f = 0.$$

The same result is true if R is replaced by an r-neighborhood $D(z_0,r)$ contained in A.

Proof. Fix $z_0 \in R$ and let γ_z be the curve joining z_0 to a general point $z \in R$, as shown in Figure 2.19. Explicitly,

$$\gamma_z \colon [0,2] \to \mathbb{C}$$

$$\gamma_z(t) = \begin{cases} x_0 + i(ty + (1-t)y_0), & t \in [0,1]; \\ (t-1)x + (2-t)x_0 + iy, & t \in [1,2]. \end{cases}$$

As Figure 2.19 indicates, we define $\tilde{\gamma}_z$ similarly.

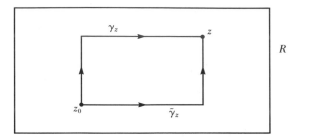

Figure 2.19   Paths for the construction of the
antiderivative F.

Let $F(z) = \int_{\gamma_z} f$. By Theorem 11 we have $F(z) = \int_{\tilde{\gamma}_z} f$, since $\int_{\gamma_z - \tilde{\gamma}_z} f = 0$.

We must show that $F'(z) = f(z)$. To do this, we first note, using $F(z) = \int_{\gamma_z} f$, that $F(x + h,y) - F(x,y) = \int_x^{x+h} f(t,y) dt$ for h sufficiently small (Why?). Therefore,

$$\frac{F(x + h,y) - F(x,y)}{h} = \frac{\int_x^{x+h} f(t,y) dt}{h}.$$

We claim that

$$\lim_{h \to 0} \frac{\int_x^{x+h} f(t,y)\,dt}{h}$$

exists and equals $f(x,y)$. The proof of this claim is the same as that of the Fundamental Theorem of Calculus. Namely, suppose that we are given $\epsilon > 0$. By the continuity of f at (x,y), we choose $\delta > 0$ such that $|t - x| < \delta$ implies that $|f(t,y) - f(x,y)| < \epsilon$. Hence for $|h| < \delta$, $h > 0$, we have

$$\left| \frac{\int_x^{x+h} f(t,y) dt}{h} - f(x,y) \right|$$

$$= \left| \frac{\int_x^{x+h} (f(t,y) - f(x,y)) dt}{h} \right| \le \frac{\int_x^{x+h} |f(t,y) - f(x,y)| dt}{|h|} \le \frac{\epsilon h}{h} = \epsilon.$$

The case where $h < 0$ is similarly dealt with, so our result is

$$\lim_{h \to 0} \frac{F(x + h,y) - F(x,y)}{h} = f(x,y).$$

Thus if we set $F(x,y) = U(x,y) + iV(x,y)$ and $f(x,y) = u(x,y) + iv(x,y)$, we have shown that $\partial U/\partial x$ and $\partial V/\partial x$ exist and that $\partial U/\partial x = u$, $\partial V/\partial x = v$. Similarly, using $F(z) = \int_{\tilde{\gamma}_z} f$, we get

$$F(x,y + h) - F(x,y) = \int_y^{y+h} f(x,t) i\, dt,$$

so

$$\frac{\partial U}{\partial y} = -v \qquad \text{and} \qquad \frac{\partial V}{\partial y} = u.$$

Therefore, since u and v are continuous, the partials of U and V are continuous. They clearly satisfy the Cauchy-Riemann equations and so, by Theorem 19 of Chapter 1, F is analytic. Also,

$$F' = \frac{\partial U}{\partial x} + i\frac{\partial V}{\partial x} = u + iv = f.$$

That $\int_\gamma f = 0$ follows at once from Theorem 4. The case in which R is replaced by a disk proceeds in exactly the same way. ∎

Using Theorem 11′ instead of Theorem 11, we get, using the same proof as for Theorem 12

Theorem 12′. *The same conclusion as that indicated by Theorem 12 holds if we assume only that f is continuous on A and analytic on $A\backslash\{z_1\}$ for some $z_1 \in R$.*

We note that continuity of f at z_1 is assumed. This assumption is stronger than the condition that $\lim_{z \to z_1} (z - z_1)f(z) = 0$ in Theorem 11. Such an assumption is needed so that the partials of F will be continuous and so that $\int_\gamma f$ is defined even if γ passes through z_1.

Homotopy and simply connected regions

To be able to extend Cauchy's Theorem to regions that are more general than rectangles, we must first clarify the concept of deforming curves, or homotopy, which was discussed informally in Section 2.2. To do so we suppose that both curves are parametrized by the same interval $[a,b]$ (which can always be done by reparametrizing if necessary) and formulate the following definition.

Definition 3. *Let $A \subset \mathbb{C}$ be a region. Two closed curves, $\gamma_1: [a,b] \to A$ and $\gamma_2: [a,b] \to A$, are called homotopic (as closed curves) in A if there exists a continuous mapping*

$$H: [a,b] \times [0,1] \to A$$

such that for each $s \in [0,1]$, $t \mapsto H(t,s)$ is a closed curve and for $s = 0$, this curve equals γ_1; for $s = 1$, it equals γ_2.

The idea of this definition is simple. As s ranges from 0 to 1, we have a family of curves that continuously change, or deform, from γ_1 to γ_2 (see Figure 2.20). Note that the curves can be self-intersecting; in fact, the constant curve $\gamma(t) = z_0$ is a closed curve.

Theorem 13 deals with a simple case in which curves may be shown to be homotopic.

Theorem 13.   *Let A be a convex region; that is, for $z_1, z_2 \in A$, the straight line joining z_1 and z_2 lies in A. Then any two closed curves $\gamma_1 : [a,b] \to A$ and $\gamma_2 : [a,b] \to A$ are homotopic as closed curves in A.*

Proof.   Define $H(t,s) = s\gamma_2(t) + (1-s)\gamma_1(t)$ for $t \in [a,b]$ and $0 \le s \le 1$. Since γ_1 and γ_2 are continuous, so is H. Also for fixed s, the map $t \mapsto H(t,s)$ is a closed curve; that is, $H(a,s) = H(b,s)$, since $s\gamma_2(a) + (1-s)\gamma_1(a) = s\gamma_2(b) + (1-s)\gamma_1(b)$ because γ_1 and γ_2 are closed curves. At $s = 0$ we get γ_1 and at $s = 1$ we get γ_2. ∎

For more complicated regions we generally rely on our geometric intuition to determine when two curves are homotopic. In other words, we try to decide whether we can continuously deform one curve to the other without leaving A. The reason is that we rarely use homotopies H explicitly in practice; they are merely theoretical tools, and we would find them complicated to write in some situations. However, we must be prepared to substantiate our geometric intuition with an explicit H in any particular situation.

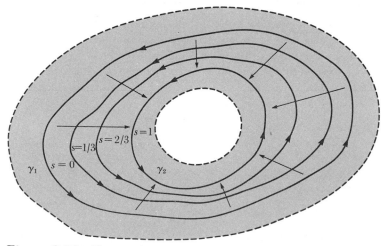

Figure 2.20   Homotopic curves.

Following is the precise definition of "simply connected" that was introduced informally in Section 2.2.

Definition 4. *A region A is called simply connected if every closed curve γ is homotopic in A to a point in A, that is, to some constant curve.*

As in Section 2.2, the informal meaning of this definition is that γ can be shrunk down to a point without leaving A. From Theorem 13 we have

Corollary 1. *A convex region A is simply connected.*

Existence of antiderivatives

Theorem 14 indicates that we can always perform integrals by finding an antiderivative or, as it is sometimes called, a primitive. Theorems 14 and 15 are the necessary technical tools that we will need to prove the general form of Cauchy's Theorem. These theorems actually reformulate the result of Theorem 12 so that it is more useful.

Theorem 14. *Let $f: A \subset \mathbb{C} \to \mathbb{C}$ be analytic on the region A and let $\gamma: [a,b] \to A$ be a curve in A. Then there is a continuous function $\lambda: [a,b] \to \mathbb{C}$ that is unique up to addition of constants and is such that the following condition holds: For any $t_0 \in [a,b]$ there is an analytic function F_{t_0} defined in a neighborhood of $\gamma(t_0)$ such that $F'_{t_0} = f$ and $F_{t_0}(\gamma(t)) = \lambda(t)$ for t sufficiently close to t_0.*
Also, if γ is piecewise differentiable, then

$$\int_\gamma f = \lambda(b) - \lambda(a).$$

Remark. In Section 2.1 it was explained that "curve" means piecewise C^1 unless otherwise stated. If the conclusion of Theorem 14 is to be valid, we actually need only assume that γ is continuous. The last statement in the theorem, then, can be used to define $\int_\gamma f$ if γ is merely continuous.
Because the proofs of this theorem and the next are somewhat technical and are not particularly instructive, they appear at the end of this section (see p. 111).
Theorem 15 is an important generalization of Theorem 14.

Theorem 15. *Let $H: [a,b] \times [0,1] \to A$ be a continuous map and let $f: A \to \mathbb{C}$ be analytic on the region A. Then there is a continuous* *function* $\lambda(t,s)$ *from* $[a,b] \times [0,1]$ *to* $\mathbb{C}$ *such that, for any* $(t_0,s_0) \in$ $[a,b] \times [0,1]$, *there is an analytic function* F *defined on a neighborhood of* $H(t_0,s_0)$ *such that* $F' = f$ *and* $F(H(t,s)) = \lambda(t,s)$ *near* (t_0,s_0).

Remark.   Note that for s fixed, $\lambda(t,s)$ satisfies the conditions of Theorem 14, so that if the curve $t \mapsto H(t,s)$ is denoted γ_s (and if γ_s is piecewise C^1), we have

$$\int_{\gamma_s} f = \lambda(b,s) - \lambda(a,s).$$

Also note that H need only be assumed to be continuous.

Deformation Theorem

Cauchy's Theorem is a special case of the Deformation Theorem. We want to be able to use the latter to transform Theorem 12 into a statement that is valid for more general regions.

Theorem 16 (Deformation Theorem).   *Let* $f: A \to \mathbb{C}$ *be analytic on the region* A *and let* γ_1 *and* γ_2 *be two closed curves that are homotopic in* A. *Then*

$$\int_{\gamma_1} f = \int_{\gamma_2} f.$$

Proof.   Let $H: [a,b] \times [0,1] \to A$ be a homotopy such that $\gamma_1(t)$ $= H(t,0)$ and $\gamma_2(t) = H(t,1)$. Construct λ as in Theorem 15 so that

$$\int_{\gamma_1} f = \lambda(b,0) - \lambda(a,0)$$

and

$$\int_{\gamma_2} f = \lambda(b,1) - \lambda(a,1)$$

as explained in the remark following Theorem 15.

Both $s \mapsto \lambda(a,s)$ and $s \mapsto \lambda(b,s)$ are curves satisfying the condition of Theorem 14 with γ taken to be the curve $s \mapsto H(a,s) = H(b,s)$. Therefore, by the uniqueness of $\lambda(t)$ up to addition of constants indicated in that theorem, we may conclude that $\lambda(a,s) - \lambda(b,s)$ is a constant, independent of s. In particular, it follows that $\lambda(b,0) - \lambda(a,0) = \lambda(b,1) - \lambda(a,1)$, and hence $\int_{\gamma_1} f = \int_{\gamma_2} f$, which is the desired result.∎

Cauchy's Theorem

Now we are ready for the statement of a general version of Cauchy's Theorem.

Theorem 17. *Let f be analytic on a region A. Let γ be a closed curve that is homotopic to a point in A. Then $\int_\gamma f = 0$.*
  In particular, if A is simply connected and γ is a closed curve in A, then

$$\int_\gamma f = 0.$$

Proof.   If $\gamma_2(t) = z_0$ is a constant curve, then $\gamma_2'(t) = 0$, so $\int_{\gamma_2} f = 0$. Setting $\gamma_1 = \gamma$, the result is thus an immediate consequence of Theorem 16. ■

Antiderivatives

The following theorem was studied informally as Theorem 9. Let us recall it here in the more precise setting of the present section.

Theorem 18. *Let f be analytic and defined on a simply connected region A. Then there exists an analytic function F on A unique up to an additive constant such that $F' = f$.*

The proof of Theorem 9 was based on Theorem 8, which was not precisely proven, but this deficiency has been remedied by the inclusion of Theorem 17. Also, the proof of the following corollary carries over unchanged from Section 2.2.

Corollary 2. *Let A be a simply connected region and assume that $0 \notin A$. Then there is an analytic function defined on A that we denote as* log *z; this function is uniquely determined up to the addition of multiples of $2\pi i$ and is such that $e^{\log z} = z$.*

Remark.   Theorems 14–18 are deduced from the property of f described in Theorem 12. (f has a local antiderivative.) Thus from Theorem 12′ we see that *the conclusions of Theorems 14–18 remain valid if we merely assume that f is continuous on A and analytic on $A \backslash \{z_1\}$.* In Section 2.4 it will be shown that this assumption actually implies that f is analytic on A, so such a weakening of the hypothesis of the theorems is only apparent; but it is necessary for the logical development of the theory.

Relationship of Cauchy's Theorem
to the Jordan Curve Theorem

An understanding of the Jordan Curve Theorem is not absolutely essential to an understanding of Cauchy's Theorem or of the material in subsequent chapters. However, the Jordan Curve Theorem is closely related to the hypotheses in Cauchy's Theorem, so it will be briefly considered here. In practical or concrete examples the result of the Jordan Curve Theorem is geometrically obvious and can usually be proven directly. The general case of the theorem is quite difficult and will not be proven here.

Theorem 19 (Jordan Curve Theorem).   *Let γ: $[a,b] \rightarrow \mathbb{C}$ be a simple, closed, continuous curve in $\mathbb{C}$. Then $\mathbb{C}\backslash\gamma([a,b])$ can be written uniquely as the disjoint union of two regions I and O such that I is bounded (that is, lies in some large disk). The region I is called the inside of γ. Region I is simply connected and γ is contractible to any point in $I \cup \gamma([a,b])$. The boundary of each of the two regions is $\gamma([a,b])$.*

The proof of this theorem uses more advanced mathematics and is beyond the scope of this book; see, for example, G. T. Whyburn, *Topological Analysis* (Princeton, N.J.: Princeton University Press, 1964).

Thus the Jordan Curve Theorem, combined with Cauchy's Theorem as stated in Theorem 17, yields the following: *If f is analytic on a region A and γ is a simple closed curve in A and the inside of γ lies in A, then $\int_\gamma f = 0$.* This is the classical way of stating Cauchy's Theorem. Although convenient in practice, it is theoretically awkward for two reasons: (1) It depends on the Jordan Curve Theorem for defining the concept of "inside"; and (2) γ is restricted to being a simple curve. The versions of the Cauchy Theorem stated in Theorems 16 and 17 do not depend on the Jordan Curve Theorem, are more general and more rigorous, and are just as easy to apply. On the other hand, the Jordan Curve Theorem reassures us that regions that we intuitively expect to be simply connected indeed are. (There is another way to describe the inside of a simple closed curve using the index, or winding number, of a curve; this method will be discussed in the next section.)

The general philosophy of this text is that we should use our geometric intuition to justify that a given region is simply connected or that two curves are homotopic, but with the realization that such knowledge is based on intuition and that to attempt to make it precise could be tedious. On the other hand, a precise argument should be used whenever possible and practical (see, for instance, the argument that a convex region is simply connected, p. 106).

Technical proofs of Theorems 14 and 15

In the proofs of these two theorems we shall need to use some ideas about compact sets that we learned in advanced calculus. In particular, we need the *Heine Borel Theorem*, which states that if $\{U_i\}$ is any collection of open sets in $\mathbb{R}$ that cover an interval $[a,b]$, then a *finite* number of these open sets cover the interval as well; see, for example, J. Marsden, *Elementary Classical Analysis* (San Francisco: W. H. Freeman and Co., 1974), Chapter 3, for the proof of this theorem.

Theorem 14.   *Let $f: A \subset \mathbb{C} \to \mathbb{C}$ be analytic on the region A and let $\gamma: [a,b] \to A$ be a curve in A. Then there is a continuous function $\lambda: [a,b] \to \mathbb{C}$ that is unique up to the addition of constants such that the following condition holds: For any $t_0 \in [a,b]$ there is an analytic function F_{t_0} defined in a neighborhood of $\gamma(t_0)$ such that $F'_{t_0} = f$ and $F_{t_0}(\gamma(t)) = \lambda(t)$ for t close to (that is, in a neighborhood of) t_0.*

  Also, if γ is piecewise differentiable, then

$$\int_\gamma f = \lambda(b) - \lambda(a).$$

Proof.   By Theorem 12 there exists, around each point $\gamma(t)$, an open disk U_t and an analytic function $F_t: U_t \to \mathbb{C}$ such that $F'_t = f$ on U_t. The open sets $\{\gamma^{-1}(U_t)\}$ form an open cover of $[a,b]$. (They are open because γ is continuous; see exercise 3, p. 60.) By compactness of $[a,b]$ we can pick a finite subcollection $\gamma^{-1}(U_{t_1}), \cdots, \gamma^{-1}(U_{t_n})$ that cover $[a,b]$. Next we choose points $a = s_0 < s_1 < s_2 \cdots < s_m = b$ such that each $\gamma([s_i,s_{i+1}])$ lies in one of the open disks U_{t_k}. We can do this because of the continuity of γ. Let us call the disk in which $\gamma([s_i,s_{i+1}])$ lies U_i (see Figure 2.21).

  We now proceed to define λ on $[a,b]$. On $[s_0,s_1] = [a,s_1]$ let us define $\lambda(t) = F_0 \circ \gamma(t)$. We observe that $U_0 \cap U_1$ is a nonempty, open, connected set (nonempty because $\gamma(s_1) \in U_0 \cap U_1$ and connected

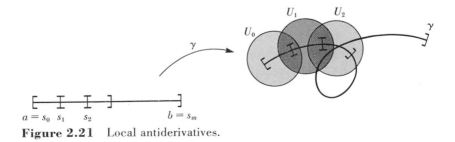

$a = s_0$ s_1 s_2 $b = s_m$

Figure 2.21   Local antiderivatives.

because U_0 and U_1 are disks). Since both F_0 and F_1 are antiderivatives of f, we conclude that $(F_0 - F_1)' = 0$ on $U_0 \cap U_1$; and since $U_0 \cap U_1$ is connected, $F_0 - F_1$ is a constant. If we adjust F_1 by adding this constant to it, we still have an antiderivative of f, and this adjusted F_1 and the function F_0 will then agree on $U_0 \cap U_1$. Let us define λ on $[s_1, s_2]$ by $\lambda(t) = F_1 \circ \gamma(t)$. Let us also adjust F_2 to agree with F_1 on $U_1 \cap U_2$ and define $\lambda(t) = F_2 \circ \gamma(t)$ on $[s_2, s_3]$. Proceeding in this way, we have λ defined on $[a,b]$. It should be clear that λ is continuous and satisfies the desired condition.

The next step is to show that λ, obeying this condition, is unique up to the addition of constants. Let λ_1 and λ_2 be two such functions. Then, by assumption, for any t_0 there is a neighborhood U_{t_0} of $\gamma(t_0)$ and antiderivatives F_1, F_2 of f defined on U_{t_0} such that $\lambda_1(t) = F_1 \circ \gamma(t)$ and $\lambda_2(t) = F_2 \circ \gamma(t)$ for t near t_0. But F_1 and F_2, both being antiderivatives of f, differ by a constant on U_{t_0}, and hence $\lambda_1 - \lambda_2$ is locally constant. But $\lambda_1 - \lambda_2$ is continuous on the connected set $[a,b]$, so $\lambda_1 - \lambda_2$ must be constant on $[a,b]$ (Why?). Thus we have uniqueness up to the addition of constants.

If γ is piecewise differentiable, we can, by performing further subdivision if necessary, assume that γ' exists on each open subinterval $]s_i, s_{i+1}[$. We compute as follows:

$$\int f = \sum_{i=0}^{m-1} \int_{s_i}^{s_{i+1}} f(\gamma(t)) \cdot \gamma'(t)\, dt$$

$$= \sum_{i=0}^{m-1} \int_{s_i}^{s_{i+1}} F_i'(\gamma(t)) \cdot \gamma'(t)\, dt$$

$$= \sum_{i=0}^{m-1} \int_{s_i}^{s_{i+1}} [F_i(\gamma(t))]'\, dt$$

$$= \sum_{i=0}^{m-1} [\lambda(s_{i+1}) - \lambda(s_i)]$$

$$= \lambda(b) - \lambda(a). \blacksquare$$

Theorem 15. *Let $H: [a,b] \times [0,1] \to A$ be a continuous map and let $f: A \to \mathbb{C}$ be analytic on the region A. Then there is a continuous function $\lambda(t,s)$ from $[a,b] \times [0,1]$ to $\mathbb{C}$ such that, for any $(t_0,s_0) \in [a,b] \times [0,1]$, there is an analytic F defined in a neighborhood of $H(t_0,s_0)$ such that $F' = f$ and $F(H(t,s)) = \lambda(t,s)$ for (t,s) near (t_0,s_0).*

Proof.   By using Theorem 12 and the compactness of $[a,b] \times [0,1]$, we can choose points $a = t_0 < t_1 < \cdots < t_n = b$ and points $0 = s_0 < s_1$

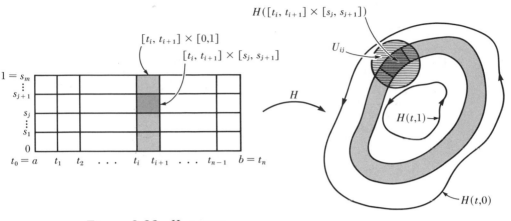

Figure 2.22 Homotopy.

$< s_2 \cdots < s_m = 1$ such that each $H([t_i, t_{i+1}] \times [s_j, s_{j+1}])$ lies in some disk U_{ij} on which f has an antiderivative F_{ij}.

We define λ in two steps. We shall piece together the $F_{ij} \circ H$ "vertically" and obtain a λ_i on $[t_i, t_{i+1}] \times [0,1]$ and then match up the λ_i's along the vertical lines $\{t_i\} \times [0,1]$ (see Figure 2.22).

The first step is to define λ_i on $[t_i, t_{i+1}] \times [s_0, s_1]$ by $\lambda_i = F_{i0} \circ H$. On $U_{i0} \cap U_{i1}$ (which is nonempty and connected), $(F_{i0} - F_{i1})' = 0$, so $F_{i0} = F_{i1} + c$ (where c is a constant) on $U_i \cap U_{i0}$. Then we adjust F_{i1} by this constant so that F_{i0} and F_{i1} become equal on $U_{i0} \cap U_{i1}$ and define $\lambda_i = F_{i1} \circ H$ on $[t_i, t_{i+1}] \times [s_1, s_2]$. We continue in this way until λ_i is defined on $[t_i, t_{i+1}] \times [0,1]$.

The second step is to define λ on $[a, t_1] \times [0,1] = [t_0, t_1] \times [0,1]$ by setting $\lambda = \lambda_0$. The curves $s \mapsto \lambda_0(t_1, s)$ and $s \mapsto \lambda_1(t_1, s)$ obey the condition of Theorem 14 for the curve $s \mapsto H(t_1, s)$ and function f. So by the uniqueness (up to the addition of constants) established in that theorem, λ_0 and λ_1 differ by a constant on $\{t_1\} \times [0,1]$. We adjust all the F_{ij}'s and λ_1 by this constant so that λ_1 agrees with λ_0 on $\{t_1\} \times [0,1]$. We define λ on $[t_1, t_2] \times [0,1]$ by $\lambda = \lambda_1$ and continue in this way to define λ on all of $[a, b] \times [0,1]$. That λ satisfies the requirements of the theorem is now easy to check (but the student should nevertheless do so). ∎

Worked Examples

1. Let A be the region bounded by the x-axis and the curve $\sigma(\theta) = Re^{i\theta}$, $0 \le \theta \le \pi$, where $R > 0$ is fixed. Let $f(z) = e^{z^2}/(2R - z)^2$. Show that for any closed curve γ in A, $\int_\gamma f = 0$.

Solution.   First observe that f fails to be analytic only when $z = 2R$ and hence f is analytic on A, since $2R$ lies outside of A (see Figure 2.23). We claim that A is simply connected. That any two points in A can be joined by a straight line lying in A (that is, that A is convex) is obvious geometrically and also is a simple matter to check (which the student should do). Hence A is simply connected by Corollary 1. By Cauchy's Theorem, $\int_\gamma f = 0$ for any closed curve in A.

2.   Let $A = \{z \in \mathbb{C} \mid 1 < |z| < 4\}$. First intuitively, then precisely, show that A is not simply connected. Also show precisely that the circles $|z| = 2$ and $|z| = 3$ are homotopic in A.

Solution.   Intuitively, the circle $|z| = 2$ cannot be contracted continuously to a point without passing over the hole in A; that is, the set $\{z \in \mathbb{C} \mid |z| \leq 1\}$ (see Figure 2.24).

Precisely, the function $1/z$ is analytic on A, and if A were simply connected, then we would have $\int_\gamma (1/z)dz = 0$ for any closed curve in A. But if we let $\gamma(t) = 2e^{it}$, $0 \leq t \leq 2\pi$, then we would obtain

$$\int_\gamma \frac{dz}{z} = \int_0^{2\pi} \frac{1}{2e^{it}} \cdot 2ie^{it}dt = 2\pi i.$$

Hence A cannot be simply connected.

Let $\gamma_1(t) = 2e^{it}$ and $\gamma_2(t) = 3e^{it}$, which represent the circles $|z| = 2$ and $|z| = 3$, respectively. Define $H(t,s) = 2e^{it} + se^{it}$; then H is a suitable homotopy between γ_1 and γ_2 in A. The effect of H is illustrated in Figure 2.24.

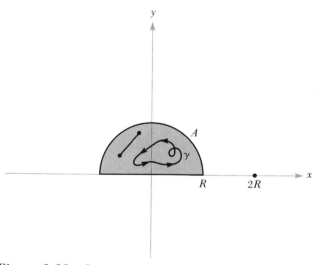

Figure 2.23   Convex region.

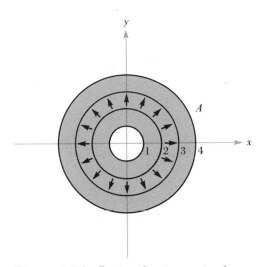

Figure 2.24 Region that is not simply connected.

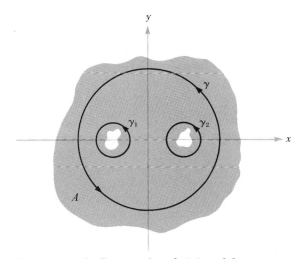

Figure 2.25 Region of analyticity of f.

3. Let γ denote the unit circle $|z| = 1$. Let γ_1 and γ_2 be two circles of radius 1/4 and centers $-1/2$ and $1/2$, respectively. Let A be a region containing γ, γ_1, and γ_2 and including the region between these curves (see Figure 2.25). For f analytic on A, show that

$$\int_\gamma f = \int_{\gamma_1} f + \int_{\gamma_2} f.$$

Solution.   We have

$$\gamma_1(t) = \frac{1}{2} + \frac{1}{4}e^{it}, \quad 0 \le t \le 2\pi,$$

and

$$\gamma_2(t) = -\frac{1}{2} + \frac{1}{4}e^{it}, \, 0 \le t \le 2\pi.$$

Let $\tilde{\gamma}$ be the curve depicted in Figure 2.26.

It is geometrically clear that γ is homotopic to $\tilde{\gamma}$ in A. The exact homotopy can be obtained easily by (1) reparametrizing $\tilde{\gamma}$ so that it has the same interval $[0,2\pi]$ as γ; (2) defining $H(t,s) = s\,\gamma(t) + (1-s)\,\tilde{\gamma}(t)$; and (3) checking that the straight line joining $\gamma(t)$ and $\tilde{\gamma}(t)$ remains in A, even though A is not convex. This method is illustrated in Figure 2.26.

By the Deformation Theorem, $\int_\gamma f = \int_{\tilde{\gamma}} f$. But $\int_{\tilde{\gamma}} f = \int_{\gamma_1} f + \int_{\gamma_2} f$, since $\tilde{\gamma} = \gamma_1 + \gamma_2 + \gamma_0 + (-\gamma_0)$ where γ_0 denotes the straight line joining $1/2$ to $-1/2$. Thus the assertion is proved. Compare this solution with that of example 3, Section 2.2, and note that here $\tilde{\gamma}$ does not have to be a simple closed curve, but merely a closed one.

Exercises

1.   Let γ_1 and γ_2 be two curves that are not necessarily closed but that have the same end points. Define the notion of homotopic curves with fixed end

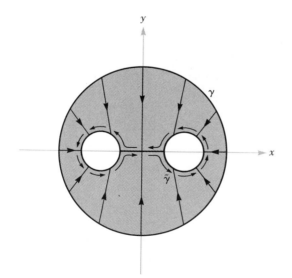

Figure 2.26   Construction of homotopy from γ to $\tilde{\gamma}$.

points in a region A. Prove that if f is analytic on A and γ_1 and γ_2 are homotopic with fixed end points in A, then

$$\int_{\gamma_1} f = \int_{\gamma_2} f.$$

Hint. Either proceed directly by modifying the Deformation Theorem or prove that $\gamma_1 + (-\gamma_2)$ is homotopic to a point in A.

2. Give a precise proof that $\mathbb{C}\backslash\{0\}$ is not simply connected.

3. A region A is called *star shaped* if there is a point $z_0 \in A$ such that for all $z \in A$, the line segment joining z_0 and z lies in A. Prove that A is simply connected.

4. Give the proof of Theorem 8.

5. Let f be continuous on a region A. Show that the conditions (a) $\int_\gamma f = 0$ for every closed curve in A, and (b) $\int_{\gamma_1} f = \int_{\gamma_2} f$ for two curves γ_1 and γ_2 with the same end points, are equivalent.

6. Evaluate the following integrals without performing an explicit computation:

   a. $\displaystyle\int_\gamma \frac{dz}{z}$; $\gamma(t) = \cos t + 2i\sin t$, $0 \leq t \leq 2\pi$.

   b. $\displaystyle\int_\gamma \frac{dz}{z^2}$; γ as in (a).

   c. $\displaystyle\int_\gamma \frac{e^z dz}{z}$; $\gamma(t) = 2 + e^{it}$, $0 \leq t \leq 2\pi$.

   d. $\displaystyle\int_\gamma \frac{dz}{z^2 - 1}$; γ is a circle of radius 1 centered at 1.

7. Evaluate $\int_\gamma dz/z$ where γ is the line segment joining 1 to i.

8. Consider $f(z) = e^{z^2}$. Find those values of θ_0 such that $f(re^{i\theta_0}) \to 0$ as $r \to \infty$.

9. Evaluate the following:

   a. $\displaystyle\int_{|z|=\frac{1}{2}} \frac{dz}{(1-z)^3}$.

   b. $\displaystyle\int_{|z+1|=\frac{1}{2}} \frac{dz}{(1-z)^3}$.

   c. $\displaystyle\int_{|z-1|=\frac{1}{2}} \frac{dz}{(1-z)^3}$.

10. a. Let γ be a curve homotopic to the unit circle in $\mathbb{C}\backslash\{0\}$. Evaluate $\int_\gamma dz/z$.

    b. Evaluate $\int_\gamma dz/z$ where γ is the curve $\gamma(t) = 3\cos t + i4\sin t$, $0 \leq t \leq 2\pi$.

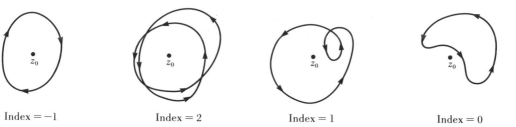

| Index $= -1$ | Index $= 2$ | Index $= 1$ | Index $= 0$ |

Figure 2.27 Index of a curve around a point.

2.4 Cauchy's Integral Formula

We are now in a position to draw important consequences from Cauchy's Theorem. In particular, we are able to show that analytic functions are infinitely differentiable and then to prove easily an otherwise difficult theorem stating that any polynomial $p(z) = a_0 + a_1 z + \cdots + a_n z^n$ has a complex root.

Index of a closed path

There is a very useful formula that expresses how many times a curve γ winds around a given point z_0 (see Figure 2.27). This number of times is called the *index* of γ with respect to z_0. The term "index" is formally defined in Definition 5.

The formula we shall use to compute the index is based on the computation that was done in example 3, Section 2.1: if γ is the unit circle $\gamma(t) = e^{it}$, $0 \le t \le 2\pi$, then

$$2\pi i = \int_\gamma \frac{dz}{z}.$$

If $\gamma(t) = e^{it}$, $0 \le t \le 2\pi n$, then γ would encircle 0 n times, and we find in the same way that

$$n = \frac{1}{2\pi i} \int_\gamma \frac{dz}{z}.$$

Now let us suppose that another closed curve $\tilde{\gamma}$ can be deformed to γ without passing through zero (that is, that $\tilde{\gamma}$ and γ are homotopic in the region $A = \mathbb{C} \setminus \{0\}$). Then again,

$$n = \frac{1}{2\pi i} \int_{\tilde{\gamma}} \frac{dz}{z} = \frac{1}{2\pi i} \int_\gamma \frac{dz}{z}$$

by the Deformation Theorem (see pp. 90 or 108). Since $\tilde{\gamma}$ and γ are homotopic in $\mathbb{C}\backslash\{0\}$, it is intuitively reasonable that they wind around 0 the same number of times.

Generally, for any point $z_0 \in \mathbb{C}$, the number of times a curve $\tilde{\gamma}$ winds around z_0 is seen to be

$$n = \frac{1}{2\pi i} \int_{\tilde{\gamma}} \frac{dz}{z - z_0},$$

by a similar argument.

Using the Jordan Curve Theorem, we can show that

$$\frac{1}{2\pi i} \int_{\gamma} \frac{dz}{z - z_0} = \begin{cases} \pm 1 & \text{if } z_0 \text{ is inside } \gamma \\ 0 & \text{if } z_0 \text{ is outside } \gamma \end{cases}$$

for a simple closed curve γ. These ideas lead to formulation of the following definition.

Definition 5.   *Let γ be a closed curve in $\mathbb{C}$; $z_0 \in \mathbb{C}$ is a point not on γ. Then the index of γ with respect to z_0 (also called the winding number of γ with respect to z_0) is defined by*

$$I(\gamma,z_0) = \frac{1}{2\pi i} \int_{\gamma} \frac{dz}{z - z_0}.$$

We say that γ "winds around z_0, $I(\gamma,z_0)$ times."

The discussion that preceded Definition 5 supports the following theorem, which is illustrated in Figure 2.28.

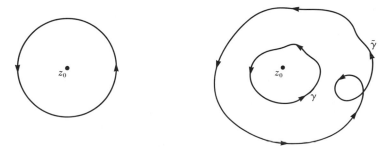

Index of circle $\tilde{\gamma}$ and γ have the same index with respect to z_0

Figure 2.28   Index (the number of times that z_0 is encircled).

Theorem 20.

i. The circle $\gamma(t) = z_0 + re^{it}$, $0 \le t \le 2\pi n$, $r > 0$, has index n with respect to z_0; the circle $-\gamma(t) = z_0 + re^{-it}$, $0 \le t \le 2\pi n$, has index $-n$.

ii. If z_0 does not lie on either $\tilde{\gamma}$ or γ and if $\tilde{\gamma}$ and γ are homotopic in $\mathbb{C}\backslash\{z_0\}$, then

$$I(\tilde{\gamma}, z_0) = I(\gamma, z_0).$$

Since homotopies are sometimes awkward to deal with directly, it is customary merely to give an intuitive geometric argument that $I(\gamma,z_0)$ has a certain value, but again the student should be prepared to give a direct proof (see example 1 at the end of this section).

Theorem 21 provides a check that $I(\gamma,z_0)$ is always an integer. (It should be if Definition 5 agrees with the ideas illustrated in Figures 2.27 and 2.28.)

Theorem 21.   Let γ be a closed curve: $[a,b] \rightarrow \mathbb{C}$; $z_0 \in \mathbb{C}$ is a point not on γ. Then $I(\gamma,z_0)$ is an integer.

Proof.   Let

$$g(t) = \int_a^t \frac{\gamma'(s)}{\gamma(s) - z_0} \, ds.$$

Then, at points where the integrand is continuous, we get

$$g'(t) = \frac{\gamma'(t)}{\gamma(t) - z_0}.$$

Thus

$$\frac{d}{dt} e^{-g(t)}(\gamma(t) - z_0) = 0$$

at points where $g'(t)$ exists and so $e^{-g(t)}(\gamma(t) - z_0)$ is piecewise constant on $[a,b]$. But $e^{-g(t)}(\gamma(t) - z_0)$ is continuous and therefore must be constant on $[a,b]$. This constant value is.

$$e^{-g(a)}(\gamma(a) - z_0),$$

so we get $e^{-g(b)}(\gamma(b) - z_0) = e^{-g(a)}(\gamma(a) - z_0)$. But $\gamma(b) = \gamma(a)$, so $e^{-g(b)} = e^{-g(a)}$. On the other hand, $g(a) = 0$; hence $e^{-g(b)} = 1$. Thus $g(b) = 2\pi n i$ for an integer n, and the theorem follows. ∎

The *inside* of a curve γ could be defined by $\{z | I(\gamma,z) \ne 0\}$ and this definition would agree with the "inside" defined by the Jordan Curve Theorem and also with the intuitive ideas illustrated in Figures 2.27 and

2.28. Thus the inside of a closed curve can be defined purely analytically without applying the Jordan Curve Theorem.

Cauchy's Integral Formula

Cauchy's Theorem will now be used to derive a very useful formula relating the value of an analytic function at z_0 to a certain integral.

Theorem 22 (Cauchy's Integral Formula).   *Let f be analytic on the region A and let γ be a closed curve in A, homotopic to a point. Let $z_0 \in A$ be a point not on γ. Then*

$$f(z_0) \cdot I(\gamma, z_0) = \frac{1}{2\pi i} \int_\gamma \frac{f(z)}{z - z_0}\, dz. \tag{1}$$

This formula* is usually applied when γ is a simple closed curve and z_0 is inside γ. Then $I(\gamma, z_0) = 1$, so that formula (1) becomes

$$f(z_0) = \frac{1}{2\pi i} \int_\gamma \frac{f(z)}{z - z_0}\, dz. \tag{1'}$$

Formula (1') is remarkable, for it says, more specifically, that the values of f on γ completely determine the values of f inside γ. In other words, the value of f is determined by its "boundary values."

Proof.   Define

$$g(z) = \begin{cases} \dfrac{f(z) - f(z_0)}{z - z_0} & \text{if } z \neq z_0; \\[2ex] f'(z_0) & \text{if } z = z_0. \end{cases}$$

Thus $g(z)$ is analytic if $z \neq z_0$ and $\displaystyle\lim_{z \to z_0} g(z)$ exists, since f is analytic. Therefore, by Cauchy's Theorem (see example 4 at the end of Section 2.2 or Theorem 17 and the remark following Corollary 2), $\int_\gamma g = 0$. But

$$\int_\gamma g(z)dz = \int_\gamma \frac{f(z)dz}{z - z_0} - \int_\gamma \frac{f(z_0)}{z - z_0}\, dz = 0$$

---

*Theorem 22 can be strengthened by requiring only that f be continuous on γ and analytic "inside" γ. This change makes little difference in solving examples. For the proof of the strengthened theorem the interested student can consult, for example, E. Hille, *Analytic Function Theory*, Vol. I (Boston: Ginn, 1959).

and

$$\int_\gamma \frac{f(z_0)dz}{z - z_0} = f(z_0) \int_\gamma \frac{1}{z - z_0}\, dz = 2\pi i f(z_0) I(\gamma, z_0),$$

so the theorem follows. ∎

Formula (1) is extremely useful for computations. For example, $\int_\gamma (e^z/z)dz = 2\pi i \cdot e^0 = 2\pi i$, where γ is the unit circle. Here $f(z) = e^z$ and $z_0 = 0$.

Note that in Theorem 22, f, not the integrand $f(z)/(z - z_0)$, is analytic on A; the integrand is analytic only on $A \backslash \{z_0\}$, so we cannot use Cauchy's Theorem to conclude that the integral is zero—in fact, the integral is generally nonzero.

Integrals of the Cauchy type

Cauchy's Integral Formula (1) is a very special formula that f satisfies, and we will use it in the proof of Theorem 24 to show that $f', f'', \cdots$, all exist. This application of formula (1) depends on the following theorem.

Theorem 23.   *Let γ be a curve in $\mathbb{C}$ and let g be a continuous function defined on $\gamma([a,b])$. Set*

$$G(z) = \frac{1}{2\pi i} \int_\gamma \frac{g(\zeta)}{\zeta - z}\, d\zeta.$$

Then G is analytic on $\mathbb{C} \backslash \gamma([a,b])$; in fact, G is infinitely differentiable with the kth derivative given by

$$G^{(k)}(z) = \frac{k!}{2\pi i} \int_\gamma \frac{g(\zeta)}{(\zeta - z)^{k+1}}\, d\zeta,\ k = 1, 2, 3, \cdots . \qquad (2)$$

Formula (2) for $G^{(k)}$ can be easily remembered if we differentiate with respect to z under the integral sign. The formal proof merely justifies this procedure and appears at the end of the section.

Existence of higher derivatives

From the preceding we are now able to deduce the following important result.

Theorem 24 (Cauchy's Integral Formula for Derivatives).
*Let f be analytic on a region A. Then all the derivatives of f exist on A.
Furthermore, for $z_0 \in A$ and any closed curve γ homotopic to a point in
A with z_0 not on γ, we have*

$$f^{(k)}(z_0) \cdot I(\gamma, z_0) = \frac{k!}{2\pi i} \int_\gamma \frac{f(\zeta)}{(\zeta - z_0)^{k+1}} \, d\zeta, \; k = 0, 1, 2, 3, \cdots, \quad (3)$$

where $f^{(k)}$ denotes the kth derivative of f.

Proof.   Around each $z_0 \in A$ we can, by the openness of A, find a
small circle $\tilde{\gamma}$ whose center is z_0 and whose interior lies in A. Then, by
formula (1), and since $I(\tilde{\gamma}, z_0) = 1$, for any z inside $\tilde{\gamma}$,

$$f(z) = \frac{1}{2\pi i} \int_{\tilde{\gamma}} \frac{f(\zeta)}{\zeta - z} \, d\zeta.$$

Therefore, by Theorem 23, f is infinitely differentiable at z_0. Also by
this theorem and by the formula

$$I(\gamma, z) = \frac{1}{2\pi i} \int_\gamma \frac{d\zeta}{\zeta - z},$$

we see that $I(\gamma, z)$ is an analytic function of z on $\mathbb{C} \setminus \gamma[a,b]$ and thus in
particular is continuous. But $I(\gamma, z)$ is an integer, so it is locally con-
stant (that is, constant in a whole neighborhood of z_0). By applying
formula (2) to formula (1), we get the desired result.∎

Cauchy's Inequalities and Liouville's Theorem

Theorem 25 (Cauchy's Inequalities).   *Let f be analytic on a
region A and let γ be a circle with radius R and center z_0 that lies in A.
Assume that the disk $\{z \mid |z - z_0| < R\}$ also lies in A. Suppose that
$|f(z)| \leq M$ for all z on γ. Then, for any $k = 0, 1, 2, \cdots$,*

$$|f^{(k)}(z_0)| \leq \frac{k!}{R^k} M. \quad (4)$$

Proof.   From formula (3) we obtain

$$f^{(k)}(z_0) = \frac{k!}{2\pi i} \int_\gamma \frac{f(\zeta)}{(\zeta - z_0)^{k+1}} \, d\zeta,$$

and hence

$$|f^{(k)}(z_0)| = \frac{k!}{2\pi} \left| \int_\gamma \frac{f(\zeta)}{(\zeta - z_0)^{k+1}} \, d\zeta \right|.$$

Now

$$\left| \frac{f(\zeta)}{(\zeta - z_0)^{k+1}} \right| \le \frac{M}{R^{k+1}}$$

since $|\zeta - z_0| = R$ for ζ on γ, and so

$$|f^{(k)}(z_0)| \le \frac{k!}{2\pi} \cdot \frac{M}{R^{k+1}} \cdot l(\gamma).$$

But $l(\gamma) = 2\pi R$, so we get our result. ∎

This result states that although the kth derivatives of f can go to infinity as $k \longrightarrow \infty$, they cannot grow too fast as $k \longrightarrow \infty$; specifically, they can grow no faster than a constant times $k!/R^k$. We can use Cauchy's inequalities to derive the following surprising result.

Theorem 26 (Liouville's Theorem). *If f is entire and there is a constant M such that $|f(z)| \le M$ for all $z \in \mathbb{C}$, then f is constant.*

This is again a quite different property than any that could possibly hold for functions of a real variable. Certainly, there are many non-constant bounded smooth functions of a real variable; for example, $f(x) = \sin x$.

Proof.   For any $z_0 \in \mathbb{C}$ we have, by formula (4), $|f'(z_0)| \le M/R$. Let $R \longrightarrow \infty$. Thus we conclude that $|f'(z_0)| = 0$ and therefore that $f'(z_0) = 0$, so f is constant. ∎

Fundamental Theorem of Algebra

Next we shall prove a result that appears to be elementary and that the student has, in the past, probably taken for granted. Algebraically, the theorem is quite difficult. However, there is a simple proof that uses the Theorem of Liouville.

Theorem 27.   *Let $a_0, a_1, \cdots, a_n$ be complex numbers and suppose that $n \ge 1$ and $a_n \ne 0$. Let $p(z) = a_0 + a_1 z + \cdots + a_n z^n$, an entire function. Then there exists a point $z_0 \in \mathbb{C}$ such that $p(z_0) = 0$.*

Note. By review exercise 16 at the end of Chapter 1, the polynomial p can have no more than n roots. It follows that p will have exactly n roots if they are counted according to their multiplicity.

Proof. Suppose that $p(z_0) \neq 0$ for all $z_0 \in \mathbb{C}$. Then $f(z) = 1/p(z)$ is entire. Now $p(z)$ and hence $f(z)$ is not constant (because $a_n \neq 0$), so it suffices, by Liouville's Theorem, to show that $f(z)$ is bounded.

To do so, we must first show that $p(z) \to \infty$ as $z \to \infty$, or, equivalently, that $f(z) \to 0$ as $z \to \infty$. In other words, we must prove that, given $M > 0$, there is a number $K > 0$ such that $|z| > K$ implies that $|p(z)| > M$. From $p(z) = a_0 + a_1 z + \cdots + a_n z^n$ we have $|p(z)| \geq |a_n|\, |z|^n - |a_0| - |a_1|\, |z| - \cdots - |a_{n-1}|\, |z|^{n-1}$. (We set $a_n z^n = p(z) - a_0 - a_1 z - \cdots - a_{n-1} z^{n-1}$ and apply the triangle inequality.) Let $a = |a_0| + |a_1| + \cdots + |a_{n-1}|$. If $|z| > 1$, then

$$|p(z)| \geq |z|^{n-1}\left(|a_n|\, |z| - \frac{|a_0|}{|z|^{n-1}} - \frac{|a_1|}{|z|^{n-2}} - \cdots - \frac{|a_{n-1}|}{1}\right)$$

$$\geq |z|^{n-1}\left(|a_n|\, |z| - a\right).$$

Let $K = \max\{1, (M + a)/a_n\}$; then, if $|z| > K$, we have $|p(z)| \geq M$.

Thus if $|z| > K$, we have $1/|p(z)| < 1/M$. But on the set of z, for which $|z| \leq K$, $1/p(z)$ is bounded in absolute value because it is continuous. If this bound for $1/p(z)$ is denoted by L, then on $\mathbb{C}$ we have $1/|p(z)| < \max(1/M, L)$, so $|f(z)|$ is bounded on $\mathbb{C}$.

Therefore, by the remark at the beginning of the proof, the theorem is proved. ∎

Another argument for showing that $f(z) \to 0$ as $z \to \infty$ that is a little simpler but accepts the validity of various limit theorems is as follows.

$$f(z) = \frac{1}{a_n z^n + a_{n-1} z^{n-1} + \cdots + a_0}$$

$$= \frac{\dfrac{1}{z^n}}{a_n + a_{n-1}\left(\dfrac{1}{z}\right) + a_{n-2}\left(\dfrac{1}{z^2}\right) + \cdots + a_0\left(\dfrac{1}{z^n}\right)}.$$

Letting $z \to \infty$, we get

$$\underset{z \to \infty}{\text{limit}}\, f(z) = \frac{0}{a_n + 0 + \cdots + 0} = 0$$

since $a_n \neq 0$.

Morera's Theorem

The following theorem is sometimes useful; it is a partial converse of Cauchy's Theorem.

Theorem 28 (Morera's Theorem). *Let f be continuous on a region A and suppose that $\int_\gamma f = 0$ for every closed curve in A. Then f is analytic on A, and $f = F'$ for some analytic function F.*

Proof.   The condition $\int_\gamma f = 0$ is equivalent to saying that the integral of f along a path joining z_1 to z_2 is independent of that path (see Theorem 8). Fix $z_0 \in A$ and let $F(z) = \int_{\gamma_z} f$ where γ_z is any path joining z_0 to z. By the proof of Theorem 9 (or Theorem 18) we know that $f = F'$ and that F is analytic. But by Theorem 24 we know that F'' exists, therefore f' exists. ∎

When applying Theorem 28 we often must choose a slightly different A than is given, since we expect $f = F'$ to hold only if A is simply connected. The following corollary is an example of such a specific application.

Corollary 1. *Let f be continuous on a region A and analytic on $A \backslash \{z_0\}$. Then f is analytic on A.*

Proof.   We restrict f to a disk $D(z_0, \epsilon)$ that is simply connected. By example 4 (Section 2.2) or Theorem 12′ we see that $\int_\gamma f = 0$ when γ is a closed curve in $D(z_0, \epsilon)$. Thus Theorem 28 can be applied to show that f is analytic on $D(z_0, \epsilon)$. Hence f is analytic on all of A. ∎

Technical proof of Theorem 23

Theorem 23. *Let γ be a curve in $\mathbb{C}$ and let g be a continuous function defined on $\gamma([a,b])$. Set*

$$G(z) = \frac{1}{2\pi i} \int_\gamma \frac{g(\zeta)}{\zeta - z}\, d\zeta.$$

Then G is analytic on $\mathbb{C} \backslash \gamma([a,b])$; in fact, G is infinitely differentiable with the kth derivative given by

$$G^{(k)}(z) = \frac{k!}{2\pi i} \int_\gamma \frac{g(\zeta)}{(\zeta - z)^{k+1}}\, d\zeta, \; k = 1, 2, 3, \cdots .$$

Proof.   We must justify differentiation under the integral sign. It is easier at this stage to proceed directly than to apply general theorems.

In this proof we shall need the following facts from advanced calculus — see, for example, J. Marsden, *Elementary Classical Analysis* (San Francisco: W. H. Freeman and Co., 1974), Chapters 3 and 4. First, $\gamma([a,b])$ is a compact set, being the continuous image of the compact set $[a,b]$. For z_0 not on γ, there is a $\delta > 0$ such that $|\zeta - z_0| \geq \delta$ for any ζ on γ; that is, z_0 is a strictly positive distance away from the curve. (In fact, $|\zeta - z_0|$ is a continuous function of ζ and thus attains its minimum value at some ζ on γ; this distance is nonzero because z_0 is not on γ.) If we let $\eta = \delta/2$ and U be the η-disk around z_0, then $z \in U$ and ζ on γ implies that $|z - \zeta| \geq \eta$ since $|z - \zeta| \geq |\zeta - z_0| - |z_0 - z| \geq 2\eta - \eta$ (see Figure 2.29).

We begin now with the case $k = 1$. We want to show that

$$\lim_{z \to z_0} \left\{ \frac{G(z) - G(z_0)}{z - z_0} - \frac{1}{2\pi i} \int_\gamma \frac{g(\zeta)}{(\zeta - z_0)^2} \, d\zeta \right\} = 0.$$

The expression in braces may be written

$$\frac{G(z) - G(z_0)}{z - z_0} - \frac{1}{2\pi i} \int_\gamma \frac{g(\zeta)}{(\zeta - z_0)^2} \, d\zeta = \frac{(z - z_0)}{2\pi i} \int_\gamma \frac{g(\zeta)}{(\zeta - z_0)^2(\zeta - z)} \, d\zeta,$$

which we get by using the identity

$$\frac{1}{z - z_0} \left\{ \frac{1}{\zeta - z} - \frac{1}{\zeta - z_0} \right\} - \frac{1}{(\zeta - z_0)^2} = \frac{z - z_0}{(\zeta - z_0)^2(\zeta - z)}.$$

Let the η-neighborhood U of z_0 be constructed as previously described and let M be the maximum of g on γ. Then $|(\zeta - z_0)^2(\zeta - z)| \geq \eta^2 \cdot \eta = \eta^3$, so $|g(\zeta)/\{(\zeta - z_0)^2(\zeta - z)\}| \leq M\eta^{-3}$, a fixed constant independent of ζ on γ, and $z, z_0 \in U$. Thus

$$\left| \frac{z - z_0}{2\pi i} \int_\gamma \frac{g(\zeta)}{(\zeta - z_0)^2(\zeta - z)} \, d\zeta \right| \leq |z - z_0| \frac{M\eta^{-3}}{2\pi} l(\gamma).$$

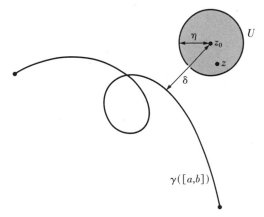

Figure 2.29   A point z_0 not on a curve γ is a positive distance from γ.

This expression approaches zero as $z \longrightarrow z_0$, so the limit is zero, as we wanted to show.

To prove the general case we proceed by induction on k. Suppose that we have shown that G can be differentiated $n - 1$ times on $\mathbb{C} \backslash \gamma([a,b])$ and that

$$G^{(n-1)}(z) = \frac{(n-1)!}{2\pi i} \int_\gamma \frac{g(\zeta)}{(\zeta - z)^n} \, d\zeta.$$

Let $z_0 \in \mathbb{C} \backslash \gamma([a,b])$. Then, using the identity

$$\frac{1}{(\zeta - z)^n} = \frac{1}{(\zeta - z)^{n-1}(\zeta - z_0)} + \frac{z - z_0}{(\zeta - z)^n(\zeta - z_0)},$$

we obtain

$$G^{(n-1)}(z) - G^{(n-1)}(z_0) = \frac{(n-1)!}{2\pi i} \left[\int_\gamma \frac{g(\zeta)}{(\zeta - z)^{n-1}(\zeta - z_0)} \, d\zeta - \int_\gamma \frac{g(\zeta)}{(\zeta - z_0)^n} d\zeta \right]$$

$$+ \frac{(n-1)!}{2\pi i} (z - z_0) \int_\gamma \frac{g(\zeta)}{(\zeta - z)^n(\zeta - z_0)} \, d\zeta. \qquad (5)$$

We can conclude from this equation that $G^{(n-1)}$ is continuous at z_0 for the following reason. By applying the induction hypothesis to $g(\zeta)/(\zeta - z_0)$, which is continuous on $\gamma([a,b])$, we see that

$$\int_\gamma \frac{g(\zeta)}{(\zeta - z)^{n-1}(\zeta - z_0)} d\zeta$$

is analytic as a function of z on the set $\mathbb{C} \backslash \gamma([a,b])$ and thus is continuous in z. Therefore,

$$\int_\gamma \frac{g(\zeta)}{(\zeta - z)^{n-1}(\zeta - z_0)} d\zeta \longrightarrow \int_\gamma \frac{g(\zeta)}{(\zeta - z_0)^n} d\zeta$$

as $z \longrightarrow z_0$. If the distance from z_0 to γ is 2η, if $|g(z)| < M$ on γ, and if $|z - z_0| < \eta$, we have

$$\left| \int_\gamma \frac{g(\zeta)}{(\zeta - z)^n(\zeta - z_0)} d\zeta \right| < \frac{M}{\eta^{n+1}} \cdot l(\gamma),$$

where $l(\gamma)$ is the length of γ. Hence

$$|z - z_0| \left| \int_\gamma \frac{g(\zeta)}{(\zeta - z)^n(\zeta - z_0)} d\zeta \right| \longrightarrow 0$$

as $z \longrightarrow z_0$, and therefore $G^{(n-1)}$ is continuous on $\mathbb{C} \backslash \gamma([a,b])$.

From equation (5) we obtain

$$\frac{G^{(n-1)}(z) - G^{(n-1)}(z_0)}{z - z_0} = \frac{(n-1)!}{2\pi i} \frac{1}{(z - z_0)}$$

$$\left[\int_\gamma \frac{g(\zeta)}{(\zeta - z)^{n-1}(\zeta - z_0)} d\zeta - \int_\gamma \frac{g(\zeta)}{(\zeta - z_0)^n} d\zeta \right]$$

$$+ \frac{(n-1)!}{2\pi i} \int_\gamma \frac{g(\zeta)}{(\zeta - z)^n(\zeta - z_0)} d\zeta. \tag{6}$$

By applying the induction hypothesis of equation (6) to $g(\zeta)/(\zeta - z_0)$, we see that the first term on the right side of equation (6) converges to

$$\frac{(n-1)(n-1)!}{2\pi i} \int_\gamma \frac{g(\zeta)}{(\zeta - z_0)^{n+1}} d\zeta$$

as $z \longrightarrow z_0$. In the paragraph following equation (5), it was shown that $G^{(n-1)}$ is continuous on $\mathbb{C} \backslash \gamma([a,b])$, and this fact applied to $g(\zeta)/(\zeta - z_0)$, instead of to $g(\zeta)$, implies that

$$\int_\gamma \frac{g(\zeta)}{(\zeta - z)^n(\zeta - z_0)} d\zeta \longrightarrow \int_\gamma \frac{g(\zeta)}{(\zeta - z_0)^{n+1}} d\zeta$$

as $z \longrightarrow z_0$. Thus we have shown that as $z \longrightarrow z_0$,

$$\frac{G^{(n-1)}(z) - G^{(n-1)}(z_0)}{z - z_0}$$

converges to

$$(n-1) \frac{(n-1)!}{2\pi i} \int_\gamma \frac{g(\zeta)}{(\zeta - z_0)^{n+1}} d\zeta + \frac{(n-1)!}{2\pi i} \int_\gamma \frac{g(\zeta)}{(\zeta - z_0)^{n+1}} d\zeta$$

$$= \frac{n!}{2\pi i} \int_\gamma \frac{g(\zeta)}{(\zeta - z_0)^{n+1}} d\zeta.$$

This concludes the induction and proves the theorem.■

Worked Examples

1. Consider the curve γ defined by $\gamma(t) = (\cos t, 3\sin t)$, $0 \le t \le 4\pi$. Show rigorously that $I(\gamma,0) = 2$.

   Solution.   Suppose that we can show that γ is homotopic in $\mathbb{C} \backslash \{0\}$ to a circle $\tilde{\gamma}$ that is centered around the origin and that is traversed twice in the counterclockwise direction (that is, $\tilde{\gamma}(t) = e^{it}$, $0 \le t \le 4\pi$). Then, by Theorem 20, $I(\gamma,0) = I(\tilde{\gamma},0) = 2$. A suitable homotopy is $H(t,s) = \cos t + i(3 - 2s)\sin t$, since H is continuous, $H(t,0) = \gamma(t)$ and $H(t,1) = \tilde{\gamma}(t)$, and H is never zero (see Figure 2.30).

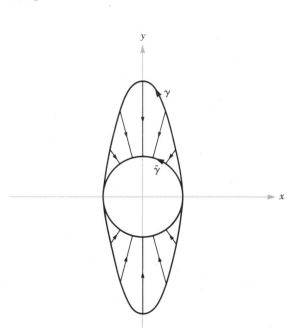

Figure 2.30 Homotopy of $\gamma(t) = (\cos t, 3\sin t)$
to $\tilde{\gamma}(t) = (\cos t, \sin t)$.

2. Evaluate

$$\int_\gamma \frac{\cos z}{z}\,dz \text{ and } \int_\gamma \frac{\sin z}{z^2}\,dz$$

where γ is the unit circle.

Solution.   The circle γ is contractible to a point in the region in which
$\cos z$ is analytic, since in fact $\cos z$ is entire. Therefore, we can apply
Cauchy's Integral Formula, observing that $I(\gamma,0) = 1$, to obtain

$$1 = \cos 0 = \frac{1}{2\pi i}\int_\gamma \frac{\cos z}{z}\,dz,$$

so that

$$\int_\gamma \frac{\cos z}{z}\,dz = 2\pi i.$$

By the Cauchy Integral Formula for Derivatives we have

$$\sin{}' (0) = \frac{1}{2\pi i}\int_\gamma \frac{\sin z}{z^2}\,dz$$

or

$$\int_\gamma \frac{\sin z}{z^2}\, dz = 2\pi i \cos 0 = 2\pi i.$$

3. Evaluate

$$\int_\gamma \frac{(e^z + z)}{z - 2}\, dz$$

where γ is (a) the unit circle; (b) a circle with radius 3 centered at 0.

Solution.
a. γ is contractible to a point in $\mathbb{C}\backslash\{2\}$ (Why?), and $(e^z + z)/(z - 2)$ is analytic on $\mathbb{C}\backslash\{2\}$. Thus by Cauchy's Theorem, $\int_\gamma (e^z+z)/(z-2)\, dz = 0$.
b. Here γ is not contractible to a point in $\mathbb{C}\backslash\{2\}$. In fact, γ winds around $2 + 0i$ exactly once, so $I(\gamma,2) = 1$. Precisely, γ is homotopic in $\mathbb{C}\backslash\{2\}$ to a circle γ centered at 2, and so $I(\gamma,2) = 1$ by Theorem 20. Thus by Cauchy's Integral Formula, which is applicable since γ is contractible to a point and $e^z + z$ is analytic on all of $\mathbb{C}$, we have

$$\int_\gamma \frac{e^z + z}{z - 2}\, dz = 2\pi i(e^2 + 2).$$

4. This example, which deals with analytic functions defined by integrals, generalizes Theorem 23. Let $f(z,w)$ be a continuous function of z,w for z in a region A and w on a curve γ. For each w on γ assume that f is analytic in z. Let

$$F(z) = \int_\gamma f(z,w)\, dw.$$

Then show that F is analytic and

$$F'(z) = \int_\gamma \frac{\partial f}{\partial z}(z,w)\, dw,$$

where $\partial f/\partial z$ denotes the derivative of f with respect to z with w held fixed.

Solution. Let $z_0 \in A$. Let γ_0 be a circle in A around z_0 whose interior also lies in A. Then for z inside γ_0,

$$f(z,w) = \frac{1}{2\pi i}\int_{\gamma_0} \frac{f(\zeta,w)}{\zeta - z}\, d\zeta$$

by Cauchy's Integral Formula. Thus

$$F(z) = \frac{1}{2\pi i}\int_\gamma \left\{\int_{\gamma_0} \frac{f(\zeta,w)}{\zeta - z}\, d\zeta\right\} dw.$$

Next we claim that we may invert the order of integration, thus obtaining

$$F(z) = \frac{1}{2\pi i} \int_{\gamma_0} \left\{ \int_\gamma \frac{f(\zeta,w)}{\zeta - z} \, dw \right\} d\zeta = \frac{1}{2\pi i} \int_{\gamma_0} \frac{F(\zeta)}{\zeta - z} \, d\zeta.$$

This procedure is justifiable because the integrand is continuous and when written out in terms of real integrals has the form

$$\int_a^b \int_\alpha^\beta h(x,y) \, dxdy + i \int_a^b \int_\alpha^\beta k(x,y) \, dxdy.$$

We know from calculus that this order can be interchanged (Fubini's Theorem); see, for instance, J. Marsden, *Elementary Classical Analysis* (San Francisco: W. H. Freeman and Co., 1974), Chapter 9.

Thus

$$F(z) = \frac{1}{2\pi i} \int_{\gamma_0} \frac{F(\zeta)}{\zeta - z} \, d\zeta,$$

so by Theorem 23, F is analytic inside γ_0 and

$$F'(z) = \frac{1}{2\pi i} \int_{\gamma_0} \frac{F(\zeta)}{(\zeta - z)^2} \, d\zeta$$

$$= \frac{1}{2\pi i} \int_{\gamma_0} \int_\gamma \frac{f(\zeta,w)}{(\zeta - z)^2} \, dwd\zeta$$

$$= \frac{1}{2\pi i} \int_\gamma \int_{\gamma_0} \frac{f(\zeta,w)}{(\zeta - z)^2} \, d\zeta \, dw$$

$$= \int_\gamma \frac{\partial f}{\partial z} (\zeta,w) \, dw,$$

again by Cauchy's Integral Formula. Since z_0 is arbitrary we obtain the desired result.

Remark. f should be analytic in z but only needs to be integrable in the w variable, as is evident from the preceding proof; we merely need an adequate hypothesis to justify interchanging the order of integration.

Exercises

1.   Prove that for closed curves γ_1, γ_2,

$$I(-\gamma_1, z_0) = -I(\gamma_1, z_0)$$

and

$$I(\gamma_1 + \gamma_2, z_0) = I(\gamma_1, z_0) + I(\gamma_2, z_0).$$

Interpret these results geometrically.

2. Prove that if the image of γ lies in a simply connected region A and if $z_0 \notin A$, then $I(\gamma, z_0) = 0$.

3. Evaluate the following integrals:

a. $\int_\gamma \dfrac{z^2}{z-1}\, dz$; γ is a circle of radius 2, centered at 0.

b. $\int_\gamma \dfrac{e^z}{z^2}\, dz$; γ is the unit circle.

c. $\int_\gamma \dfrac{z^2 - 1}{z^2 + 1}\, dz$; γ is a circle of radius 2, centered at 0.

4. Let f be analytic "inside and on" a simple closed curve γ. Suppose that $f = 0$ on γ. Show that $f = 0$ inside γ.

5. Prove that $\int_0^\pi e^{\cos \theta} \cos(\sin \theta)\, d\theta = \pi$ by considering $\int_\gamma (e^z/z)\, dz$ where γ is the unit circle.

6. Let f be analytic on a region A and let γ be a closed curve in A. For any $z_0 \in A$ not on γ, show that

$$\int_\gamma \frac{f'(\zeta)}{\zeta - z_0}\, d\zeta = \int_\gamma \frac{f(\zeta)}{(\zeta - z_0)^2}\, d\zeta.$$

Can you think of a way to generalize this result?

7. Suppose that f is entire and that $\lim\limits_{z \to \infty} f(z)/z = 0$. Prove that f is constant.

8. Let f be entire. If $|f(z)| \leq M|z|^n$ for large $|z|$, for a constant M and for an integer n, show f is a polynomial of degree $\leq n$.

9. Prove that if γ is a circle, $\gamma(t) = z_0 + re^{it}$, $0 \leq t \leq 2\pi$, then for every z inside γ (that is, $|z - z_0| < r$), $I(\gamma, z) = 1$.

10. Use example 4 to show that

$$F(z) = \int_0^1 e^{-z^2 x^2}\, dx$$

is analytic in z. What is $F'(z)$?

11. Show that if F is analytic on A, then so is f where

$$f(z) = \frac{F(z) - F(z_0)}{z - z_0}$$

if $z \neq z_0$ and $f(z_0) = F'(z_0)$ where z_0 is some point in A.

12. Evaluate the following integrals:

a. $\int_\gamma \dfrac{dz}{z^3}$; γ is the unit square (with vertices $-1 - i$, $1 - i$, $1 + i$, $-1 + i$).

b. $\int_\gamma \dfrac{\sin z}{z^4}\, dz$; γ is the unit circle.

c. $\displaystyle\int_\gamma \frac{\sin{(e^z)}}{z}\,dz$; γ is the unit circle.

13. Use example 3 of Section 2.1 (where appropriate) and Cauchy's Integral Formula to evaluate the following integrals. γ is the circle $|z| = 2$ in each case.

a. $\displaystyle\int_\gamma \frac{dz}{z^2 - 1}$.

b. $\displaystyle\int_\gamma \frac{dz}{z^2 + z + 1}$.

c. $\displaystyle\int_\gamma \frac{dz}{z^2 - 8}$.

d. $\displaystyle\int_\gamma \frac{dz}{z^2 + 2z - 3}$.

14. Suppose that we know that $f(z)$ is analytic on $|z| < 1$ and that $|f(z)| \leq 1$. What estimate can be made about $|f'(0)|$?

15. Consider the function $f(z) = 1/z^2$.
   a. It satisfies $\int_\gamma f(z)dz = 0$ for all closed contours γ (not passing through the origin) but is not analytic at $z = 0$. Does this statement contradict Morera's Theorem?
   b. It is bounded as $z \to \infty$ but is not a constant. Does this statement contradict Liouville's Theorem?

16. Evaluate

$$\int_C \frac{|z|e^z}{z^2}\,dz$$

where C is the circumference of the circle of radius 2 around the origin.

17. Does

$$\int_{|z|=1} \frac{e^z}{z^2}\,dz = 0?$$

Does

$$\int_{|z|=1} \frac{\cos z}{z^2}\,dz = 0?$$

18. Let $f(z)$ be entire and let $|f(z)| \geq 1$ on the whole complex plane. Prove that f is constant.

19. Evaluate

a. $\displaystyle\int_{|z-1|=2} \frac{dz}{(z^2 - 2i)}$.

b. $\displaystyle\int\limits_{|z|=2} \frac{dz}{z^2(z^2+16)}.$

20. Let f be analytic inside and on the circle γ: $|z-z_0| = R$. Prove that

$$\frac{f(z_1)-f(z_2)}{z_1-z_2} - f'(z_0) = \frac{1}{2\pi i} \int\limits_{\gamma} \left[\frac{1}{(z-z_1)(z-z_2)} - \frac{1}{(z-z_0)^2}\right] f(z)dz$$

for z_1, z_2 inside γ.

2.5   Maximum Modulus Theorem and Harmonic Functions

In this section some more consequences will be drawn from the Cauchy Integral Formula. The main consequence is called the Maximum Modulus Theorem. It states that if f is analytic on a region A, then $|f(z)|$ takes its maximum on the boundary of A (that is, there are no strict local maxima of f inside A). This theorem and the Cauchy Integral Formula will be applied to harmonic functions to develop some of their basic properties.

Maximum Modulus Theorem

Theorem 29 is a preliminary statement of the general form of the Maximum Modulus Theorem (Theorem 30).

Theorem 29.   *Let f be analytic on a region A. Suppose that $|f|$ has a relative maximum at $z_0 \in A$ (that is, $|f(z)| \le |f(z_0)|$ for all z in some neighborhood of z_0). Then f is constant in some neighborhood of z_0.*

The proof of Theorem 29 is based on the following.

Mean Value Property.   *Let f be analytic inside and on a circle of radius r and center z_0. (In other words, f is analytic on a region containing the circle and its interior.) Then*

$$f(z_0) = \frac{1}{2\pi} \int\limits_{0}^{2\pi} f(z_0 + re^{i\theta})d\theta. \tag{1}$$

Equation (1) states that the value of f at the center of the circle is the average of f on the circumference.

Proof.   By Cauchy's Integral Formula,

$$f(z_0) = \frac{1}{2\pi i} \int_\gamma \frac{f(z)}{z - z_0}\, dz$$

where $\gamma(\theta) = z_0 + re^{i\theta}$, $0 \leq \theta \leq 2\pi$. But by definition of the integral,

$$\frac{1}{2\pi i} \int_\gamma \frac{f(z)}{z - z_0}\, dz = \frac{1}{2\pi i} \int_0^{2\pi} \frac{f(z_0 + re^{i\theta})}{re^{i\theta}}\, rie^{i\theta}\, d\theta$$

$$= \frac{1}{2\pi} \int_0^{2\pi} f(z_0 + re^{i\theta})\, d\theta. \blacksquare$$

We can use the Mean Value Property to prove Theorem 29.

Proof of Theorem 29.   Let us first prove a special case of Theorem 29 in which we assume that $f(z_0)$ is real and ≥ 0.

Let $r_0 > 0$ be such that the open disk $D(z_0, r_0)$ is contained in the neighborhood of z_0 on which $|f(z_0)| \geq |f(z)|$. Choose any $r < r_0$. Our first step is to show that $f(z) = f(z_0)$ for every z on the circle $|z - z_0| = r$, which obviously implies that $f(z_0) = f(z)$ on the disk $D(z_0, r_0)$. We observe from the Mean Value Property that

$$f(z_0) = |f(z_0)| = \frac{1}{2\pi} \int_0^{2\pi} f(z_0 + re^{i\theta})\, d\theta \leq \sup\{|f(z_0 + re^{i\theta})| \mid 0 \leq \theta \leq 2\pi\}$$

$$= \sup \{|f(z)| \mid |z - z_0| = r\} \leq f(z_0).$$

Here, sup S denotes the supremum or least upper bound of a set $S \subset \mathbb{R}$. The preceding inequalities imply that $f(z_0) = \sup \{|f(z)| \mid |z - z_0| = r\}$. Let us consider the continuous function $g(z) = f(z_0) - \text{Re } f(z)$ defined on the circle $|z - z_0| = r$. Since $\text{Re } f(z) \leq |f(z)| \leq f(z_0)$, we know that $g(z) \geq 0$. But by the Mean Value Property, $\int_0^{2\pi} g(z_0 + re^{i\theta})\, d\theta = 2\pi f(z_0) - \int_a^{2\pi} \text{Re } f(z_0 + re^{i\theta})\, d\theta = 2\pi f(z_0) - \text{Re}(2\pi f(z_0)) = 0$. We know from real variable calculus that if h is a continuous real-valued function defined on $[a, b]$ with $h(t) \geq 0$ for all $t \in [a, b]$ and $\int_a^b h(s)\, ds = 0$, then $h(t) = 0$ for all $t \in [a, b]$. This implies that g is identically zero on the circle $|z - z_0| = r$ and hence that $f(z_0) = \text{Re } f(z)$ on the circle.

Thus on $D(z_0, r_0)$, $\text{Re } f(z)$ is constant and equal to $f(z_0)$. We can use the Cauchy-Riemann equations to show that if f is analytic on a region ($D(z_0, r_0)$ in this case) and $f = u + iv$, then u constant implies that v is constant. Indeed, $\partial v/\partial x = - \partial u/\partial y = 0$ and $\partial v/\partial y = \partial u/\partial x = 0$, so v is constant. Thus f is constant. Hence $f(z) = f(z_0)$ on $D(z_0, r_0)$. This completes the proof for the case in which $f(z_0)$ is real and ≥ 0.

Let us suppose that $w = f(z_0)$ is not real and ≥ 0 (specifically, $w \neq 0$) and consider the function $\tilde{f}(z) = w^{-1}f(z)$. This function $\tilde{f}$ still obeys the condition of Theorem 29 and in addition, $\tilde{f}(z_0) = w^{-1}f(z_0) = 1$, which is real and ≥ 0. Thus $w^{-1}f(z)$ is constant in a neighborhood of z_0 and hence so is $f(z)$. ∎

If, instead of proving that $f(z)$ is constant, we merely wished to prove that $|f(z)|$ is constant, the preceding argument could have been shortened considerably, as follows. From the Mean Value Property and Theorem 3, Section 2.1, we can immediately conclude that

$$0 < \frac{1}{2\pi} \int\limits_{0}^{2\pi} \{|f(z_0 + re^{i\theta})| - |f(z_0)|\}\, d\theta,$$

and since the integrand is not positive, we can conclude that $|f(z)|$ is locally constant. But to prove Theorem 29, we would still have to deduce that f itself is constant. (This requirement is what necessitated the extra work in the preceding proof.)

This argument that $|f(z)|$ is constant can be used to give a quick proof of the fact that z_0 cannot be a *strict* local maximum; the student can easily fill in the details.

At this point we shall need a few additional terms to enable us to formulate a more general version of Theorem 29.

Definition 6.   *The closure of a set A, denoted by $\mathrm{cl}(A)$ or by $\bar{A}$, consists of A together with the limit points of all convergent sequences from A. A set A is called closed iff $A = \mathrm{cl}(A)$. The boundary of A is defined by $\mathrm{bd}(A) = \mathrm{cl}(A) \cap \mathrm{cl}(\mathbb{C} \backslash A)$. Finally, a set is bounded when it can be enclosed in a disk of finite radius.*

It is not difficult to see that a set is closed exactly when its complement is open and that if A is open, then $\mathrm{bd}(A) = \mathrm{cl}(A) \backslash A$. The interested student can work out these relations for himself. He is reminded of the intuitive idea of closure and boundary as illustrated in Figure 2.31.

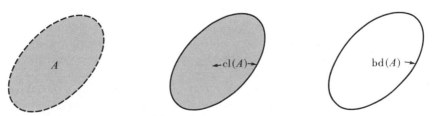

Figure 2.31   Closure and boundary.

Now we are ready for a statement of the general form of the Maximum Modulus Theorem.

Theorem 30 (Maximum Modulus Theorem). *Let A be an open, connected, and bounded set and let $\text{cl}(A)$ denote its closure. Let f: $\text{cl}(A) \to \mathbb{C}$ be continuous on $\text{cl}(A)$ and analytic on A. Let $\text{bd}(A)$ be the boundary of A and let M be the maximum of $|f(z)|$ on $\text{bd}(A)$; that is, M is the least upper bound, or supremum of $|f(z)|$ as z ranges through $\text{bd}(A)$. In symbols, $M = \sup \{|f(z)| \mid z \in \text{bd}(A)\}$. Then (i) $|f(z)| \leq M$ for all $z \in A$, and (ii) if $|f(z)| = M$ for some $z \in A$, then f is constant on A.*

Theorem 30 states that the maximum of f occurs on the boundary of A and if that maximum is attained on A itself, then f must be constant. This is a rather striking result and is certainly a very special property of analytic functions. If A is not bounded, the result of Theorem 30 is not true; see review exercise 28 at the end of Chapter 3.

In applications of this theorem, A often will be the inside of a simple closed curve γ, so $\text{cl}(A)$ will be A union γ and $\text{bd}(A)$ will be γ.

From calculus we know that a continuous real-valued function on a closed bounded set attains a maximum on that set. It therefore follows that if $M' = \sup\{|f(z)| \mid z \in \text{cl}(A)\}$, then $M' = |f(a)|$ for some $a \in \text{cl}(A)$. (Continuity of $|f(z)|$ was established in exercise 1, Section 1.4.) Some additional facts about connected sets that we need are summarized in the Appendix to this section. These results from calculus are used in the following proof.

Proof.   The first step is to show that $M = M'$, where $M' = \sup\{|f(z)| \mid z \in \text{cl}(A)\}$ and $M = \sup\{|f(z)| \mid z \in \text{bd}(A)\}$.

In this first step, there are two cases. First, let us suppose that there is no $a \in A$ such that $|f(a)| = M'$. Hence there must be an $a \in \text{bd}(A)$ such that $|f(a)| = M'$ (because we know that there is some $a \in \text{cl}(A) = A \cup \text{bd}(A)$ such that $|f(a)| = M'$). But then we must have $|f(a)| = M' = M$. Second, let us suppose that there is an $a \in A$ such that $|f(a)| = M'$. By Theorem 29, the set $B = \{z \in A \mid f(z) = f(a)\}$ is open, since every $z \in B$ has a neighborhood on which f is the constant value $f(z) = f(a)$. On the other hand, B is the inverse image of the closed set $\{f(a)\}$ by the continuous map f restricted to A; therefore, B is a closed set in A (consult exercise 3, Section 1.4). Thus B is both open and closed in A and, of course, nonempty, so by the basic facts about connectedness we have $B = A$ (see the Appendix at the end of this section). Hence f is the constant value $f(a)$ on A and so, by the continuity of f on $\text{cl}(A)$, the value of f equals the constant value $f(a)$ on $\text{cl}(A)$ as well (Why?). Thus in this case we also have $M = M'$.

Since $M = M'$, we obviously have $|f(z)| \leq M$ for all $z \in A$ and thus

(i) is proved. If $|f(z)| = M = M'$ for some $z \in A$, corresponding to the second case, then, as shown there, f is constant on A, and thus (ii) is proved. ■

Schwarz Lemma

Theorem 31 is an example of an application of the Maximum Modulus Theorem. This result is not one of the most basic results of the theory, but it further indicates the type of severe restrictions that analyticity imposes.

Theorem 31 (Schwarz Lemma). *Let f be analytic on $A = \{z \in \mathbb{C} \mid |z| < 1\}$ (the open unit disk) and suppose that $|f(z)| \leq 1$ for $z \in A$ and $f(0) = 0$. Then $|f(z)| \leq |z|$ for all $z \in A$ and $|f'(0)| \leq 1$. If $|f(z_0)| = |z_0|$ for some $z_0 \in A$, $z_0 \neq 0$, then $f(z) = cz$ for all $z \in A$ for some constant c, $|c| = 1$.*

Proof. Let

$$g(z) = \begin{cases} \dfrac{f(z)}{z} & \text{if } z \neq 0; \\ f'(0) & \text{if } z = 0. \end{cases}$$

Then g is analytic on A because it is continuous on A and analytic on $A\backslash\{0\}$ (see example 4 of Section 2.2 or the remarks following Corollary 2, Section 2.3). Let $A_r = \{z \mid |z| \leq r\}$ for $0 < r < 1$ (see Figure 2.32). Then g is analytic on A_r, and on $|z| = r$, $|g(z)| = |f(z)/z|$

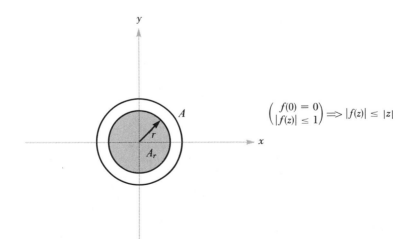

Figure 2.32 Schwarz Lemma.

$\leq 1/r$. By Theorem 30, $|g(z)| \leq 1/r$ on all of A_r; that is, $|f(z)| \leq |z|/r$ on A_r. But by holding $z \in A$ fixed, we can let $r \to 1$ to obtain $|f(z)| \leq |z|$. Clearly, $|g(0)| \leq 1$, or $|f'(0)| \leq 1$.

If $|f(z_0)| = |z_0|$, $z_0 \neq 0$, then $|g(z_0)| = 1$ is maximized in A_r, where $|z_0| < r < 1$, so g is constant on A_r. The constant is independent of r (Why?), and the theorem is proved. ∎

The Schwarz Lemma is a tool for many elegant and occasionally useful geometric results of complex analysis. A generalization that is useful for obtaining accurate estimates of bounds for functions is known as the *Lindelöf Principle*, which is as follows.

Suppose that f and g are analytic on $|z| < 1$, that g maps $|z| < 1$ one-to-one onto a set G, that $f(0) = g(0)$, and that the range of f is contained in G. Then $|f'(0)| \leq |g'(0)|$ and the image of $|z| < r$, for $r < 1$, under f is contained in its image under g.

This principle is made particularly useful by the convenient availability of linear fractional transformations for the role of g; these have the form $g(z) = (az + b)/(cz + d)$. As will be proved in Chapter 5, they take circles into circles, and so the g image of the disk $|z| < r$ is usually easy to find (see exercise 3 for further details).

For a useful survey of some of the more geometric results and a bibliography, see T. H. MacGregor, "Geometric Problems in Complex Analysis," *American Mathematical Monthly* (May, 1972), page 447.

Harmonic functions and harmonic conjugates

A function $u: A \to \mathbb{R}$ on a region $A \subset \mathbb{C}$ that is twice continuously differentiable is called *harmonic* if

$$\nabla^2 u = \frac{\partial^2 u}{\partial x^2} + \frac{\partial^2 u}{\partial y^2} = 0. \tag{2}$$

The expression $\nabla^2 u$ is called the *Laplacian of u* and is one of the most basic operations in mathematics and physics. Harmonic functions play a fundamental role in the physical examples discussed later in Chapters 5 and 8. For the moment let us study them from the mathematical point of view.

If f is analytic on A and $f = u + iv$, we know that u and v are infinitely differentiable and are harmonic (by Theorem 24 and example 5, Section 1.4). Let us now show that the converse is also true.

Theorem 32. *Let A be a region and let u be a twice continuously differentiable, harmonic function on A. Then u is C^∞, and in a neighborhood of each point $z_0 \in A$, u is the real part of some analytic function. If A is simply connected, there is an analytic function f on A such that $u = \mathrm{Re}\, f$.*

Thus harmonic functions always occur as the real part of an analytic function f (or the imaginary part of the analytic function if) at least locally, and on all of the domain of that function if the domain is simply connected.

Proof.   We prove the last statement of the theorem first. Let us consider the function $g = (\partial u/\partial x) - i(\partial u/\partial y)$. We claim that g is analytic. Setting $g = U + iV$ where $U = \partial u/\partial x$ and $V = -\partial u/\partial y$, we must check that U and V have continuous first partials and that they satisfy the Cauchy-Riemann equations (Theorem 5, Section 1.4). Indeed, the functions $\partial U/\partial x = \partial^2 u/\partial x^2$ and $\partial V/\partial y = -\partial^2 u/\partial y^2$ are continuous by assumption and are equal since $\nabla^2 u = 0$. Also,

$$\frac{\partial U}{\partial y} = \frac{\partial^2 u}{\partial y \partial x} = \frac{\partial^2 u}{\partial x \partial y} = -\frac{\partial V}{\partial x}.$$

Thus we conclude that g is analytic. Furthermore, there is an analytic function f on A such that $f' = g$ by Theorem 18 (or Theorem 9). Let $f = \tilde{u} + i\tilde{v}$. Then $f' = (\partial \tilde{u}/\partial x) - i(\partial \tilde{u}/\partial y)$, so $\partial \tilde{u}/\partial x = \partial u/\partial x$ and $\partial \tilde{u}/\partial y = \partial u/\partial y$. Thus $\tilde{u}$ differs from u by a constant. Adjusting f by subtracting this constant, we get $u = \operatorname{Re} f$.

Now we prove the first statement. If D is a disk around z_0 in A, it is simply connected. Therefore, as a result of what we have just proven, we can write $u = \operatorname{Re} f$ for some analytic f on D. Thus since f is C^∞, u is also C^∞ on a neighborhood of each point in A, and so is C^∞ on A.∎

When there is an analytic function f such that u and v are related by $f = u + iv$, we say that u and v are *harmonic conjugates*. Since if is analytic, $-v$ and u are also harmonic conjugates.

Theorem 32 says that *on any simply connected region A, any harmonic function has a harmonic conjugate $v = \operatorname{Im} f$.* Since the Cauchy-Riemann equations ($\partial u/\partial x = \partial v/\partial y$ and $\partial u/\partial y = -\partial v/\partial x$) must hold, v is uniquely determined up to the addition of a constant. These equations may be used as a practical method for finding v when u is given (see example 3).

Theorem 33 describes a property of harmonic conjugates that has a very important physical interpretation that will be used in Chapter 5.

Theorem 33.   *Let u and v be harmonic conjugates on a region A. Suppose that*

$$u(x,y) = constant = c_1$$

and

$$v(x,y) = constant = c_2$$

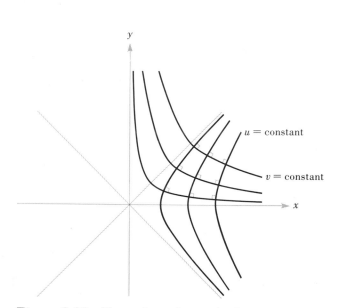

Figure 2.33 Harmonic conjugates in the case
$u = x^2 - y^2$, $v = 2xy$, $f(z) = z^2$.

define smooth curves. Then these curves intersect orthogonally (see
Figure 2.33).

Note. We shall accept from calculus the fact that $u(x,y) = c_1$ defines
a smooth curve if the gradient grad $u(x,y) = (\partial u/\partial x, \partial u/\partial y) = (\partial u/\partial x)$
$+ i(\partial u/\partial y)$ is nonzero for x and y satisfying $u(x,y) = c_1$. (The student
should be aware of this fact even though it is a technical point that
does not play a major role in concrete examples.) It is also true that
the vector grad u is perpendicular to that curve (see Figure 2.34).
This can be explained as follows. If $(x(t), y(t))$ is the curve, then
$u(x(t), y(t)) = c_1$, so

$$\frac{d}{dt} \cdot u(x(t), y(t)) = 0,$$

or

$$\frac{\partial u}{\partial x} \cdot x'(t) + \frac{\partial u}{\partial y} \cdot y'(t) = 0;$$

that is,

$$\left(\frac{\partial u}{\partial x}, \frac{\partial u}{\partial y} \right) \cdot (x'(t), y'(t)) = 0.$$

Now we prove Theorem 33.

Proof.   By the above note, it suffices to show that grad u and grad v are perpendicular. Their inner product is

$$\text{grad } u \cdot \text{grad } v = \frac{\partial u}{\partial x} \cdot \frac{\partial v}{\partial x} + \frac{\partial u}{\partial y} \cdot \frac{\partial v}{\partial y},$$

which is zero by the Cauchy-Riemann equations. ∎

Mean Value Property and Maximum Principle for Harmonic Functions

One reason why Theorem 32 is important is that it enables us to deduce properties of harmonic functions from corresponding properties of analytic functions. This is done in the next theorem.

Theorem 34.   *Let u be harmonic on a region containing a circle of radius r around $z_0 = x_0 + iy_0$ and its interior. Then*

$$u(x_0, y_0) = \frac{1}{2\pi} \int_0^{2\pi} u(z_0 + re^{i\theta}) \, d\theta. \tag{3}$$

Proof.   By Theorem 32 there is an analytic function f defined on a region containing this circle and its interior such that $u = \text{Re } f$. This containing region may be chosen to be a slightly larger disk. The existence of a slightly larger circle in A is intuitively clear; the precise proof is left for those students wishing to practice their topology in the plane (see the Appendix at the end of this section). By the Mean Value Property for f,

$$f(z_0) = \frac{1}{2\pi} \int_0^{2\pi} f(z_0 + re^{i\theta}) \, d\theta.$$

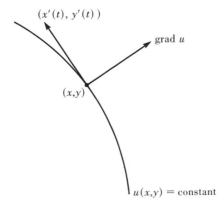

$(x'(t), y'(t))$

grad u

(x,y)

$u(x,y) = $ constant   **Figure 2.34**   Gradient.

Taking the real part of both sides of the equation gives the desired result. ■

From Theorem 34 we can deduce, in a way similar to the way we deduced Theorem 29, the following theorem.

Theorem 35. *Let u be harmonic on a region A. Suppose that u has a relative maximum at $z_0 \in A$ (that is, $u(z) \le u(z_0)$ for z near z_0). Then u is constant in a neighborhood of z_0.*

In this theorem "maximum" can be replaced by "minimum" (see exercise 5).

Instead of actually going through a proof for $u(z)$ similar to the proof of Theorem 29, we can use the result of Theorem 29 to give a quick proof of Theorem 35.

Proof. On a disk around z_0, $u = \mathrm{Re}\, f$ for some analytic f. Then $e^{f(z)}$ is analytic and $|e^{f(z)}| = e^{u(z)}$. Thus since e^x is strictly increasing in x for all real x, the maxima of u are the same as those of $|e^f|$. By Theorem 29, e^f is constant in a neighborhood of z_0; therefore, e^u and hence u are also (again because e^x is strictly increasing for x real). ■

From Theorem 35 we deduce, exactly as Theorem 30 was deduced from Theorem 29, the following more useful version of Theorem 35.

Theorem 36 (Maximum Principle for Harmonic Functions).
Let $A \subset \mathbb{C}$ be open, connected, and bounded. Let u: $\mathrm{cl}(A) \to \mathbb{R}$ be continuous and harmonic on A. Let M be the maximum of u on $\mathrm{bd}(A)$. Then
i. *$u(x,y) \le M$ for all $(x,y) \in A$.*
ii. *if $u(x,y) = M$ for some $(x,y) \in A$, then u is constant on A.*

There is a corresponding result for the minimum. Let m denote the minimum of u on $\mathrm{bd}(A)$. Then
i. $u(x,y) \ge m$ for $(x,y) \in A$.
ii. if $u(x,y) = m$ on A, then u is constant.

The Minimum Principle for Harmonic Functions may be deduced from the Maximum Principle for Harmonic Functions by applying the latter to $-u$.

Dirichlet Problem for the disk and Poisson's Formula

There is a very important problem in mathematics and physics called the *Dirichlet Problem*. It is this: Let A be an open bounded region and let u_0 be a given continuous function on $\mathrm{bd}(A)$. Find a real-valued function u on $\mathrm{cl}(A)$ that is continuous on $\mathrm{cl}(A)$ and harmonic on A and that equals u_0 on $\mathrm{bd}(A)$.

There do exist theorems stating that if the boundary $\mathrm{bd}(A)$ is "sufficiently smooth," then there always is a solution u. These theorems are quite difficult. However, we can easily show that the solution is always unique.

Theorem 37.   *The solution to the Dirichlet Problem is unique (assuming that there is a solution).*

Proof.   Let u and $\tilde{u}$ be two solutions. Let $\varphi = u - \tilde{u}$. Then φ is harmonic and $\varphi = 0$ on $\mathrm{bd}(A)$. We must show that $\varphi = 0$.

By the Maximum Principle for Harmonic Functions, $\varphi(x,y) \leq 0$ inside A. Similarly, from the corresponding Minimum Principle, $\varphi(x,y) \geq 0$ on A. Thus $\varphi = 0$. ∎

We want to find the solution to the Dirichlet Problem for the case where the region is an open disk. To do so we derive a formula that explicitly expresses the values of the solution in terms of its values on the boundary of the disk.

Theorem 38 (Poisson's Formula).   *If u is defined and continuous on the closed disk $\overline{D(0,r)} = \{z \mid |z| \leq r\}$ and is harmonic on the open disk $D(0,r) = \{z \mid |z| < r\}$, then for $\rho < r$,*

$$u(\rho e^{i\varphi}) = \frac{(r^2 - \rho^2)}{2\pi} \int_0^{2\pi} \frac{u(re^{i\theta})}{r^2 - 2r\rho\cos(\theta - \varphi) + \rho^2} \, d\theta. \tag{4}$$

Note.   The technical parts of the following proof require an acquaintance with the idea of *uniform convergence*. The student who has not studied uniform convergence in advanced calculus may wish to reread this proof after studying Section 3.1, where the relevant ideas are discussed.

Proof.   Since u is harmonic on $D(0,r)$ and $D(0,r)$ is simply connected, there is an analytic function f defined on $D(0,r)$ such that $u = \mathrm{Re}\, f$.

Let $0 < s < r$ and let γ_s be the circle $|z| = s$. Then, by Cauchy's Integral Formula, we have

$$f(z) = \frac{1}{2\pi i} \int_{\gamma_s} \frac{f(\zeta)}{\zeta - z} \, d\zeta$$

for all z such that $|z| < s$. We can manipulate this expression into a form suitable for taking real parts.

Let $\tilde{z} = s^2/\bar{z}$, which is called the *reflection* of z in the circle $|z| = s$. Reflection is pictured geometrically in Figure 2.35.

Thus if z lies inside the circle, then $\tilde{z}$ lies outside the circle, and therefore

$$\frac{1}{2\pi i} \int_{\gamma_s} \frac{f(\zeta)\, d\zeta}{\zeta - \tilde{z}} = 0$$

for $|z| < s$. We may subtract this integral from

$$f(z) = \frac{1}{2\pi i} \int_{\gamma_s} \frac{f(\zeta)}{\zeta - z}\, d\zeta$$

to obtain

$$f(z) = \frac{1}{2\pi i} \int_{\gamma_s} f(\zeta) \left(\frac{1}{\zeta - z} - \frac{1}{\zeta - \tilde{z}} \right) d\zeta.$$

Observing that $|\zeta| = s$, we can simplify

$$\frac{1}{\zeta - z} - \frac{1}{\zeta - \tilde{z}} = \frac{1}{\zeta - z} + \frac{1}{\zeta - \dfrac{|\zeta|^2}{\bar{z}}} = \frac{1}{\zeta - z} - \frac{\bar{z}}{\zeta(\bar{z} - \bar{\zeta})}$$

$$= \frac{-\zeta\bar{z} + |\zeta|^2 + \zeta\bar{z} - |z|^2}{\zeta\,|\zeta - z|^2} = \frac{|\zeta|^2 - |z|^2}{\zeta\,|\zeta - z|^2}.$$

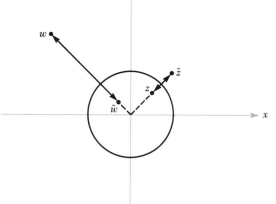

Figure 2.35   Reflection of a complex number in a circle.

Hence we have

$$f(z) = \frac{1}{2\pi i} \int_{\gamma_s} \frac{f(\zeta)\,(|\zeta|^2 - |z|^2)}{\zeta\,|\zeta - z|^2}\, d\zeta$$

or

$$f(\rho e^{i\varphi}) = \frac{1}{2\pi} \int_0^{2\pi} \frac{f(s e^{i\theta})\,(s^2 - \rho^2)}{|s e^{i\theta} - \rho e^{i\varphi}|^2}\, d\theta$$

where $\rho < s$. Noting that $|s e^{i\theta} - \rho e^{i\varphi}|^2 = s^2 + \rho^2 - 2s\rho\cos(\theta - \varphi)$ and taking the real parts on both sides of the equations, we obtain

$$u(\rho e^{i\varphi}) = \frac{1}{2\pi} \int_0^{2\pi} \frac{u(s e^{i\theta})\,(s^2 - \rho^2)\,d\theta}{s^2 + \rho^2 - 2s\rho\cos(\theta - \varphi)}.$$

Keeping ρ and φ fixed, we observe that this formula is true for any s such that $\rho < s < r$. Since u is continuous on $\overline{D(0,r)}$ and since $s^2 + \rho^2 - 2s\rho\cos(\theta - \varphi)$ is never zero when $s > \rho$, we conclude that for $s > \rho$, $[u(s e^{i\theta})(s^2 - \rho^2)]/[s^2 + \rho^2 - 2s\rho\cos(\theta - \varphi)]$ is a continuous function of s and θ and hence (with ρ, φ fixed) is uniformly continuous on the compact set, $0 \le \theta \le 2\pi$, $(r + \rho)/2 \le s \le r$. Consequently,

$$\frac{u(s e^{i\theta})\,(s^2 - \rho^2)}{s^2 + \rho^2 - 2s\rho\cos(\theta - \varphi)} \to \frac{u(r e^{i\theta})\,(r^2 - \rho^2)}{r^2 + \rho^2 - 2r\rho\cos(\theta - \varphi)}$$

as $s \to r$, uniformly in θ, which implies that

$$\frac{1}{2\pi} \int_0^{2\pi} \frac{u(s e^{i\theta})\,(s^2 - r^2)}{s^2 + \rho^2 - 2s\rho\cos(\theta - \varphi)}\, d\theta \to \frac{1}{2\pi} \int_0^{2\pi} \frac{u(r e^{i\theta})\,(r^2 - \rho^2)}{r^2 + \rho^2 - 2r\rho\cos(\theta - \varphi)}\, d\theta$$

as $s \to r$. (See Theorem 7, Section 3.1, if you are not familiar with this result about the convergence of integrals.) Thus

$$u(\rho e^{i\varphi}) = \frac{r^2 - \rho^2}{2\pi} \int_0^{2\pi} \frac{u(r e^{i\theta})}{r^2 + \rho^2 - 2r\rho\cos(\theta - \varphi)}\, d\theta. \ \blacksquare$$

Actually, this formula also enables us to settle the Dirichlet Problem for the case where the region is an open disk. Suppose that we are given the continuous function u_0 defined on the circle $|z| = r$. We define u by formula (4):

$$u(\rho e^{i\varphi}) = \frac{r^2 - \rho^2}{2\pi} \int_0^{2\pi} \frac{u_0(r e^{i\theta})}{r^2 - 2r\rho\cos(\theta - \varphi) + \rho^2}\, d\theta$$

for $\rho < r$ and $u(re^{i\varphi}) = u_0(re^{i\varphi})$. It is relatively simple to show that u is harmonic in $D(0,r)$ (see exercise 13), but it is more difficult to show that u is continuous on $\overline{D}(0,r)$. The expression must be examined in the critical case in which $\rho \to r$ and the integrand takes the values $0/0$ near $\theta = \varphi$. A proof will not be given here.

Appendix to Chapter 2:
On Connectedness

The proof of Theorem 30 used a certain property of connected sets that is learned in advanced calculus. This property is stated in the following theorem.

Theorem.   *Let $A \subset \mathbb{R}^n$ be (path) connected. If $C \subset A$, $C \neq \phi$, and C is both open and closed relative to A (that is, if $C = U \cap A$ where U is an open set and $C = F \cap A$ where F is a closed set), then $C = A$.*

Proof.   If $C \neq A$, there is a $z_1 \in C$ and a $z_2 \in A \backslash C$. Let $c : [a,b] \longrightarrow A$ be a continuous path joining z_1 to z_2. Let $B = c^{-1}(C)$. Then $B \subset [a,b]$ and B is both open relative to $[a,b]$ and closed relative to $[a,b]$. This is because c is continuous and the inverse images of open (or closed) sets under a continuous mapping are open (or closed — see exercise 1, Section 1.4). Now $B \neq \phi$ since $a \in B$, and $[a,b]/B \neq \phi$ since it contains b.

This argument restricts us to proving the theorem for the case $A = [a,b]$. Such a proof requires use of the least upper bound property. Let $x = \sup B$ (that is, the least upper bound of B). We find that $x \in B$ since B is closed. But since B is open there is a whole neighborhood around x lying in B (remember that $x \neq b$ since $b \in [a,b] \backslash B$). This means that for some $\epsilon > 0$, $x + \epsilon \in B$. Thus x cannot be the least upper bound. This contradiction shows that such a set B cannot exist. ∎

The proof of the following theorem utilizes some other techniques from advanced calculus.

Theorem. *Let $A \subset \mathbb{C}$ be open, $z_0 \in A$, and suppose that $B_r = \{z \mid |z - z_0| \leq r\} \subset A$. Then there is a $\rho > r$ such that $D(z_0, \rho) \subset A$.*

Proof.   We know from calculus that a continuous real-valued function on a closed bounded set in $\mathbb{C}$ assumes its maximum and minimum at some point of the set. Let $f(z) = \inf\{|z - w| \mid w \in \mathbb{C} \backslash A\}$ where $z \in B_r$. Here inf means greatest lower bound. In other words, $f(z)$ is the distance from $z \in B_r$ to the complement of A. Since A is open, $f(z) > 0$ for $z \in B_r$. We can also verify that f is continuous. Thus f assumes its minimum at, say, $z_1 \in B_r$. Let $\rho = f(z_1) + r$ and check that this ρ has the desired properties.■

Worked Examples

1. Let f be analytic and nonzero in a region A. Show that $|f|$ has no strict local minima in A. If f has zeros in A, show by example that this conclusion does not hold.

    Solution.   Since f is analytic and nonzero in A, $1/f$ is analytic in A. By the Maximum Modulus Theorem, $1/|f|$ can have no local maxima in A unless $1/|f|$ is constant. Thus $1/|f|$ has no strict local maxima in A. Hence $|f|$ can have no strict local minima in A. The identity function $I: z \mapsto z$ is analytic on the unit disk and $|I|$ has a strict minimum at the origin.

2. Find the maximum of $|\sin z|$ on $[0, 2\pi] \times [0, 2\pi]$.

    Solution.   Since $\sin z$ is entire we can apply the Maximum Modulus Theorem, which tells us that the maximum occurs on the boundary of this square.
        Now $|\sin z|^2 = \sinh^2 y + \sin^2 x$ since $\sin (x + iy) = \sin x \cosh y + i \cos x \cdot \sinh y$ and $\sin^2 x + \cos^2 x = 1$, and $\cosh^2 y - \sinh^2 y = 1$.
        On the boundary $y = 0$, $|\sin z|^2$ has maximum 1; for $x = 0$ the maximum is $\sinh^2 2\pi$ since $\sinh y$ increases with y; for $x = 2\pi$ the maximum is again $\sinh^2 2\pi$; for $y = 2\pi$ the maximum is $\sinh^2 2\pi + 1$. Thus the maximum of $|\sin z|^2$ occurs at $x = \pi/2$, $y = 2\pi$, and is $\sinh^2 2\pi + 1 = \cosh^2 2\pi$. Therefore, the maximum of $|\sin z|$ on $[0, 2\pi] \times [0, 2\pi]$ is $\cosh 2\pi$.

3. Find the harmonic conjugates of the following harmonic functions on $\mathbb{C}$:
    a. $u(x,y) = x^2 - y^2$.
    b. $u(x,y) = \sin x \cosh y$.

    Remark.   The student probably immediately recognizes these expressions as the real parts of z^2 and $\sin z$, $z = x + iy$. From this observation it follows that the conjugates, up to addition of constants, are $2xy$ and $\sinh y \cos x$.

It is, however, instructive to solve the problem using the Cauchy-Riemann equations, because the student might not always recognize the $f(z)$.

Solution.

a. Let v be the harmonic conjugate of u (which we know exists because u is harmonic on the simply connected region $\mathbb{C}$). Then $\partial v/\partial x = 2y$ and $\partial v/\partial y = 2x$. Therefore, $v = 2yx + g_1(y)$ and $v = 2xy + g_2(x)$. Hence $g_1(y) = g_2(x) = $ constant, so $v(x,y) = 2yx + $ constant.

b. $\partial v/\partial x = -\sin x \sinh y$ and $\partial v/\partial y = \cos x \cosh y$. The first equation implies that $v = \cos x \sinh y + g_1(y)$ and the second equation implies that $v = \cos x \sinh y + g_2(x)$. Hence $g_1(y) = g_2(x) = $ constant. So $v(x,y) = \cos x \sinh y + $ constant.

4. Find the maximum of $u(x,y) = \sin x \cosh y$ on the unit square $[0,1] \times [0,1]$.

Solution. $u(x,y)$ is a harmonic function and is not constant, so the maximum of u on the unit square $[0,1] \times [0,1]$ occurs on the boundary. The maximum of $\sin x \cosh y$ is $\sin(1) \cosh(1)$, since both sin and cosh are increasing on the interval $[0,1]$.

Exercises

1. Let f be entire and let $|f(z)| \leq M$ for z on the circle $|z| = R$; let R be fixed. Then prove that

$$|f^{(k)}(re^{i\theta})| \leq \frac{k! \, M}{(R-r)^k}, \quad k = 0,1,2, \cdots$$

for all $0 \leq r < R$.

2. Prove: If f is analytic and $|f|$ is constant on a r-neighborhood of z_0, then so is $f(z)$.

3. a. Show that the mapping

$$T: z \longmapsto \frac{R(z - z_0)}{R^2 - \bar{z}_0 z}$$

takes the open disk of radius R one-to-one onto the disk of radius 1 and takes z_0 to the origin, where $|z_0| < R$.

Hint. Use the Maximum Modulus Theorem and verify that $z_0 \longmapsto 0$ and that $|z| = R$ implies that $|Tz| = 1$.

b. Suppose that f is analytic on the open disk $|z| < R$ and that $|f(z)| < M$ for $|z| < R$. Suppose also that $f(z_0) = w_0$. Show that

$$\left| \frac{M(f(z) - w_0)}{M^2 - \bar{w}_0 f(z)} \right| \leq \left| \frac{R(z - z_0)}{R^2 - \bar{z}_0 z} \right|.$$

(This is a generalization of the Schwarz Lemma.)

4. Let f and g be continuous on $\mathrm{cl}(A)$ and analytic on A, where A is an open, connected, bounded region. If $f = g$ on $\mathrm{bd}(A)$, show that $f = g$ on all of $\mathrm{cl}(A)$.

5. Let u be harmonic on the bounded region A and continuous on $\mathrm{cl}(A)$. Then show that u takes its minimum only on $\mathrm{bd}(A)$ unless u is constant. (Compare this exercise with example 1.)

Hint.   Consider $-u$.

6. Let $f: A \rightarrow \mathbb{C}$ be analytic and let $w: B \rightarrow \mathbb{R}$ be harmonic with $f(A) \subset B$. Show that $w \circ f: A \rightarrow \mathbb{R}$ is harmonic.

7. Find the maximum of $|e^{z^2}|$ on the unit disk.

8. Find the maximum of $u = \mathrm{Re}\, z^3$ on the unit square $[0,1] \times [0,1]$.

9. Find the harmonic conjugates for each of the following functions (specify a region in each case):
a. $u(x,y) = \sinh x \sin y$.
b. $u(x,y) = \log \sqrt{x^2 + y^2}$.
c. $u(x,y) = e^x \cos y$.

10. If u is harmonic, show that, in terms of polar coordinates,

$$r^2 \frac{\partial^2 u}{\partial r^2} + r \frac{\partial u}{\partial r} + \frac{\partial^2 u}{\partial \theta^2} = 0.$$

Hint.   Use the Cauchy-Riemann equations in polar form.

11. Verify directly that the level curves of $\mathrm{Re}\, e^z$ and $\mathrm{Im}\, e^z$ intersect orthogonally.

12. a. Prove that

$$\int_0^{2\pi} \frac{R^2 - r^2}{R^2 - 2rR\cos(\theta - \varphi) + r^2}\, d\theta = 2\pi$$

for $R > r$ and any φ.

Hint.   Use uniqueness of the solution to the Dirichlet Problem.

b. Solve the Dirichlet Problem directly for the boundary condition $u_0(z) = x$. Deduce that for $r < 1$,

$$r\cos \varphi = \frac{1}{2\pi} \int_0^{2\pi} \frac{(1 - r^2)\cos \theta}{1 - 2r\cos(\theta - \varphi) + r^2}\, d\theta.$$

13. Show that in Theorem 38,

$$u(z) = \mathrm{Re}\left[\frac{1}{2\pi i} \int_{\gamma_r} \frac{\zeta + z}{\zeta - z}\, u(\zeta)\, \frac{d\zeta}{\zeta} \right];$$

then write a formula for the harmonic conjugate of u. Use this formula to show that if $u(\zeta)$ is continuous on the boundary, then $u(z)$ is harmonic inside it.

Hint.   Use Theorem 25.

14. Prove *Hadamard's Three-Circle Theorem*: Let f be analytic on a region containing the set R in Figure 2.36; let $R = \{z \mid r_1 \le |z| \le r_3\}$; and let $0 < r_1 < r_2 < r_3$. Let M_1, M_2, M_3 be the maxima of $|f|$ on the circles $|z| = r_1, r_2, r_3$, respectively. Then $M_2{}^{\log(r_3/r_1)} \le M_1{}^{\log(r_3/r_2)} M_3{}^{\log(r_2/r_1)}$.

Hint.   Let $\lambda = -\log(M_3/M_1)/\log(r_3/r_1)$ and consider $g(z) = z^\lambda f(z)$. Apply the Maximum Principle to g; be careful about the domain of analyticity of g.

15. Prove: If u is continuous and satisfies the Mean Value Property, then u is C^∞ and is harmonic.

Hint.   Think about Poisson's Formula.

16. Let g be analytic on $\{z \mid |z| < 1\}$ and assume that $|g(z)| = |z|$ for all $|z| < 1$. Show that $g(z) = e^{i\theta}z$ for some constant $\theta \in [0,2\pi[$.

Hint.   Use Theorem 31.

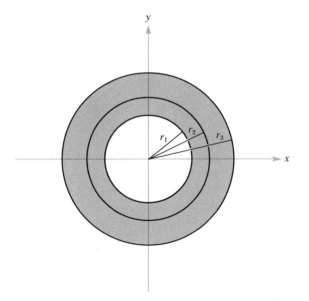

Figure 2.36 Hadamard's Three-Circle Theorem.

17. Evaluate $\int_\gamma dz/(z^2 - 1)$ where γ is the circle $|z| = 2$.

18. The function $f(z)$ is analytic over the whole complex plane and $\mathrm{Im}\, f < 0$. Prove that f is a constant.

19. If $u(z)$ is harmonic and $v(z)$ is harmonic, then
a. is $u(v(z))$ harmonic?
b. is $u(z) \cdot v(z)$ harmonic?
c. is $u(z) + v(z)$ harmonic?

20. Consider the function $f(z) = 1/z$. Draw the contours $u = \mathrm{Re}\, f = $ constant and $v = \mathrm{Im}\, f = $ constant. How do they intersect? Is it always true that grad u is parallel to the curve $v = $ constant?

Review Exercises for Chapter 2

1. Let f be analytic and let $f'(z) \neq 0$ on a region A. Let $z_0 \in A$ and assume that $f(z_0) \neq 0$. Given $\epsilon > 0$, show that there exists $z \in A$ and $\zeta \in A$ such that $|z - z_0| < \epsilon$, $|\zeta - z_0| < \epsilon$, and

$$|f(z)| > |f(z_0)|;$$
$$|f(\zeta)| < |f(z_0)|.$$

Hint.   Use the Maximum Modulus Theorem.

2. Let f and g be analytic in a region A. Prove: If $|f| = |g|$ on A, and $f \neq 0$ except at isolated points then $f(z) = e^{i\theta} g(z)$ on A for some constant θ, $0 \leq \theta < 2\pi$.

3. Let f be analytic on $\{z \mid |z| > 1\}$. Show that if γ_r is the circle of radius $r > 1$ and center 0, then $\int_{\gamma_r} f$ is independent of r.

4. Let $f(z) = P(z)/Q(z)$ where P and Q are polynomials and Q has a degree of at least two more than that of P.
a. Argue that if R is sufficiently large, there is a constant M such that

$$\left| \frac{P(z)}{Q(z)} \right| \leq \frac{M}{|z|^2} \text{ for } |z| \geq R.$$

b. If γ is a circle of radius r, and center 0 with r large enough so f is analytic outside γ, prove that $\int_\gamma f(z)dz = 0$.

Hint.   Use exercise 3 and let $r \to \infty$.

c. Evaluate $\int_\gamma dz/(1 + z^2)$ where γ is a circle of radius 2 and center 0.

5. Evaluate $\int_\gamma f$ where $f(x + iy) = x^2 + iy^2$ and γ is the line joining 1 to i.

6. Let $f = u + iv$ be analytic on a region A. Indicate which of the following are analytic on A:
a. $u - iv$.
b. $-u - iv$.
c. $iu - v$.

7. Let f be analytic on the open connected set A and suppose that there is a $z_0 \in A$ such that $|f(z)| \le |f(z_0)|$ for all $z \in A$. Then show that f is constant on A.

8. Evaluate the following integrals:

   a. $\displaystyle\int_\gamma \sin z \, dz;$ γ is the unit circle.

   b. $\displaystyle\int_\gamma \frac{\sin z}{z} \, dz;$ γ is the unit circle.

   c. $\displaystyle\int_\gamma \frac{\sin z}{z^2} \, dz;$ γ is the unit circle.

   d. $\displaystyle\int_{|z|=1} \frac{\sin(e^z)}{z^2} \, dz.$

9. Find a harmonic conjugate for

$$u(x,y) = \frac{x^2 + y^2 - x}{(x-1)^2 + y^2}$$

   on a suitable domain.

10. Let f be analytic on A and let $f'(z_0) \ne 0$. Show that if γ is a sufficiently small circle centered at z_0, then

$$\frac{2\pi i}{f'(z_0)} = \int_\gamma \frac{dz}{f(z) - f(z_0)}.$$

   Hint.   Use the Inverse Function Theorem.

11. Evaluate $\int_0^{2\pi} e^{-i\theta} e^{e^{i\theta}} \, d\theta.$

12. Let f and g be analytic in a region A and let $g'(z) \ne 0$ for all $z \in A$; let g be one-to-one and let γ be a closed curve in A. Then for z not on γ, prove that

$$f(z) \, I(\gamma, z) = \frac{g'(z)}{2\pi i} \int_\gamma \frac{f(\zeta)}{g(\zeta) - g(z)} \, d\zeta.$$

   Hint.   Apply the Cauchy Integral Theorem to

$$h(\zeta) = \begin{cases} \dfrac{f(\zeta)(\zeta - z)}{g(\zeta) - g(z)} & z \ne \zeta; \\[2ex] \dfrac{f(\zeta)}{g'(\zeta)} & z = \zeta. \end{cases}$$

   Apply this result to the case in which $g(z) = e^z$.

13. Simplify: $e^{\log i}$; $\log i$; $\log (-i)$; $i^{\log(-1)}$.

14. Let $A = \mathbb{C}$ minus the negative real axis and zero. Show that $\log z = \int_{\gamma_z} d\zeta/\zeta$ where γ_z is any curve in A joining 1 to z. Is A simply connected?

15. Let f be analytic on a region A and let f be nonzero. Let γ be a curve homotopic to a point in A. Show that

$$\int_\gamma \frac{f'(z)}{f(z)}\, dz = 0.$$

16. Let f be analytic on and inside the unit circle. Suppose that the image of the unit circle $|z| = 1$ lies in the disk $D = \{z\mid |z - z_0| < r\}$. Show that the image of the whole inside of the unit circle lies in D. Illustrate with e^z.

17. Is $\int_\gamma x\,dx + x\,dy$ always zero if γ is a closed curve?

18. Compute the cube roots of $8i$.

19. If f is analytic on and inside the unit disk, then show that

$$f(re^{i\varphi}) = \frac{1}{2\pi}\int_0^{2\pi} \frac{f(e^{i\theta})}{1 - re^{i(\varphi - \theta)}}\, d\theta;\ r < 1.$$

20. Prove *Harnack's Inequality*: If u is harmonic and nonnegative for $|z| \le R$, then

$$u(0)\frac{R - |z|}{R + |z|} \le u(z) \le u(0)\frac{R + |z|}{R - |z|}.$$

21. Discuss the following sketch of a proof for Cauchy's Theorem: Suppose f is analytic in a convex region G containing 0 and γ is a closed curve in G. Define $F(t) = t \int_\gamma f(t_z)\, dz$ for $0 \le t \le 1$. Cauchy's Theorem is that $F(1) = 0$. Compute that $F'(t) = \int_\gamma f(tz)\, dz + t \int_\gamma zf'(tz)\, dz$, and integrate the second integral by parts to obtain:

$$F'(t) = \int_\gamma f(tz)\, dz + t\left\{\left[\frac{zf(tz)}{t}\right]_\gamma - \frac{1}{t}\int_\gamma f(tz)\, dz\right\} = 0$$

so $F(1) = F(0) = 0$. [See: Philip M. Morse and Herman Feshback, *Methods of Mathematical Physics, Part I.*, McGraw-Hill Book Co., New York, 1953, pp. 364–365.]

Chapter 3

Series Representation of Analytic Functions

There is an important alternate way to define an analytic function. A function f is analytic iff it is locally representable as a convergent power series. This series is called the *Taylor series* of f. We also want to investigate the series representation of a function that is analytic on a deleted neighborhood; that is, a function that has an isolated singularity. The resulting series, called the *Laurent series*, yields valuable information about the behavior of functions near singularities and this behavior is the key to the subject of residues and its subsequent applications.

This chapter begins with an analysis of the basic properties of convergent and uniformly convergent series and then concentrates on the special case of power series.

3.1 Convergent Series of Analytic Functions

We shall use the Cauchy Integral Formula to determine when the limit of a convergent sequence or series of analytic functions is an analytic function and when the derivative (or integral) of the limit is the limit of the derivative (or integral) of the terms in the sequence or series. The basic type of convergence studied in this chapter is uniform convergence; the Weierstrass M Test is the basic tool used to determine such convergence.

We shall be especially interested in the special case of power series (studied in Section 3.2), but we should be aware that some important functions are convergent series other than power series, such as the Riemann zeta function (see example 1).

The proofs of Theorems 1 to 4 are slightly technical, and since they are analogous to the case of real series, they appear at the end of the section.

Convergence of sequences and series

Definition 1. *A sequence z_n of complex numbers is said to converge to a complex number z_0 if, for all $\epsilon > 0$, there is an integer N such that $n \geq N$ implies that $|z_n - z_0| < \epsilon$. Convergence of z_n to z_0 is denoted by $z_n \rightarrow z_0$. An infinite series $\Sigma_{k=1}^{\infty} a_k$ is said to converge to S and we write $\Sigma_{k=1}^{\infty} a_k = S$ if the sequence of partial sums $s_n = \Sigma_{k=1}^{n} a_k$ converge to S.*

The limit z_0 is unique; that is, a sequence can converge to only one point z_0. (This and other properties of limits were discussed in Section 1.4.) A sequence z_n converges iff it is a *Cauchy sequence*; in other words, if, for every $\epsilon > 0$, there is an N such that $n,m \geq N$ implies that $|z_n - z_m| < \epsilon$. (Equivalently, the definition of Cauchy sequence can read: For every $\epsilon > 0$ there is an N such that $n \geq N$ implies that $|z_n - z_{n+p}| < \epsilon$ for every integer $p = 0, 1, 2, \cdots$.) This property of $\mathbb{C} = \mathbb{R}^2$ follows easily from the corresponding property of $\mathbb{R}$ and we shall accept it from advanced calculus.

Corresponding statements for the series $\Sigma_{k=1}^{\infty} a_k$ can be made if we consider the sequence of partial sums $s_n = \Sigma_{k=1}^{n} a_k$. Since $s_{n+p} - s_n = \Sigma_{k=n+1}^{n+p} a_k$, the Cauchy criterion becomes: $\Sigma_{k=1}^{\infty} a_k$ *converges* iff, *for every $\epsilon > 0$, there is an N such that $n \geq N$ implies that $|\Sigma_{k=n+1}^{n+p} a_k| < \epsilon$ for all $p = 1, 2, 3, \cdots$*. As a particular case of the Cauchy criterion, with $p = 1$ we see that *if $\Sigma_{k=1}^{\infty} a_k$ converges, then $a_k \rightarrow 0$.* The converse is not necessarily true (see exercise 1).

As with real series, a complex series $\Sigma_{k=1}^{\infty} a_k$ is said to *converge absolutely* if $\Sigma_{k=1}^{\infty} |a_k|$ converges. From the Cauchy criterion we get

Theorem 1. *If $\Sigma_{k=1}^{\infty} a_k$ converges absolutely, then it converges. (The converse is generally not true.)*

The proof of this theorem is found at the end of the section.

This result is important because $\Sigma_{k=1}^{\infty} |a_k|$ is a *real* series, and the usual tests for real series that we know from calculus can be applied. Some of those tests are included in Theorem 2 (again the proof appears at the end of the section).

Theorem 2.

i. *If $|r| < 1$, then $\Sigma_{n=0}^{\infty} r^n$ converges to $1/(1 - r)$ and diverges (does not converge) if $|r| \geq 1$.*

ii. *Comparison test: If $\Sigma_{k=1}^{\infty} b_k$ converges, $b_k \geq 0$, and $0 \leq a_k \leq b_k$, then $\Sigma_{k=1}^{\infty} a_k$ converges; if $\Sigma_{k=1}^{\infty} c_k$ diverges, $c_k \geq 0$, and $0 \leq c_k \leq d_k$, then $\Sigma_{k=1}^{\infty} d_k$ diverges.*

iii. *p-series test: $\Sigma_{n=1}^{\infty} n^{-p}$ converges if $p > 1$ and diverges to ∞ (that is, the partial sums increase without bound) if $p \leq 1$.*

iv. *Ratio test: Suppose that $\lim\limits_{n \to \infty} \left| \dfrac{a_{n+1}}{a_n} \right|$ exists and is strictly less than 1. Then $\Sigma_{n=1}^{\infty} a_n$ converges absolutely. If the limit is strictly greater than 1, the series diverges. If the limit equals 1, the test fails.*

v. *Root test: Suppose that $\lim\limits_{n \to \infty} (|a_n|)^{1/n}$ exists and is strictly less than 1. Then $\Sigma_{n=1}^{\infty} a_n$ converges absolutely. If the limit is strictly greater than 1, the series diverges; if the limit equals 1, the test fails.*

Uniform convergence

Suppose that $f_n: A \to \mathbb{C}$ is a sequence of functions all defined on the set A. The sequence is said to *converge pointwise* iff, for each $z \in A$, the sequence $f_n(z)$ converges. The limit defines a new function $f(z)$ on A.

A more important kind of convergence is called uniform convergence and is defined as follows.

Definition 2.   *A sequence $f_n: A \to \mathbb{C}$ of functions defined on a set A is said to converge uniformly to a function f if, for any $\epsilon > 0$, there is an N such that $n \geq N$ implies that $|f_n(z) - f(z)| < \epsilon$ for all $z \in A$. This is written $f_n \to f$ uniformly on A.*

A series $\Sigma_{k=1}^{\infty} g_k(z)$ is said to converge pointwise if the corresponding partial sums $s_n(z) = \Sigma_{k=1}^{n} g_k(z)$ converge pointwise. A series $\Sigma_{k=1}^{\infty} g_k(z)$ is said to converge uniformly iff $s_n(z)$ converges uniformly.

Obviously, uniform convergence implies pointwise convergence. The difference between uniform and pointwise convergence is as follows. For pointwise convergence, given $\epsilon > 0$, the N required is allowed to vary from point to point, whereas for uniform convergence we must be able to find a *single N* that works for all z.

The Cauchy criterion for uniform convergence is stated in Theorem 3.

Theorem 3.

i. *A sequence $f_n(z)$ converges uniformly on A iff, for any $\epsilon > 0$, there is an N such that $n \geq N$ implies that $|f_n(z) - f_{n+p}(z)| < \epsilon$ for all $z \in A$ and all $p = 1, 2, 3, \cdots$.*

ii. *A series $\sum_{k=1}^{\infty} g_k(z)$ converges uniformly on A iff, for all $\epsilon > 0$, there is is an N such that $n \geq N$ implies that*

$$\left| \sum_{k=n+1}^{n+p} g_k(z) \right| < \epsilon$$

for all $z \in A$, and all $p = 1, 2, \cdots$.

Theorem 4 states a basic property of uniform convergence.

Theorem 4.   *If f_n are continuous functions defined on A and if $f_n \to f$ uniformly, then f is continuous. Similarly, if $g_k(z)$ are continuous and $g(z) = \sum_{k=1}^{\infty} g_k(z)$ converges uniformly on A, g is continuous on A.*

The Weierstrass M Test

This test is probably the most useful theoretical and practical tool for testing whether a series converges uniformly. It does not always work, but it is effective in a great majority of cases.

Theorem 5 (Weierstrass M Test).   *Let g_n be a sequence of functions on a set $A \subset \mathbb{C}$. Suppose that there is a sequence of real constants $M_n \geq 0$ such that*
 i. $|g_n(z)| \leq M_n$ *for all $z \in A$, and*
 ii. $\sum_{n=1}^{\infty} M_n$ *converges.*
Then $\sum_{n=1}^{\infty} g_n$ converges absolutely and uniformly on A.

Proof.   Since $\sum M_n$ converges, for any $\epsilon > 0$ there is an N such that $n \geq N$ implies that $\sum_{k=n+1}^{n+p} M_k < \epsilon$ for all $p = 1, 2, 3, \cdots$. (Absolute value bars are not needed on $\sum_{k=n+1}^{n+p} M_k$ because $M_n \geq 0$.) Thus $n \geq N$ implies that

$$\left| \sum_{k=n+1}^{n+p} g_k(z) \right| \leq \sum_{k=n+1}^{n+p} |g_k(z)| \leq \sum_{k=n+1}^{n+p} M_k < \epsilon,$$

so by Theorem 3(ii) we have the desired result. ∎

Example.   Consider the series $g(z) = \sum_{n=1}^{\infty} z^n/n$. It will be shown that this series converges uniformly on the sets $A_r = \{z \mid |z| \leq r\}$ for each $0 \leq r < 1$. (We cannot let $r = 1$.) Here $g_n(z) = z^n/n$ and $|g_n(z)| = |z|^n/n \leq r^n/n$ since $|z| \leq r$. Therefore, let $M_n = r^n/n$. But $r^n/n \leq r^n$ and $\sum_{n=1}^{\infty} r^n$ converges for $0 \leq r < 1$. Thus $\sum M_n$ converges, so by the Weierstrass M Test, the given series converges uniformly on A_r. It converges pointwise on $A = \{z \mid |z| < 1\}$ since each $z \in A$ lies in some A_r for r close enough to 1 (see Figure 3.1).

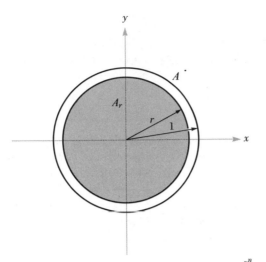

Figure 3.1 Region of convergence of $\sum \dfrac{z^n}{n}$:
uniformly on A_r, pointwise on A.

This series does not, however, converge uniformly on A. Indeed if it did, $\sum x^n/n$ would converge uniformly on $[0,1[$. Suppose that this were true. Then for any $\epsilon > 0$ there would be an N such that $n \geq N$ would imply that

$$\frac{x^n}{n} + \frac{x^{n+1}}{n+1} + \cdots + \frac{x^{n+p}}{n+p} < \epsilon$$

for all $x \in [0,1[$ and $p = 0, 1, 2, \cdots$. But

$$\frac{1}{N} + \frac{1}{N+1} + \cdots$$

diverges to infinity (that is, the partial sums $\longrightarrow \infty$) so we can choose p such that

$$\frac{1}{N} + \cdots + \frac{1}{N+p} > 2\epsilon.$$

Next, we choose x so close to one that $x^{N+p} > 1/2$. Then

$$\frac{x^N}{N} + \cdots + \frac{x^{N+p}}{N+p} > x^{N+p}\Big(\frac{1}{N} + \cdots + \frac{1}{N+p}\Big) > \epsilon,$$

a contradiction.

However, note that $g(z)$ is still continuous on A because it is continuous at each z, since each z lies in some A_r on which we do have uniform convergence.

Series of analytic functions

Theorem 6 is one of the main theorems concerning the convergence of analytic functions. The theorem was formulated by Karl Weierstrass in approximately 1860.

Theorem 6.

i. *Let A be a region in $\mathbb{C}$ and let f_n be a sequence of analytic functions defined on A. If $f_n \rightarrow f$ uniformly on every closed disk contained in A, then f is analytic. Furthermore, $f_n' \rightarrow f'$ pointwise on A and uniformly on every closed disk in A (see Figure 3.2).*
ii. *If g_k are analytic functions on A and $g(z) = \Sigma_{k=1}^{\infty} g_k(z)$ converges uniformly on every closed disk in A, then g is analytic on A and $g'(z) = \Sigma_{k=1}^{\infty} g_k'(z)$ pointwise on A and also uniformly on every closed disk contained in A.*

(See exercise 3 for an alternate way to state the conditions of this theorem and example 1 for a nontrivial application of the theorem.)

Once again we have a theorem that reveals a remarkable property of analytic functions that is not shared by functions of a real variable (compare Section 2.4). Uniform convergence usually is not sufficient to justify differentiation of a series term by term, but for analytic functions it is sufficient.

The proof of Theorem 6 depends on Morera's Theorem and Cauchy's Integral Formula, which were studied in Section 2.4. To prepare for this proof let us first analyze a result concerning integration of sequences and series.

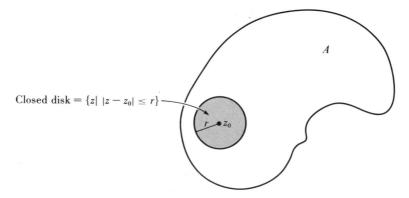

Closed disk $= \{z|\ |z - z_0| \leq r\}$

Figure 3.2   Convergence of analytic functions.

Theorem 7.   *Let* γ: $[a,b] \rightarrow A$ *be a curve in a region A and let* f_n *be a sequence of continuous functions defined on* $\gamma([a,b])$ *converging uniformly to f on* $\gamma([a,b])$. *(Thus f is continuous by Theorem 4.) Then*

$$\int_\gamma f_n \rightarrow \int_\gamma f.$$

Similarly, if $\Sigma_{n=1}^\infty g_n(z)$ *converges uniformly on* γ, *then*

$$\int_\gamma \left(\sum_{n=1}^\infty g_n(z) \right) dz = \sum_{n=1}^\infty \int_\gamma g_n(z) \ dz.$$

Proof of Theorem 7.   Given $\epsilon > 0$, we can choose N such that $n \geq N$ implies that $|f_n(z) - f(z)| < \epsilon$ for all z on γ. Then, by Theorem 3, Chapter 2,

$$\left| \int_\gamma f_n - \int_\gamma f \right| = \left| \int_\gamma f_n - f \right| \leq \int_\gamma |f_n(z) - f(z)| \ |dz| < \epsilon l(\gamma),$$

from which the first assertion follows. The second is obtained by applying the first to the partial sums. (The student should write out the details.)■

Proof of Theorem 6.   As usual, it suffices to prove (i). Let $z_0 \in A$ and let $\{z \mid |z - z_0| \leq r\}$ be a closed disk around z_0 entirely contained in A. (Why does such a disk always exist?) Consider $D(z_0,r)$ $= \{z \mid |z - z_0| < r\}$, which is a simply connected region because it is convex. Since $f_n \rightarrow f$ uniformly on $\{z \mid |z - z_0| \leq r\}$, it is clear that $f_n \rightarrow f$ uniformly on $D(z_0,r)$. We wish to show that f is analytic on $D(z_0,r)$. To do this we use Morera's Theorem (Theorem 28, Chapter 2). By Theorem 4, f is continuous on $D(z_0,r)$. Let γ be any closed curve in $D(z_0,r)$. Since f_n is analytic, $\int_\gamma f_n = 0$ by Cauchy's Theorem and by the fact that $D(z_0,r)$ is simply connected. But by Theorem 7, $\int_\gamma f_n \rightarrow \int_\gamma f$ and so $\int_\gamma f = 0$. Thus by Morera's Theorem f is analytic on $D(z_0,r)$.

We must still show that $f_n' \rightarrow f'$ uniformly on closed disks. To do this we use Cauchy's Integral Formula. Let $B = \{z \mid |z - z_0| \leq r\}$ be a closed disk in A. We can draw a circle γ of radius $\rho > r$ centered at z_0 that contains B entirely in its interior (see p. 150 and Figure 3.3).

For any $z \in B$,

$$f_n'(z) = \frac{1}{2\pi i} \int_\gamma \frac{f_n(\zeta)}{(\zeta - z)^2} \ d\zeta \quad \text{and} \quad f'(z) = \frac{1}{2\pi i} \int_\gamma \frac{f(\zeta)}{(\zeta - z)^2} \ d\zeta$$

by the Cauchy Integral Formula. By hypothesis, $f_n \rightarrow f$ uniformly on the closed disk $\{z \mid |z - z_0| \leq \rho\}$, which lies entirely in A. Then,

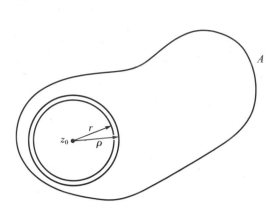

Figure 3.3 A closed disk in an open set
can be slightly enlarged.

given $\epsilon > 0$, we pick N such that $n \geq N$ implies that $|f_n(z) - f(z)| < \epsilon$
for all z in this disk (which we can do by hypothesis). Since γ is the
boundary of this disk, $n \geq N$ implies that $|f_n(\zeta) - f(\zeta)| < \epsilon$ on γ. Let
us note that

$$|f_n'(z) - f'(z)| = \left| \frac{1}{2\pi i} \int_\gamma \frac{f_n(\zeta) - f(\zeta)}{(\zeta - z)^2} \, d\zeta \right|$$

and observe that for ζ on γ and $z \in B$, $|\zeta - z| \geq \rho - r$. Hence $n \geq N$
implies that

$$|f_n'(z) - f'(z)| \leq \frac{1}{2\pi} \cdot \frac{\epsilon}{(\rho - r)^2} \cdot l(\gamma) = \frac{\epsilon \rho}{(\rho - r)^2}.$$

Since ρ and r are fixed constants that are independent of $z \in B$, we
get the desired result.∎

Applying Theorem 6 repeatedly we see that the kth derivatives $f_n^{(k)}$
converge to $f^{(k)}$ uniformly on closed disks in A.

Notice that Theorem 6 does not assume uniform convergence on all
of A. For example, $\sum_{n=1}^{\infty} z^n/n$ on $A = \{z \mid |z| < 1\}$ converges uniformly on
the sets $A_r = \{z \mid |z| \leq r\}$ for $0 \leq r < 1$ (as we saw in the preceding
example) and hence converges uniformly on all closed disks in A. Thus
we can conclude that $\sum z^n/n$ is analytic on A and that the derivative is
$\sum z^{n-1}$, which also converges on A. However, as that example demon-
strated, we do have pointwise but not uniform convergence on A;
convergence is uniform only on each closed subdisk in A.

Technical proofs of Theorems 1–4

Theorem 1. *If $\sum_{k=1}^{\infty} a_k$ converges absolutely, then it converges. (The
converse is generally not true.)*

Proof.   By the Cauchy criterion, given $\epsilon > 0$ there is an N such that $n \geq N$ implies that

$$\sum_{k=n+1}^{n+p} |a_k| < \epsilon, \, p = 1, 2, \, \cdots \, .$$

But

$$\left| \sum_{k=n+1}^{n+p} a_k \right| \leq \sum_{k=n+1}^{n+p} |a_k| < \epsilon$$

by the triangle inequality (see Section 1.2). Thus by the Cauchy criterion, $\Sigma_{k=1}^{\infty} a_k$ converges. ∎

Theorem 2.

i. *If $|r| < 1$, then $\Sigma_{n=0}^{\infty} r^n$ converges to $1/(1-r)$ and diverges (does not converge) if $|r| \geq 1$.*

ii. *Comparison test: If $\Sigma_{k=1}^{\infty} b_k$ converges, $b_k \geq 0$, and $0 \geq a_k \geq b_k$, then $\Sigma_{k=1}^{\infty} a_k$ converges; if $\Sigma_{k=1}^{\infty} c_k$ diverges, $c_k \geq 0$, and $0 \leq c_k \leq d_k$, then $\Sigma_{k=1}^{\infty} d_k$ diverges.*

iii. *p-series test: $\Sigma_{n=1}^{\infty} n^{-p}$ converges if $p > 1$ and diverges to ∞ (that is, the partial sums increase without bound) if $p \leq 1$.*

iv. *Ratio test: Suppose that $\lim\limits_{n \to \infty} \left| \dfrac{a_{n+1}}{a_n} \right|$ exists and is strictly less than 1.*

   Then $\Sigma_{n=1}^{\infty} a_n$ converges absolutely. If the limit is strictly greater than 1, the series diverges. If the limit equals 1, the test fails.

v. *Root test: Suppose that $\lim\limits_{n \to \infty} (|a_n|)^{1/n}$ exists and is strictly less than 1.*

   Then $\Sigma_{n=1}^{\infty} a_n$ converges absolutely. If the limit is strictly greater than 1, the series diverges; if the limit equals 1, the test fails.

Proof.

i. We have, by elementary algebra, that

$$1 + r + r^2 + \cdots + r^n = \frac{1 - r^{n+1}}{1 - r}$$

if $r \neq 1$. Clearly, $r^{n+1} \to 0$ as $n \to \infty$ if $|r| < 1$, and $|r|^{n+1} \to \infty$ if $|r| > 1$. Thus we have convergence if $|r| < 1$ and divergence if $|r| > 1$. Obviously, $\Sigma_{n=0}^{\infty} r^n$ diverges if $|r| = 1$, since $r^n \not\to 0$.

ii. The partial sums of the series $\Sigma_{k=1}^{\infty} b_k$ form a Cauchy sequence and thus the partial sums of the series $\Sigma_{k=1}^{\infty} a_k$ also form a Cauchy sequence, since for any k and p we have $a_k + a_{k+1} + \cdots + a_{k+p} \leq b_k + b_{k+1} + \cdots + b_{k+p}$. Hence $\Sigma_{k=1}^{\infty} a_k$ converges. A positive series can diverge only to $+\infty$, so given $M > 0$, we can find k_0 such that $k \geq k_0$ implies that $c_1 + c_2 + \cdots + c_k \geq M$. Therefore, for $k \geq k_0$, $d_1 + d_2 + \cdots + d_k \geq M$, so $\Sigma_{k=1}^{\infty} d_k$ also diverges to ∞.

iii. First suppose that $p \leq 1$; in this case $1/n^p \geq 1/n$ for all $n = 1$, 2, $\cdots$. Therefore, by (ii), $\Sigma_{n=1}^{\infty} 1/n^p$ will diverge if $\Sigma_{n=1}^{\infty} 1/n$ diverges. (We recall the proof of this from calculus.*) If $s_k = 1/1 + 1/2 + \cdots + 1/k$, then s_k is a strictly increasing sequence of positive real numbers. Write s_{2^k} as follows:

$$s_{2^k} = 1 + \frac{1}{2} + \left(\frac{1}{3} + \frac{1}{4}\right) + \left(\frac{1}{5} + \frac{1}{6} + \frac{1}{7} + \frac{1}{8}\right)$$

$$+ \cdots \left(\frac{1}{2^{k-1}+1} + \cdots + \frac{1}{2^k}\right)$$

$$\geq 1 + \frac{1}{2} + \left(\frac{1}{2}\right) + \left(\frac{1}{2}\right) + \cdots + \left(\frac{1}{2}\right) = 1 + \frac{k}{2}.$$

Hence s_k can be made arbitrarily large if k is made sufficiently large; thus $\Sigma_{n=1}^{\infty} 1/n$ diverges.

Now suppose that $p > 1$. If we let

$$s_k = \frac{1}{1^p} + \frac{1}{2^p} + \frac{1}{3^p} + \cdots + \frac{1}{k^p},$$

then s_k is an increasing sequence of positive real numbers. On the other hand,

$$s_{2^k-1} = \frac{1}{1^p} + \left(\frac{1}{2^p} + \frac{1}{3^p}\right) + \left(\frac{1}{4^p} + \frac{1}{5^p} + \frac{1}{6^p} + \frac{1}{7^p}\right) \cdots$$

$$+ \left(\frac{1}{(2^{k-1})^p} + \cdots + \frac{1}{(2^k-1)^p}\right) \leq \frac{1}{1^p} + \frac{2}{2^p} + \frac{4}{4^p} + \frac{2^{k-1}}{(2^{k-1})^p}$$

$$= \frac{1}{1^{p-1}} + \frac{1}{2^{p-1}} + \frac{1}{4^{p-1}} + \cdots \frac{1}{(2^{k-1})^{p-1}} < \frac{1}{1 - \dfrac{1}{2^{p-1}}}$$

(Why?). Thus the sequence $\{s_k\}$ is bounded from above by $1/(1 - 1/2^{p-1})$; hence $\Sigma_{n=1}^{\infty} 1/n^p$ converges.

iv. Suppose that $\lim\limits_{n \to \infty} \left|\dfrac{a_{n+1}}{a_n}\right| = r < 1$. Choose r' such that $r < r' < 1$ and let N be such that $n \geq N$ implies that

$$\left|\frac{a_{n+1}}{a_n}\right| < r'.$$

Then $|a_{N+p}| < |a_N|(r')^p$. Consider the series $|a_1| + \cdots + |a_N| + |a_N| r' + |a_N| (r')^2 + |a_N| (r')^3 + \cdots$. This converges to

$$|a_1| + \cdots + |a_{N-1}| + \frac{|a_N|}{1 - r'}.$$

*We can also prove (iii) by using the integral test for positive series (see any calculus text). The demonstration given here also proves the *Cauchy Condensation Test*: Let Σa_n be a series of positive terms with $a_{n+1} \leq a_n$. Then Σa_n converges iff $\Sigma_{j=1}^{\infty} 2^j a_{2^j}$ converges (see G. J. Porter, "An Alternative to the Integral Test for Infinite Series," *American Mathematical Monthly* 79 (1972), page 634).

By (ii) we can conclude that $\Sigma_{k=1}^{\infty} |a_k|$ converges. If $\lim\limits_{n \to \infty} \left| \frac{a_{n+1}}{a_n} \right|$
$= r > 1$, choose r' such that $1 < r' < r$ and let N be such that
$n \geq N$ implies that $\left| \frac{a_{n+1}}{a_n} \right| > r'$. Hence $|a_{N+p}| > (r')^p |a_N|$, so
$\lim\limits_{n \to \infty} |a_N| = \infty$, whereas the limit would have to be zero if the sum
converged (see exercise 1). Thus $\Sigma_{k=1}^{\infty} a_k$ diverges. To see that the
test fails if $\lim\limits_{n \to \infty} \left| \frac{a_{n+1}}{a_n} \right| = 1$, consider the series $1+1+1+\cdots$,
and $\Sigma_{n=1}^{\infty} 1/n^p$ for $p > 1$. In both cases $\lim\limits_{n \to \infty} \left| \frac{a_{n+1}}{a_n} \right| = 1$, but the
first series diverges and the second converges.

v. Suppose that $\lim\limits_{n \to \infty} (|a_n|)^{1/n} = r < 1$. Choose r' such that $r < r' < 1$
and N such that $n \geq N$ implies that $|a_n|^{1/n} < r'$; in other words,
$|a_n| < (r')^n$. The series $|a_1| + |a_2| + \cdots + |a_{N-1}| + (r')^N$
$+ (r')^{N+1} + \cdots$ converges to $|a_1| + |a_2| + \cdots + |a_{N-1}|$
$+ (r')^N/(1 - r')$, so by (ii), $\Sigma_{k=1}^{\infty} a_k$ converges. If $\lim\limits_{n \to \infty} (|a_n|)^{1/n}$
$= r > 1$, choose $1 < r' < r$ and N such that $n \geq N$ implies that
$|a_n|^{1/n} > r'$ or, in other words, $|a_n| > (r')^n$. Hence $\lim\limits_{n \to \infty} |a_n| = \infty$.
Therefore, $\Sigma_{k=1}^{\infty} a_k$ diverges.

To show that the test fails when $\lim\limits_{n \to \infty} (|a_n|)^{1/n} = 1$, observe that,
by elementary analysis,

$$\lim_{n \to \infty} \left(\frac{1}{n} \right)^{1/n} = 1 \quad \text{and} \quad \lim_{n \to \infty} \left(\frac{1}{n^2} \right)^{1/n} = 1$$

(take logarithms and use the fact that $(\log x)/x \to 0$ as $x \to \infty$).
But $\Sigma_{n=1}^{\infty} 1/n$ diverges and $\Sigma_{n=1}^{\infty} 1/n^2$ converges. ∎

Theorem 3.

i. *A sequence $f_n(z)$ converges uniformly on A iff, for any $\epsilon > 0$, there is
an N such that $n \geq N$ implies that $|f_n(z) - f_{n+p}(z)| < \epsilon$ for all $z \in A$
and all $p = 1, 2, 3, \cdots$.*

ii. *A series $\Sigma_{k=1}^{\infty} g_k(z)$ converges uniformly on A iff, for all $\epsilon > 0$, there is
an N such that $n \geq N$ implies that*

$$\left| \sum_{k=n+1}^{n+p} g_k(z) \right| < \epsilon$$

for all $z \in A$, and $p = 1, 2, \cdots$.

Proof.

i. Suppose that the Cauchy criterion holds. Let $f(z) = \lim\limits_{n \to \infty} f_n(z)$,
which exists because $f_n(z)$ converges pointwise. We wish to show
that $f_n \to f$ uniformly on A. Given $\epsilon > 0$, choose N such that

$|f_n(z) - f_{n+p}(z)| < \epsilon/2$, for $n \geq N$ and all $p \geq 1$. The first step is to show that for any z and any $n \geq N$, $|f_n(z) - f(z)| < \epsilon$. For $z \in A$, choose p large enough so that $|f_{n+p}(z) - f(z)| < \epsilon/2$, which is possible by pointwise convergence. Then, by the triangle inequality, $|f_n(z) - f(z)| \leq |f_n(z) - f_{n+p}(z)| + |f_{n+p}(z) - f(z)| < \epsilon/2 + \epsilon/2 = \epsilon$. (Notice that although p depends on z, N does not.)

Conversely, if $f_n \to f$ uniformly, given $\epsilon > 0$ choose N such that $n \geq N$ implies that $|f_n(z) - f(z)| < \epsilon/2$ for all z. Since $n + p \geq N$, $|f_n(z) - f_{n+p}(z)| \leq |f_n(z) - f(z)| + |f(z) - f_{n+p}(z)| < \epsilon/2 + \epsilon/2 = \epsilon$.

By applying (i) to the partial sums, we can immediately deduce (ii). ■

Theorem 4. *If f_n are continuous on A and $f_n \to f$ uniformly, then f is continuous. Similarly, if $g_k(z)$ are continuous and $g(z) = \Sigma_{k=1}^{\infty} g_k(z)$ converges uniformly on A, g is continuous on A.*

Proof.   It suffices to prove the assertion for sequences. We wish to show that for $z_0 \in A$, given $\epsilon > 0$, there is a $\delta > 0$ such that $|z - z_0| < \delta$ implies that $|f(z) - f(z_0)| < \epsilon$. Choose N such that $|f_N(z) - f(z)| < \epsilon/3$ for all $z \in A$. Since f_N is continuous, there is a $\delta > 0$ such that $|f_N(z) - f_N(z_0)| < \epsilon/3$ if $|z - z_0| < \delta$. Thus $|f(z) - f(z_0)| \leq |f(z) - f_N(z)| + |f_N(z) - f_N(z_0)| + |f_N(z_0) - f(z_0)| < \epsilon/3 + \epsilon/3 + \epsilon/3 = \epsilon$. ■

Note that in the last step we need an N that is independent of z so we can conclude that both $|f_N(z) - f(z)| < \epsilon/3$ and $|f_N(z_0) - f(z_0)| < \epsilon/3$.

Worked Examples

1. Show that the *Riemann ζ function*, defined by

$$\zeta(z) = \sum_{n=1}^{\infty} n^{-z},$$

is analytic on the region $A = \{z|\ \mathrm{Re}\ z > 1\}$. Compute $\zeta'(z)$ on that set.

Solution.   We use Theorem 6. We must be careful to try to prove uniform convergence only on closed disks in A and not on all of A. In this example we do not in fact have uniform convergence on all of A (see exercise 9).

Let B be a closed disk in A and let δ be its distance from the line $\mathrm{Re}\ z = 1$ (Figure 3.4).

We shall show that $\Sigma_{n=1}^{\infty} n^{-z}$ converges uniformly on B. Here $n^{-z} = e^{-z \log n}$, where $\log n$ means the usual log of real numbers. Now $|n^{-z}| = |e^{-z \log n}| = e^{-x \log n} = n^{-x}$. But $x \geq 1 + \delta$ if $z \in B$, so $|n^{-z}| \leq n^{-(1+\delta)}$ for all $z \in B$. Let us, therefore, choose $M_n = n^{-(1+\delta)}$.

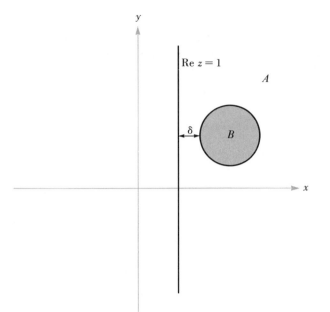

Figure 3.4

By Theorem 2 (iii), $\Sigma_{n=1}^{\infty} M_n$ converges. Thus by the Weierstrass M Test, our series $\Sigma_{n=1}^{\infty} n^{-z}$ converges uniformly on B. Hence by Theorem 6, ζ is analytic on A. Also by Theorem 6, we can differentiate term by term to give

$$\zeta'(z) = -\sum_{n=1}^{\infty} (\log n)n^{-z},$$

which we know must also converge on A (and uniformly on closed disks in A).

2. Show that

$$f(z) = \sum_{n=1}^{\infty} \frac{z^n}{n^2}$$

is analytic on $A = \{z| \ |z| < 1\}$. Write a series for $f'(z)$.

Solution. Again we use Theorem 6 (although this is a power series that can be dealt with more easily after the student has read Section 3.2).

In this case we actually have uniform convergence on all of A. Let $M_n = 1/n^2$. Clearly, ΣM_n converges and $|z^n/n^2| < 1/n^2 = M_n$ for all $z \in A$. Thus, by the Weierstrass M Test, $\Sigma_{n=1}^{\infty} z^n/n^2$ converges uniformly on A; therefore, the series converges on closed disks in A. By Theorem 6 the sum $\Sigma_{n=1}^{\infty} z^n/n^2$ is an analytic function on A. Furthermore,

$$f'(z) = \sum_{n=1}^{\infty} \frac{nz^{n-1}}{n^2} = \sum_{n=1}^{\infty} \frac{z^{n-1}}{n}.$$

Note.   This series for $f'(z)$ does not converge for $z = 1$, so f cannot be extended to be analytic on any region containing the closed unit disk.

3. Compute

$$\int_\gamma \left(\sum_{n=-1}^{\infty} z^n \right) dz$$

where γ is a circle of radius 1/2.

Solution.   Let B be a closed disk in $A = \{z \mid |z| < 1\}$ a distance of δ from the circle $|z| = 1$. For $z \in B$, $|z^n| = (|z|)^n \leq (1 - \delta)^n$ for $n \geq 0$. We choose $M_n = (1 - \delta)^n$ and note that $\Sigma \, M_n$ is convergent. Therefore, $\Sigma_{n=0}^{\infty} z^n$ is uniformly convergent on B, so by Theorem 6, $\Sigma_{n=0}^{\infty} z^n$ is analytic on A. Therefore,

$$\int_\gamma \left(\sum_{n=-1}^{\infty} z^n \right) dz = \int_\gamma \frac{1}{z} \, dz + \int_\gamma \left(\sum_{n=0}^{\infty} z^n \right) dz = 2\pi i$$

by example 3 (Section 2.1) and Cauchy's Theorem.

Exercises

1. If $\Sigma_{k=1}^{\infty} a_k$ converges, prove that $a_k \to 0$. If $\Sigma_{k=1}^{\infty} g_k(z)$ converges uniformly, show that $g_k \to 0$ uniformly.

2. If $\Sigma_{k=1}^{\infty} g_k(z)$ is a uniformly convergent series of continuous functions and if $z_n \to z$, show that

$$\lim_{n \to \infty} \sum_{k=1}^{\infty} g_k(z_n) = \sum_{k=1}^{\infty} g_k(z).$$

3. Prove: $f_n \to f$ uniformly on every closed disk in a region A iff $f_n \to f$ uniformly on every compact (closed and bounded) subset of A.

4. Show that $\Sigma_{n=1}^{\infty} (\log n)^k n^{-z}$ is analytic on $\{z \mid \text{Re } z > 1\}$.

Hint.   Use the result of example 1.

5. Let f_n be analytic on a bounded region A and continuous on $\text{cl}(A)$. Suppose that f_n converge uniformly on $\text{bd}(A)$. Then prove that f_n converge uniformly to an analytic function on A.

Hint.   Use the Maximum Modulus Theorem.

6. Show that $\Sigma_{n=1}^{\infty} e^{-n} \sin nz$ is analytic in the region $A = \{z \mid -1 < \text{Im } z < 1\}$.

7. Find a suitable region in which

$$\sum_{n=1}^{\infty} \frac{(z - 1)^n}{n}$$

is analytic.

8.   Show that

$$\sum_{n=1}^{\infty} \frac{1}{n!z^n}$$

is analytic on $\mathbb{C}\backslash\{0\}$. Compute its integral around the unit circle.

9.   Prove that

$$\zeta(z) = \sum_{n=1}^{\infty} \frac{1}{n^z}$$

does not converge uniformly on $A = \{z \mid \text{Re } z > 1\}$.

10.   If $f_n(z) \rightarrow f(z)$ uniformly on a region A, and if f_n is analytic on A, is it true that $f_n'(z) \rightarrow f'(z)$ uniformly on A?

Hint.   See example 2.

11.   Show that

$$\sum_{n=1}^{\infty} \frac{1}{z^n}$$

is analytic on $A = \{z \mid |z| > 1\}$.

12.   Let f be an analytic function on the disk $D(0,2)$ such that $|f(z)| \leq 7$ for all $z \in D(0,2)$. Prove that there exists a $\delta > 0$ such that if $z_1, z_2 \in D(0,1)$, and if $|z_1 - z_2| < \delta$, then $|f(z_1) - f(z_2)| < 1/10$. Find a numerical value of δ independent of f that has this property.

Hint.   Use the Cauchy Integral Formula.

13.   What is the limit of the sequence $f_n(x) = (1+x)^{1/n}$, $x \geq 0$? Does it converge uniformly?

14.   Do the following sequences converge, and if so, what is their limit?

a. $z_n = (-1)^n + \dfrac{i}{n+1}$.

b. $z_n = \dfrac{n!}{n^n} i^n$.

15.   Test the following series for absolute convergence and convergence:

a. $\displaystyle\sum_{2}^{\infty} \frac{i^n}{\log n}$.

b. $\displaystyle\sum_{1}^{\infty} \frac{i^n}{n}$.

16.   Prove that the series

$$\sum_{n=1}^{\infty} \frac{z^n}{1 + z^{2n}}$$

converges in both the interior and exterior of the unit circle and represents an analytic function in each region.

3.2   Power Series and Taylor's Theorem

This section will consider a special kind of series called power series. These series have the form $\sum_{n=0}^{\infty} a_n(z - z_0)^n$. We shall examine their convergence properties and show that a function is analytic iff it is locally representable as a convergent power series. To obtain this representation, we first need to establish Taylor's Theorem, which asserts that if f is analytic on an open disk centered at z_0, then the Taylor series of f,

$$\sum_{n=0}^{\infty} \frac{f^{(n)}(z_0)}{n!}\, (z - z_0)^n,$$

converges on the disk and

$$f(z) = \sum_{n=0}^{\infty} \frac{f^{(n)}(z_0)}{n!}\, (z - z_0)^n$$

everywhere on that disk.

Again this is a remarkable property that is not shared by functions of a real variable; a function $f(x)$, $x \in \mathbb{R}$, even if infinitely differentiable, need not have a convergent Taylor series representation. For example:

$$f(x) = \begin{cases} e^{-1/x^2}, & x \neq 0; \\ 0, & x = 0. \end{cases}$$

We can show that f is infinitely differentiable and that all the derivatives of f are zero at $x = 0$ but f is not identically zero, so f does not equal its Taylor series.

In proving the results of this section we shall use the techniques developed in Section 3.1 and Cauchy's Integral Formula. Many of the results concerning the convergence of power series can be proved without using the theory developed in Chapter 2, but this method is more difficult.

Convergence of power series

A *power series* is a series of the form $\sum_{n=0}^{\infty} a_n(z - z_0)^n$. (Here a_n and $z_0 \in \mathbb{C}$ are fixed complex numbers.) Each term $a_n(z - z_0)^n$ is entire, so in proving that the sum is analytic on a region we can use Theorem 6. The basic fact to remember about power series is that the appropriate domain of analyticity is the interior of a circle centered at z_0. This is established in Theorem 8.

Theorem 8.   *Let $\sum_{n=0}^{\infty} a_n(z - z_0)^n$ be a given power series. There is a unique number $R \geq 0$, possibly $+\infty$, called the radius of convergence* *such that if $|z - z_0| < R$, the series converges, and if $|z - z_0| > R$, the series diverges. Furthermore, the convergence is uniform and absolute on every closed disk in $A = \{z \in \mathbb{C} \mid |z - z_0| < R\}$. No general statement about convergence can be made if $|z - z_0| = R$.*

Thus on the region $A = \{z \in \mathbb{C} \mid |z - z_0| < R\}$, the series converges and we have divergence at z if $|z - z_0| > R$. The circle $|z - z_0| = R$ is called the *circle of convergence* of the given power series (see Figure 3.5). Practical methods of calculating R are described in Theorem 11.

Proof.   Set $R = \sup\{r \geq 0 \mid \Sigma_{n=0}^{\infty} |a_n| \, r^n$ converges$\}$ where sup means the least upper bound of that set of real numbers. We shall show that R has the desired properties. The following lemma is useful in this regard.

Lemma 1 (Abel-Weierstrass Lemma).   *Suppose that $r_0 \geq 0$ and that $|a_n| \, r_0^n \leq M$ for all n, where M is some constant. Then for $r < r_0$, $\Sigma_{n=0}^{\infty} a_n(z - z_0)^n$ converges uniformly and absolutely on the closed disk $A_r = \{z \mid |z - z_0| \leq r\}$.*

Proof.   For $z \in A_r$ we have

$$|a_n(z - z_0)^n| \leq |a_n| \, r^n \leq M\left(\frac{r}{r_0}\right)^n.$$

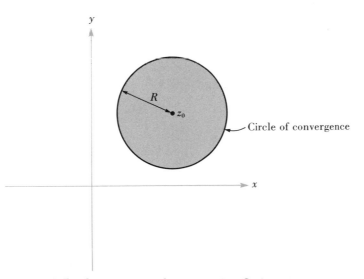

Figure 3.5   Convergence of power series. Series converges within circle; series diverges outside circle.

Let

$$M_n = M\left(\frac{r}{r_0}\right)^n.$$

Since $r/r_0 < 1$, $\Sigma\, M_n$ converges. Thus by the Weierstrass M Test (Theorem 5), the series converges uniformly and absolutely on A_r. ∎

Now we can prove the first part of Theorem 8. Let $r_0 < R$. By the definition of R there is an $r_1, r_0 < r_1 \leq R$ such that $\Sigma\, |a_n|\, r_1^n$ converges. Then $\Sigma_{n=0}^{\infty} |a_n|\, r_0^n$ converges, by the comparison test. The terms $|a_n|\, r_0^n$ are bounded (in fact $\to 0$), so by Lemma 1 the series converges uniformly and absolutely on A_r for any $r < r_0$. Since any z with $|z - z_0| < R$ lies in some A_r and since we can always choose r_0 such that $r < r_0 < R$, we have convergence at z.

Now suppose that $|z_1 - z_0| > R$ and $\Sigma\, a_n(z_1 - z_0)^n$ converges. We shall derive a contradiction. The terms $a_n(z_1 - z_0)^n$ are bounded in absolute value because they approach zero. Thus, by Lemma 1, if $R < r < |z_1 - z_0|$, then $\Sigma\, a_n(z_1 - z_0)^n$ converges absolutely if $z_1 \in A_r$. Therefore, $\Sigma\, |a_n|\, r^n$ converges. But this would mean, by definition of R, that $R < R$.

We have proved that the convergence is uniform and absolute on every *strictly smaller* closed disk A_r and hence on any closed disk in A. ∎

From Theorems 6 and 8 we may immediately deduce the following theorem.

Theorem 9. *A power series $\Sigma_{n=0}^{\infty} a_n(z - z_0)^n$ is an analytic function on the inside of its circle of convergence.*

We also know (by Theorem 6) that we can differentiate term by term. In fact, we have:

Theorem 10. *Let $f(z) = \Sigma_{n=0}^{\infty} a_n(z - z_0)^n$ be the analytic function defined on the inside of the circle of convergence of the given power series. Then $f'(z) = \Sigma_{n=1}^{\infty} na_n(z - z_0)^{n-1}$, and this series has the same circle of convergence as $\Sigma\, a_n(z - z_0)^n$. Furthermore, the coefficients a_n are given by $a_n = f^{(n)}(z_0)/n!$.*

Proof.   We know from Theorem 6 that $f'(z) = \Sigma_{n=0}^{\infty} na_n(z - z_0)^{n-1}$ converges on $A = \{z \in \mathbb{C}|\ |z - z_0| < R\}$. To show that the derived series has the same circle of convergence as the original series, we need only show that it diverges for $|z_1 - z_0| > R$. But if it converged, $na_n r_0^{n-1}$ would be bounded where $r_0 = |z_1 - z_0|$. Thus $a_n r_0^n$ would be bounded as well, so by the Abel-Weierstrass Lemma, $\Sigma\, a_n(z - z_0)^n$ would converge if $R < |z - z_0| < r_0$, thus contradicting the property of R in Theorem 8. In the formula defining $f(z)$, we set $z = z_0$ so that $f(z_0) = a_0$. We obtain $f'(z_0) = a_1$ in the same way, from the formula for $f'(z)$. Proceeding inductively we obtain

$$f^{(k)}(z) = \sum_{n=k}^{\infty} n(n-1) \cdots (n-k+1)a_n(z-z_0)^{n-k},$$

and setting $z = z_0$ we get $f^{(k)}(z_0) = k!\, a_k.$ ∎

We will now obtain some practical methods of computing the radius of convergence R. (The method of defining R given in the proof of Theorem 8 is not useful for computing R in specific examples.)

Theorem 11.   *Consider the power series $\Sigma_{n=0}^{\infty} a_n(z - z_0)^n$.*
i. *If*

$$\lim_{n \to \infty} \frac{|a_n|}{|a_{n+1}|}$$

exists, it equals R, the radius of convergence of the series.
ii. *If $\rho = \lim\limits_{n \to \infty} \sqrt[n]{|a_n|}$ exists, then $R = 1/\rho$ is the radius of convergence.*
(Set $R = \infty$ if $\rho = 0$; set $R = 0$ if $\rho = \infty$.)

We shall refer to (i) of Theorem 11 as the "ratio test" for the radius of convergence; (ii) of that theorem will be referred to as the "root test" for the radius of convergence.

Proof.   To prove both cases we must show that $R = \sup\{r \geq 0 \mid |\Sigma_{n=0}^{\infty}\, |a_n|\, r^n < \infty\}$.
i. By Theorem 2 we know that $\Sigma_{n=0}^{\infty}\, |a_n|\, r^n$ converges or diverges as

$$\lim_{n \to \infty} \frac{|a_{n+1} r^{n+1}|}{|a_n r^n|} < 1 \text{ or } > 1,$$

that is, according to whether

$$\lim_{n \to \infty} \frac{|a_n|}{|a_{n+1}|} > r$$

or

$$\lim_{n \to \infty} \frac{|a_n|}{|a_{n+1}|} < r.$$

Thus, by the characterization of R in Theorem 8, the limit equals R.

ii. By Theorem 2 we know that $\Sigma_{n=0}^{\infty} |a_n| \, r^n$ converges or diverges as $\lim_{n\to\infty} (|a_n| \, r^n)^{1/n} < 1$ or > 1; that is, according to whether $r < 1/\lim_{n\to\infty} |a_n|^{1/n}$ or $r > 1/\lim_{n\to\infty} |a_n|^{1/n}$. The result follows as in (i). ∎

For example:

The series $\Sigma_{n=0}^{\infty} z^n$ has radius of convergence 1 since $a_n = 1$, and thus $\lim_{n\to\infty} |a_n/a_{n+1}| = 1$.

The series $\Sigma_{n=0}^{\infty} z^n/n!$ has radius of convergence $R = +\infty$ (that is, the function is entire), since $a_n = 1/n!$, and so $|a_n/a_{n+1}| = n + 1 \to \infty$.

The series $\Sigma_{n=0}^{\infty} n! \, z^n$ has radius of convergence $R = 0$ because $|a_n/a_{n+1}| = 1/(n + 1) \to 0$. (This function thus does not have a region of analyticity.)

Remark.   By refining the root test, we can show that $R = 1/\rho$ where $\rho = \lim_{n\to\infty} \sup \sqrt[n]{|a_n|}$, which always exists ($\lim_{n\to\infty} \sup c_n = \lim_{n\to\infty} (\sup \{c_n, c_{n+1}, \cdots\})$, by definition). There is no analogous refinement for the ratio test.

Taylor's Theorem

It is obvious from the preceding computations that if $f: A \to \mathbb{C}$ equals, in a small disk around each $z_0 \in A$, a convergent power series, then f is analytic. The converse is also true: If f is analytic it equals, on every disk in its domain, a convergent power series. This is made clear by Theorem 12.

Theorem 12.   *Let f be analytic on a region A. Let $z_0 \in A$ and let $A_r = \{z \mid |z - z_0| < r\}$ be contained in A — usually the largest open disk possible is used: If $r = \infty$, $A_r = A = \mathbb{C}$ (see Figure 3.6). Then for every $z \in A_r$,*

$$\sum_{n=0}^{\infty} \frac{f^{(n)}(z_0)}{n!} (z - z_0)^n$$

converges on A_r (that is, has a radius of convergence $\geq r$), and we have

$$f(z) = \sum_{n=0}^{\infty} \frac{f^{(n)}(z_0)}{n!} (z - z_0)^n. \tag{1}$$

(We use the convention $0! = 1$.) The series (1) is called the Taylor series of f around the point z_0.

Before proving this result let us study an example that illustrates its usefulness. Consider $f(z) = e^z$. Here f is analytic, and $f^{(n)}(z) = e^z$ for all

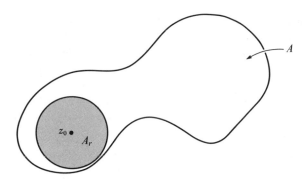

Figure 3.6 Taylor's Theorem.

n, so $f^{(n)}(0) = 1$ and thus

$$e^z = \sum_{n=0}^{\infty} \frac{z^n}{n!}, \tag{2}$$

which is valid for all $z \in \mathbb{C}$, since e^z is entire. Formula (2) is very useful. Table 3.1 lists the Taylor series of some common elementary functions. The proofs will be left for the examples and exercises since they are similar to the derivation of formula (2). The Taylor series around the point $z_0 = 0$ is sometimes called the *Maclaurin series*.

Table 3.1 Some Common Expansions

Function	*Taylor Series around 0*	*Where Valid*
e^z	$\displaystyle\sum_{n=0}^{\infty} \frac{z^n}{n!}$	all z
$\sin z$	$z - \dfrac{z^3}{3!} + \dfrac{z^5}{5!} - \cdots = \displaystyle\sum_{n=1}^{\infty} (-1)^{n+1}\dfrac{z^{2n-1}}{(2n-1)!}$	all z
$\cos z$	$1 - \dfrac{z^2}{2} + \dfrac{z^4}{4!} - \dfrac{z^6}{6!} + \cdots = \displaystyle\sum_{n=0}^{\infty} (-1)^n \dfrac{z^{2n}}{(2n)!}$	all z
$\log(1+z)$ (principal branch)	$\displaystyle\sum_{n=1}^{\infty} \frac{(-1)^{n-1}}{n} z^n$	$|z| < 1$
$(1+z)^\alpha$ (principal branch with $\alpha \in \mathbb{C}$ fixed)	$\displaystyle\sum_{n=0}^{\infty} \binom{\alpha}{n} z^n$ (binomial series), where $\binom{\alpha}{n} = \dfrac{\alpha(\alpha-1)\,\cdots\,(\alpha-n+1)}{n!}$, take to be zero if α is an integer $< n$ and $\binom{\alpha}{0} = 1$.	$|z| < 1$

The binomial series in this table is a useful one to know and may be familiar from high school algebra (where it is often used without proof).

The proof of Theorem 12, which follows, is based on the Cauchy Integral Formula.

Proof.   Let $0 < \sigma < r$ and let γ be the circle $\gamma(t) = z_0 + \sigma e^{it}, 0 \le t \le 2\pi$. By Cauchy's Integral Formula, if z is any point on the inside of γ, then

$$f(z) = \frac{1}{2\pi i} \int_\gamma \frac{f(\zeta)}{\zeta - z} \, d\zeta.$$

We shall next rewrite $1/(\zeta - z)$ as a power series in $z - z_0$. To do so we must interrupt this proof to prove the following lemma.

Lemma 2.   *The power series $\Sigma_{n=0}^{\infty} z^n$ converges uniformly on and inside any circle of radius $\rho < 1$ to $1/(1 - z)$.*

Proof.   We use the Weierstrass M Test (Theorem 5). Let $A_\rho = \{z| \; |z| \le \rho\}$. Let $M_n = \rho^n$ so that as $0 \le \rho < 1$, $\Sigma \, M_n$ converges and $|z^n| = |z|^n \le \rho^n$ for $z \in A_\rho$. Thus the series $\Sigma \, z^n$ converges uniformly on A_ρ. The sum is $(1 - z)^{-1}$ as in Theorem 2(i).

Continuing with the proof of Theorem 12, we note that

$$\frac{1}{\zeta - z} = \frac{1}{\zeta - z_0} \cdot \frac{1}{\left(1 - \dfrac{z - z_0}{\zeta - z_0}\right)} = \frac{1}{\zeta - z_0} \sum_{n=0}^{\infty} \left(\frac{z - z_0}{\zeta - z_0}\right)^n$$

by Lemma 2; we have convergence since z is a point inside γ, and thus

$$\left|\frac{z - z_0}{\zeta - z_0}\right| < 1.$$

Also by the lemma, the convergence of

$$\sum_{n=0}^{\infty} \left(\frac{z - z_0}{\zeta - z_0}\right)^n$$

is uniform in ζ with ζ traversing the circle γ and z fixed. Next we note that on γ, $f(\zeta)/(\zeta - z_0)$ is a continuous function in ζ, and so $f(\zeta)/(\zeta - z_0)$ is bounded. It follows that

$$\sum_{n=0}^{\infty} \frac{f(\zeta)(z - z_0)^n}{(\zeta - z_0)^{n+1}}$$

converges on γ uniformly in ζ to $f(\zeta)/(\zeta - z)$. (The student is asked to verify this in exercise 21.) By Theorem 7 we thus have

$$\int_\gamma \frac{f(\zeta)}{\zeta - z} \, d\zeta = \sum_{n=0}^\infty \int_\gamma \frac{f(\zeta)(z - z_0)^n}{(\zeta - z_0)^{n+1}} \, d\zeta.$$

Hence

$$f(z) = \frac{1}{2\pi i} \int_\gamma \frac{f(\zeta)}{\zeta - z} \, d\zeta = \sum_{n=0}^\infty \frac{(z - z_0)^n}{2\pi i} \int_\gamma \frac{f(\zeta)}{(\zeta - z_0)^{n+1}} \, d\zeta,$$

which, by the Cauchy Integral Formula for Derivatives, equals

$$\sum_{n=0}^\infty (z - z_0)^n \frac{f^{(n)}(z_0)}{n!}.$$

Since the radius σ of the circle γ was arbitrary, this representation of $f(z)$ is valid in the largest open disk around z_0 contained in the region A. ■

The following consequence of Theorem 12 was mentioned informally at the beginning of this section.

Corollary 1.   *Let A be a region in $\mathbb{C}$ and let $f: A \subset \mathbb{C} \to \mathbb{C}$ be defined on A. Then f is analytic on A iff, for each $z_0 \in A$, there is a disk $D(z_0,r) = \{z \mid |z - z_0| < r\}$ (for some fixed r) contained in A such that f equals a convergent power series on $D(z_0,r)$.*

Proof.   Theorem 12 proves that every analytic function equals a power series (the Taylor series) on every disk in A. On the other hand, if $f(z)$ equals a convergent power series on $D(z_0,r)$, then $D(z_0,r)$ must be contained in the inside of the circle of convergence of the power series and so f must be analytic on $D(z_0,r)$. Since there is such a disk and convergent power series for each $z_0 \in A$, it follows from Theorem 9 that f is analytic on A. ■

The condition in Corollary 1 thus may be taken as an alternative definition of "analytic."

Cauchy's Theorem, Cauchy's integral formulas, and Taylor's Theorem are among the most fundamental theorems of complex analysis.

In attempting to find the Taylor series in concrete examples in which the derivatives of f are complicated, it is sometimes easiest to search for a convergent series that represent f directly rather than computing the derivatives $f^{(k)}(z_0)$. If $f(z) = \Sigma_{n=0}^\infty a_n(z - z_0)^n$ *and the series converges, it must be the Taylor series by Theorem 10.* A few such tricks are found in example 4.

Worked Examples

1. Can a power series $\Sigma\, a_n(z - 2)^n$ converge at $z = 0$ but diverge at $z = 3$?

Solution.   No. If it converges at $z = 0$ this means, by Theorem 8, that the radius of convergence is $R \geq 2$. But $z = 3$ lies inside that circle, so the series would converge there (see Figure 3.7).

2. Suppose that $\Sigma\, a_n z^n$ and $\Sigma\, b_n z^n$ have radii of convergence $\geq r_0$. Define $c_n = \Sigma_{k=0}^{n} a_k b_{n-k}$. Prove that $\Sigma\, c_n z^n$ has radius of convergence $\geq r_0$ and that inside this circle of radius r_0, we have

$$\sum_{n=0}^{\infty} c_n z^n = \left(\sum_{n=0}^{\infty} a_n z^n \right)\left(\sum_{n=0}^{\infty} b_n z^n \right).$$

Solution.   This way of multiplying out two power series is a generalization of the manner in which polynomials are multiplied. A direct proof can be given but would be somewhat lengthy. If we use Taylor's Theorem, the proof is fairly simple. Let $f(z) = \Sigma_{n=0}^{\infty} a_n z^n$, $g(z) = \Sigma_{n=0}^{\infty} b_n z^n$, and $A = \{z|\ |z| < r_0\}$. Then f and g are analytic on A, so fg is also analytic on A. By Taylor's Theorem we can write

$$(f \cdot g)(z) = \sum_{n=0}^{\infty} \frac{(f \cdot g)^{(n)}(0)}{n!}\, z^n$$

for all z in A. It is a simple exercise (as in calculus) to show by induction that the nth derivative of the product $f(z)\, g(z)$ is given by

$$(f \cdot g)^{(n)}(z) = \sum_{k=0}^{n} \binom{n}{k} f^{(k)}(z)\, g^{(n-k)}(z)$$

where

$$\binom{n}{k} = \frac{n!}{k!(n-k)!}.$$

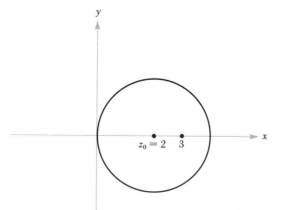

Figure 3.7

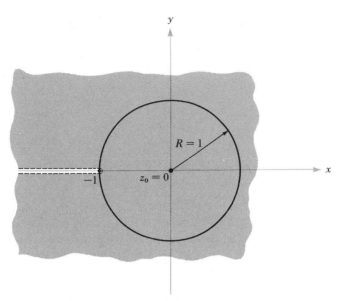

Figure 3.8 Taylor series of $\log(1 + z)$.

Hence

$$\frac{(f \cdot g)^{(n)}(0)}{n!} = \sum_{k=0}^{n} \frac{1}{k!(n-k)!} f^{(k)}(0)\, g^{(n-k)}(0)$$

$$= \sum_{k=0}^{n} a_k\, b_{n-k}.$$

Thus $\sum_{n=0}^{\infty} c_n z^n$ converges on A (and therefore, by Taylor's Theorem, the radius of convergence is $\geq r_0$) and on A

$$\sum_{n=0}^{\infty} c_n z^n = (f \cdot g)(z) = \left(\sum_{n=0}^{\infty} a_n z^n \right)\left(\sum_{n=0}^{\infty} b_n z^n \right).$$

3. Find the Taylor series of $\log(1 + z)$ around $z = 0$ and give its radius of convergence (see Table 3.1).

Solution. We use the principal branch of log so that the function $f(z) = \log(1 + z)$ is defined at $z = 0$. Since f is analytic on the region $A = \mathbb{C}\backslash\{x + iy \,|\, y = 0,\ x \leq -1\}$ shown in Figure 3.8, the radius of convergence of the Taylor series will be ≥ 1 by Theorem 12. That it is exactly one can be shown as follows.

We know that

$$f(0) = \log 1 = 0;$$

$$f'(z) = \frac{1}{z+1}, \text{ so } f'(0) = 1;$$

$$f''(z) = -\frac{1}{(z+1)^2}, \text{ so } f''(0) = -1;$$

and

$$f'''(z) = \frac{2}{(z+1)^3}, \text{ so } f'''(0) = 2.$$

Inductively, we see that

$$f^{(n)}(z) = \frac{(n-1)!(-1)^{n-1}}{(z+1)^n},$$

so $f^{(n)}(0) = (n-1)!(-1)^{n-1}$. Thus the Taylor series is

$$\sum_{n=0}^{\infty} \frac{f^{(n)}(0)}{n!} z^n = \sum_{n=1}^{\infty} \frac{(-1)^{n-1}}{n} z^n$$

(in agreement with Table 3.1). When $z = -1$, it is the harmonic series that always diverges, so the radius of convergence is ≤ 1 and thus is exactly 1. (A general procedure to follow for determining the exact radius of convergence without computing the series is found in exercise 19.)

4. Compute the Taylor series around $z = 0$ and give the radii of convergence for

    a. $\dfrac{z}{z-1}$.

    b. $\dfrac{e^z}{1-z}$. (Compute the first few terms only.)

    Solution.
    a. From the binomial expansion (Table 3.1) we have $(1-z)^{-1} = 1 + z + z^2 + \cdots$ valid for $|z| < 1$. Hence for $|z| < 1$, $z/(z-1) = -z(1+z+z^2 + \cdots) = -z - z^2 - z^3 - z^4 - \cdots$. By the uniqueness of representation by power series (Theorem 10), this is the Taylor series of $z/(z-1)$ around 0. By observing that $z/(z-1)$ is analytic on the open disk $|z| < 1$, we know by the Taylor Theorem that the Taylor series must have a radius of convergence ≥ 1. Of course, a close analysis of the series $-z - z^2 - z^3 - z^4 - \cdots$, using the ratio test or the root test, shows that the radius of convergence is exactly 1.
    b. $1/(1-z) = 1 + z + z^2 + \cdots$ for $|z| < 1$ and $e^z = 1 + z + z^2/2 + \cdots$ for all z. Thus by example 2 we get the series for the product by formally multiplying the two series out as if they were polynomials; the result must still converge for $|z| < 1$. We get

$$\frac{e^z}{1-z} = (1 + z + z^2 + z^3 + \cdots)\left(1 + z + \frac{z^2}{2} + \frac{z^3}{3!} + \cdots\right)$$

$$= 1 + (z + z) + \left(\frac{z^2}{2} + z^2 + z^2\right) + \left(\frac{z^3}{6} + \frac{z^3}{2} + z^3 + z^3\right) + \cdots$$

$$= 1 + 2z + \frac{5z^2}{2} + \frac{8z^3}{3} + \cdots.$$

In this last case the general term has no simple form. Note that this method is faster than computing $f^{(k)}(0)$ when the value of k increases.

Exercises

1. Prove: If $\sum_{n=0}^{\infty} a_n z^n = \sum_{n=0}^{\infty} b_n z^n$ (both converging on an open disk A centered at 0), then $a_n = b_n$ for all n.

2. Find the radius of convergence of each of the following power series:

    a. $\displaystyle\sum_{n=0}^{\infty} n z^n.$

    b. $\displaystyle\sum_{n=0}^{\infty} \frac{z^n}{e^n}.$

    c. $\displaystyle\sum_{n=0}^{\infty} n! \frac{z^n}{n^n}.$

    d. $\displaystyle\sum_{n=1}^{\infty} \frac{z^n}{n}.$

3. Establish the Taylor series for $\sin z$, $\cos z$, and $(1 + z)^\alpha$ in Table 3.1.

4. Compute the Taylor series of the following functions around the indicated points and state on which set the series converges:
    a. e^z, $z_0 = 1$.

    b. $\dfrac{1}{z}$, $z_0 = 1$.

5. Compute the Taylor series of the following around the indicated point:
    a. $\sin z^2$, $z_0 = 0$.
    b. e^{2z}, $z_0 = 0$.

6. Compute the Taylor series of the following. (Give only the first few terms where appropriate.)
    a. $\dfrac{\sin z}{z}$; $z_0 = 1$.
    b. $z^2 e^z$; $z_0 = 0$.
    c. $e^z \sin z$; $z_0 = 0$.

7. Compute the first few terms in the Taylor expansion of $\sqrt{z^2 - 1}$ around 0.

8. What is the flaw in the following reasoning? Since $e^z = \sum_{n=0}^{\infty} z^n/n!$, we get $e^{1/z} = \sum_{n=0}^{\infty} 1/n! z^n$. Since this converges (because e^z is entire) and since the Taylor expansion is unique, the Taylor expansion of $f(z) = e^{1/z}$ around $z = 0$ is $\sum_{n=0}^{\infty} z^{-n}/n!$.

9. Let $f(z) = \sum_{n=0}^{\infty} a_n z^n$ and $g(z) = \sum_{n=0}^{\infty} b_n z^n$ converge for $|z| < R$. Let γ be a circle of radius $r < R$ and define

$$F(z) = \frac{1}{2\pi i} \int_{\gamma} \frac{f(\zeta)}{\zeta} g\left(\frac{z}{\zeta}\right) d\zeta.$$

    Show that $F(z) = \sum_{n=0}^{\infty} a_n b_n z^n$.

    Hint.   Use example 4, Section 2.4.

10. Let $f(z) = \sum_{n=0}^{\infty} a_n z^n$ converge for $|z| < R$. If $0 < r < R$, show that $f(z) = \sum_{n=0}^{\infty} a_n r^n e^{in\theta}$ where $z = re^{i\theta}$ and

$$a_n = \frac{1}{2\pi r^n} \int_0^{2\pi} f(re^{i\theta}) e^{-in\theta} \, d\theta, \tag{3}$$

and that

$$\frac{1}{2\pi} \int_0^{2\pi} |f(re^{i\theta})|^2 \, d\theta = \sum_{n=0}^{\infty} |a_n|^2 \, r^{2n}. \tag{4}$$

(The result of this exercise is referred to as *Parseval's Theorem*.)

Note. Formula (3) expresses the Taylor series as a *Fourier series*.

Hint. Use the Cauchy Integral Formula for a_n and expand $f\bar{f}$ in a series, then integrate term by term.

11. Show that

$$\sinh z = \sum_{n=1}^{\infty} \frac{z^{2n-1}}{(2n-1)!}$$

and that

$$\cosh z = \sum_{n=0}^{\infty} \frac{z^{2n}}{(2n)!}.$$

12. If $\sum_{n=0}^{\infty} a_n z^n$ has radius of convergence R, show that $\sum_{n=0}^{\infty} (\text{Re } a_n) z^n$ has radius of convergence $\geq R$.

13. Let $f(z) = \sum a_n z^n$ have radius of convergence R and let $A = \{z \mid |z| < R\}$. Let $z_0 \in A$ and let $\tilde{R}$ be the radius of convergence of the Taylor series of f around z_0. Prove that $R - |z_0| \leq \tilde{R} \leq R + |z_0|$.

14. Differentiate the series for $1/(1-z)$ to obtain expansions for

$$\frac{1}{(1-z)^2} \quad \text{and} \quad \frac{1}{(1-z)^3}.$$

Give the radius of convergence.

15. Let $f(z) = \sum a_n z^n$ be a power series with radius of convergence $R > 0$. For any closed curve γ in $A = \{z \mid |z| < R\}$ show that $\int_\gamma f = 0$ by
a. using Cauchy's Theorem.
b. justifying term-by-term integration.

16. Compute the Taylor expansion of $\zeta(z) = \sum_{n=1}^{\infty} n^{-z}$ around $z = 2$ (see example 1, Section 3.1).

17. In what region does

$$\sum_{n=1}^{\infty} \frac{\sin nz}{2^n}$$

represent an analytic function? What about

$$\sum_{n=1}^{\infty} \frac{\sin nz}{n^2}?$$

18. Find the first few terms of the Taylor expansion of $\tan z = \sin z/\cos z$ around $z = 0$.

Hint.   We know that such an expansion exists. Write

$$\frac{\sin z}{\cos z} = a_0 + a_1 z + a_2 z^2 + \cdot \cdot \cdot \cdot$$

Multiply by

$$\cos z = 1 - \frac{z^2}{2!} + \frac{z^4}{4!} - \cdot \cdot \cdot$$

and use example 2 to solve for a_0, a_1, a_2.

19. Let f be analytic on the region A, let $z_0 \in A$, and let D be the largest open disk centered at z_0 and contained in A.
   a. If f is unbounded on D, then show that the radius of D equals the radius of convergence of the Taylor series for f at z_0.
   b. If there exists no analytic extension of f (that is, there is no $\tilde{f}$ and A' such that $\tilde{f}$ is analytic on A', $A' \supset A$, $A' \neq A$, and $f = \tilde{f}|A$), then show by an example that the radius of convergence of the Taylor series of f at z_0 can still be greater than the radius of D.

Hint.   Use the principal branch of $\log(1 + z)$ with $z_0 = -2 + i$.

20. Prove: A power series converges absolutely everywhere or nowhere on its circle of convergence. Give an example to show that each case can occur.

21. If $\Sigma g_n(z)$ converges uniformly on a set $B \subset \mathbb{C}$ and $h(z)$ is a bounded function on B, prove that $\Sigma(h(z) \cdot g_n(z))$ converges uniformly on B to $h(z)\{\Sigma g_n(z)\}$.

3.3   Laurent's Series and Classification of Singularities

The Taylor series enables us to find a convergent power series expansion around z_0 for $f(z)$ when f is analytic in a whole disk around z_0. Thus the Taylor expansion does not apply to functions like $f(z) = 1/z$ or e^z/z^2, around $z_0 = 0$, that fail to be analytic at $z = 0$. For such functions there is another expansion, called the *Laurent expansion* (formulated in approximately 1840), that uses inverse powers of z rather than powers of z. This expansion is particularly important in the study of singular points of functions and leads to another fundamental result of complex analysis, the Residue Theorem, which is studied in the next chapter.

Laurent's Theorem

Theorem 13.   *Let $r_1 \geq 0$, $r_2 > r_1$, and $z_0 \in \mathbb{C}$, and consider the region $A = \{z \in \mathbb{C} \mid r_1 < |z - z_0| < r_2\}$ (see Figure 3.9). We can allow $r_1 = 0$ or $r_2 = \infty$ (or both). Let f be analytic on the region A. Then we can write*

$$f(z) = \sum_{n=0}^{\infty} a_n(z - z_0)^n + \sum_{n=1}^{\infty} \frac{b_n}{(z - z_0)^n} \tag{1}$$

(the Laurent expansion), where both series on the right side of the equation converge absolutely on A and uniformly on any set of the form $B_{\rho_1,\rho_2} = \{z \mid \rho_1 \leq |z - z_0| \leq \rho_2\}$ where $r_1 < \rho_1 < \rho_2 < r_2$. If γ is a circle around z_0 with radius r, $r_1 < r < r_2$, then the coefficients are given by

$$a_n = \frac{1}{2\pi i} \int_\gamma \frac{f(\zeta)}{(\zeta - z_0)^{n+1}} \, d\zeta, \, n = 0, 1, 2, \cdots$$

and $\tag{2}$

$$b_n = \frac{1}{2\pi i} \int_\gamma f(\zeta)(\zeta - z_0)^{n-1} \, d\zeta, \, n = 1, 2, \cdots .$$

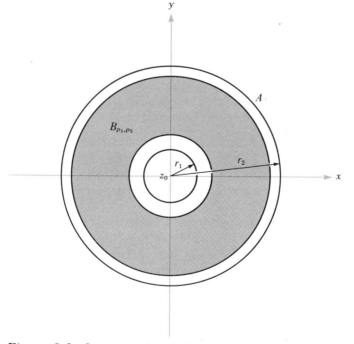

Figure 3.9   Laurent series, with $z_0 = 0$.

(If we set $b_n = a_{-n}$, then the first formula covers both cases.) The series for f is called the Laurent series around z_0 in the annulus A. Any point-wise convergent expansion of f of this form equals the Laurent expansion. (In other words, the Laurent expansion is unique.)

Note.   We *cannot* set $a_n = f^{(n)}(z_0)/n!$ as we did with the Taylor expansion. Indeed, $f^{(n)}(z_0)$ is not even defined since $z_0 \notin A$.

Formula (2) for a_n and b_n is not very practical for computing the Laurent series of a given function f. Tricks can be used to obtain some expansions of the desired form, and the uniqueness of the expansion indicates that it is the desired one. Techniques are given in the following text and in the worked examples.

In the following proof we shall see that the power series part of f, that is,

$$\sum_{n=0}^{\infty} a_n(z - z_0)^n,$$

converges *inside* the circle $|z - z_0| = r_2$, whereas the singular part,

$$\sum_{n=1}^{\infty} \frac{b_n}{(z - z_0)^n},$$

converges *outside* $|z - z_0| = r_1$.

The student is also cautioned that uniqueness is dependent on the choice of A. For example, if $A = \{z \mid |z| > 1\}$, then $f(z) = 1/[z(z-1)]$ has Laurent expansion

$$f(z) = \frac{1}{z(z - 1)} = \frac{1}{z}\left(\frac{1}{z\left(1 - \frac{1}{z}\right)}\right) = \frac{1}{z^2}\left\{1 + \frac{1}{z} + \frac{1}{z^2} + \cdots\right\} = \frac{1}{z^2} + \frac{1}{z^3} + \cdots$$

(valid if $|z| > 1$), whereas on $A = \{z \mid 0 < |z| < 1\}$, it has expansion

$$f(z) = \frac{1}{z(z - 1)} = -\frac{1}{z}(1 + z + z^2 + \cdots) = -\left(\frac{1}{z} + 1 + z + z^2 + \cdots\right)$$

(valid for $0 < |z| < 1$). By uniqueness these are *the* Laurent expansions for the appropriate regions.

Proof.   As with the proof of Taylor's Theorem, we begin with Cauchy's Integral Formula. We will first show uniform convergence of the stated series on B_{ρ_1,ρ_2}, where a_n, b_n are defined by formula (2). Since all the curves γ of radius r are homotopic in A as long as $r_1 < r < r_2$ (Why?), the numbers a_n, b_n are independent of r, so

$$a_n = \frac{1}{2\pi i} \int_{\gamma_1} \frac{f(\zeta)}{(\zeta - z_0)^{n+1}} \, d\zeta$$

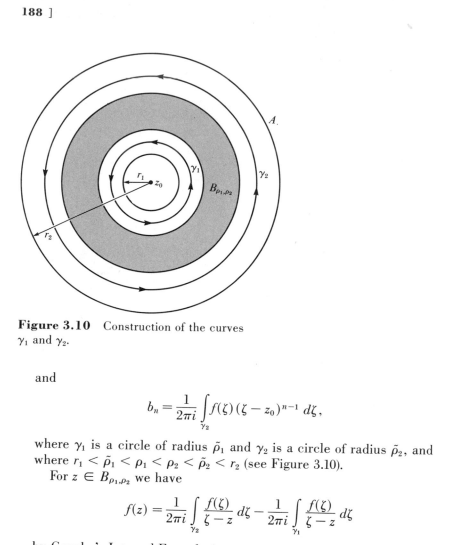

Figure 3.10 Construction of the curves
γ_1 and γ_2.

and

$$b_n = \frac{1}{2\pi i} \int_{\gamma_2} f(\zeta)(\zeta - z_0)^{n-1}\, d\zeta,$$

where γ_1 is a circle of radius $\tilde{\rho}_1$ and γ_2 is a circle of radius $\tilde{\rho}_2$, and where $r_1 < \tilde{\rho}_1 < \rho_1 < \rho_2 < \tilde{\rho}_2 < r_2$ (see Figure 3.10).
For $z \in B_{\rho_1,\rho_2}$ we have

$$f(z) = \frac{1}{2\pi i} \int_{\gamma_2} \frac{f(\zeta)}{\zeta - z}\, d\zeta - \frac{1}{2\pi i} \int_{\gamma_1} \frac{f(\zeta)}{\zeta - z}\, d\zeta$$

by Cauchy's Integral Formula (see exercise 1).
As in Taylor's Theorem, for ζ on γ_2 (and z fixed inside γ_2),

$$\frac{1}{\zeta - z} = \frac{1}{\zeta - z_0} + \frac{z - z_0}{(\zeta - z_0)^2} + \frac{(z - z_0)^2}{(\zeta - z_0)^3} + \cdots,$$

which converges uniformly in ζ on γ_2.
We may integrate term by term (by Theorem 7 and the fact that $f(\zeta)$ is bounded—see exercise 21, Section 3.2) and thus obtain

$$\frac{1}{2\pi i} \int_{\gamma_2} \frac{f(\zeta)}{\zeta - z}\, d\zeta = \sum_{n=0}^{\infty} \frac{1}{2\pi i} \left(\int_{\gamma_2} \frac{f(\zeta)}{(\zeta - z_0)^{n+1}}\, d\zeta \right)(z - z_0)^n$$

$$= \sum_{n=0}^{\infty} a_n (z - z_0)^n.$$

Since this power series converges for z inside γ_2, it converges uniformly on strictly smaller discs (in particular, on B_{ρ_1,ρ_2}).

Similarly,

$$\frac{-1}{\zeta - z} = \frac{1}{(z - z_0)\left(1 - \dfrac{\zeta - z_0}{z - z_0}\right)} = \frac{1}{z - z_0} + \frac{\zeta - z_0}{(z - z_0)^2} + \frac{(\zeta - z_0)^2}{(z - z_0)^3} + \cdots$$

converges uniformly in ζ on γ_1. Thus

$$\frac{-1}{2\pi i} \int_{\gamma_1} \frac{f(\zeta)}{\zeta - z}\, d\zeta = \sum_{n=1}^{\infty} \frac{1}{2\pi i} \left(\int_{\gamma_1} f(\zeta) \cdot (\zeta - z_0)^{n-1}\, d\zeta \right) \cdot \frac{1}{(z - z_0)^n}$$

$$= \sum_{n=1}^{\infty} \frac{b_n}{(z - z_0)^n}.$$

This series converges for z outside γ_1, so the convergence is uniform outside any strictly larger circle. This fact can be proved in the same way as the analogous fact for power series by using the Abel-Weierstrass Lemma in Section 3.2. Another method is to make the transformation $w = 1/(z - z_0)$ and apply the power series result to $\sum_{n=1}^{\infty} b_n w^n$. The student is asked to do this in exercise 13.

We have now proved the existence of the Laurent expansion. To show uniqueness, let us suppose that we have an expansion for f:

$$f(z) = \sum_{n=0}^{\infty} a_n(z - z_0)^n + \sum_{n=1}^{\infty} \frac{b_n}{(z - z_0)^n}.$$

If this converges in A it will, by the preceding remarks, do so uniformly on the circle γ, so we can form

$$\frac{f(z)}{(z - z_0)^{k+1}} = \sum_{n=0}^{\infty} a_n(z - z_0)^{n-k-1} + \sum_{n=1}^{\infty} \frac{b_n}{(z - z_0)^{n+k+1}},$$

which also converges uniformly. We then integrate term by term. By example 3, Section 2.1,

$$\int_{\gamma} (z - z_0)^m\, dz = \begin{cases} 0, & m \neq -1; \\ 2\pi i, & m = -1. \end{cases}$$

Thus if $k \geq 0$, all terms in the preceding series are zero except one, the term with $n = k$, and therefore,

$$\int_{\gamma} \frac{f(z)}{(z - z_0)^{k+1}}\, dz = 2\pi i\, a_k.$$

Similarly, if $k \leq -1$ we find that all terms are zero except for $n = -k$ and

$$b_{-k} = \frac{1}{2\pi i} \int_{\gamma} f(z)(z - z_0)^{k+1} \, dz,$$

so the proof is complete. ∎

Isolated singularities: classification of singular points

We want to look at the special case of Theorem 13 when $r_1 = 0$ in more detail. Then f is analytic on $\{z \mid 0 < |z - z_0| < r_2\}$, which is a deleted r_2-neighborhood of z_0 (see Figure 3.11). In this case we say that z_0 is an *isolated singularity* of f. Thus we can expand f in a Laurent series as follows:

$$f(z) = \cdots + \frac{b_n}{(z - z_0)^n} + \cdots$$

$$+ \frac{b_1}{z - z_0} + a_0 + a_1(z - z_0) + a_2(z - z_0)^2 + \cdots \quad (3)$$

(valid for $0 < |z - z_0| < r_2$).

Definition 3. *If f is analytic on a region A and $z_0 \notin A$ but some deleted ϵ-neighborhood of z_0 lies in A, then z_0 is called an isolated singularity. (Thus the preceding Laurent expansion (formula (3)) is valid in such a deleted ϵ-neighborhood.)*

If z_0 is an isolated singularity of f and if all but a finite number of the b_n in formula (3) are zero, then z_0 is called a pole of f. If k is the highest integer such that $b_k \neq 0$, it is called a pole of order k. (To emphasize that $k \neq \infty$, we sometimes say a pole of finite order k.) If z_0 is a first-order pole,

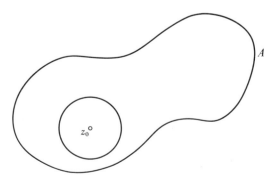

Figure 3.11   Isolated singularity.

we also say it is a simple pole. If an infinite number of b_k's are nonzero, z_0 is called an essential singularity. (Sometimes this z_0 is called a pole of infinite order, but that terminology will not be used here.) "Pole" shall always mean a pole of finite order.

We call b_1 the residue of f at z_0.

If all the b_k's are zero, we say that z_0 is a removable singularity.

A function that is analytic in a region A, except for poles in A, is called meromorphic in A. The phrase "f is a meromorphic function" means that f is meromorphic in $\mathbb{C}$.

Thus f has a pole of order k iff its Laurent expansion has the form

$$\frac{b_k}{(z-z_0)^k} + \cdots + \frac{b_1}{z-z_0} + a_0 + a_1(z-z_0) + \cdots$$

The part

$$\frac{b_k}{(z-z_0)^k} + \cdots + \frac{b_1}{z-z_0},$$

often called the *principal part* of f at z_0, tells just "how singular" f is at z_0.

If f has a removable singularity, then

$$f(z) = \sum_{n=0}^{\infty} a_n(z-z_0)^n$$

is a convergent power series. Thus if we set $f(z_0) = a_0$, f will be analytic at z_0. In other words, *f has a removable singularity at z_0 iff f can be defined at z_0 in such a way that f becomes analytic at z_0.*

As we shall see in Chapter 4, finding the Laurent expansion is not as important as being able to compute the residue b_1, and this computation can often be done without computing the Laurent series. Techniques for doing so will be studied in Section 4.1.

The important property of b_1 not shared by other coefficients is stated in Theorem 14.

Theorem 14.   Let f be analytic on a region A and have an isolated singularity at z_0 with residue b_1 at z_0. If γ is any circle around z_0 whose interior, except for the point z_0, lies in A, then

$$\int_\gamma f(z)\,dz = b_1 \cdot 2\pi i. \tag{4}$$

This conclusion follows immediately from the formula for b_1 in Theorem 13. The point is that we can compute b_1 by methods other than formula (4) and therefore we can use formula (4) to compute $\int_\gamma f$. For example, if $z \neq 0$, then

$$e^{1/z} = 1 + \frac{1}{z} + \frac{1}{2!z^2} + \frac{1}{3!z^3} + \cdots$$

(Why?), so $e^{1/z}$ has an essential singularity at $z = 0$ and $b_1 = 1$. Thus $\int_\gamma e^{1/z}\, dz = 2\pi i$ for any circle γ around 0.

The following theorem tells us when we have the various types of singularities. It is due, in part, to Riemann.

Theorem 15. *Let f be analytic on a region A and have an isolated singularity at z_0.*

  i. *z_0 is a removable singularity iff any one of the following conditions holds: (1) f is bounded in a deleted neighborhood of z_0; (2) $\lim\limits_{z \to z_0} f(z)$ exists; or (3) $\lim\limits_{z \to z_0} (z - z_0) f(z) = 0$.* (Note that it is not immediately evident that these three conditions are equivalent but the assertion is that they are and that each is equivalent to the condition that f has a removable singularity.)

  ii. *z_0 is a simple pole iff $\lim\limits_{z \to z_0} (z - z_0)f(z)$ exists and is unequal to zero.*

  This limit equals the residue of f at z_0.

 iii. *z_0 is a pole of order $\leq k$ (or possibly a removable singularity) iff any one of the following conditions holds: (1) There is a constant $M > 0$ and an integer $k \geq 1$ such that*
$$|f(z)| \leq \frac{M}{|z - z_0|^k}$$
 in a deleted neighborhood of z_0; (2) $\lim\limits_{z \to z_0} (z - z_0)^{k+1}f(z) = 0$; or

 (3) $\lim\limits_{z \to z_0} (z - z_0)^k f(z)$ exists.

 iv. *z_0 is a pole of order $k \geq 1$ iff there is an analytic function φ defined on a neighborhood U of z_0 such that $U \setminus \{z_0\} \subset A$, such that $\varphi(z_0) \neq 0$, and such that $f(z) = \varphi(z)/(z - z_0)^k$ for $z \in U$, $z \neq z_0$.*

Proof.
  i. If z_0 is a removable singularity, then in a deleted neighborhood of z_0 we have $f(z) = \sum_{n=0}^{\infty} a_n(z - z_0)^n$. Since this series represents an analytic function in an undeleted neighborhood of z_0, obviously conditions (1), (2), and (3) hold. Conditions (1) and (2) each obviously implies condition (3), so it remains to be shown that (3) implies that z_0 is a removable singularity for f. We must prove that each b_k in the Laurent expansion of f around z_0 is 0. Now $b_k = \frac{1}{2\pi i} \int_{\gamma_r} f(\zeta)(\zeta - z_0)^{k-1}\, d\zeta$, where γ_r is a circle whose interior

(except for z_0) lies in A. Let $\epsilon > 0$ be given. By condition (3) we can choose $r > 0$ with $r < 1$ such that on γ_r we have $|f(\zeta)| < \epsilon/|\zeta - z_0| = \epsilon/r$. Then

$$|b_k| \leq \frac{1}{2\pi} \int_{\gamma_r} |f(\zeta)| \, |\zeta - z_0|^{k-1} \, |d\zeta|$$

$$\leq \frac{1}{2\pi} \frac{\epsilon}{r} r^{k-1} \int_{\gamma_r} |d\zeta|$$

$$= \frac{1}{2\pi} \frac{\epsilon}{r} r^{k-1} 2\pi r$$

$$= \epsilon \, r^{k-1} \leq \epsilon.$$

Thus $|b_k| \leq \epsilon$. Since ϵ was arbitrary, $b_k = 0$. We shall use (iii) to prove (ii), so (iii) is proved next.

iii. This proof follows immediately by applying (i) to the function $(z - z_0)^k f(z)$, which is analytic on A. (The student should write out the details of the proof.)

ii. If z_0 is a simple pole, then in a deleted neighborhood of z_0 we have

$$f(z) = \frac{b_1}{z - z_0} + \sum_{n=0}^{\infty} a_n (z - z_0)^n = \frac{b_1}{z - z_0} + h(z),$$

where h is analytic at z_0 and where $b_1 \neq 0$ by the Laurent expansion. Hence $\lim_{z \to z_0} (z - z_0)f(z) = \lim_{z \to z_0} (b_1 + (z - z_0)h(z)) = b_1$. On the other hand, suppose that $\lim_{z \to z_0} (z - z_0)f(z)$ exists and is unequal to zero. Thus $\lim_{z \to z_0} (z - z_0)^2 f(z) = 0$. By the result obtained in (iii), this says that

$$f(z) = \frac{b_1}{(z - z_0)} + \sum_{n=0}^{\infty} a_n (z - z_0)^n = \frac{b_1}{z - z_0} + h(z)$$

for some constant b_1 and analytic function h where b_1 may or may not be zero. But then $(z - z_0)f(z) = b_1 + (z - z_0)h(z)$, so $\lim_{z \to z_0} (z - z_0)f(z) = b_1$. Thus, in fact, $b_1 \neq 0$, and therefore f has a simple pole at z_0.

iv. z_0 is a pole of order $k \geq 1$ iff

$$f(z) = \frac{b_k}{(z - z_0)^k} + \frac{b_{k-1}}{(z - z_0)^{k-1}} + \cdots + \frac{b_1}{(z - z_0)} + \sum_{n=0}^{\infty} a_n (z - z_0)^n$$

$$= \frac{1}{(z - z_0)^k} \{ b_k + b_{k-1}(z - z_0) + \cdots + b_1 (z - z_0)^{k-1}$$

$$+ \sum_{n=0}^{\infty} a_n (z - z_0)^{n+k} \}$$

(where $b_k \neq 0$). This expansion is valid in a deleted neighborhood of z_0. Let $\varphi(z) = b_k + b_{k-1}(z - z_0) + \cdots + b(z - z_0)^{k-1} + \Sigma_{n=0}^{\infty} a_n(z - z_0)^{n+k}$. Then $\varphi(z)$ is analytic in the corresponding undeleted neighborhood (since it is a convergent power series) and $\varphi(z_0) = b_k \neq 0$. Conversely, given such a φ, we can retrace these steps to show that z_0 is a pole of order $k \geq 1$.∎

Zeros of order k

Let f be analytic on a region A and let $z_0 \in A$. We say that f has a zero *of order* k *at* z_0 iff $f(z_0) = 0, \cdots, f^{(k-1)}(z_0) = 0, f^{(k)}(z_0) \neq 0$.

From the Taylor expansion

$$f(z) = \sum_{n=0}^{\infty} \frac{f^{(n)}(z_0)}{n!} (z - z_0)^n,$$

we see that f has a zero of order k iff, in a neighborhood of z_0, we can write $f(z) = (z - z_0)^k g(z)$ where $g(z)$ is analytic at z_0 and $g(z_0) = f^{(k)}(z_0)/k! \neq 0$. Thus from Theorem 15(iv), we get (letting $\varphi(z) = g(z)^{-1}$) Theorem 16.

Theorem 16. *If f is analytic in a neighborhood of z_0, then f has a zero of order k at z_0 iff $1/f(z)$ has a pole of order k at z_0. If h is analytic and $h(z_0) \neq 0$, then $h(z)/f(z)$ also has a pole of order k.*

Obviously, if z_0 is a zero of f and f is not identically equal to zero in a neighborhood of z_0, then z_0 has some finite order k. (Otherwise the Taylor series would be identically zero.)

Essential singularities

In practical examples we usually are dealing with a pole. It is not hard to show that if $f(z)$ has a pole (of finite order k) at z_0, then $|f(z)| \rightarrow \infty$ as $z \rightarrow z_0$ (see exercise 2). However, in case of an essential singularity, $|f|$ will not, in general, approach ∞ as $z \rightarrow z_0$. In fact, we have the following result proven by E. Picard in 1879.

Theorem 17 (Picard Theorem). *Let f have an essential singularity at z_0 and let U be any (arbitrarily small) deleted neighborhood of z_0. Then for all $w \in \mathbb{C}$, except perhaps one value, the equation $f(z) = w$ has infinitely many solutions z in U.*

Theorem 17 actually belongs in a more advanced course. However, we can easily prove a simple version.

Theorem 18 (Casorati-Weierstrass Theorem). Let f have an essential singularity at z_0 and let $w \in \mathbb{C}$. Then there exist $z_1, z_2, z_3, \cdots,$ $z_n \to z_0$, such that $f(z_n) \to w$.

Proof. If the assertion were false, there would be a deleted neighborhood U of z_0 and an $\epsilon > 0$ such that $|f(z) - w| \geq \epsilon$ for all $z \in U$ (Why?). Let $g(z) = 1/(f(z) - w)$. Thus on U, g is analytic, and since $g(z)$ is bounded on U ($|g(z)| \leq \epsilon^{-1}$), z_0 is a removable singularity by Theorem 15(i). Let k be the order of the zero of g at z_0 (set $k = 0$ if $g(z_0) \neq 0$). (The order must be finite because otherwise, as mentioned previously, by the Taylor Theorem, g would be zero in a neighborhood of z_0, whereas g is 0 nowhere on U.) Thus $f(z) - w + 1/g(z)$ is either analytic (if $k = 0$) or has a pole of order k by Theorem 16. This conclusion contradicts our assumption that f has an essential singularity. ∎

Worked Examples

1. Find the Laurent expansions of the following functions (with z_0, r_1, r_2 as indicated):

    a. $\dfrac{z+1}{z}$; $z_0 = 0$, $r_1 = 0$, $r_2 = \infty$.

    b. $\dfrac{z}{z^2+1}$; $z_0 = i$, $r_1 = 0$, $r_2 = 2$.

    Solution.

    a. $\dfrac{z+1}{z} = \dfrac{1}{z} + 1$.

        This equation is in the form of the Laurent expansion so, by uniqueness (see Theorem 13), it equals it; that is, $b_k = 0$ for $k > 1$, $b_1 = 1$, $a_0 = 1$, $a_k = 0$ for $k \geq 1$.

    b. $\dfrac{z}{z^2+1} = \dfrac{z}{(z+i)(z-i)}$.

        Because $1/(z+i)$ is analytic near $z = i$, it can be expanded as a power series in $z - i$ as follows:

        $$\frac{1}{z+i} = \frac{1}{2i + (z-i)} = \frac{1}{2i\left(1 + \dfrac{z-i}{2i}\right)}$$

        $$= \frac{1}{2i}\left\{1 - \frac{(z-i)}{2i} + \left[\frac{(z-i)}{2i}\right]^2 - \cdots\right\}.$$

        This is valid for $|z - i| < 2$ (see Figure 3.12).

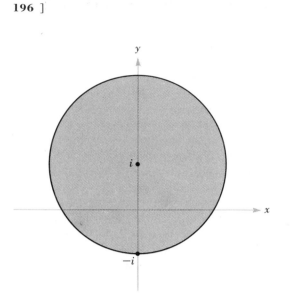

Figure 3.12 Region of convergence for the expansion of $\dfrac{1}{z+i}$.

Thus, after simplifying,

$$\frac{z}{z^2+1} = \frac{1}{2(z-i)} + \frac{1}{4i}\left\{ 1 - \frac{(z-i)}{2i} + \left[\frac{(z-i)}{2i} \right]^2 - \cdots \right\}$$

is the Laurent expansion.

The following method of partial fractions is sometimes useful. Write

$$\frac{z}{(z+i)(z-i)} = \left[\frac{(z-i)+i}{2i} \right]\left\{ \frac{1}{z-i} - \frac{1}{z+i} \right\}$$

and

$$\frac{1}{z+i} = \frac{1}{2i}\left(1 - \frac{(z-i)}{2i} + \left[\frac{(z-i)}{2i} \right]^2 - \cdots \right).$$

Gathering terms produces the same result.

2. Determine the order of the pole of each of the following functions at the indicated singularity:

a. $\dfrac{\cos z}{z^2}$; $z_0 = 0$.

b. $\dfrac{e^z - 1}{z^2}$; $z_0 = 0$.

c. $\dfrac{z+1}{z-1}$; $z_0 = 0$.

Solution.

a. z^2 has a zero of order 2 and $\cos 0 = 1$, so $\cos z/z^2$ has a pole of order 2 by Theorem 16. Alternatively,

$$\frac{\cos z}{z^2} = \frac{1}{z^2}\left\{1 - \frac{z^2}{2!} + \frac{z^4}{4!} - \cdots\right\} = \frac{1}{z^2} - \frac{1}{2!} + \frac{z^2}{4!} - \cdots,$$

so again the pole is of order 2.

b. The numerator vanishes at 0 so Theorem 16 does not apply. (We could use exercise 4, however.) But

$$\frac{e^z - 1}{z^2} = \frac{1}{z^2}\left\{1 + z + \frac{z^2}{2!} + \cdots - 1\right\} = \frac{1}{z} + \frac{1}{2!} + \frac{z}{3!} + \frac{z^2}{4!} + \cdots,$$

so the pole is simple.

c. There is no pole; it is analytic at 0.

3. Determine which of the following functions have removable singularities at $z_0 = 0$:

a. $\dfrac{\sin z}{z}$.

b. $\dfrac{e^z}{z}$.

c. $\dfrac{(e^z - 1)^2}{z^2}$.

d. $\dfrac{z}{e^z - 1}$.

Solution.

a. $\lim\limits_{z \to 0} z \cdot (\sin z)/z = \lim\limits_{z \to 0} \sin z = 0$, so the singularity is removable (by Theorem 15(i)). Alternatively,

$$\frac{\sin z}{z} = \frac{1}{z}\left\{z - \frac{z^3}{3!} + \cdots\right\} = 1 - \frac{z^2}{3!} + \frac{z^4}{5!} - \cdots.$$

b. $\lim\limits_{z \to 0} z \cdot e^z/z = 1$ so the pole is simple (the singularity is not removable).

c. $(e^z - 1)/z$ has a removable singularity, since $\lim\limits_{z \to 0} z \cdot (e^z - 1)/z = 0$, so $[(e^z - 1)/z]^2$ also has a removable singularity.

d. $\lim\limits_{z \to 0} z/(e^z - 1) = 1$ because $(e^z - 1)/z = 1 + z/2 + z^2/3! + \cdots \to 1$ as $z \to 0$. Thus $z/(e^z - 1)$ has a removable singularity (by Theorem 15(i)).

Exercises

1. Let γ_1 and γ_2 be two concentric circles around z_0 of radii R_1 and R_2, $R_1 < R_2$. If z lies between the circles and f is analytic on a region containing γ_1, γ_2 and the region between them, show that

$$f(z) = \frac{1}{2\pi i}\int_{\gamma_2} \frac{f(\zeta)}{\zeta - z}\,d\zeta - \frac{1}{2\pi i}\int_{\gamma_1} \frac{f(\zeta)}{\zeta - z}\,d\zeta.$$

2. Let f have a pole at z_0 of order $k \geq 1$. Prove that $f(z) \to \infty$ as $z \to z_0$.

Hint.   Use Theorem 15 (iv).

3. Find the Laurent expansions of the following functions around $z_0 = 0$:

a. $\sin\left(\dfrac{1}{z}\right)$.

b. $\dfrac{1}{z(z+1)}$.

c. $\dfrac{z}{z+1}$.

d. $\dfrac{e^z}{z^2}$.

4. Show that if g and f are analytic at z_0 and g has a zero of order l, f has a zero of order k, $k \geq l$, then g/f has a pole of order $k - l$. What if $k < l$?

5. Which of the following functions have removable singularities at the indicated points:

a. $\dfrac{\cos(z-1)}{z^2}$; $z_0 = 0$.

b. $\dfrac{z}{z-1}$; $z_0 = 1$.

c. $\dfrac{f(z)}{(z-z_0)^k}$ if f has a zero at z_0 of order k.

6. Prove, using the Taylor series, the following complex version of *l'Hopital's Rule:* Let $f(z)$ and $g(z)$ be analytic, both having zeros of order k at z_0. Then $f(z)/g(z)$ has a removable singularity and

$$\underset{z \to z_0}{\text{limit}} \frac{f(z)}{g(z)} = \frac{f^{(k)}(z_0)}{g^{(k)}(z_0)}.$$

7. If f is analytic on a region containing a circle γ and its interior and has a zero of order 1 only at z_0 inside or on γ, show that

$$z_0 = \frac{1}{2\pi i} \int_\gamma \frac{z f'(z)}{f(z)} \, dz.$$

Hint.   Let $f(z) = (z - z_0)\varphi(z)$ and apply the Cauchy Integral Formula.

8. Expand $\dfrac{1}{z(z-1)(z-2)}$ in a Laurent series in the following regions:

a. $0 < |z| < 1$.
b. $1 < |z| < 2$.

9. Find the first few terms in the Laurent expansion of $1/(e^z - 1)$ around $z = 0$.

Hint.   Show that since $1/(e^z - 1)$ has a simple pole,

$$\frac{1}{e^z - 1} = \frac{b_1}{z} + a_0 + a_1 z + a_2 z^2 + \cdots$$

Then cross multiply (using example 2, Section 3.2) and solve for b_1, a_0, a_1.

10. Use the Hint of problem 9 to find the first few terms in the Laurent expansion of $\cot z = \cos z/\sin z$ around $z = 0$.

11. For f as in Theorem 13, show that if $r_1 < r < r_2$, then

$$\int_0^{2\pi} |f(z_0 + re^{i\theta})|^2 \, d\theta = 2\pi \sum_{n=0}^{\infty} |a_n|^2 \, r^{2n} + 2\pi \sum_{n=1}^{\infty} |b_n|^2 \, r^{-2n}.$$

12. Define a branch of $\sqrt{z^2 - 1}$ that is analytic except for the segment $[-1,1]$ on the real axis. Determine the first few terms in the Laurent expansion that is valid for $|z| > 1$.

13. If

$$\sum_{n=1}^{\infty} \frac{b_n}{(z - z_0)^n}$$

converges for $|z - z_0| > R$, prove that it converges uniformly on the set $F_r = \{z \mid |z - z_0| > r\}$ for $r > R$.

Hint.   Adapt the Abel-Weierstrass Lemma and the Weierstrass M Test to this case.

14. Let z_0 be an essential singularity of f and let U be any deleted neighborhood of z_0. Prove that $\text{cl}(f(U)) = \mathbb{C}$. (cl$=$closure; see definition on p. 137.)

15. Find the residues of the following functions at the indicated points:

a. $\dfrac{1}{z^2 - 1}$; $z = 1$.

b. $\dfrac{z}{z^2 - 1}$; $z = 1$.

c. $\dfrac{e^z - 1}{z^2}$; $z = 0$.

d. $\dfrac{e^z - 1}{z}$; $z = 0$.

16. Discuss the singularities of $1/\cos(1/z)$.

Review Exercises for Chapter 3

1. Find the Taylor expansion of $\log z$ (principle branch of the logarithm) around $z = 1$.

2. Where are the poles of $1/\cos z$ and what are their orders?

3. Find the Laurent expansion of $1/(z^2 + z^3)$ around $z = 0$.

4.   Expand $z^2 \sin(z^2)$ in a Taylor series around $z = 0$.

5.   The $2n$th derivative of $f(z) = e^{z^2}$ at $z = 0$ is given by $(2n)!/n!$. Prove this without actually computing the $2n$th derivative.

6.   Verify the Picard Theorem (Theorem 17) for the function $e^{1/z}$.

7.   If f is analytic and nonconstant on any open set in a region A, prove that the zeros of f are *isolated*. (In other words, prove that if z_0 is a zero, there is a neighborhood of z_0 in which there are no other zeros.)

8.   Let $\exp(t(z - 1/z)/2) = \sum_{n=-\infty}^{\infty} J_n(t)z^n$ be the Laurent expansion for each fixed $t \in \mathbb{R}$. $J_n(t)$ is called the *Bessel function* of order n. Show that

a.
$$J_n(t) = \frac{1}{\pi} \int_0^{\pi} \cos(t \sin\theta - n\theta) \, d\theta.$$

b.
$$J_{-n}(t) = (-1)^n J_n(t).$$

9.   Let $\sum_{n=0}^{\infty} a_n(z - z_0)^n$ be a power series with radius of convergence $R > 0$. If $0 < r < R$, show that there is a constant M such that $|a_n| \le M \, r^{-n}$, $n = 0, 1, 2, \cdots$.

10.   Let f have a zero at z_0 of multiplicity k. Show that the residue of f'/f at z_0 is k.

11.   Suppose that f is analytic on the open unit disk $|z| < 1$ and that there is a constant M such that $|f^{(k)}(0)| \le M^k$ for all k. Show that f can be extended to an entire function.

12.   Suppose that f is analytic in a region containing the unit disk $|z| \le 1$, that $f(0) = 0$, and that $|f(z)| < 1$ if $|z| = 1$. Show that there are no $z \ne 0$ with $|z| < 1$ and $f(z) = z$.

Hint.   Use the Schwarz Lemma.

13.   What is the radius of convergence of the Taylor expansion of

$$f(z) = \frac{e^z}{(z - 1)(z + 1)(z - 2)(z - 3)}$$

when expanded around $z = i$?

14.   Evaluate

$$\int_\gamma \frac{z^2 + e^z}{z(z - 3)} \, dz$$

where γ is the unit circle.

15.   Suppose that $\sum_{n=0}^{\infty} a_n$ converges but that $\sum_{n=0}^{\infty} |a_n|$ diverges. Show that $\sum_{n=0}^{\infty} a_n z^n$ has a radius of convergence equal to 1. Answer the same question but assume that $\sum_{n=0}^{\infty} a_n$ converges and that $\sum_{n=0}^{\infty} n|a_n|$ diverges.

16.   Find the Laurent expansion of

$$f(z) = \frac{1}{z(z^2 + 1)}$$

that is valid for

a. $0 < |z| < 1$.

b. $1 < |z|$.

17. Let f be analytic on $\mathbb{C}\backslash\{0\}$. Show the coincidence of the Laurent expansions of f valid in

a. $|z| > 0$.

b. $|z| > 1$.

18. Find the radii of convergence of

a. $\displaystyle\sum_{n=0}^{\infty} \frac{2^n}{n^2} z^n$.

b. $\displaystyle\sum_{n=0}^{\infty} z^{n!}$.

19. Let f be entire and let $g(z) = \sum_{n=0}^{\infty} a_n z^n$ have radius of convergence R. Can you find another power series $\sum b_n z^n$ with radius of convergence $\geq R$ such that

$$\sum_{n=0}^{\infty} b_n z^n = f\left(\sum_{n=0}^{\infty} a_n z^n\right)?$$

20. Let f be entire and suppose that $f(z) \longrightarrow \infty$ as $z \longrightarrow \infty$. Show that f is a polynomial.

Hint.   Show that $f(1/z)$ has a pole of finite order at $z = 0$.

21. Let f have an isolated singularity at z_0. Show that if $f(z)$ is bounded in a deleted neighborhood of z_0, then $\lim_{z \to z_0} f(z)$ exists.

22. Let f be analytic on $|z| < 1$. Show that the inequality $|f^{(k)}(0)| \geq k!\, 5^k$ cannot hold for all k.

23. Evaluate $\int_0^{2\pi} e^{e^{i\theta}}\, d\theta$.

24. Evaluate $\int_\gamma e^z/z^2\, dz$ where γ is the unit circle.

25. Determine the order of the poles of the following functions at their singularities:

a. $\dfrac{e^z(z - 3)}{(z - 1)(z - 5)}$.

b. $\dfrac{e^z - 1}{z}$.

c. $\dfrac{e^z - 2}{z}$.

d. $\dfrac{\cos z}{1 - z}$.

26. Let $f(z)$ be entire and satisfy these two conditions:

a. $f'(z) = f(z)$.

b. $f(0) = 1$.

Show that $f(z) = e^z$. If you replace (a) by $f(z_1 + z_2) = f(z_1)f(z_2)$, show that $f(z) = e^{az}$ for some constant a.

27. a. Show by example that the Mean Value Theorem for analytic functions is not true. In other words, let f be defined on the region A and let $z_1, z_2 \in A$ be such that the straight line joining z_1 to z_2 lies in A. Show that there need not be a z_0 on this straight line such that

$$f'(z_0) = \frac{f(z_1) - f(z_2)}{z_1 - z_2}.$$

b. If, however, $|f'(z_0)| \leq M$ on this line, prove that $|f(z_1) - f(z_2)| \leq M |z_1 - z_2|$ and generally that if $|f'(z_0)| \leq M$ on a curve γ joining z_1 to z_2, then $|f(z_1) - f(z_2)| \leq M l(\gamma)$.

28. Show that the Maximum Modulus Theorem is false if A is unbounded even if $|f|$ is bounded on $\mathrm{bd}(A)$.

29. Let f be analytic on a region A containing $\{z \mid |z - z_0| < R\}$ so

$$f(z) = \sum_{n=0}^{\infty} \frac{f^{(n)}(z_0)(z - z_0)^n}{n!}.$$

Let $R_n(z)$ equal $f(z)$ minus the nth partial sum. (R_n is thus the remainder.) Let $\rho < R$ and let M be the maximum of f on $\{z \mid |z - z_0| = R\}$. Show that $|z - z_0| \leq \rho$ implies that

$$|R_n(z)| \leq M \left(\frac{\rho}{R}\right)^{n+1} \left\{ \frac{1}{1 - \left(\frac{\rho}{R}\right)} \right\}.$$

30. If $f(z) = f(-z)$ and $f(z) = \sum_{n=0}^{\infty} a_n z^n$ is convergent on a disk $|z| < R, R > 0$, show that $a_n = 0$ (for $n = 1, 3, 5, 7, \cdots$).

31. If f is entire and is bounded on the real axis, then f is constant. Prove or give a counterexample.

32. Find the radius of convergence of the power series $\sum_{n=0}^{\infty} 2^n z^{n^2}$.

33. The Bernoulli numbers B_n are related to the coefficients of the power series of $z/(e^z - 1)$ by the formula

$$\frac{z}{e^z - 1} = \sum_{n=0}^{\infty} \frac{B_n}{n!} z^n.$$

a. Determine the radius of convergence of the series

$$\sum_{n=0}^{\infty} \frac{B_n}{n!} z^n.$$

b. Using the Cauchy integral formulas and the contour $|z| = 1$, find an integral expression for B_n of the form

$$B_n = \int_0^{2\pi} g_n(\theta) \, d\theta$$

(for suitable functions $g_n(\theta)$; $0 \leq \theta \leq 2\pi$).

34. The *Legendre Polynomials* are defined as the coefficients of z^n in the Taylor development

$$(1 - 2\alpha z + z^2)^{-1/2} = \sum_{n=0}^{\infty} P_n(\alpha) z^n.$$

Prove that $P_n(\alpha)$ is a polynomial of degree n and find P_1, P_2, P_3, P_4.

35. Prove:

a. $\left(\dfrac{z^n}{n!}\right)^2 = \dfrac{1}{2\pi i} \displaystyle\int \dfrac{z^n e^{zt}}{n! t^n} \dfrac{dt}{t}$; where γ is the unit circle.

b. $\displaystyle\sum_{n=0}^{\infty} \left(\dfrac{z^n}{n!}\right)^2 = \dfrac{1}{2\pi} \int_0^{2\pi} e^{2z\cos\theta} \, d\theta.$

36. Find a power series that solves the functional equation $f(z) = z + f(z^2)$ and prove that this is the only power series that solves the equation.

37. a. What is wrong with the following argument? Consider

$$f(z) = \cdots + \frac{1}{z^3} + \frac{1}{z^2} + \frac{1}{z} + 1 + z + z^2 + \cdots$$

Note that

$$z + z^2 + \cdots = \frac{z}{1 - z}$$

whereas

$$1 + \frac{1}{z} + \frac{1}{z^2} + \cdots = \frac{1}{1 - 1/z} = \frac{-z}{1 - z}.$$

Hence $f(z) = 0$.

b. Is f in fact the zero function?

Chapter 4

Calculus of Residues

This chapter focuses on the Residue Theorem, which states that the integral of an analytic function f around a closed contour equals $2\pi i$ times the sum of the residues of f inside the contour. We shall use this theorem in our first main application of complex analysis, the evaluation of definite integrals. So that the student will gain ample facility in handling residues, the chapter begins with the techniques for computing residues of functions at isolated singularities.

4.1 Calculation of Residues

We recall from Section 3.3 that if f has an isolated singularity at z_0, then f admits a Laurent expansion that is valid in a deleted neighborhood of z_0:

$$f(z) = \cdots + \frac{b_2}{(z - z_0)^2} + \frac{b_1}{(z - z_0)} + a_0 + a_1(z - z_0) + \cdots$$

and b_1 is called the *residue* of f at z_0. This is written

$$b_1 = \mathrm{Res}(f, z_0).$$

We want to develop techniques for computing the residue without having to find the Laurent expansion. Of course, if the Laurent expansion is known there is no problem. For example, since

$$e^{1/z} = 1 + \frac{1}{z} + \frac{1}{2z^2} + \cdots + \frac{1}{n! z^n} + \cdots,$$

we have residue 1 at $z_0 = 0$.

Particularly important is the case of a pole (in contrast to an essential singularity). For this case we have fairly straightforward techniques that are easy to apply if the order of the pole is not too large.

If we are given an f defined on a region A with an isolated singularity at z_0, then we proceed in the following way to find the residue. First we decide whether we can easily find the first few terms in the Laurent expansion. If so, the residue of f at z_0 will be the coefficient of $1/(z-z_0)$ in the expansion. If not, then we guess the order of singularity, verify it according to the rules that will be developed in this section (rules were also developed in Theorem 15), and calculate the residue according to these rules. (The rules are summarized in Table 4.1, p. 213.) If we have any doubt as to what order to guess, we should work systematically by first guessing removable singularity, then simple pole, and so on, checking against Table 4.1 until we obtain the correct answer.

As each rule is introduced, it will be illustrated with an example.

Removable singularity

It was shown in Section 3.3 that f has a removable singularity at z_0 iff $\lim_{z \to z_0} (z - z_0)f(z) = 0$. The following theorem covers many important cases and may sometimes be easier to use than the criterion of $\lim_{z \to z_0} (z - z_0) f(z) = 0$.

Theorem 1. *If $g(z)$ and $h(z)$ are analytic and have zeros at z_0 of the same order, then $f(z) = g(z)/h(z)$ has a removable singularity at z_0.*

Proof. By Theorem 16, p. 194, $g(z) = (z - z_0)^k \tilde{g}(z)$ where $\tilde{g}(z_0) \neq 0$ and $h(z) = (z - z_0)^k \tilde{h}(z)$ where $\tilde{h}(z_0) \neq 0$ and $\tilde{g}$ and $\tilde{h}$ are analytic and nonzero at z_0. Thus $f(z) = \tilde{g}(z)/\tilde{h}(z)$ is analytic at z_0. ∎

Examples.
  i. $e^z/(z - 1)$ has no singularity at $z_0 = 0$.
  ii. $(e^z - 1)/z$ has a removable singularity at 0 because $e^z - 1$ and z have zeros of order 1. (They vanish but their derivatives do not.)
  iii. $z^2/\sin^2 z$ has a removable singularity at $z_0 = 0$ because both the numerator and the denominator have zeros of order 2.

The preceding discussion is summarized in lines one and two of Table 4.1.

Simple poles

By Theorem 15, if $\lim\limits_{z \to z_0} (z - z_0) f(z)$ exists and is nonzero, then there must be a simple pole at z_0 and this limit equals the residue. Let us apply this result to obtain the following theorem, which is probably the most useful method of all for computing residues.

Theorem 2. *Let g and h be analytic at z_0 and assume that $g(z_0) \neq 0$, $h(z_0) = 0$, and $h'(z_0) \neq 0$. Then $f(z) = g(z)/h(z)$ has a simple pole at z_0 and*

$$\mathrm{Res}(f, z_0) = g(z_0)/h'(z_0). \tag{1}$$

Proof.   We know that

$$\lim_{z \to z_0} \frac{h(z) - h(z_0)}{z - z_0} = \lim_{z \to z_0} \frac{h(z)}{z - z_0} = h'(z_0) \neq 0,$$

so

$$\lim_{z \to z_0} \frac{z - z_0}{h(z)} = \frac{1}{h'(z_0)}.$$

Thus

$$\lim_{z \to z_0} (z - z_0) \frac{g(z)}{h(z)} = \frac{g(z_0)}{h'(z_0)}$$

exists and therefore equals the residue. ∎

Alternate Proof. $h(z_0) = 0$ and $h'(z_0) \neq 0$ imply that $h(z) = \tilde{h}(z)(z - z_0)$ where $\tilde{h}(z)$ is analytic at z_0, and that $\tilde{h}(z_0) = h'(z_0) \neq 0$. Thus $g(z)/h(z) = g(z)/\{\tilde{h}(z)(z - z_0)\}$ and $g(z)/\tilde{h}(z)$ is analytic at z_0. Hence we have a Taylor series $g(z)/\tilde{h}(z) = \sum_{n=0}^{\infty} a_n (z - z_0)^n$, where $a_0 = g(z_0)/\tilde{h}(z_0)$. Therefore, $g(z)/\{\tilde{h}(z)(z - z_0)\} = \sum_{n=0}^{\infty} a_n (z - z_0)^{n-1}$ is the Laurent expansion and thus $a_0 = g(z_0)/\tilde{h}(z_0) = g(z_0)/h'(z_0)$ is the residue. ∎

Generally, if $g(z)$ has a zero of order k, and h has a zero of order l with $l > k$, then $g(z)/h(z)$ has a pole of order $l - k$. This follows because $g(z) = (z - z_0)^k \tilde{g}(z)$ and $h(z) = (z - z_0)^l \tilde{h}(z)$ where $\tilde{g}(z_0) \neq 0$ and $\tilde{h}(z_0) \neq 0$. Thus

$$\frac{g(z)}{h(z)} = \frac{\varphi(z)}{(z - z_0)^{l-k}}$$

where $\varphi(z) = \tilde{g}(z)/\tilde{h}(z)$, which is analytic at z_0 because $\tilde{h}(z_0) \neq 0$ (and hence $\tilde{h}(z) \neq 0$ in a neighborhood of z_0). Thus our assertion follows from Theorem 15. If $l = k + 1$, we have a simple pole and can obtain the residue from Theorem 3. This generalizes the result in Theorem 2.

Theorem 3. *Let us suppose that $g(z)$ has a zero of order k at z_0 and that $h(z)$ a zero of order $k + 1$. Then $g(z)/h(z)$ has a simple pole with residue given by*

$$\operatorname{Res}\left(\frac{g}{h}, z_0\right) = (k + 1) \frac{g^{(k)}(z_0)}{h^{(k+1)}(z_0)}. \tag{2}$$

Proof.   By Taylor's Theorem and the fact that $g(z_0), \cdots, g^{(k-1)}(z_0) = 0$, we can write

$$g(z) = \frac{(z - z_0)^k}{k!} g^{(k)}(z_0) + (z - z_0)^{k+1} \tilde{g}(z)$$

where $\tilde{g}$ is analytic. Similarly,

$$h(z) = \frac{(z - z_0)^{k+1}}{(k + 1)!} h^{(k+1)}(z_0) + (z - z_0)^{k+2} \tilde{h}(z).$$

Thus

$$(z - z_0) \frac{g(z)}{h(z)} = \frac{\dfrac{g^{(k)}(z_0)}{k!} + (z - z_0)\tilde{g}(z)}{\dfrac{h^{(k+1)}(z_0)}{(k + 1)!} + (z - z_0)\tilde{h}(z)}.$$

As $z \to z_0$, this converges to (by the Quotient Theorem for Limits)

$$(k + 1) \frac{g^{(k)}(z_0)}{h^{(k+1)}(z_0)},$$

which proves our assertion. ∎

Note.   We could also prove

$$\lim_{z \to z_0} (z - z_0) \frac{g(z)}{h(z)} = (k + 1) \frac{g^{(k)}(z_0)}{h^{(k+1)}(z_0)}$$

by using l'Hopital's Rule (see exercise 6, Section 3.3) and observing that both $(z - z_0) g(z)$ and $h(z)$ are analytic at z_0 with z_0 a zero of order $(k + 1)$.

Examples.
   i. e^z/z at $z = 0$. In this case 0 is not a zero of e^z but is a first-order zero of z, so the residue at 0 is $1 \cdot e^0/1 = 1$. Clearly, Theorem 2 also applies.
   ii. $e^z/\sin z$ at 0. e^z has a zero of order zero at the point $z = 0$, and, since $\cos 0 = 1$, 0 is a first-order 0 of $\sin z$. Thus the residue is $e^0/\cos 0 = 1$.
   iii. $z/(z^2 + 1)$ at $z = i$. Here $g(z) = z$, $h(z) = z^2 + 1$. Therefore, $g(i) = i \neq 0$ and $h(i) = 0$, $h'(i) = 2i \neq 0$. Thus the residue at i is $g(i)/h'(i) = 1/2$.

iv. $z/(z^4-1)$ at $z=1$. Here $g(z)=z$, and $h(z)=z^4-1$. Thus $g(1)=1$ $\neq 0$ and $h(1)=0$, $h'(1)=4\neq 0$, so the residue is 1/4.

v. $z/(1-\cos z)$ at $z=0$. Here $g(0)=0$ and $g'(z)=1\neq 0$, so 0 is a simple zero of g. Also $h(0)=0$, $h'(0)=\sin 0=0$, and $h''(0)=\cos 0$ $=1\neq 0$, so 0 is a double zero of h. Thus, by formula (2) (see also line 5 of Table 4.1), the residue at 0 is

$$2 \cdot \frac{g'(0)}{h''(0)} = 2 \cdot \frac{1}{1} = 2.$$

Double poles

As the order of poles increases the formulas become more complicated and the residues become more laborious to obtain. However, for second-order poles the situation is still relatively simple. Probably the most useful formula for finding the residue in this case is formula (3) of the following theorem.

Theorem 4.   *Let g and h be analytic at z_0 and let $g(z_0)\neq 0$, $h(z_0)=0$, $h'(z_0)=0$, and $h''(z_0)\neq 0$. Then $g(z)/h(z)$ has a second-order pole at z_0 and the residue is*

$$\mathrm{Res}\left(\frac{g}{h}, z_0\right) = 2\,\frac{g'(z_0)}{h''(z_0)} - \frac{2}{3}\,\frac{g(z_0)h'''(z_0)}{[h''(z_0)]^2}. \tag{3}$$

Proof.   Since g has no zero and h has a second-order zero, we know that the pole is of second order (see the remark preceding Theorem 3). Thus we may write the Laurent series in the form

$$\frac{g(z)}{h(z)} = \frac{b_2}{(z-z_0)^2} + \frac{b_1}{z-z_0} + a_0 + a_1(z-z_0) + a_2(z-z_0)^2 + \cdots,$$

and we want to compute b_1. We can write

$$g(z) = g(z_0) + g'(z_0)(z-z_0) + \frac{g''(z_0)}{2}(z-z_0)^2 + \cdots$$

and

$$h(z) = \frac{h''(z_0)}{2}(z-z_0)^2 + \frac{h'''(z_0)}{6}(z-z_0)^3 + \cdots.$$

Therefore,

$$g(z) = h(z)\left\{\frac{b_2}{(z-z_0)^2} + \frac{b_1}{z-z_0} + a_0 + a_1(z-z_0) + \cdots\right\}$$

$$= \left(\frac{h''(z_0)}{2} + \frac{h'''(z_0)}{6}(z-z_0) + \cdots\right)$$

$$\cdot (b_2 + b_1(z-z_0) + a_0(z-z_0)^2 + \cdots).$$

We know we can multiply out these two convergent power series as if they were polynomials (see example 2, Section 3.2). The result is

$$g(z) = \frac{b_2 \, h''(z_0)}{2} + \left\{ \frac{b_2 \, h'''(z_0)}{6} + \frac{b_1 \, h''(z_0)}{2} \right\} (z - z_0) + \cdots$$

Since these two power series are equal we can conclude that the coefficients are equal. Therefore,

$$g(z_0) = \frac{b_2 \cdot h''(z_0)}{2}, \; g'(z_0) = \frac{b_2 \, h'''(z_0)}{6} + \frac{b_1 \, h''(z_0)}{2}.$$

Solving for b_1 yields the theorem. ∎

The following theorem may be proved in an analogous manner.

Theorem 5. *Let g and h be analytic at z_0 and let $g(z_0) = 0$, $g'(z_0) \neq 0$, $h(z_0) = 0$, $h'(z_0) = 0$, $h''(z_0) = 0$, and $h'''(z_0) \neq 0$. Then g/h has a second-order pole at z_0 with residue*

$$\frac{3g''(z_0)}{h'''(z_0)} - \frac{3}{2} \frac{g'(z_0) h^{(iv)}(z_0)}{[h'''(z_0)]^2}. \tag{4}$$

The proof is left for the student as exercise 3.

Examples.
 i. $e^z/(z-1)^2$ has a second-order pole at $z_0 = 1$; $g(z) = e^z, h(z) = (z-1)^2$ and $g(1) = e \neq 0$, $h(z_0) = 0$, $h'(z_0) = 2(z_0 - 1) = 0$ and $h''(z_0) = 2 \neq 0$. Therefore, by formula (3), the residue at 1 is $(2 \cdot e)/2 - 2/3 \cdot (e \cdot 0)/2^2 = e$.
 ii. $(e^z - 1)/\sin^3 z$ with $z_0 = 0$. $g(z) = e^z - 1$, $h(z) = \sin^3 z$ and $g(0) = 0$, $g'(0) \neq 0$, $h(0) = 0$, $h'(0) = 3\sin^2 0 \cos 0 = 0$, $h''(z) = 6\sin z \cdot \cos^2 z - 3\sin^3 z$, which also is zero at 0, and $h'''(z) = 6\cos^3 z - 12\sin^2 z \cdot \cos z - 9\sin^2 z \cdot \cos z$, which is 6 at $z = 0$. We also compute $h^{(iv)}(0) = 0$. Thus by formula (4) the residue is $3 \cdot 1/6 = 1/2$.

Note.   For a second-order pole of the form $g(z)/(z - z_0)^2$ where $g(z_0) \neq 0$, formula (3) simplifies to $g'(z_0)$.

Higher-order poles

For poles of order greater than 2 we could develop formulas in the same manner in which we developed the preceding ones, but they would be quite complicated. Instead two general methods can be used. The first is described in Theorem 6.

Theorem 6. *Let f have an isolated singularity at z_0 and let k be the smallest integer ≥ 0 such that $\lim\limits_{z \to z_0} (z - z_0)^k f(z)$ exists. Then $f(z)$ has a pole of order k at z_0 and, if we let $\varphi(z) = (z - z_0)^k f(z)$, then φ can be defined uniquely at z_0 so that φ is analytic at z_0 and*

$$\text{Res}(f, z_0) = \frac{\varphi^{(k-1)}(z_0)}{(k-1)!}. \tag{5}$$

Proof.   Since $\lim\limits_{z \to z_0} (z - z_0)^k f(z)$ exists, $\varphi(z) = (z - z_0)^k f(z)$ has a removable singularity at z_0, by Theorem 15, Section 3.3. Thus $\varphi(z) = (z - z_0)^k f(z) = b_k + b_{k-1}(z - z_0) + \cdots + b_1(z - z_0)^{k-1} + a_0(z - z_0)^k + \cdots$ in a neighborhood of z_0, and so

$$f(z) = \frac{b_k}{(z - z_0)^k} + \frac{b_{k-1}}{(z - z_0)^{k-1}}$$

$$+ \cdots + \frac{b_1}{(z - z_0)} + a_0 + a_1(z - z_0) + \cdots .$$

If $b_k = 0$, then $\lim\limits_{z \to z_0} (z - z_0)^{k-1} f(z)$ exists, which contradicts the hypothesis about k. Thus z_0 is a pole of order k. If we differentiate the expansion for $\varphi(z)$ $k-1$ times at z_0, we obtain $\varphi^{(k-1)}(z_0) = (k-1)! \, b_1$. ∎

In this theorem it is the formula

$$\text{Res}(f, z_0) = \frac{\varphi^{(k-1)}(z_0)}{(k-1)!}$$

that is important rather than the test for the order. It usually is easier to test the order by writing (if possible) $f = g/h$ and showing that h has a zero of order k greater than g. Then we have a pole of order k (as was explained in the text preceding Theorem 3).

Let us now suppose that the form of f makes application of Theorem 6 inconvenient. (For example, consider $e^z/\sin^4 z$ with $z_0 = 0$. Here $k = 4$ since the numerator has no zero and the denominator has a zero of order 4.) There is an alternate method that generalizes Theorem 4. Suppose that $f(z) = g(z)/h(z)$ and that $h(z)$ has a zero of order k more than g at z_0; therefore, f has a pole of order k. We may write

$$\frac{g(z)}{h(z)} = \frac{b_k}{(z - z_0)^k} + \cdots + \frac{b_1}{(z - z_0)} + \rho(z)$$

where ρ is analytic. Also suppose that z_0 is a zero of order m for $g(z)$ and a zero of order $m + k$ for $h(z)$. Then

$$g(z) = \sum_{n=m}^{\infty} \frac{g^{(n)}(z_0)(z - z_0)^n}{n!}$$

and

$$h(z) = \sum_{n=m+k}^{\infty} \frac{h^{(n)}(z_0)(z-z_0)^n}{n!}.$$

Thus we can write

$$\sum_{n=m}^{\infty} \frac{g^{(n)}(z_0)(z-z_0)^n}{n!} = \left\{ \sum_{n=m+k}^{\infty} \frac{h^{(n)}(z_0)(z-z_0)^n}{n!} \right\}$$
$$\cdot \left\{ \frac{b_k}{(z-z_0)^k} + \cdots + \frac{b_1}{(z-z_0)} + \rho(z) \right\}.$$

We can then multiply out the right side of the equation as if the factors were polynomials (because of example 2, Section 3.2) and compare the coefficients of $(z-z_0)^m$, $(z-z_0)^{m+1}$, $\cdots$, $(z-z_0)^{m+k-1}$ to obtain k equations in $b_1, b_2, \cdots, b_k$. Finally, we can solve these equations for b_1. This method may sometimes be more practical than that of Theorem 6. When $m = 0$ (that is, when $g(z_0) \neq 0$), the explicit formula contained in Theorem 7 can be used. (The student should prove this theorem by using the procedure just described.)

Theorem 7. *Let g and h be analytic at z_0, with $g(z_0) \neq 0$, and let $h(z_0) = 0 = \cdots = h^{(k-1)}(z_0)$, $h^{(k)}(z_0) \neq 0$. Then g/h has a pole of order k and the residue is given by*

$$\text{Res } (g/h, z_0) = \left[\frac{k!}{h^{(k)}(z_0)} \right]^k \times$$

$$\begin{vmatrix}
\dfrac{h^{(k)}(z_0)}{k!} & 0 & 0 & \cdots & 0 & g(z_0) \\[2ex]
\dfrac{h^{(k+1)}(z_0)}{(k+1)!} & \dfrac{h^{(k)}(z_0)}{k!} & 0 & \cdots & 0 & g^{(1)}(z_0) \\[2ex]
\dfrac{h^{(k+2)}(z_0)}{(k+2)!} & \dfrac{h^{(k+1)}(z_0)}{(k+1)!} & \dfrac{h^{(k)}(z_0)}{k!} & \cdots & 0 & \dfrac{g^{(2)}(z_0)}{2!} \\[2ex]
\vdots & \vdots & \vdots & & & \vdots \\[2ex]
\dfrac{h^{(2k-1)}(z_0)}{(2k-1)!} & \dfrac{h^{(2k-2)}(z_0)}{(2k-2)!} & \dfrac{h^{(2k-3)}(z_0)}{(2k-3)!} & \cdots & \dfrac{h^{(k+1)}(z_0)}{(k+1)!} & \dfrac{g^{(k-1)}(z_0)}{(k-1)!}
\end{vmatrix} \quad (6)$$

where $|\cdot|$ indicates the determinant of the $k \times k$ matrix.

Examples.
   i. $z^2/[(z-1)^3(z+1)]$ at $z_0 = 1$. The pole is of order 3. We use formula 5. In this case,

$$\varphi(z) = \frac{z^2}{z+1},$$

and so

$$\varphi'(z) = \frac{(z+1) \cdot 2z - z^2}{(z+1)^2} = \frac{z^2 + 2z}{(z+1)^2} = 1 - \frac{1}{(z+1)^2},$$

and

$$\varphi''(z) = \frac{2}{(z+1)^3} \quad \text{so} \quad \varphi''(1) = \frac{1}{4}.$$

Thus since $k = 3$, the residue is $1/2 \cdot 1/2^2 = 1/8$.

ii. $e^z/\sin^3 z$ at $z = 0$. Here $k = 3$ and we shall use formula (6), with $g(z) = e^z$ and $h(z) = \sin^3 z$. We need to compute $h'''(0)$, $h^{(iv)}(0)$, and $h^{(v)}(0)$. These are, by straightforward but laborious computation, $h'''(0) = 6$, $h^{(iv)}(0) = 0$, and $h^{(v)}(0) = -60$; so $h'''/3! = 1$, $h^{(iv)}/4! = 0$, and $h^{(v)}/5! = -1/2$. Also $g^{(l)}(0)/l! = 1/l!$, so the residue is, by formula (6),

$$\left(\frac{3!}{6}\right)^3 \times \begin{vmatrix} 1 & 0 & 1 \\ 0 & 1 & 1 \\ \dfrac{-1}{2} & 0 & \dfrac{1}{2} \end{vmatrix} = \begin{vmatrix} 0 & 0 & 1 \\ -1 & 1 & 1 \\ -1 & 0 & \dfrac{1}{2} \end{vmatrix} = 1.$$

(The last column is subtracted from the first.)

Essential singularities

In the case of an essential singularity there are no simple formulas like the preceding ones, so we must rely on our ability to find the Laurent expansion.

Example.

$$f(z) = e^{(z + 1/z)} = e^z \cdot e^{1/z} = \left(1 + z + \frac{z^2}{2!} + \cdots\right)\left(1 + \frac{1}{z} + \frac{1}{2!z^2} + \cdots\right).$$

Gathering terms involving $1/z$, we get

$$\frac{1}{z}\left\{1 + \frac{1}{2!} + \frac{1}{2!3!} + \frac{1}{3!4!} + \cdots\right\}.$$

(We multiply out as in the procedure of example 2, Section 3.2, a method that is justified by a proof of the more general result that is outlined in exercise 10.) The residue is thus

$$\text{Res}(f,0) = 1 + \frac{1}{2!} + \frac{1}{2!3!} + \frac{1}{3!4!} + \cdots .$$

We cannot sum the series explicitly.

Table 4.1 Techniques for Finding Residues

In this table g and h are analytic at z_0 and f has an isolated singularity. The most useful and common tests are indicated by an asterisk.

Function	Test	Type of Singularity	Formula for Residue at z_0
1. $f(z)$	$\lim_{z \to z_0}(z - z_0)f(z) = 0$	removable	residue $= 0$
*2. $\dfrac{g(z)}{h(z)}$	g and h have zeros of same order	removable	residue $= 0$
*3. $f(z)$	$\lim_{z \to z_0}(z - z_0)f(z)$ exists and is $\neq 0$	simple pole	residue $= \lim_{z \to z_0}(z - z_0)f(z)$
*4. $\dfrac{g(z)}{h(z)}$	$g(z_0) \neq 0,\ h(z_0) = 0,\ h'(z_0) \neq 0$	simple pole	residue $= \dfrac{g(z_0)}{h'(z_0)}$
5. $\dfrac{g(z)}{h(z)}$	g has zero of order k, h has zero of order $k + 1$	simple pole	residue $= (k + 1)\dfrac{g^{(k)}(z_0)}{h^{(k+1)}(z_0)}$
*6. $\dfrac{g(z)}{h(z)}$	$g(z_0) \neq 0$ $h(z_0) = 0 = h'(z_0)$ $h''(z_0) \neq 0$	second-order pole	residue $= 2\dfrac{g'(z_0)}{h''(z_0)} - \dfrac{2}{3}\dfrac{g(z_0)h'''(z_0)}{[h''(z_0)]^2}$
*7. $\dfrac{g(z)}{(z - z_0)^2}$	$g(z_0) \neq 0$	second-order pole	residue $= g'(z_0)$
8. $\dfrac{g(z)}{h(z)}$	$g(z_0) = 0,\ g'(z) \neq 0,\ h(z_0) = 0 = h'(z_0)$ $= h''(z_0),\ h'''(z_0) \neq 0$	second-order pole	residue $= 3\dfrac{g''(z_0)}{h'''(z_0)} - \dfrac{3}{2}\dfrac{g'(z_0)h^{(iv)}(z_0)}{[h'''(z_0)]^2}$
9. $f(z)$	k is the smallest integer such that $\lim_{z \to z_0} \varphi(z_0)$ exists where $\varphi(z) = (z - z_0)^k f(z)$	pole of order k	residue $= \lim_{z \to z_0}\dfrac{\varphi^{(k-1)}(z)}{(k-1)!}$
*10. $\dfrac{g(z)}{h(z)}$	g has zero of order l, h has zero of order $k + l$	pole of order k	residue $= \lim_{z \to z_0}\dfrac{\varphi^{(k-1)}(z)}{(k-1)!}$ where $\varphi(z) = (z - z_0)^k \dfrac{g}{h}$
11. $\dfrac{g(z)}{h(z)}$	$g(z_0) \neq 0,\ h(z_0) = \cdots = h^{k-1}(z_0)$ $= 0,\ h^k(z_0) \neq 0$	pole of order k	residue given by formula (6) in Theorem 7

Worked Examples

1. Find the residues at all singularities of

$$\tan z = \frac{\sin z}{\cos z}.$$

Solution.   The singularities of $\tan z$ occur when $\cos z = 0$. The zeros of $\cos z$ occur at

$$z = \pm\frac{\pi}{2}, \pm\frac{3\pi}{2}, \pm\frac{5\pi}{2}, \cdots$$

The student should recall that these points are the only zeros of $\cos z$. We conclude that the singularities of $\tan z$ occur at the points $z_n = (2n + 1)\pi/2$, where n is an integer. We choose $g(z) = \sin z$ and $h(z) = \cos z$. At any z_n, $h'(z_n) = \pm 1 \neq 0$, so each z_n is a simple pole of $\tan z$. Thus we may use formula 4 of Table 4.1 to obtain

$$\text{Res}(\tan z, z_n) = \frac{g(z_n)}{h'(z_n)} = -1.$$

2. Evaluate the residue of $(z^2 - 1)/(z^2 + 1)^2$ at $z = i$.

Solution. $(z^2 + 1)^2$ has a zero order 2 at i and $i^2 - 1 \neq 0$, so $(z^2 - 1)/(z^2 + 1)^2$ has a pole of order 2; thus to find the residue we use formula 6 of Table 4.1. We choose $g(z) = z^2 - 1$, which, at $z = i$, is -2, and $g'(i) = 2i$. We also take $h(z) = (z^2 + 1)^2$ and note that $h'(z) = 4z(z^2 + 1)$, so $h(i) = h'(i) = 0$. Also, $h''(z) = 4(z^2 + 1) + 8z^2 = 12z^2 + 4$, so $h''(i) = -8$ and $h'''(z) = 24z$; therefore, $h'''(i) = 24i$.
Thus the residue is

$$\frac{2 \cdot 2i}{-8} - \frac{2}{3} \cdot \frac{(-2) \cdot 24i}{64} = 0.$$

3. Compute the residue of $z^2/\sin^2 z$ at $z = 0$.

Solution.   Since both numerator and denominator have a zero of order two, the singularity is removable, so the residue is zero.

Exercises

1. Find the residues of the following functions at the indicated points:

a. $\dfrac{e^{z^2}}{z - 1}$, $z_0 = 1$.

b. $\dfrac{e^{z^2}}{(z - 1)^2}$, $z_0 = 0$.

c. $\left(\dfrac{\cos z - 1}{z}\right)^2$, $z_0 = 0$.

d. $\dfrac{z^2}{z^4 - 1}$, $z_0 = e^{\frac{i\pi}{2}}$.

e. $\dfrac{e^z - 1}{\sin z}$, $z_0 = 0$.

f. $\dfrac{1}{e^z - 1}$, $z_0 = 0$.

g. $\dfrac{z + 2}{z^2 - 2z}$, $z_0 = 0$.

h. $\dfrac{1 + e^z}{z^4}$, $z_0 = 0$.

i. $\dfrac{e^z}{(z^2 - 1)^2}$, $z_0 = 1$.

2. If $f(z)$ has residue b_1 at $z = z_0$, show by example that $[f(z)]^2$ does not have residue b_1^2 at $z = z_0$.

3. Deduce Theorem 5 from Theorem 4.

4. Explain what is wrong with the following reasoning. Let

$$f(z) = \frac{1 + e^z}{z^2} + \frac{1}{z}.$$

$f(z)$ has a pole at $z = 0$ and the residue at that point is the coefficient of $1/z$, namely 1. Compute the residue correctly.

5. Find all singular points of the following functions and compute the residues at those points:

a. $\dfrac{1}{e^z - 1}$.

b. $\sin\left(\dfrac{1}{z}\right)$.

c. $\dfrac{1}{z^3(z + 4)}$.

d. $\dfrac{1}{z^2 + 2z + 1}$.

e. $\dfrac{1}{z^3 - 3}$.

6. Find the residue of $1/(z^2\sin z)$ at $z = 0$.

7. Complete the proof of Theorem 7.

8. If f_1 and f_2 have residues r_1 and r_2 at z_0, show that the residue of $f_1 + f_2$ at z_0 is $r_1 + r_2$.

9. If f_1 and f_2 have simple poles at z_0, show that $f_1 f_2$ has a second-order pole at z_0. Derive a formula for the residue.

10.   Let

$$f = \cdots + \frac{b_k}{(z - z_0)^k} + \cdots + \frac{b_1}{z - z_0} + a_0 + a_1(z - z_0) + \cdots$$

and

$$g = \cdots + \frac{d_k}{(z - z_0)^k} + \cdots + \frac{d_1}{z - z_0} + c_0 + c_1(z - z_0) + \cdots$$

be Laurent expansions for f and g valid for $0 < |z - z_0| < r$. Show that the Laurent expansion for fg is obtained by formally multiplying these series. Do this by proving the following result: If $\sum_{n=0}^{\infty} a_n$ and $\sum_{n=0}^{\infty} b_n$ are absolutely convergent, then

$$\left(\sum_{n=0}^{\infty} a_n \right) \left(\sum_{n=0}^{\infty} b_n \right) = \sum_{n=0}^{\infty} c_n$$

where $c_n = \sum_{j=0}^{n} a_j b_{n-j}$; moreover, the series $\sum_{n=0}^{\infty} c_n$ is absolutely convergent.

Hint.   Show that

$$\sum_{j=0}^{n} |c_j| \leq \sum_{j=0}^{n} \sum_{k=0}^{n} |a_j| \, |b_{k-j}| \leq \left(\sum_{j=0}^{n} |a_j| \right) \left(\sum_{k=0}^{n} |b_k| \right)$$

and use this to deduce that $\sum c_n$ converges absolutely. Estimate the error between

$$\sum_{k=0}^{n} c_k, \qquad \left(\sum_{j=0}^{n} a_j \right) \left(\sum_{k=0}^{n} b_k \right), \qquad \text{and} \qquad \left(\sum_{j=0}^{\infty} a_j \right) \left(\sum_{k=0}^{\infty} b_k \right).$$

11.   Compute the residues of the following functions at their singularities:

a. $\dfrac{1}{(1 - z)^3}.$

b. $\dfrac{e^z}{(1 - z)^3}.$

c. $\dfrac{1}{z(1 - z)^3}.$

d. $\dfrac{e^z}{z(1 - z)^3}.$

4.2   The Residue Theorem

The Residue Theorem, which is proved in this section, includes Cauchy's Theorem and Cauchy's Integral Formula as special cases. It is one of the main results of complex analysis and leads quickly to interesting applications, some of which are considered in the next section. The main tools needed to prove the theorem are Cauchy's

Theorem (see Theorems 5 and 17 of Chapter 2) and the Laurent expansion (see Theorem 13, Section 3.3).

The precise proof of the Residue Theorem is preceded by two intuitive proofs that are based only on the material in Section 2.2 and the following property of the residue at z_0:

$$2\pi i \; \text{Res}(f,z_0) = \int_\gamma f,$$

where γ is a small circle around z_0 (see Theorem 14, Section 3.3). For most practical examples the intuitive proofs are in fact perfectly adequate, but, as was evident in Section 2.2, it is difficult to formulate a general theorem to which the argument rigorously applies.

Residue Theorem

Theorem 8 (Residue Theorem). *Let A be a region and let $z_1, \cdots ,$ $z_n \in A$, be n distinct points in A. Let f be analytic on $A\backslash\{z_1, z_2, \cdots , z_n\}$; that is, f is analytic on A except for isolated singularities at $z_1, \cdots , z_n$. Let γ be a closed curve in A homotopic to a point in A. Assume that no z_i lies on γ. Then*

$$\int_\gamma f = 2\pi i \sum_{i=1}^{n} \{\text{Res}(f,z_i) \cdot I(\gamma,z_i)\}, \qquad (1)$$

where Res (f,z_i) is the residue of f at z_i (see the preceding section) and $I(\gamma,z_i)$ is the index (or winding number) of γ with respect to z_i (see Section 2.4).

In practical examples, γ will be a simple closed curve traversed counterclockwise, and thus $I(\gamma,z_i)$ will be 1 or 0 according to whether z_i lies inside γ or outside γ. (This is illustrated in Figure 4.1.)

The same policy regarding computation of the index $I(\gamma,z)$ as was used in Section 2.4 will be followed in this section. An intuitive proof is acceptable so long as such statements can be substantiated with

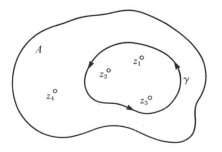

Figure 4.1   Residue Theorem:

$$\int_\gamma f = 2\pi i \{\text{Res}(f,z_1) + \text{Res}(f,z_2)$$

$$+ \text{Res}(f,z_3)\}.$$

homotopy arguments. For example, $I(\gamma,z_0) = +1$ if γ can be shown to be homotopic, in $\mathbb{C}\backslash\{z_0\}$, to $\gamma(t) = z_0 + re^{i\theta}$, $0 \le \theta \le 2\pi$.

Assuming that we accept the Jordan Curve Theorem, a general formulation of the Residue Theorem for simple closed curves may be stated as follows: *If γ is a simple closed curve in the region A whose inside lies in A and if f is analytic on $A\backslash\{z_1, \cdots, z_n\}$, then $\int_\gamma f$ is $2\pi i$ times the sum of the residues of f inside γ when γ is traversed in the counterclockwise direction.* This is the classical way of stating the Residue Theorem, but the statement in Theorem 8 is preferable because it does not restrict us to simple closed curves and does not rely on the difficult Jordan Curve Theorem.

Following are two short intuitive proofs of Theorem 8 for simple closed curves. They will be illustrated by an example showing that, in practical cases, such proofs can be made quite precise.

First Intuitive Proof of Theorem 8 for Simple Closed Curves.   Since γ is contractible in A to a point in A, the inside of γ lies in A. Suppose that each z_i lies in the inside of γ. Around each z_i draw a circle γ_i that also lies in the inside of γ. Then apply example 3 of Section 2.2 (the generalized Deformation Theorem) to obtain $\int_\gamma f = \sum_{i=1}^n \int_{\gamma_i} f$, since f is analytic on $\gamma,\gamma_1, \cdots, \gamma_n$ and the region between them (Figure 4.2). Suppose that $\gamma,\gamma_1, \cdots, \gamma_n$ are all traversed in the counterclockwise direction. As was proved in Theorem 14, Section 3.3, $\int_{\gamma_i} f = 2\pi i$ Res(f,z_i), so $\int_\gamma f = 2\pi i \sum_{i=1}^n$ Res(f,z_i), which is the statement of Theorem 8 since $I(\gamma,z) = 1$ for z inside γ and $I(\gamma,z) = 0$ for z outside γ. ∎

Second Intuitive Proof of Theorem 8 for Simple Closed Curves.   This proof proceeds in the same manner as the preceding one except that

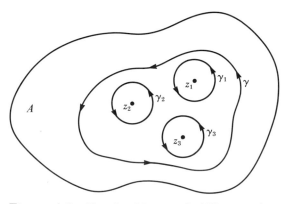

Figure 4.2   First intuitive proof of Theorem 8.

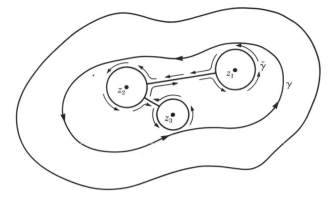

Figure 4.3   Second intuitive proof of Theorem 8.

a different justification is given that $\int_\gamma f = \Sigma_{i=1}^n \int_{\gamma_i} f$. The circles are connected as shown in Figure 4.3, to obtain a new curve $\tilde{\gamma}$. Thus γ and $\tilde{\gamma}$ are homotopic in $A\backslash\{z_1, \cdot \cdot \cdot, z_n\}$, so, by the Deformation Theorem, $\int_\gamma f = \int_{\tilde{\gamma}} f$. But $\int_{\tilde{\gamma}} f = \Sigma_{i=1}^n \int_{\gamma_i} f$, since the portions along the connecting curves cancel out. ∎

It is important that the student clearly understand why these proofs fail to be precise proofs of Theorem 8. First, we assumed that γ is simple. Second, we used the Jordan Curve Theorem (which was not proved) to be able to discuss the inside and outside of γ and the fact that $I(\gamma,z) = 1$ for z inside γ and that $I(\gamma,z) = 0$ for z outside γ. Finally, in the first intuitive proof we used example 3 of Section 2.2, which was established only informally, to justify that $\int_\gamma f = \Sigma_{i=1}^n \int_{\gamma_i} f$.

Example.   Consider $\int_\gamma dz/(z^2 - 1)$, where γ is a circle with center 0 and radius 2. The function $1/(z^2 - 1)$ has simple poles at -1, 1 (see Figure 4.4). Note that in this example we can discuss the inside and outside of γ, and we know that -1 and 1 have an index $+1$ with respect to γ. Thus the Jordan Curve Theorem is not needed in such a concrete example (see Section 2.4).

We draw two circles γ_1 and γ_2 of radius $1/4$ around -1 and 1, respectively. The only statement in the proofs of Theorem 8 that was not precise was

$$\int_\gamma f = \int_{\gamma_1} f + \int_{\gamma_2} f.$$

The exact justification of $\int_\gamma f = \int_{\gamma_1} f + \int_{\gamma_2} f$ in terms of homotopies was explained in example 3, Section 2.3.

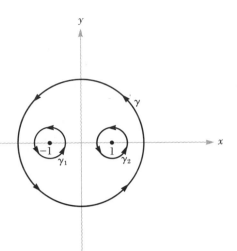

Figure 4.4

Note that in this case

$$
\int_\gamma \frac{dz}{z^2 - 1} = 2\pi i \left\{ \text{Res}\left(\frac{1}{z^2 - 1}, -1\right) + \text{Res}\left(\frac{1}{z^2 - 1}, 1\right) \right\}
$$

$$
= 2\pi i \left\{ \frac{1}{2(-1)} + \frac{1}{2 \cdot 1} \right\} = 0.
$$

We could also justify $\int_\gamma f = \int_{\gamma_1} f + \int_{\gamma_2} f$ by considering the curve in Figure 4.5(i) and showing that it is homotopic in $\mathbb{C}\backslash\{1, -1\}$ to a point. This is geometrically clear; a homotopy is indicated in Figure 4.5(ii) and (iii).

Precise Proof of Theorem 8.   Since z_i is an isolated singularity of f, we can write a Laurent series expansion

$$
f(z) = \sum_{n=0}^\infty a_n(z - z_i)^n + \sum_{m=1}^\infty \frac{b_m}{(z - z_i)^m}
$$

in some deleted neighborhood of z_i of the form $\{z \mid r > |z - z_i| > 0\}$ for some $r > 0$. Recall from Theorem 13, Section 3.3, that

$$
\sum_{m=1}^\infty \frac{b_m}{(z - z_i)^m}
$$

is called the singular part of the Laurent series expansion and that it converges on $\mathbb{C}\backslash\{z_i\}$ and uniformly outside any circle $|z - z_i| = \epsilon > 0$. Hence

$$
\sum_{m=1}^\infty \frac{b_m}{(z - z_i)^m}
$$

is analytic on $\mathbb{C}\backslash\{z_i\}$ (see Theorem 6, Section 3.1). Denote the singular part of the Laurent expansion of f around z_i by $S_i(z)$.

Consider the function

$$g(z) = f(z) - \sum_{i=1}^{n} S_i(z).$$

Since f is analytic on $A\backslash\{z_1, \cdots, z_n\}$ and since each $S_i(z)$ is analytic on $\mathbb{C}\backslash\{z_i\}$, g is analytic on $A\backslash\{z_1, \cdots, z_n\}$.

All the z_i's are removable singularities of g because on a deleted neighborhood $\{z|\ r > |z - z_i| > 0\}$, which does not contain any of the singularities, we have

$$f(z) = \sum_{n=0}^{\infty} a_n(z - z_i)^n + S_i(z),$$

and so

$$g(z) = \sum_{n=0}^{\infty} a_n(z - z_i)^n - \sum_{j=1}^{i-1} S_j(z) - \sum_{j=i+1}^{n} S_j(z).$$

Since the functions S_j, $j \neq i$, are analytic on $\mathbb{C}\backslash\{z_j\}$, we know that $\lim_{z \to z_i} g(z)$ exists and equals $a_0 - \Sigma_{\substack{j=i \\ j \neq i}}^{n} S_j(z_i)$. Consequently, z_i is a removable singularity of g.

Because g can be defined at the points z_i in such a way that g is analytic on all of A, we may apply the Cauchy Theorem (Theorem 17, Section 2.3) to obtain $\int_\gamma g = 0$. Hence

$$\int_\gamma f = \sum_{i=1}^{n} \int_\gamma S_i.$$

Next consider the integral $\int_\gamma S_i$. The function $S_i(z)$ is of the form

$$\sum_{m=1}^{\infty} \frac{b_m}{(z - z_i)^m},$$

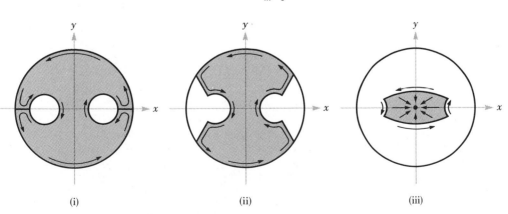

(i)   (ii)   (iii)

Figure 4.5   A curve that is homotopic to a point.

which, as we have noted, converges uniformly outside a small disk centered at z_j. Thus the convergence is uniform on γ. (Since $\mathbb{C}\backslash\gamma[a,b]$ is an open set, each z_j has a small disk around it not meeting γ.) By Theorem 7, Section 3.1,

$$\int_\gamma S_i = \sum_{m=1}^\infty \int_\gamma \frac{b_m}{(z-z_i)^m}\, dz.$$

But for $m > 1$,

$$\frac{1}{(z-z_i)^m} = \frac{d}{dz}\left\{\frac{(z-z_i)^{1-m}}{1-m}\right\} \quad (z \neq z_i),$$

so by Theorem 4, Section 2.1, and the fact that γ is a closed curve, all terms are zero except the term in which $m = 1$. Thus

$$\int_\gamma S_i = \int_\gamma \frac{b_1}{z-z_i}\, dz = b_1 \int_\gamma \frac{1}{z-z_i}\, dz.$$

But this equals, by Definition 5 of Section 2.4 for the index, $b_1 \cdot 2\pi i \cdot I(\gamma, z_i) = 2\pi i\, \mathrm{Res}(f, z_i)\, I(\gamma, z_i)$. Thus

$$\int_\gamma f = \sum_{i=1}^n \int_\gamma S_i = \sum_{i=1}^n 2\pi i \mathrm{Res}(f, z_i) I(\gamma, z_i),$$

and the theorem is proved. ■

Worked Examples

1.   Let γ be a circle of radius $1/2$; $\gamma(t) = e^{it}/2$, $0 \le t \le 2\pi$. Evaluate

$$\int_\gamma \frac{dz}{z^2 + z + 1}.$$

Solution.   The singularities occur at the points for which the denominator vanishes. These points are

$$z_1 = \frac{-1 + \sqrt{1-4}}{2} = \frac{-1 + \sqrt{3}i}{2}$$

and

$$z_2 = \frac{-1 - \sqrt{1-4}}{2} = \frac{-1 - \sqrt{3}i}{2}.$$

It is easy to check that $|z_1| = |z_2| = 1$, so both z_1 and z_2 lie outside the circle of radius $1/2$. Now γ is homotopic to 0 in $\mathbb{C}\backslash\{z_1\}$ and $1/(z-z_1)$ is analytic on $\mathbb{C}\backslash\{z_1\}$, so by Cauchy's Theorem, $I(\gamma, z_1) = 0$. Similarly, $I(\gamma, z_2) = 0$. Therefore, by the Residue Theorem,

$$\int_\gamma \frac{dz}{z^2 + z + 1} = 0.$$

Alternatively, we could note that γ is homotopic to 0 in $\mathbb{C}\setminus\{z_1, z_2\}$ and $1/(z^2 + z + 1)$ is analytic on $\mathbb{C}\setminus\{z_1, z_2\}$; thus, by Cauchy's Theorem,

$$\int_\gamma \frac{dz}{z^2 + z + 1} = 0$$

(see Figure 4.6).

2. Evaluate

$$\int_\gamma \frac{dz}{z^4 + 1}$$

where γ consists of the portion of the x-axis from -2 to $+2$ and the semi-circle in the upper half plane from 2 to -2 centered at 0.

Solution.   The singular points of the integrand occur at the fourth roots of -1: $e^{\pi i/4}$, $e^{(\pi + 2\pi)i/4} = e^{3\pi i/4}$, $e^{(\pi + 4\pi)i/4} = e^{5\pi i/4}$, and $e^{(\pi + 6\pi)i/4} = e^{7\pi i/4}$ (see Figure 4.7). Thus, by the Residue Theorem,

$$\int_\gamma \frac{dz}{z^4 + 1} = 2\pi i \left\{ \operatorname{Res}\left(\frac{1}{z^4 + 1}, e^{\frac{\pi i}{4}}\right) I(\gamma,\, e^{\frac{\pi i}{4}}) \right.$$

$$+ \operatorname{Res}\left(\frac{1}{z^4 + 1}, e^{\frac{3\pi i}{4}}\right) I(\gamma,\, e^{\frac{3\pi i}{4}})$$

$$+ \operatorname{Res}\left(\frac{1}{z^4 + 1}, e^{\frac{5\pi i}{4}}\right) I(\gamma,\, e^{\frac{5\pi i}{4}})$$

$$\left. + \operatorname{Res}\left(\frac{1}{z^4 + 1}, e^{\frac{7\pi i}{4}}\right) I(\gamma,\, e^{\frac{7\pi i}{4}}) \right\}.$$

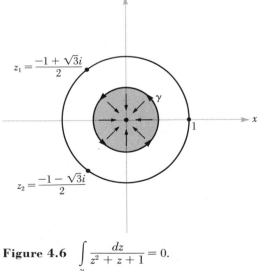

Figure 4.6 $\displaystyle\int_\gamma \frac{dz}{z^2 + z + 1} = 0.$

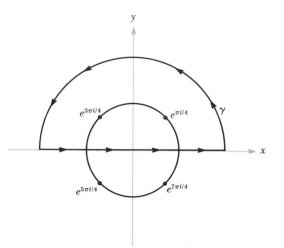

Figure 4.7

It is intuitively clear that $I(\gamma, e^{\pi i/4}) = 1$, $I(\gamma,\ e^{3\pi i/4}) = 1$, whereas the other two indexes are zero.

This can be more carefully justified as follows. γ is homotopic to a circle $\tilde{\gamma}$ around $e^{\pi i/4}$ traversed counterclockwise. To see this, reparametrize γ so that it is defined on the interval $[0, 2\pi]$. A suitable homotopy is then $H(s,t) = (1-t)\gamma(s) + t\tilde{\gamma}(s)$. This homotopy is illustrated in Figure 4.8. We know that $I(\tilde{\gamma},\ e^{\pi i/4}) = 1$ by example 3, Section 2.1, and that $I(\tilde{\gamma},\ e^{\pi i/4}) = I(\gamma, e^{\pi i/4})$ by the Deformation Theorem. Thus $I(\gamma, e^{\pi i/4}) = 1$ and, similarly, $I(\gamma, e^{3\pi i/4}) = 1$. Furthermore, γ can be contracted to the origin along the radii of the semicircle, so by Cauchy's Theorem, $I(\gamma, e^{5\pi i/4}) = 0$ and $I(\gamma, e^{7\pi i/4}) = 0$.

To calculate

$$\operatorname{Res}\left(\frac{1}{z^4 + 1},\ e^{\frac{\pi i}{4}}\right),$$

observe that $e^{\pi i/4}$ is a simple pole of the function $1/(z^4 + 1)$, so we may use formula 4 of Table 4.1 to obtain

$$\operatorname{Res}\left(\frac{1}{z^4 + 1},\ e^{\frac{\pi i}{4}}\right) = \frac{1}{4\left(e^{\frac{\pi i}{4}}\right)^3} = \frac{e^{\frac{\pi i}{4}}}{4e^{\pi i}} = -\frac{e^{\frac{\pi i}{4}}}{4}.$$

Similarly,

$$\operatorname{Res}\left(\frac{1}{z^4 + 1},\ e^{\frac{3\pi i}{4}}\right) = \frac{1}{4\left(e^{\frac{3\pi i}{4}}\right)^3} = \frac{e^{\frac{-\pi i}{4}}}{4}.$$

Therefore,

$$\int_\gamma \frac{dz}{z^4 + 1} = \frac{2\pi i}{4}\ \{e^{\frac{-\pi i}{4}} - e^{\frac{\pi i}{4}}\} = \pi \sin \frac{\pi}{4} = \frac{\pi\sqrt{2}}{2}.$$

Remark.   We do not actually have to use such a detailed method to calculate the indexes (winding numbers). We simply use our intuition to estimate the number of times the curve in question winds around the given point in the counterclockwise direction. But we must bear in mind that the justification for this intuition consists of an argument like the preceding one.

3. Evaluate

$$\int_\gamma \frac{1+z}{1-\cos z}\, dz$$

where γ is the circle of radius 7 around zero.

Solution.   The singularities of $(1+z)/(1-\cos z)$ occur where $1-\cos z = 0$. But $(e^{iz}+e^{-iz})/2 = 1$ implies that $(e^{iz})^2 - 2(e^{iz})+1 = 0$ and hence $e^{iz} = 1$. Therefore, the singularities occur at $z = 2\pi n$ for $n = \cdots, -2, -1, 0, 1, 2, 3, \cdots$. The only singularities of $(1+z)/(1-\cos z)$ that lie inside the circle of radius 7 are $z_1 = 0$, $z_2 = 2\pi$, and $z_3 = -2\pi$ (see Figure 4.9). Furthermore, $d(1-\cos z)/dz = \sin z$, which is zero at $0, -2\pi, +2\pi$; and $d^2(1-\cos z)/dz^2 = \cos z$, which is nonzero at $0, -2\pi$, and 2π; so these singularities are poles of order 2.

The residue at one of these poles z_0 is, by formula 6 of Table 4.1,

$$2\,\frac{g'(z_0)}{h''(z_0)} - \frac{2}{3}\,\frac{g(z_0)\, h'''(z_0)}{[h''(z_0)]^2}.$$

In this case $g(z) = 1 + z$, so $g'(z) = 1$ and $h(z) = 1 - \cos z$; therefore, $h'(z) = \sin z$, $h''(z) = \cos z$, $h'''(z) = -\sin z$. Thus $h'''(z) = 0$ for $z = z_1, z_2, z_3$, so the formula becomes $2g'(z_0)/h''(z_0)$.

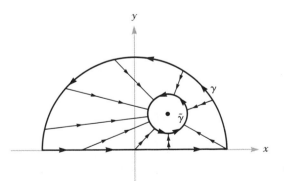

Figure 4.8   Homotopy between γ and $\tilde{\gamma}$.

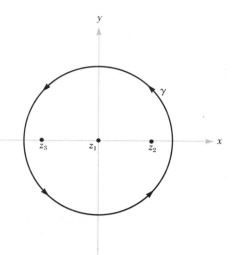

Figure 4.9

Hence

$$\text{Res}(f,z_1) = \frac{2}{\cos 0} = 2;$$

$$\text{Res}(f,z_2) = \frac{2}{\cos(2\pi)} = 2;$$

$$\text{Res}(f,z_3) = \frac{2}{\cos(-2\pi)} = 2.$$

Thus, by the Residue Theorem,

$$\int_\gamma \frac{1+z}{1-\cos z}\, dz = 2\pi i\{\text{Res}(f,z_1) + \text{Res}(f,z_2) + \text{Res}(f,z_3)\}$$

$$= 12\pi i.$$

Notice that we have implicitly used the fact that $I(\gamma,z) = 0$ for z outside γ and $I(\gamma,z) = 1$ for z inside γ. If the student does not believe his intuition he should write out a careful proof of this fact.

Exercises

1. Deduce Cauchy's Integral Formula from the Residue Theorem.
2. Evaluate

$$\int_\gamma \frac{dz}{(z+1)^3}$$

where (a) γ is a circle of radius 2, center 0; (b) γ is a square with vertices $0, 1, 1+i, i$.

3. Evaluate

$$\int_\gamma \frac{z}{z^2 + 2z + 5}\, dz$$

where γ is the unit circle.

4. Evaluate

$$\int_\gamma \frac{1}{e^z - 1}\, dz$$

where γ is the circle of radius 9 and center 0.

5. Show that

$$\int_\gamma \frac{5z - 2}{z(z - 1)}\, dz = 10\pi i$$

where γ is any circle of radius greater than 1 and center 0.

6. Evaluate

$$\int_\gamma \tan z\, dz$$

where γ is the circle of radius 8 centered at 0.

7. Evaluate

$$\int_\gamma \frac{e^{-z^2}}{z^2}\, dz$$

where (a) γ is the square with vertices $-1 - i$, $1 - i$, $-1 + i$, and $1 + i$; (b) γ is the ellipse $\gamma(t) = a\cos t + ib\sin t$, where $a,b > 0$, $0 \le t \le 2\pi$.

8. Let f be analytic on $\mathbb{C}$ except for poles at 1 and -1. Assume that $\mathrm{Res}(f,1) = -\mathrm{Res}(f,-1)$. Let $A = \{z\mid z \notin [-1,1]\}$. Show that there is an analytic function h on A such that $h'(z) = f(z)$.

9. Let $f\colon A \to B$ be analytic, one-to-one, and onto and let $f'(z) \ne 0$ for $z \in A$. Let γ be a curve in A and let $\tilde\gamma = f \circ \gamma$. Also let g be continuous on $\tilde\gamma$. Show that

$$\int_\gamma (g\circ f) \cdot f' = \int_{\tilde\gamma} g.$$

What does this become in the case where $f(z) = 1/z$?

10. Suppose that f is analytic except for a finite number of poles in $\mathbb{C}$. Define the *residue of* f *at* ∞ to be minus (the residue of $g(z) = (1/z^2)f(1/z)$ at $z = 0$).
    a. Show that $\mathrm{Res}\,(f,\infty) = (-1/2\pi i)\int_\Gamma f$ where Γ is a sufficiently large circle.

    Hint.   Use exercise 9 and the Laurent expansion of g.

b. Show that if $\underset{z \to \infty}{\text{limit}} - \{zf(z)\}$ exists, it equals the residue of f at ∞.

c. Find the residue of $(z-1)^3/(z(z/2)^3)$ at $z = \infty$.

d. Show informally that if γ is a simple closed curve described counter-clockwise, then

$$\int_\gamma f = -2\pi i \ \Sigma \ \{\text{residues of } f \text{ outside } \gamma \text{ including } \infty\}.$$

Hint.   Apply the Residue Theorem to the curve in Figure 4.10.

Note.   This formulation of residues at infinity is convenient for pur-poses of computation but the student may find it strange from the theoretical point of view. This can be remedied by introducing the point at ∞ as a legitimate point on the Riemann sphere (see Section 5.1).

e. Show two methods of evaluating

$$\int_\gamma \frac{(z-1)^3}{z(z+2)^3} \, dz$$

where γ is the circle with center 0 and radius 3.

11. Choose a branch of $\sqrt{z^2 - 1}$ that is analytic on $\mathbb{C}$ except for the segment $[-1,1]$ on the real axis. Evaluate

$$\int_\gamma \sqrt{z^2 - 1} \, dz$$

where γ is the circle of radius 2 centered at 0.

12. Evaluate the following integrals:

a. $\displaystyle\int_{|z| = \frac{1}{2}} \frac{dz}{(1-z)^3}.$

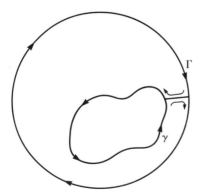

Figure 4.10   Residue at infinity.

b. $\displaystyle\int\limits_{|z+1|=\frac{1}{2}} \frac{dz}{(1-z)^3}.$

c. $\displaystyle\int\limits_{|z-1|=\frac{1}{2}} \frac{dz}{(1-z)^3}.$

d. $\displaystyle\int\limits_{|z-1|=\frac{1}{2}} \frac{e^z}{(1-z)^3}\, dz.$

e. $\displaystyle\int\limits_{|z|=\frac{1}{2}} \frac{dz}{z(1-z)^3}.$

f. $\displaystyle\int\limits_{|z|=\frac{1}{2}} \frac{e^z\, dz}{z(1-z)^3}.$

4.3   Evaluation of Definite Integrals

In this section some systematic methods will be described whereby the Residue Theorem can be used to evaluate certain types of integrals. These techniques are summarized in Table 4.2 (p. 253). Some examples of special devices that can be used to evaluate integrals of "multi-valued" (multiple-valued) functions will also be given. The worked examples that would normally appear at the end of this section are incorporated in the text.

Integrals of the type $\displaystyle\int_{-\infty}^{\infty} f(x)\, dx$

The first and simplest type of real integral that can be evaluated using complex analysis is dealt with by the following theorem.

Theorem 9.   *Let f(z) be analytic on* $\mathbb{C}$ *except for a finite number of poles, none of which is on the real axis. Suppose that there is a constant M and a number R such that*

$$|f(z)| \le M/|z|^2$$

for $|z| \ge R$. (More generally, we could assume that $|f(z)| \le M/|z|^\alpha$ for some $\alpha > 1$.) Then

$$\int_{-\infty}^{\infty} f(x)\, dx = 2\pi i \, \Sigma \, \{residues\ of\ f\ in\ the\ upper\ half\ plane\}$$

$$= -2\pi i \, \Sigma \, \{residues\ of\ f\ in\ the\ lower\ half\ plane\}. \quad (1)$$

In particular, formula (1) holds if $f = P/Q$ where P and Q are polynomials, the degree of Q is ≥ 2 plus the degree of P, and Q has no zeros on the real axis.

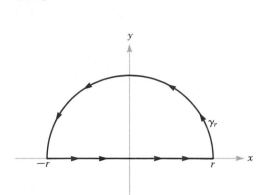

Figure 4.11 The curve γ_r.

Proof.   Let $r > R$ and consider the curve γ_r shown in Figure 4.11. Choose r large enough that all the poles of f in the upper half plane lie inside γ_r.

Then, by the Residue Theorem,

$$\int_{\gamma_r} f = 2\pi i \; \Sigma \; \{\text{residues of } f \text{ in upper half plane}\}.$$

But

$$\int_{\gamma_r} f = \int_{-r}^{r} f(x) \; dx + \int_{0}^{\pi} f(re^{i\theta}) i r e^{i\theta} \; d\theta.$$

Suppose that we could show that as $r \rightarrow \infty$, the last term approaches zero. Then we would have

$$\underset{r \rightarrow \infty}{\text{limit}} \int_{-r}^{r} f(x) \; dx = 2\pi i \; \Sigma \; \{\text{residues of } f \text{ in upper half plane}\}.$$

From calculus, we know that $f(x)$ is integrable as a function of $x \in \mathbb{R}$ (because f is continuous on $\mathbb{R}$ and because of the condition $|f(z)| \le M/|x|^{\alpha}$ for $|x| \ge R$); therefore,

$$\underset{r \rightarrow \infty}{\text{limit}} \int_{-r}^{r} f(x) \; dx = \int_{-\infty}^{\infty} f(x) \; dx.$$

Thus it remains to be proved that

$$\underset{r \rightarrow \infty}{\text{limit}} \int_{0}^{\pi} f(re^{i\theta}) \; r i e^{i\theta} d\theta = 0.$$

But

$$\left| \int_0^\pi f(re^{i\theta}) \, rie^{i\theta} \, d\theta \right| \le \pi \cdot \frac{M}{r^\alpha} \cdot r = \frac{\pi M}{r^{\alpha-1}},$$

which approaches zero as $r \to \infty$ since $\alpha > 1$. The formula concerning the residues in the lower half plane follows in a similar way.

Finally, if $f = P/Q$, we shall establish that $|f(z)| \le M/|z|^2$ for $|z|$ large and thus complete the proof. If P is of degree n and Q is of degree $n + p$, $p \ge 2$, we know there is an $M_1 > 0$ such that $|P(z)| \le M_1 |z|^n$ for $|z| \ge 1$ and an $M_2 > 0$ such that $|Q(z)| \ge M_2 |z|^{n+p}$ if $|z| \ge R$ for some $R > 1$ (see the proof of Theorem 27, Section 2.4). Thus

$$\left| \frac{P(z)}{Q(z)} \right| \le \frac{M_1}{M_2} \cdot \frac{1}{|z|^p} \le \frac{M_1}{M_2} \cdot \frac{1}{|z|^2}$$

for $|z| \ge R$, since $p \ge 2$. Therefore, we can let $M = M_1/M_2$ in this case. ∎

Example 1.   Evaluate

$$\int_{-\infty}^\infty \frac{dx}{x^4 + 1}.$$

Solution.   Here $P(x) = 1$, $Q(x) = x^4 + 1$, so the conditions of Theorem 9 are met. The poles of P/Q are located at the fourth roots of -1, namely, $e^{\pi i/4}$, $e^{3\pi i/4}$, $e^{5\pi i/4}$, $e^{7\pi i/4}$. These poles are simple and only the first two lie in the upper half plane. The residue at such a point z_0 is $1/4z_0^3 = -z_0/4$ (see Table 4.1, formula 4), so

$$\text{Res}\left(f, e^{\frac{\pi i}{4}}\right) + \text{Res}\left(f, e^{\frac{3\pi i}{4}}\right) = -\frac{1}{4}\left(e^{\frac{\pi i}{4}} + e^{\frac{3\pi i}{4}}\right) = -\frac{1}{4}e^{\frac{\pi i}{4}}\left(1 + e^{\frac{\pi i}{2}}\right)$$

$$= -\frac{1}{4}\left(\frac{1+i}{\sqrt{2}}\right)(1 + i) = -\frac{1}{4} \cdot \frac{2i}{\sqrt{2}}$$

$$= -\frac{i}{2\sqrt{2}}.$$

Thus the answer is $(2\pi i)(-i)/2\sqrt{2} = \pi/\sqrt{2}$.

Some simple checks such as determining that the answer must be real and positive (because $1/(x^4 + 1) \ge 0$ on $\mathbb{R}$) can often detect basic computational errors.

The integral

$$\int_{-\infty}^\infty \frac{dx}{x^4 + 1}$$

could have been evaluated by using the method of partial fractions. However, later in this section we shall encounter integrals that can be evaluated by using residues but for which the method of partial fractions (and, for that matter, all the elementary techniques of integration) fails.

Some important integrals cannot be evaluated by using Theorem 9. For example, consider the formula

$$\sqrt{\pi} = \int_{-\infty}^{\infty} e^{-x^2} \, dx.$$

Theorem 9 does not apply. The trouble with this integral is that e^{-z^2} does not have the desired limiting behavior as $z \longrightarrow \infty$. In fact, along the lines $\arg z = \pm\pi/4$ and $\arg z = \pm 3\pi/4$, $|e^{-z^2}| = 1$; and along $\arg z = \pm\pi/2, 0, \pi$, we have $|e^{-z^2}| \longrightarrow 0$ faster than the reciprocal of any polynomial. This particular integral can be evaluated by other means, however; see J. Marsden, *Elementary Classical Analysis* (San Francisco: W. H. Freeman and Co., 1974), Chapter 9, and Section 7.1 of this text. For a way to evaluate this using residues, see G. J. O. Jameson, *A First Course on Complex Functions* (N.Y.: Halsted, 1970.)

We shall now consider a technique for evaluating integrals of the form $\int_{-\infty}^{\infty} \cos ax \, f(x) \, dx$ and $\int_{-\infty}^{\infty} \sin ax \, f(x) \, dx$.

Theorem 10. *Suppose that $f(z)$ is analytic on $\mathbb{C}$ except for a finite number of poles, none of which lies on the real axis, and suppose that there exists $M, R > 0$ such that $|f(z)| \le M/|z|$ for $|z| \ge R$. Then, for any $a > 0$, the integral $\int_{-\infty}^{\infty} e^{iax} f(x) \, dx$ exists in the conditional sense (that is,* $\underset{c \to \infty}{\text{limit}} \int_0^c e^{iax} f(x) \, dx$ *and* $\underset{c \to -\infty}{\text{limit}} \int_c^0 e^{iax} f(x) \, dx$ *both exist) and*

$$\int_{-\infty}^{\infty} e^{iax} f(x) \, dx = 2\pi i \, \Sigma \, \{\text{residues of } f(z) \, e^{iaz} \text{ in upper half plane}\}. \tag{2}$$

Formula (2) holds if $f = P/Q$ for polynomials P and Q, Q has no real zeros, and $\deg Q \ge 1 + \deg P$. (deg Q stands for the degree of Q.)

If f is real-valued on the real axis, we also have

$$\int_{-\infty}^{\infty} \cos ax \, f(x) \, dx = \text{Re}[2\pi i \, \Sigma \, \{\text{residues of } f(z) e^{iaz} \text{ in upper half plane}\}] \tag{3}$$

and

$$\int_{-\infty}^{\infty} \sin ax \, f(x) \, dx = \text{Im}[2\pi i \, \Sigma \, \{\text{residues of } f(z) e^{iaz} \text{ in upper half plane}\}]. \tag{4}$$

If $a < 0$, we replace "upper half plane" by "lower half plane" and change the sign of formula (2) to "equals $-2\pi i \, \Sigma$ {residues of $f(z)e^{iaz}$ in lower half plane}."

Note.   The conditions on f are weaker than those in Theorem 9; we demand only that $|f| \leq M/|z|$ and not $|f| \leq M/|z|^\alpha$ for $\alpha > 1$. If $f = P/Q$, this restriction allows Q to have only degree one greater than P rather than two greater as in Theorem 9.* However, the extra factor e^{iax} is essential if this method is to work. (Theorem 9 is not true if we set $\alpha = 1$.)

Proof.   Let γ be the curve shown in Figure 4.12, where $y_1 > x_1, x_2 > R$ and x_1, x_2, y_1 are chosen large enough that γ contains all the poles of f in the upper half plane. By the Residue Theorem, $\int_\gamma e^{iaz} f(z) \, dz = 2\pi i \, \Sigma$ {residues of $f(z) \, e^{iaz}$ in upper half plane}.

Now estimate the absolute values of the three integrals

$$I_1 = \int_0^{y_1} e^{ia(x_2 + yi)} f(x_2 + yi) i \, dy,$$

$$I_2 = \int_{x_2}^{-x_1} e^{ia(x + iy_1)} f(x + iy_1) \, dx,$$

$$I_3 = \int_{y_1}^0 e^{ia(-x_1 + yi)} f(-x_1 + iy) i \, dy.$$

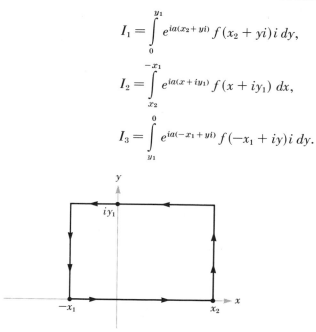

Figure 4.12   The curve γ used for the proof of Theorem 10.

---

*The assumptions on f in Theorem 10 can be weakened even further, as the following proof demonstrates. "$|f(z)| \leq M/|z|$" can be replaced by "$f(z) \to 0$ as $z \to \infty$ uniformly in arg z, $0 \leq$ arg $z \leq \pi$." This refinement makes little difference in the examples.

First,

$$|I_1| \le \int_0^{y_1} e^{-ay}|f(x_2 + yi)| \, dy \le \frac{M}{x_2} \int_0^{y_1} e^{-ay} \, dy = \frac{M(1 - e^{-ay_1})}{ax_2} \le \frac{M}{ax_2}.$$

Similarly,

$$|I_3| \le \frac{M}{ax_1}.$$

Finally,

$$|I_2| \le -\int_{x_2}^{-x_1} e^{-ay_1}|f(x + iy_1)| \, dx \le -\frac{M}{y_1} \int_{x_2}^{-x_1} e^{ay_1} \, dx = Me^{-ay_1} \frac{(x_2 + x_1)}{y_1}.$$

This last expression approaches zero as $y_1 \to \infty$. Hence, since

$$\int_\gamma e^{iaz} f(z)dz = I_1 + I_2 + I_3 + \int_{-x_1}^{x_2} e^{iax} f(x) \, dx,$$

we have

$$\left| \int_{-x_1}^{x_2} e^{iax} f(x) \, dx - 2\pi i \, \Sigma \, \{\text{residues of } f(z)e^{iaz} \text{ in upper half plane}\} \right|$$

$$\le \frac{M}{a}\left(\frac{1}{x_1} + \frac{1}{x_2}\right) + Me^{-ay_1} \frac{(x_2 + x_1)}{y_1}.$$

If we let $y_1 \to \infty$, we can drop the last term. Thus

$$\lim_{\substack{x_1 \to \infty \\ x_2 \to \infty}} \int_{-x_1}^{x_2} e^{iax} f(x) \, dx$$

exists and is the required value. The student is left the exercise of proving the fact from calculus that the existence of this double limit is the same as the existence of

$$\lim_{x_2 \to \infty} \int_0^{x_2} e^{iax} f(x) \, dx$$

and

$$\lim_{x_1 \to \infty} \int_{-x_1}^{0} e^{iax} f(x) \, dx.$$

Thus we have proved the existence (in the conditional sense) of

$$\int_{-\infty}^{\infty} e^{iax} f(x) \, dx$$

and that $\int_{-\infty}^{\infty} e^{iax} f(x)\ dx = 2\pi i\ \Sigma$ {residues of $f(z)e^{iaz}$ in upper half plane}.

That Theorem 10 holds for $f(x) = P(x)/Q(x)$, where P and Q are polynomials with deg $Q(x) \geq 1 + \deg P(x)$, follows as in the proof of Theorem 9: For $|z| \geq 1$ there is an $M_1 > 0$ such that $|P(z)| \leq M_1 |z|^n$ where $n = \deg P(x)$, and there is an $R > 1$ and an $M_2 > 0$ such that for $|z| \geq R$ we have $|Q(z)| \geq M_2 |z|^{n+1}$. Hence

$$\left| \frac{P(z)}{Q(z)} \right| \leq \frac{M_1}{M_2 |z|}$$

for $|z| \geq R$ and the condition of the theorem is satisfied.

The proof of the last statement of the theorem (the case where $a < 0$) is similar to the preceding proof, except that the appropriate curve is a rectangle in the lower half plane. ∎

Note that $\int_{-\infty}^{\infty} \cos ax\, f(x)\ dx$ is *not* $2\pi i\ \Sigma$ {residues of $(\cos az)f(z)$ in upper half plane}. (This formula is actually false.) The hypotheses of Theorems 9 and 10 simply do not apply, even if $|f(z)| \leq M/|z|^2$.

There is another method of proving Theorem 10 that the interested student can work out. It is based on *Jordan's Lemma*: If $f(z) \rightarrow 0$ as $|z| \rightarrow \infty$, uniformly in arg z, $0 \leq \arg z \leq \pi$, and if $f(z)$ is analytic when $|z| > c$, c a constant, $0 \leq \arg z \leq \pi$, then $\int_{\gamma_\rho} e^{iaz} f(z)\ dz \longrightarrow 0$ as $\rho \longrightarrow \infty$, where $\gamma_\rho(\theta) = \rho e^{i\theta}$, $0 \leq \theta \leq \pi$. Consult E. T. Whittaker and G. N. Watson, *A Course of Modern Analysis* (New York: Cambridge University Press, 1927), p. 115.

Example 2.   Show that

$$\int_0^{\infty} \frac{\cos x}{x^2 + b^2}\ dx = \frac{\pi e^{-b}}{2b};\ b > 0.$$

Solution.   Since $\cos x/(x^2 + b^2)$ is an even function, we have

$$\int_0^{\infty} \frac{\cos x}{x^2 + b^2}\ dx = \frac{1}{2} \int_{-\infty}^{\infty} \frac{\cos x}{x^2 + b^2}\ dx.$$

We must find the residues of $e^{iz}/(z^2 + b^2)$ in the upper half plane. The only pole in the upper half plane is at bi and the pole is simple, so

$$\mathrm{Res}\left(\frac{e^{iz}}{z^2 + b^2}, ib \right) = \frac{e^{-b}}{2ib}$$

(Table 4.1, formula 4), and hence

$$\int_{-\infty}^{\infty} \frac{\cos x}{x^2 + b^2}\ dx = \mathrm{Re}\left(2\pi i \left\{ \frac{e^{-b}}{2ib} \right\} \right) = \frac{\pi e^{-b}}{b}.$$

Thus we have the desired result.

Trigonometric integrals

Theorem 11.   *Let R(x,y) be a rational function of x,y having no pole on the unit circle. Then*

$$\int_0^{2\pi} R(\cos\theta,\,\sin\theta)\,d\theta = 2\pi i\,\Sigma\,\{residues\ of\ f(z)\ inside\ unit\ circle\}$$

where

$$f(z) = \frac{R\left\{\frac{1}{2}\left(z+\frac{1}{z}\right),\,\frac{1}{2i}\left(z-\frac{1}{z}\right)\right\}}{iz}.$$

Proof.   Since R has no poles on the unit circle, neither does f, so if γ is the unit circle, we have, by the Residue Theorem,

$$\int_\gamma f = 2\pi i\,\Sigma\,\{residues\ of\ f\ inside\ \gamma\}.$$

Therefore,

$$\int_0^{2\pi} R(\cos\theta,\,\sin\theta)d\theta = \int_0^{2\pi} R\left(\frac{e^{i\theta}+e^{-i\theta}}{2},\,\frac{e^{i\theta}-e^{-i\theta}}{2i}\right)\frac{ie^{i\theta}}{ie^{i\theta}}\,d\theta$$

$$= \int_0^{2\pi} f(e^{i\theta})\,ie^{i\theta}\,d\theta = \int_\gamma f,$$

which proves the theorem. ■

If the student forgets the formula for f, he can always begin with $\int_0^{2\pi} R(\cos\theta,\,\sin\theta)\,d\theta$ and follow the method of the preceding proof (that is, write $\cos\theta = (e^{i\theta} + e^{-i\theta})/2 \cdots$). The formula for f will then be apparent.

Example 3.   Evaluate

$$I = \int_0^{2\pi} \frac{d\theta}{1 + a^2 - 2a\cos\theta};\ a > 0,\,a \neq 1.$$

Solution.   By Theorem 11,

$$I = \int_\gamma \frac{dz}{iz\left(1 + a^2 - \frac{2a}{2}\left(z + \frac{1}{z}\right)\right)}$$

$$= \int_\gamma \frac{dz}{i\{-az^2 + (1 + a^2)z - a\}}$$

$$= \int_\gamma \frac{idz}{(z - a)(az - 1)}.$$

The poles of the integrand are at $z = a$ and $z = 1/a$. First, suppose that $a < 1$; the pole inside the circle thus is at $z = a$. The residue is

$$\frac{i}{a^2 - 1}.$$

If we suppose that $a > 1$, the integrand has a pole at $z = 1/a$ and the residue is

$$\frac{i}{a \cdot \left(\dfrac{1}{a} - a\right)} = \frac{i}{1 - a^2}.$$

Thus

$$I = \begin{cases} \dfrac{2\pi}{1 - a^2} & \text{if } a < 1; \\[2ex] \dfrac{2\pi}{a^2 - 1} & \text{if } a > 1. \end{cases}$$

Theorems 9, 10, and 11 take care of some of the more common types of integrals. Now we come to more specialized integrals. The first involves the multiple-valued function $z \longmapsto z^a$.

Integrals of the type $\int_0^\infty x^{a-1} f(x)\, dx$

Theorem 12. *Let f be analytic on $\mathbb{C}$ except for a finite number of poles, none of which lies on the strictly positive real axis (that is, all the poles lie in the complement of the set $\{x - iy \mid y = 0 \text{ and } x > 0\}$). Let $a > 0$ with the restriction that a is not an integer and suppose that (i) there exist constants $M_1, R_1 > 0$ and $b > a$ such that for $|z| \geq R_1$, $|f(z)| \leq M_1/|z|^b$; and (ii) there exist constants $M_2, R_2 > 0$, and $0 < d < a$, such that for $0 < |z| \leq R_2$, $|f(z)| \leq M_2/|z|^d$.*

Then the integral $\int_0^\infty x^{a-1} f(x)\, dx$ exists (in the sense of being absolutely integrable) and

$$\int_0^\infty x^{a-1} f(x)dx = -\frac{\pi e^{-\pi a i}}{\sin(a\pi)} \sum \begin{Bmatrix} \textit{residues of } z^{a-1} f(z) \textit{ at the poles of } f, \\ \textit{excluding the residue at } 0 \end{Bmatrix}. \qquad (2)$$

Here $z^{a-1} = e^{(a-1)\log z}$ using the branch $0 < \arg z < 2\pi$.

The proof of Theorem 12 is typical of the approach taken when dealing with branches.

Proof.   The existence, in the sense of absolute integrability, of $\int_0^\infty x^{a-1} f(x) dx$ follows if we use the assumed conditions $|f(x)| \leq M_1/x^b$ for x large and $|f(x)| \leq M_2/x^d$ for x small, together with the comparison test for integrals. The curve that we shall use, $\gamma = \gamma_1 + \gamma_2 + \gamma_3 + \gamma_4$, is illustrated in Figure 4.13. The radius of the incomplete circle γ_1 is r; the radius of the incomplete circle γ_3 is $\epsilon > 0$; and γ_4, γ_2 each make an angle of $\eta > 0$ with the positive x-axis. We choose ϵ small enough, r large enough, and η small enough so that (i) $\epsilon \leq R_2$; (ii) $r \geq R_1$; and (iii) $\gamma = \gamma_1 + \gamma_2 + \gamma_3 + \gamma_4$ encloses all the poles of $f(z)$ excluding the pole at 0.

By z^{a-1} we mean $e^{(a-1)\log z}$ where $\log z$ denotes the branch of log with $0 < \arg z < 2\pi$. Let $I = \int_\gamma z^{a-1} f(z) dz$; $I_1 = \int_{\gamma_1} z^{a-1} f(z) dz$; $I_2 = \int_{\gamma_2} z^{a-1} f(z) dz$; $I_3 = \int_{\gamma_3} z^{a-1} f(z) dz$; and $I_4 = \int_{\gamma_4} z^{a-1} f(z) dz$ so that $I = I_1 + I_2 + I_3 + I_4$. By the Residue Theorem,

$$I = 2\pi i \, \Sigma \, \{\text{residues of } z^{a-1} f(z), \text{ excluding the residue at } 0\}.$$

Clearly, the poles of $z^{a-1} f(z)$ are the same as the poles of $f(z)$, except possibly 0.

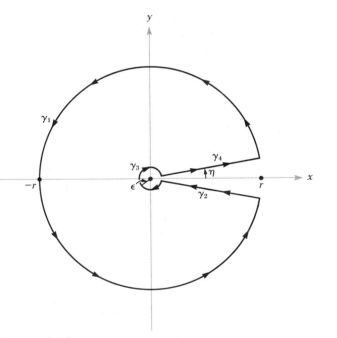

Figure 4.13 $\gamma = \gamma_1 + \gamma_2 + \gamma_3 + \gamma_4$.

First we shall show that $I_3 \to 0$ as $\epsilon \to 0$ and $I_1 \to 0$ as $r \to \infty$ independently of η. Recall that when a is real, $|z^{a-1}| = |z|^{a-1}$, so we get $|I_3| \leq \int_{\gamma_3} |z^{a-1}| \, |f(z)| \, |dz| \leq M_2 \int_{\gamma_3} |z|^{a-1}/|z|^d \, |dz| = M_2 \, \epsilon^{a-d-1} \, l(\gamma_3) < 2\pi M_2 \, \epsilon^{a-d}$. Also, $|I_1| \leq \int_{\gamma_1} |z^{a-1}| \, |f(z)| \, |dz| \leq M_1 \cdot \int_{\gamma_1} |z|^{a-1}/|z|^b \, |dz| < 2\pi M_1 \, r^{a-b}$. These estimates show that $I_3 \to 0$ as $\epsilon \to 0$ and $I_1 \to 0$ as $r \to \infty$, both independently of η.

Next we shall study the limiting behavior of I_2 and I_4 for fixed ϵ and r but with $\eta \to 0$. Notice that we have no reason to expect I_2 and $-I_4$ to converge to the same value as $\eta \to 0$ (and in fact they do not); this is because of the discontinuity in the function z^{a-1} as we cross the positive x-axis. (The positive x-axis is a branch line.) By definition, $I_2 = \int_r^\epsilon (te^{i(2\pi-\eta)})^{a-1} f(te^{i(2\pi-\eta)}) e^{i(2\pi-\eta)} \, dt$.

The student is left the exercise of proving that as $\eta \to 0$, we have $(te^{i(2\pi-\eta)})^{a-1} f(te^{i(2\pi-\eta)}) e^{i(2\pi-\eta)} \to t^{a-1} e^{2\pi i a} f(t)$ uniformly on $[\epsilon, r]$; hence as $\eta \to 0$, $I_2 \to \int_r^\epsilon t^{a-1} e^{2\pi i a} f(t) \, dt$; similarly, $I_4 \to \int_\epsilon^r t^{a-1} \cdot f(t) \, dt$; hence $I_2 + I_4 \to (1 - e^{2\pi i a}) \int_\epsilon^r t^{a-1} f(t) \, dt = -2i e^{\pi i a} \sin \pi a \cdot \int_\epsilon^r t^{a-1} f(t) \, dt$.

Suppose that we are given $\delta > 0$. We can choose r sufficiently large and $\epsilon > 0$ sufficiently small that

$$\left| \int_\epsilon^r t^{a-1} f(t) dt - \int_0^\infty t^{a-1} f(t) dt \right| < \delta/4.$$

Furthermore, we can choose r larger and ϵ smaller, if necessary, so that

$$\frac{|I_1|}{|2i e^{\pi i a} \sin \pi a|} < \frac{\delta}{4}, \quad \frac{|I_3|}{|2i e^{\pi i a} \sin \pi a|} < \frac{\delta}{4}.$$

Also, we can choose η small enough that

$$\left| \frac{I_2 + I_4}{(-2i e^{\pi i a} \sin \pi a)} - \int_\epsilon^r t^{a-1} f(t) dt \right| < \frac{\delta}{4}.$$

This implies that

$$\left| \int_0^\infty t^{a-1} f(t) dt - \frac{I}{(-2i e^{\pi i a} \sin \pi a)} \right| < \delta$$

(Why?). Since δ was arbitrary, we have

$$\int_0^\infty t^{a-1} f(t) dt = \frac{I}{(-2i e^{\pi i a} \sin \pi a)}$$

$$= \frac{-\pi e^{-\pi i a}}{\sin \pi a} \Sigma \, \{\text{residues of } z^{a-1} f(z) \text{ excluding the residue at } 0\}. \blacksquare$$

The student should verify the following corollary.

Corollary 1.   *Theorem 12 applies when $f(z) = P(z)/Q(z)$ for polynomials P of degree p and Q of degree q and when the following two conditions hold:*
i. $0 < a < q - p$.
ii. *If n_Q is of the order of the zero of Q at $z = 0$ (with the convention that $n_Q = 0$ if $Q(0) \neq 0$) and if n_p is the order of the zero of P at $z = 0$, then $n_Q - n_P < a$. (This condition holds, for instance, if $n_Q = 0$.)*

Example 4.   Prove that

$$\int_0^\infty \frac{x^{a-1}}{1+x^2}\, dx = \frac{\pi}{2\sin\left(\frac{a\pi}{2}\right)}, \quad 0 < a < 2.$$

Solution.   With this restriction on a, Corollary 1 holds (here $q = 2$, $p = 0$), so

$$\int_0^\infty \frac{x^{a-1}}{1+x^2}\, dx = \frac{-\pi e^{-\pi a i}}{\sin(a\pi)} \sum \left\{ \text{residues of } \frac{z^{a-1}}{1+z^2} \right\}.$$

The poles of $1/(1+z^2)$ are at $\pm i$ and are simple. Thus

$$\text{Res}\left(\frac{z^{a-1}}{1+z^2}, i\right) = \frac{i^{a-1}}{2i}$$

and

$$\text{Res}\left(\frac{z^{a-1}}{1+z^2}, -i\right) = -\frac{(-i)^{a-1}}{2i},$$

and the sum is

$$\frac{i^{a-1} - (-i)^{a-1}}{2i}.$$

We compute $i^{a-1} = e^{(a-1)\log i} = e^{(a-1)\pi i/2}$ and $(-i)^{a-1} = e^{(a-1)(3\pi i/2)}$. (Remember that we must choose $\arg(-i)$ such that $0 < \arg(-i) < 2\pi$.) Thus

$$\frac{i^{a-1} - (-i)^{a-1}}{2i} = \frac{1}{2i}\left\{ e^{(a-1)\frac{\pi i}{2}} - e^{(a-1)\frac{3\pi i}{2}} \right\}$$

$$= -\frac{1}{2}\left\{ e^{\frac{a\pi i}{2}} + e^{\frac{3a\pi i}{2}} \right\}$$

$$= -\frac{1}{2}e^{a\pi i}\left(e^{\frac{a\pi i}{2}} + e^{\frac{-a\pi i}{2}} \right)$$

$$= -e^{a\pi i}\cos\left(\frac{a\pi}{2}\right).$$

Hence

$$\int\limits_{0}^{\infty} \frac{x^{a-1}}{1+x^2}\, dx = \frac{\pi \cos\left(\dfrac{a\pi}{2}\right)}{\sin(a\pi)}.$$

But $\sin(2\varphi) = 2\sin\varphi \cdot \cos\varphi$, so this becomes $\pi/2\sin(a\pi/2)$, as required.

Cauchy Principal Value

Suppose that $f(x)$ is continuous on the real line $\mathbb{R}$ except at the point x_0. Then $\int_{-\infty}^{\infty} f(x)\,dx$ possibly might not be defined. We recall from calculus a way to make a meaningful definition. Consider

$$\int\limits_{-\infty}^{x_0-\epsilon} f(x)\,dx + \int\limits_{x_0+\eta}^{\infty} f(x)\,dx \text{ for } \epsilon > 0 \text{ and } \eta > 0$$

(assuming that both integrals are convergent for every such ϵ and η) and let $\epsilon \to 0$ and $\eta \to 0$. If each limit exists, we say the integral is *convergent*. We must exercise care when using this definition to solve examples. For instance, suppose that we consider

$$\int\limits_{-\infty}^{-\epsilon} \frac{1}{x^3}\,dx + \int\limits_{\epsilon}^{\infty} \frac{1}{x^3}\,dx = -\frac{1}{2\epsilon^2} + \frac{1}{2\epsilon^2} = 0.$$

On the other hand,

$$\int\limits_{-\infty}^{-\epsilon} \frac{1}{x^3}\,dx + \int\limits_{2\epsilon}^{\infty} \frac{1}{x^3}\,dx = -\frac{1}{2\epsilon^2} + \frac{1}{2(2\epsilon)^2} = -\frac{3}{8\epsilon^2} \to -\infty$$

as $\epsilon \to 0$. Thus we see that we can get different values for $\int_{-\infty}^{\infty} 1/x^3\,dx$ depending on how we let ϵ and η approach zero; in this example the integrals $\int_0^{\infty} 1/x^3\,dx$ and $\int_{-\infty}^{0} 1/x^3\,dx$ are not convergent.

We shall choose a particular, more restrictive, way of letting ϵ and η approach zero: the symmetric way, taking $\epsilon = \eta$. (We will find that we can readily apply the Residue Theorem to the evaluation of such integrals.) The definition that follows is slightly more general than the preceding one in that any finite number of discontinuities are allowed on the real axis. Let f be continuous on $\mathbb{R}$ except for a finite number of points $x_1 < x_2 \cdots < x_n$. If $\int_{-\infty}^{x_1-\epsilon} f(x)\,dx$ is convergent for every $\epsilon > 0$, if $\int_{x_n+\epsilon}^{\infty} f(x)\,dx$ is convergent for every $\epsilon > 0$, and if

$$\lim_{\epsilon \to 0} \int\limits_{-\infty}^{x_1-\epsilon} f(x)\,dx + \int\limits_{x_1+\epsilon}^{x_2-\epsilon} f(x)\,dx + \cdots + \int\limits_{x_{n-1}+\epsilon}^{x_n-\epsilon} f(x)\,dx + \int\limits_{x_n+\epsilon}^{\infty} f(x)\,dx)$$

exists and is finite, then we call this limit the *Cauchy Principal Value*, abbreviated P.V. $\int_{-\infty}^{\infty} f(x)\ dx$. Observe that if the integral is convergent at all these points, we recover the usual value for the integral $\int_{-\infty}^{\infty} f(x)\ dx$. However, as the previous example shows, the Cauchy Principal Value can exist even though the integrals are not convergent in the usual sense.

The general technique to be used for evaluating a large class of such Cauchy Principal Value integrals is the following. We require that $f(z)$ be defined and analytic on $\mathbb{C}$ with a finite number of poles, some of which may lie on the real axis, say, at points $x_1, \cdots, x_n$ where $x_1 < x_2 < \cdots < x_n$. The remaining poles lie off the real axis. Let us consider the curve γ shown in Figure 4.14. In this figure, the radius r of the large semicircle is chosen sufficiently large and the radius $\epsilon > 0$ of the small semicircles is chosen sufficiently small that γ encloses all the poles of f in the upper half plane, not including those on the real axis. We also require that the integral around the large semicircle approach zero as $r \longrightarrow \infty$ and that the limits of the integrals around the small semicircles exist and are finite as $\epsilon \longrightarrow 0$. These requirements, together with the fact that $\int_{\gamma} f = 2\pi i \sum$ {residues at poles in upper half plane, off the real axis}, ensure the existence of P.V. $\int_{-\infty}^{\infty} f(x)\,dx$ and enable us to calculate its value.

These requirements are met in Theorem 13, in which an explicit formula is derived for P.V. $\int_{-\infty}^{\infty} f(x)\,dx$. This theorem reduces to Theorems 9 and 10 in the event that no poles lie on the real axis.

Theorem 13.   *Let $f(z)$ be analytic except for a finite number of poles. Let $x_1, \cdots, x_m$ be the poles of f that lie on the real axis and suppose that they are all simple poles. Assume that either of the following conditions is obeyed.*

i. *There exist R and $M > 0$ such that for z with $\operatorname{Im} z \geq 0$ and $|z| \geq R$, we have $|f(z)| \leq M/|z|^2$.*

ii. *$f(z) = e^{iaz}g(z)$, where $a > 0$, and there exist R and $M > 0$ such that for z with $\operatorname{Im} z \geq 0$ and $|z| \geq R$, we have $|g(z)| \leq M/|z|$ (see Theorems 9 and 10 for cases in which these conditions hold).*

Then P.V. $\int_{-\infty}^{\infty} f(x)\,dx$ exists and

$$\text{P.V.} \int_{-\infty}^{\infty} f(x)\,dx = 2\pi i \sum \{\text{residues in upper half plane}\}$$

$$+ \pi i \sum \{\text{residues on x-axis}\}. \tag{3}$$

Remark.   The following proof assumes that condition (i) holds. The proof that assumes condition (ii) differs in one way: The three sides of a large rectangle in the upper half plane would be more suitable than the large semicircle as in Theorem 10. Note that $\int_{-\infty}^{x_1 - \epsilon} f(x)\ dx$

and $\int_{x_n+\epsilon}^{\infty} f(x)\ dx$ are convergent integrals for every $\epsilon > 0$ in the event that condition (ii) is fulfilled. This conclusion follows from the proof of Theorem 10.

If we consider the curve γ in Figure 4.14, we see that to prove Theorem 13 we need to know how to handle integrals over small circles. This result is provided by the following lemma.

Lemma 1.   *Let $f(z)$ have a simple pole at z_0 and let γ_ϵ be a portion of a circular arc of radius ϵ ánd angle α (see Figure 4.15). Then*

$$\lim_{\epsilon \to 0} \int_{\gamma_\epsilon} f = \alpha i\ \mathrm{Res}(f, z_0). \qquad (4)$$

Proof of Lemma 1.   Near z_0 we can write $f(z) = b_1/(z - z_0) + h(z)$ where h is analytic and $b_1 = \mathrm{Res}(f, z_0)$ (Why?). Thus

$$\int_{\gamma_\epsilon} f = \int_{\gamma_\epsilon} \frac{b_1}{z - z_0}\ dz + \int_{\gamma_\epsilon} h(z)\,dz.$$

Therefore,

$$\int_{\gamma_\epsilon} \frac{b_1}{z - z_0}\ dz = b_1 \int_{\alpha_0}^{\alpha_0 + \alpha} \frac{\epsilon i e^{i\theta}}{\epsilon e^{i\theta}}\ d\theta = b_1 \alpha i.$$

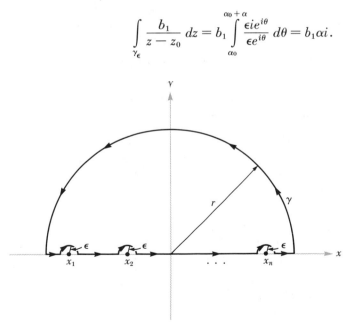

Figure 4.14   The curve γ for evaluating the Cauchy Principal Value.

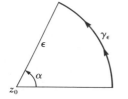

Figure 4.15 The curve γ_ϵ.

Here, $\gamma_\epsilon(\theta) = z_0 + \epsilon e^{i\theta}$, $\alpha_0 \le \theta \le \alpha_0 + \alpha$. Also, since h is analytic, it is bounded near z_0, say, by M, so

$$\left| \int_{\gamma_\epsilon} h(z)\,dz \right| \le Ml(\gamma_\epsilon) = M\alpha\epsilon \longrightarrow 0$$

as $\epsilon \longrightarrow 0$. The lemma follows. ∎

Proof of Theorem 13.   Let $\gamma = \gamma_r + \gamma_1 + \cdots + \gamma_m + \tilde{\gamma}$ where γ_r is the semicircular portion of radius r; $\gamma_1, \cdots, \gamma_m$ are the semicircular portions of radius ϵ; and $\tilde{\gamma}$ consists of the straight-line portions of γ along the real axis.

By the Residue Theorem, we have $\int_\gamma f = 2\pi i \Sigma$ {residues in upper half plane}. Next we observe that $\int_{-\infty}^{x_1-\epsilon} f(x)\,dx$ and $\int_{x_m+\epsilon}^{\infty} f(x)\,dx$ are convergent integrals for every $\epsilon > 0$ by the comparison test for integrals, together with the condition that $|f(z)| \le M/|z|^2$ for large $|z|$. Thus $\lim_{r \to \infty} \int_{\tilde{\gamma}} f$ exists. Therefore, $2\pi i \Sigma$ {residues in upper half plane} $= \int_\gamma f = \int_{\tilde{\gamma}} f + \int_{\gamma_r} f + \Sigma_{i=1}^m \int_{\gamma_i} f$. As in Theorem 9, $\int_{\gamma_r} f \to 0$ as $r \to \infty$ because of the condition $|f(z)| \le M/|z|^2$ for large $|z|$ where Im $z \ge 0$. Thus $\Sigma_{i=1}^m \int_{\gamma_i} f + \lim_{r \to \infty} \int_{\tilde{\gamma}} f = 2\pi i \Sigma$ {residues in upper half plane}. By Lemma 1, $\lim_{\epsilon \to 0} \int_{\gamma_j} f = -\pi i \, \text{Res}(f, x_j)$, $j = 1, 2, \cdots, m$. Hence $\lim_{\epsilon \to 0} (\lim_{r \to \infty} \int_{\tilde{\gamma}} f)$ exists and equals $2\pi i \Sigma$ {residues in upper half plane} $+ \pi i \Sigma$ {residues on real axis}. But by definition the Cauchy Principal Value is precisely $\lim_{\epsilon \to 0} (\lim_{r \to \infty} \int_{\tilde{\gamma}} f)$ and the theorem is proved. ∎

Example 5.   Consider $(\sin x)/x$, which is defined to have value 1 when $x = 0$. Then $(\sin x)/x$ is defined and continuous on $\mathbb{R}$. Show that $\int_0^\infty (\sin x)/x \, dx$ exists and compute its value.

Solution.   The integral

$$I = \text{P.V.} \int_{-\infty}^{\infty} \frac{e^{ix}}{x} \, dx$$

exists by Theorem 13. Therefore,

$$\text{Im } I = \text{P.V.} \int_{-\infty}^{\infty} \frac{\sin x}{x} \, dx$$

exists. Then

$$\text{P.V.} \int_{-\infty}^{\infty} \frac{\sin x}{x} \, dx = \int_{-\infty}^{\infty} \frac{\sin x}{x} \, dx$$

by the continuity of $(\sin x)/x$ at $x=0$. The existence of $\int_{-\infty}^{\infty} (\sin x)/x \, dx$ implies the existence of $\int_{0}^{\infty} (\sin x)/x \, dx$, and thus

$$\text{Im } I = 2 \int_{0}^{\infty} \frac{\sin x}{x} \, dx.$$

From formula (3) of Theorem 13 we have $I = \pi i \, \Sigma \, \{\text{residue of } e^{iz}/z \text{ at } z = 0\} = \pi i$. Hence

$$\int_{0}^{\infty} \frac{\sin x}{x} \, dx = \frac{\pi}{2}.$$

Further analysis of integrals involving multiple-valued functions

By the result of Theorem 12, we have seen that when dealing with multiple-valued functions, we must choose a curve suitable to a branch of the function. This requirement can be illustrated further with an example.

Example 6.   Use residues to prove that

$$\int_{1}^{\infty} \frac{dx}{x\sqrt{x^2 - 1}} = \frac{\pi}{2}.$$

This integral cannot be evaluated by any of the formulas that we have developed. The basic techniques used in the solution of this problem, however, are similar to those we have already applied. Following is an outline, the details of which are left to the student.

Solution.   Recall that a suitable domain of $\sqrt{z^2 - 1}$ consists of $\mathbb{C}$ minus the half lines $x \geq 1$ and $x \leq -1$. Consider the curve γ in Figure 4.16, consisting of the incomplete circles of radius r around 0 and radius ϵ around 1 and -1 and horizontal lines a distance δ from the real axis. The function $1/z\sqrt{z^2 - 1}$ is defined and analytic in the

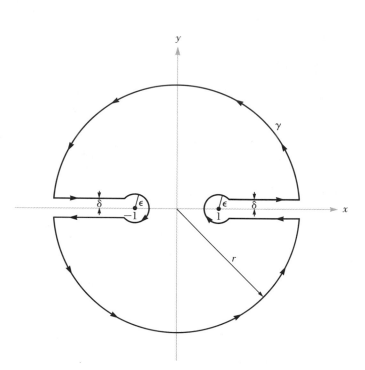

Figure 4.16   The curve γ.

region $\mathbb{C}$ minus the half lines $x \geq 1$ and $x \leq -1$ except for a simple pole at 0. By the Residue Theorem,

$$\int_\gamma \frac{dz}{z\sqrt{z^2 - 1}} = 2\pi i \; \mathrm{Res}\left(\frac{1}{z\sqrt{z^2 - 1}}, 0\right) = 2\pi.$$

The student should verify that (a) the integral over the incomplete circle of radius r approaches zero as $r \rightarrow \infty$ (the integrand is less than or equal to $M/|z|^2$ for $|z|$ large); (b) the integral over the incomplete circles of radius ϵ approaches zero as $\epsilon \rightarrow 0$ (the integral is bounded by a constant times $\epsilon/\sqrt{\epsilon} = \sqrt{\epsilon}$ on those circles); and (c) for fixed ϵ and r the integral over the horizontal lines approaches

$$4\int_{1+\epsilon}^{r} \frac{dx}{x\sqrt{x^2 - 1}}.$$

These three facts, together with the fact that

$$\int_\gamma \frac{dz}{z\sqrt{z^2 - 1}} = 2\pi,$$

show that

$$\int_1^\infty \frac{dx}{x\sqrt{x^2-1}}$$

exists and equals $\pi/2$.

Theorems 14 and 15 are useful in evaluating series. For instance, we shall see that by applying these theorems we can easily prove that

$$\sum_{n=1}^\infty \frac{1}{n^2} = \frac{\pi^2}{6}.$$

This is a famous formula of Leonhard Euler, who discovered it in the eighteenth century by using other techniques.

We shall develop a general method for evaluating series of form $\sum_{n=-\infty}^\infty f(n)$, where f is a given function. Suppose that we restrict f to being a meromorphic function with a finite number of poles, none of which are integers. Suppose that $G(z)$ is a meromorphic function whose only poles are simple poles at the integers, where the residues are all 1. Thus at the integers, the residues of $f(z)G(z)$ are $f(n)$. Then if γ is a closed curve enclosing $-N, -N+1, \cdots, 0, 1, \cdots, N$, we have, by the Residue Theorem,

$$\int_\gamma G(z)f(z)\,dz = 2\pi i \left\{ \left[\sum_{n=-N}^N f(n) \right] \right.$$

$$\left. + \sum \{\text{residues of } G(z)f(z) \text{ at poles of } f\} \right\}.$$

If $\int_\gamma G(z)f(z)\,dz$ exhibits a controllable limiting behavior as γ becomes large, we will have information about the limiting behavior of $\sum_{n=-N}^N f(n)$ as $N \to \infty$ in terms of the residues of $G(z)f(z)$ at the poles of f. A suitable $G(z)$ is $\pi\cot \pi z$. The last example of this section will show how, by slightly modifying Theorem 14, we can also handle the case in which some of the poles of f are at the integers.

Theorem 14. *Let f be a function meromorphic in $\mathbb{C}$, with a finite number of poles, none of which are at the integers. Let C_N be a rectangle with vertices at $(N + 1/2)(\pm 1 \pm i)$, $N = 1, 2, 3, \cdots$ (Figure 4.17). If N is sufficiently large that C_N encloses all the poles of f, then $\int_{C_N} \pi\cot \pi z \cdot f(z)\,dz$ is defined. Suppose that $\int_{C_N} \pi\cot \pi z\, f(z)\,dz \to 0$ as $N \to \infty$. Then $\lim_{N\to\infty} \sum_{n=-N}^N f(n)$ exists and is finite and*

$$\lim_{N\to\infty} \sum_{n=-N}^N f(n) = -\sum \{\text{residues of } \pi\cot \pi z \cdot f(z) \text{ at the poles of } f\}. \quad (5)$$

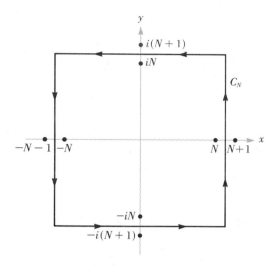

Figure 4.17 Contour for evaluating $\sum\limits_{-\infty}^{\infty} f(n)$.

Proof.   By the Residue Theorem,

$$\int_{C_N} \pi\cot \pi z\cdot f(z)\,dz = 2\pi i \sum \left\{ \begin{array}{l} \text{residues of } \pi\cot \pi z\cdot f(z) \text{ at the} \\ \text{integers } -N, -N+1, \cdots, 0, 1, \cdots, N \end{array} \right\}$$

$$+ 2\pi i \sum \left\{ \begin{array}{l} \text{residues of } \pi\cot \pi z\cdot f(z) \\ \text{at the poles of } f \end{array} \right\}$$

for N sufficiently large so that C_N encloses all the poles of f. Since $\cot \pi z = (\cos \pi z)/(\sin \pi z)$ and $(\sin \pi z)' \neq 0$ at $z = n$, we see that n is a simple pole of $\cot \pi z$ and that Res $(\cot \pi z, n) = \cos \pi n/\pi\cos \pi n = 1/\pi$ (use formula 4 of Table 4.1). Therefore, Res $(\pi\cot \pi z\cdot f(z), n) = \pi f(n)$ Res $(\cot \pi z, n) = f(n)$. Thus Σ {residues of $\pi\cot \pi z\cdot f(z)$ at the integers $-N, -N + 1, \cdots, 0, 1, \cdots, N$} $= \Sigma_{n=-N}^{N} f(n)$. Taking limits on both sides of the preceding equation for $\int_{C_N} \pi\cot \pi z \cdot f(z)\,dz$ and using the fact that $\int_{C_N} \pi\cot \pi z\cdot f(z)\,dz \rightarrow 0$ as $N \rightarrow \infty$, we obtain

$$\lim_{N\to\infty} \sum_{n=-N}^{N} f(n) = -\sum \{\text{residues of } \pi\cot \pi z\cdot f(z) \text{ at the poles of } f\}. \blacksquare$$

The following theorem establishes a criterion by which f can be judged to satisfy the hypotheses of Theorem 14.

Theorem 15.   *Suppose that f is a meromorphic function with no poles at the integers. Also suppose that there are constants R and $M > 0$ such* *that for $|z| > R$ we have $|zf(z)| \leq M$. Then the hypotheses of Theorem 14 are satisfied.*

Proof. Since $|zf(z)|$ is bounded, $f(z)$ cannot have any poles in the region $|z| > R$. Thus all the poles of f lie in the region $|z| \leq R$. Since, by the definition, the poles of a meromorphic function are isolated, there must be a finite number of them (Why?). Furthermore, $|f(1/z)/z|$ is bounded by M in the region $|z| < 1/R$, so 0 is a removable singularity of $f(1/z) \cdot 1/z$ and we can therefore write $f(1/z) \cdot 1/z = a_0 + a_1 z + a_2 z^2 + \cdots$ for $|z| < 1/R$; hence

$$f(z) = \frac{a_0}{z} + \frac{a_1}{z^2} + \frac{a_2}{z^3} + \cdots$$

for $|z| > R$.

Consider the integral

$$\int_{C_N} \frac{\pi \cot \pi z}{z}\, dz.$$

By the Residue Theorem,

$$\int_{C_N} \frac{\pi \cot \pi z}{z}\, dz = 2\pi i \cdot \left\{ \text{residue of } \frac{\pi \cot \pi z}{z} \text{ at } z = 0 \right\}$$

$$+ 2\pi i \sum \left\{ \begin{matrix} \text{residue of } \dfrac{\pi \cot \pi z}{z} \text{ at} \\ n = \pm 1, \pm 2,\ \cdots,\ \pm N \end{matrix} \right\}.$$

Since the pole at 0 is of order 2, we can write

$$\frac{\pi \cot \pi z}{z} = \frac{b_{-2}}{z^2} + \frac{b_{-1}}{z} + b_0 + b_1 z + b_2 z^2 + \cdots.$$

Because $(\pi \cot \pi z)/z$ is an even function of z (that is, $[\pi \cot \pi (-z)]/(-z) = (\pi \cot \pi z)/z$), we have, by uniqueness of the Laurent expansion, that coefficients of odd powers of z are zero; in particular, $b_{-1} = 0$. But b_{-1} is exactly $\text{Res}[(\pi \cot \pi z)/z, 0]$. (Instead of this trick we could have used formula 9 of Table 4.1.) Also, $\text{Res}[(\pi \cot \pi z)/z, n] = 1/n$ for $n = \pm 1, \pm 2, \cdots, \pm N$ (Why?); so Σ {residues of $(\pi \cot \pi z)/z$ at $n = \pm 1, \pm 2, \cdots, \pm N$} $= 0$. Consequently,

$$\int_{C_N} \frac{\pi \cot \pi z}{z}\, dz = 0.$$

Thus we can write

$$\int_{C_N} \pi \cot \pi z \cdot f(z)\, dz = \int_{C_N} \pi \cot \pi z \left(f(z) - \frac{a_0}{z} \right) dz.$$

To estimate this latter integral, we observe that

$$f(z) - \frac{a_0}{z} = \frac{a_1}{z^2} + \frac{a_2}{z^3} + \cdots$$

for $|z| > R$. Since $a_1 + a_2 z + a_3 z^2 + \cdots$ represents an analytic function for $|z| < 1/|R|$, it is bounded, say, by M' on the closed disk $|z| \leq 1/R'$, where $R' > R$. This implies that

$$\left| f(z) - \frac{a_0}{z} \right| \leq \frac{M'}{|z|^2}$$

for $|z| \geq R'$. Suppose that N is sufficiently large so that all points on C_N satisfy $|z| \geq R'$. Then

$$\left| \int_{C_N} \pi \cot \pi z \left(f(z) - \frac{a_0}{z} \right) dz \right| \leq \frac{\pi M' \cdot 8 \left(N + \frac{1}{2} \right)}{\left(N + \frac{1}{2} \right)^2} \left\{ \sup_{z \text{ on } C_N} \left| \cot \pi z \right| \right\}. \quad (6)$$

The student should verify that

$$\sup \{ |\cot \pi z| \mid z \text{ lies on } C_N \} = \frac{e^{2\pi(N+1/2)} + 1}{e^{2\pi(N+1/2)} - 1}$$

(noting that on the vertical sides, $|\cot \pi z| \leq 1$; on the horizontal sides, the maximum occurs at $x = 0$). Hence for all N sufficiently large we have $\sup_{z \text{ on } C_N} |\cot \pi z| \leq 2$. Inequality (6) thus shows that

$$\int_{C_N} \pi \cot \pi z \cdot \left(f(z) - \frac{a_0}{z} \right) dz$$

approaches zero as $N \to \infty$, which, in turn, shows that

$$\int_{C_N} \pi \cdot \cot \pi z \, f(z) \, dz \to 0 \text{ as } N \to \infty. \blacksquare$$

This result will now be applied to an important example. (The result of this example will be used in Chapter 7).

Example 7.   Let z be any complex number not equal to an integer. Show that both

$$\sum_{n=1}^{\infty} \left(\frac{1}{z-n} + \frac{1}{n} \right)$$

and

$$\sum_{n=1}^{\infty} \left(\frac{1}{z+n} - \frac{1}{n} \right)$$

are absolutely convergent series and that

$$\pi \cot \pi z = \frac{1}{z} + \sum_{n=1}^{\infty} \left(\frac{1}{z-n} + \frac{1}{n} \right) + \sum_{n=1}^{\infty} \left(\frac{1}{z+n} - \frac{1}{n} \right).$$

This equation can also be written

$$\frac{1}{z} + {\sum_{n=-\infty}^{\infty}}' \left(\frac{1}{z-n} + \frac{1}{n} \right);$$

the prime indicates that the term corresponding to $n = 0$ is omitted.

Solution.   For n sufficiently large, $|z - n| > n/2$. Therefore,

$$\left| \frac{1}{z-n} + \frac{1}{n} \right| = \left| \frac{z}{(z-n)n} \right| \leq \frac{2|z|}{n^2}.$$

By comparison with the convergent series,

$$2|z| \cdot \left(\frac{1}{n^2} + \frac{1}{(n+1)^2} + \cdots \right),$$

we see that

$$\sum_{n=1}^{\infty} \left(\frac{1}{z-n} + \frac{1}{n} \right)$$

is absolutely convergent. Similarly,

$$\sum_{n=1}^{\infty} \left(\frac{1}{z+n} - \frac{1}{n} \right)$$

is absolutely convergent. Fix z and consider the function $f(w) = 1/(w - z)$. This function is meromorphic; its only pole is at z, which is not an integer, and it is easy to see that $|wf(w)|$ is bounded for w sufficiently large (as in Theorem 9). By Theorem 15, we see that the hypotheses of Theorem 14 are satisfied, so

$$\lim_{N \to \infty} \sum_{n=-N}^{N} \frac{1}{n-z} = - \left\{ \text{residue of } \frac{\pi \cot \pi w}{w-z} \text{ at } w = z \right\} = -\pi \cot \pi z.$$

We note that

$$\sum_{n=-N}^{N} \frac{1}{z-n} = \frac{1}{z} + \sum_{n=1}^{N} \left(\frac{1}{z-n} + \frac{1}{n} \right) + \sum_{n=1}^{N} \left(\frac{1}{z+n} - \frac{1}{n} \right),$$

so

$$\frac{1}{z} + \sum_{n=1}^{\infty} \left(\frac{1}{z-n} + \frac{1}{n} \right) + \sum_{n=1}^{\infty} \left(\frac{1}{z+n} - \frac{1}{n} \right) = \pi \cot \pi z.$$

The final example in this section illustrates the power of the methods for summing series developed in Theorems 14 and 15.

Example 8.   Show that

$$\sum_{n=1}^{\infty} \frac{1}{n^2} = \frac{\pi^2}{6}.$$

Solution.   We cannot apply Theorem 14 directly since $f(z) = 1/z^2$ has a pole at 0. But the proof of Theorem 15 shows that

$$\int_{C_N} \frac{\pi \cot \pi z}{z^2}\, dz \to 0$$

as $N \to \infty$. By the Residue Theorem,

$$\int_{C_N} \frac{\pi \cot \pi z}{z^2}\, dz = 2\pi i \left(\sum \left\{ \begin{matrix} \text{residues of } \frac{\pi \cot \pi z}{z^2} \text{ at} \\ z = 0,\ \pm 1,\ \pm 2,\ \cdots,\ \pm N \end{matrix} \right\} \right).$$

As before,

$$\sum \left\{ \text{residues of } \frac{\pi \cot \pi z}{z^2} \text{ at } z = \pm 1, \pm 2, \cdots, \pm N \right\} = 2 \sum_{n=1}^{N} \frac{1}{n^2}.$$

Since $\tan z$ has a simple zero at $z = 0$, $\cot z$ has a simple pole at 0, so we can write, by the Laurent expansion, $\cot z = b_1/z + a_0 + a_1 z + \cdots$. By writing

$$\left(1 - \frac{z^2}{2!} + \frac{z^4}{4!} + \cdots \right) = \left(z - \frac{z^3}{3!} + \frac{z^5}{5!} + \cdots \right) \cdot \left(\frac{b_1}{z} + a_0 + a_1 z + \cdots \right),$$

multiplying out, and equating coefficients of like powers of z, we find that $b_1 = 1$, $a_0 = 0$, and $a_1 = -1/3$. Therefore,

$$\frac{\pi \cot \pi z}{z^2} = \frac{\pi \left(\frac{1}{\pi z} - \frac{\pi z}{3} + \cdots \right)}{z^2} = \frac{1}{z^3} - \frac{\pi^2}{z} \cdot \frac{1}{3} + \cdots.$$

Hence

$$\mathrm{Res}\left(\frac{\pi \cot \pi z}{z^2}, 0 \right) = \frac{-\pi^2}{3}.$$

We conclude that

$$\sum_{n=1}^{\infty} \frac{1}{n^2} = \frac{\pi^2}{6}.$$

Table 4.2 Evaluation of Definite Integrals

Type of Integral	Conditions	Formula
1. $\displaystyle\int_{-\infty}^{\infty} f(x)\,dx$	No poles of $f(z)$ on real axis; finite number of poles on $\mathbb{C}$; $|f(z)| \le M/|z|^2$ for large $|z|$.	$\displaystyle\int_{-\infty}^{\infty} f(x)\,dx = 2\pi i \sum \left\{\begin{array}{l}\text{residues of } f \text{ in}\\ \text{upper half plane}\end{array}\right\}$.
2. $\displaystyle\int_{-\infty}^{\infty} \frac{P(x)}{Q(x)}\,dx$	P, Q polynomials; $\deg Q \ge 2 + \deg P$; no real zeros of Q.	$\displaystyle\int_{-\infty}^{\infty} \frac{P(x)}{Q(x)}\,dx = 2\pi i \sum \left\{\begin{array}{l}\text{residues of } P/Q \text{ in}\\ \text{upper half plane}\end{array}\right\}$.
3a. $\displaystyle\int_{-\infty}^{\infty} e^{iax} f(x)\,dx$	$a > 0$; $|f(z)| \le M_f|z|$ for $|z|$ large and no poles of f on real axis; or $f(z) = P(z)/Q(z)$ where $\deg Q(z) \ge 1 + \deg P(z)$ and Q has no real zeros.	$\displaystyle\int_{-\infty}^{\infty} e^{iax} f(x)\,dx = I$ $\displaystyle= 2\pi i \left(\sum \left\{\begin{array}{l}\text{residues of } e^{iaz} f(z)\\ \text{in upper half plane}\end{array}\right\}\right)$. If $a < 0$, use $-\sum\{$residues in lower half plane$\}$.
b. $\displaystyle\int_{-\infty}^{\infty} \cos(ax) f(x)\,dx$ $\displaystyle\int_{-\infty}^{\infty} \sin(ax) f(x)\,dx$	f real on real axis.	$\displaystyle\int_{-\infty}^{\infty} \cos(ax) f(x)\,dx = \operatorname{Re} I$. $\displaystyle\int_{-\infty}^{\infty} \sin(ax) f(x)\,dx = \operatorname{Im} I$. If $a < 0$, use lower half plane as above.
4. $\displaystyle\int_{0}^{2\pi} R(\cos\theta, \sin\theta)\,d\theta$	R rational and $R(\cos\theta, \sin\theta)$ is continuous in θ. (No poles on unit circle.)	$\displaystyle\int_{0}^{2\pi} R(\cos\theta, \sin\theta)\,d\theta = 2\pi i \sum \left\{\begin{array}{l}\text{residues of } f\\ \text{inside unit circle}\end{array}\right\}$. $\displaystyle f(z) = \frac{1}{iz} R\left(\frac{1}{2}\left(z + \frac{1}{z}\right), \frac{1}{2i}\left(z - \frac{1}{z}\right)\right)$.

Table 4.2 *continued*

Type of Integral	Conditions	Formula
5. $\displaystyle\int_{0}^{\infty} x^{a-1} f(x)\,dx$	$a > 0$ and f has finite number of poles, none on positive real axis: $\lvert f(z)\rvert \leq M\lvert z\rvert^{b}$, $b > a$, for $\lvert z\rvert$ large; and $\lvert f(z)\rvert \leq M\lvert z\rvert^{d}$, $d < a$, for $\lvert z\rvert \to 0$. *or* $f = P/Q$, and Q has no zeros on positive real axis; $0 < a < \deg Q - \deg P$ and $n_Q - n_P < a$, where $n_Q =$ order of zero of Q at 0 and $n_P =$ order of 0 of P at 0.	$\displaystyle\int_{0}^{\infty} x^{a-1} f(x)\,dx = \frac{-\pi e^{-\pi a i}}{\sin(\pi a)} \sum \left\{ \begin{array}{l} \text{residues of } z^{a-1}f(z) \text{ at} \\ \text{poles of } f \text{ excluding } 0 \end{array} \right\},$ using the branch $0 < \arg z < 2\pi$.
6. $\displaystyle\int_{-\infty}^{\infty} f(x)\,dx$	Same as (1) except that simple poles are allowed on x-axis.	$\displaystyle\int_{-\infty}^{\infty} f(x)\,dx = 2\pi i \sum \{\text{residues in upper half plane}\}$ $\quad + \pi i \sum \{\text{residues on } x\text{-axis}\}.$
7. $\displaystyle\int_{-\infty}^{\infty} \frac{P(x)}{Q(x)}\,dx$	Same as (2) except that simple poles are allowed on x-axis.	$\displaystyle\int_{-\infty}^{\infty} \frac{P(x)}{Q(x)}\,dx = 2\pi i \sum \{\text{residues in upper half plane}\}$ $\quad + \pi i \sum \{\text{residues on } x\text{-axis}\}.$
8a. $\displaystyle\int_{-\infty}^{\infty} e^{iax} f(x)\,dx$	Same as (3) except that simple poles are allowed on x-axis.	i. $a > 0.$ $\displaystyle\int_{-\infty}^{\infty} e^{iax} f(x)\,dx = I$ $\quad = 2\pi i \sum \left\{ \begin{array}{l} \text{residues of } e^{iaz}f(z) \\ \text{in upper half plane} \end{array} \right\}$ $\quad + \pi i \sum \left\{ \begin{array}{l} \text{residues of } e^{iaz}f(z) \\ \text{on } x\text{-axis} \end{array} \right\}.$

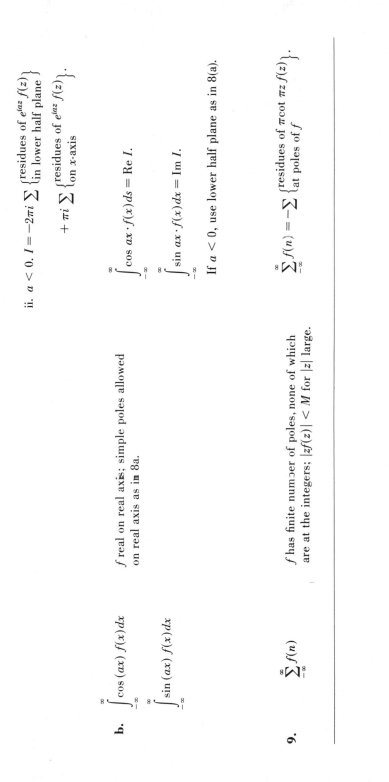

ii. $a < 0$. $I = -2\pi i \sum \begin{Bmatrix} \text{residues of } e^{iaz} f(z) \\ \text{in lower half plane} \end{Bmatrix}$

$+ \pi i \sum \begin{Bmatrix} \text{residues of } e^{iaz} f(z) \\ \text{on } x\text{-axis} \end{Bmatrix}.$

b. $\displaystyle\int_{-\infty}^{\infty} \cos(ax)\, f(x)\, dx$ $\quad$ f real on real axis; simple poles allowed on real axis as in 8a.

$\displaystyle\int_{-\infty}^{\infty} \cos ax \cdot f(x)\, ds = \text{Re } I.$

$\displaystyle\int_{-\infty}^{\infty} \sin(ax)\, f(x)\, dx$

$\displaystyle\int_{-\infty}^{\infty} \sin ax \cdot f(x)\, dx = \text{Im } I.$

If $a < 0$, use lower half plane as in 8(a).

9. $\displaystyle\sum_{-\infty}^{\infty} f(n)$ $\quad$ f has finite number of poles, none of which are at the integers; $|zf(z)| < M$ for $|z|$ large.

$\displaystyle\sum_{-\infty}^{\infty} f(n) = -\sum \begin{Bmatrix} \text{residues of } \pi \cot \pi z\, f(z) \\ \text{at poles of } f \end{Bmatrix}.$

Exercises

1. Evaluate

$$\int_{-\infty}^{\infty} \frac{dx}{x^2 - 2x + 4}.$$

2. Prove that

$$\int_{0}^{\infty} \frac{\sin^2 x}{x^2}\, dx = \frac{\pi}{2}.$$

Hint.   Consider

$$\int_{-\infty}^{\infty} \frac{1 - e^{2ix}}{x^2}\, dx$$

and apply Theorem 13.

3. Evaluate

$$\int_{0}^{\pi} \frac{d\theta}{(a + b\cos \theta)^2},\ 0 < b < a.$$

4. Evaluate

$$\int_{0}^{\infty} \frac{dx}{1 + x^6}.$$

5. Evaluate

$$\int_{0}^{\infty} \frac{\cos mx}{1 + x^4}\, dx.$$

6. Evaluate

$$\int_{0}^{\infty} \frac{x\sin x}{1 + x^2}\, dx.$$

7. Evaluate

$$\int_{0}^{\infty} \frac{x^{a-1}}{1 + x^3}\, dx,\ 0 < a < 3.$$

8. a. Prove that

$$\int_{-\infty}^{\infty} \frac{\cos x}{e^x + e^{-x}}\, dx = \frac{\pi}{e^{\frac{\pi}{2}} + e^{\frac{-\pi}{2}}}$$

by integrating $e^{iz}/(e^z + e^{-z})$ around the rectangle with vertices $-r$, r, $r + \pi i$, $-r + \pi i$; let $r \to \infty$.

b. Use the same technique to show that

$$\int_{-\infty}^{\infty} \frac{e^{-x}}{1 + e^{-2\pi x}} \, dx = \frac{1}{2\sin\left(\dfrac{1}{2}\right)}.$$

9. Evaluate

$$\text{P.V.} \int_{-\infty}^{\infty} \frac{dx}{(x - a)^2(x - 1)}$$

where Im $a > 0$.

10. Show that

$$\int_{0}^{\infty} \frac{\cos ax}{(x^2 + b^2)^2} \, dx = \frac{\pi}{4b^3} (1 + ab)e^{-ab}, \; a > 0, \, b > 0.$$

11. Show that

$$\int_{0}^{\pi} \sin^{2n} \theta \, d\theta = \frac{\pi(2n)!}{(2^n n!)^2}.$$

12. Show that

$$\int_{0}^{\infty} \frac{1}{x^b(x + 1)} \, dx = \frac{\pi}{\sin(b\pi)} \text{ for } 0 < b < 1.$$

13. Prove that

$$\int_{0}^{\infty} \frac{\log x}{(x^2 + 1)^2} \, dx = -\frac{\pi}{4}.$$

14. Find

$$\text{P.V.} \int_{-\infty}^{\infty} \frac{dx}{x(x^2 - 1)}.$$

15. Find

$$\int_{0}^{1} \frac{dx}{\sqrt{x^2 - 1}}$$

by (a) changing the variables to $y = 1/(x + \sqrt{x^2 - 1})$, and (b) considering the curve in Figure 4.18 and finding the residue of a branch of $1/\sqrt{z^2 - 1}$ at ∞ (see exercise 10 of Section 4.2).

16. Let $P(z)$ and $Q(z)$ be polynomials with degree $Q(z) \geq 2 + $ degree $P(z)$. Show that the sum of the residues of $P(z)/Q(z)$ is zero.

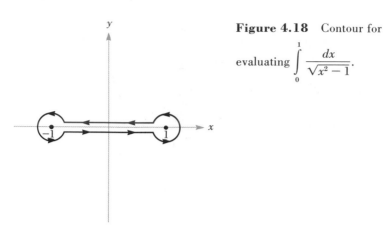

Figure 4.18 Contour for evaluating $\int_0^1 \dfrac{dx}{\sqrt{x^2-1}}$.

17. Prove:

a.
$$\sum_{n=1}^{\infty} \frac{1}{n^4} = \frac{\pi^4}{90}.$$

b.
$$\sum_{n=1}^{\infty} \frac{(-1)^{n-1}}{(2n-1)^3} = \frac{\pi^3}{32}.$$

c.
$$\sum_{n=0}^{\infty} \frac{1}{(n^2+a^2)} = \frac{\pi}{2a} \coth(\pi a) + \frac{1}{2a^2}, \; a > 0.$$

18. Show that

$$\frac{\pi^2}{\sin^2 \pi z} = \sum_{n=-\infty}^{\infty} \frac{1}{(z-n)^2}.$$

Hint. Start with the expansion for $\pi \cot \pi z$ discussed in the example on p. 250.

19. Develop a method for evaluating series of the form $\sum_{n=-\infty}^{\infty} (-1)^n f(n)$ where f is a meromorphic function in $\mathbb{C}$ with a finite number of poles none of which lie at the integers. In other words, develop theorems analogous to Theorems 14 and 15.

Hint. $\pi/\sin \pi z$ has poles at the integers with $\operatorname{Res}(\pi/\sin \pi z, n) = (-1)^n$. Discuss how you would handle the summation if some of the poles of f do lie at the integers; see the last example in this section.

20. Let $f(z)$ be as in formula 5 of Table 4.2 except allow f to have a finite number of simple poles on the (strictly) positive real axis. Show that:

$$\text{P.V.} \int_0^\infty x^{a-1} f(x)dx = \frac{-\pi e^{-\pi ai}}{\sin(\pi a)} \sum \{\text{residues of } (-z)^{a-1}f(z) \text{ at poles of } f$$
$$\text{off the nonnegative real axis}\}$$

$$+ \frac{\pi e^{-\pi ai}\cos \pi a}{\sin \pi a} \sum \{\text{residues of } (-z)^{a-1}f(z) \text{ on the positive real axis}\}.$$

21. Use exercise 20 to show that

$$\text{P.V.} \int_0^\infty \frac{x^{a-1}}{1-x} dx = -\pi \cot(\pi a) \text{ for } 0 < a < 1.$$

22. Establish the following formulas:

a.
$$\int_0^\pi \frac{d\theta}{1 + \sin^2 \theta} = \frac{\pi}{\sqrt{2}}$$

b.
$$\int_0^\infty \frac{x^2 dx}{(x^2 + a^2)^2} = \frac{\pi}{4a}, \ a > 0.$$

c.
$$\int_{-\infty}^\infty \frac{x^3 \sin x}{(x^2 + 1)^2} dx = \frac{\pi}{2} e^{-1}.$$

d.
$$\int_0^\infty \frac{x \sin x}{x^4 + 1} dx = \frac{\pi}{2} e^{\frac{-1}{\sqrt{2}}} \sin\left(\frac{1}{\sqrt{2}}\right).$$

23. Let $f(z)$ be meromorphic with simple poles at $a_1, a_2, a_3, \cdots$ where $0 < |a_1| < |a_2| < \cdots$. Let $b_1, b_2, \cdots$ be the residues of f at a_1, $a_2, \cdots$. Suppose that f is uniformly bounded on a sequence of circles whose radii approach infinity. Prove that

$$f(z) = f(0) + \sum_{n=1}^\infty \left(\frac{b_n}{z - a_n} + \frac{b_n}{a_n}\right).$$

Note. This is a special case of a general result known as the *Mittag-Leffler Theorem.*

24. Use exercise 23 to prove that

$$\cot z = \frac{1}{z} + \sum'\left(\frac{1}{z - n\pi} + \frac{1}{n\pi}\right)$$

where $\sum'$ means the sum is over all $n \neq 0$.

25. Prove that

$$1 - \frac{1}{2^2} + \frac{1}{3^2} - \frac{1}{4^2} + \cdots = \frac{\pi^2}{12}.$$

Review Exercises for Chapter 4

1. Evaluate

$$\int_0^{2\pi} \frac{d\theta}{2 - \sin\theta}.$$

2. Evaluate

$$\int_{-\infty}^{\infty} \frac{x^2}{x^4 + 1}\, dx.$$

3. Evaluate

$$\int_{-\infty}^{\infty} \frac{\cos bx}{x^2 + a^2}\, dx, \ a > 0, \ b > 0.$$

4. Let $f(z)$ have a zero of order k at z_0. Show that Res $(f'/f, z_0) = k$. Can you find Res $(f''/f', z_0)$?

5. Compute

$$\int_{-\infty}^{\infty} \frac{\sin x}{x(x+1)(x^2+1)}\, dx.$$

6. Find the residues of the following at each singularity:

   a. $\dfrac{z}{1 - e^{z^2}}$.

   b. $\dfrac{\sin(z^2)}{(\sin z)^2}$.

   c. $\sin\left(e^{\frac{1}{z}}\right)$.

7. Evaluate

$$\int_0^{\pi} \frac{d\theta}{2\cos\theta + 3}.$$

8. Evaluate

$$\int_{|z|=1} \frac{\cos(e^{-z})}{z^2}\, dz.$$

9. Evaluate

$$\int_0^{2\pi} e^{e^{i\theta}}\, d\theta.$$

10. Show that

$$\frac{2}{\pi} \int\limits_{0}^{\infty} \frac{\sin(kt)}{t} \, dt = \begin{cases} 1, k > 0; \\ 0, k = 0; \\ -1, k < 0. \end{cases}$$

11. Let f be analytic on a region containing the upper half plane $\{z | \text{Im } z \geq 0\}$. Suppose that for some $a > 0$, $|f(z)| \leq M/|z|^a$ for $|z|$ large. Show that for $\text{Im } z > 0$,

$$f(z) = \frac{1}{2\pi i} \int\limits_{-\infty}^{\infty} \frac{f(x)}{x - z} \, dx.$$

12. Show that

$$\int\limits_{0}^{\infty} \frac{x^{m-1}}{1 + x^n} \, dx = \frac{\pi}{n \sin\left(\dfrac{m\pi}{n}\right)}$$

where $0 < m < n$.

13. Find the Laurent expansions of

$$f(z) = \frac{1}{(z - 1)(z - 2)}$$

that are valid for $0 < |z| < 1$; for $|z| > 2$.

14. What is the radius of convergence of the Taylor series of $1/\cos z$ around $z = 0$?

15. Let f be entire and suppose that $\text{Re } f$ is a polynomial in x, y. Prove that f is a polynomial.

16. Explain what is wrong with the following reasoning. We know that $a^z = e^{z \log a}$, so $da^z/dz = (\log a)a^z$. On the other hand, $da^z/dz = za^{z-1}$. Thus $za^{z-1} = a^z(\log a)$, so $z = a \log a$.

17. Verify the Maximum Principle for Harmonic Functions and the Minimum Principle for Harmonic Functions for the harmonic function $u(x,y) = x^2 - y^2$ on $[0,1] \times [0,1]$.

18. Where is $\sum_{n=0}^{\infty} z^n e^{-izn}$ analytic?

19. Explain what is wrong with the following argument, then compute the residue correctly. The expansion

$$\frac{1}{z(z - 1)^2} = \frac{1}{(z - 1)^2} \cdot \frac{1}{[1 + (z - 1)]} = \cdots + \frac{1}{(z - 1)^5} - \frac{1}{(z - 1)^4} + \frac{1}{(z - 1)^3}$$

is the Laurent expansion; since there is no term in $1/(z - 1)$, the residue at $z = 1$ is zero.

20. Evaluate

$$\int_\gamma \frac{1}{z(z-1)(z-2)}\, dz$$

where γ is the circle centered at 0 with radius 3/2.

21. Determine the radius of convergence of the following series:

a. $\sum_1^\infty \frac{\log(n^n)}{n!}\, z^n.$

b. $\sum_1^\infty \left(1-\frac{1}{n}\right)^n z^n.$

22. Expand the following in Laurent series as indicated: $f(z) = \left(\frac{1}{1-z}\right)^3$

a. for $|z| < 1$.
b. for $|z| > 1$.
c. for $|z+1| < 2$.
d. for $0 < |z-1| < \infty$.

23. Establish the following formulas:

a.
$$\int_0^\infty \frac{\sin^3 x}{x^3}\, dx = \frac{3\pi}{8}.$$

b.
$$\int_0^\infty \frac{x^a}{x^2+b^2}\, dx = \frac{\pi b^{a-1}}{2\cos\left(\dfrac{\pi a}{2}\right)}, \quad -1 < a < 1.$$

24. Establish the following:

a.
$$\int_0^\infty \frac{\sinh ax}{\sinh \pi x}\, dx = \frac{1}{2}\tan\frac{a}{2}, \quad -\pi < a < \pi.$$

Hint.   Integrate $e^{az}/\sinh(\pi z)$ over a "square" with sides $y = 0$, $y = 1$, $x = -R$, $x = +R$, and circumvent the singularities at 0, i.

b.
$$\int_0^{\pi/2} \frac{d\theta}{(a+\sin^2\theta)^2} = \frac{\pi(2a+1)}{4(a^2+a)^{3/2}}, \quad a > 0.$$

c.
$$\int_0^\infty \cos(x^2)\, dx = \int_0^\infty \sin(x^2)\, dx = \frac{1}{2}\sqrt{\frac{\pi}{2}}.$$

(These are known as the *Fresnel integrals*.)

Hint.   Integrate e^{-z^2} over the curve γ with sides $z = x$, $(0 \le x \le R)$, $z = Re^{i\theta}(0 \le \theta \le \pi/4)$, $z = re^{i\pi/4}(0 \le r \le R)$.

25. Attempt to sum the series $\Sigma_{n=1}^{\infty} 1/n^3$.†

26. Prove that

$$\tan z = 2z \sum_{0}^{\infty} \frac{1}{\left(n + \frac{1}{2}\right)^2 \pi^2 - z^2}, \quad z \neq \left(n + \frac{1}{2}\right)\pi.$$

Hint.   Start with the identity for $\cot z$ on p. 251 and use $\tan z = \cot z - 2 \cot 2z$.

27. Let $f(z)$ be analytic inside and on a simple closed contour γ. For z_0 on γ, and γ differentiable near z_0, show that

$$f(z) = \frac{1}{\pi i} \text{ P.V. } \int_{\gamma} \frac{f(\zeta)}{\zeta - z_0} \, d\zeta.$$

28. Use exercise 27 to find sufficient conditions under which

$$f(x,0) = \frac{1}{\pi i} \text{ P.V. } \int_{-\infty}^{\infty} \frac{f(\zeta,0)}{\zeta - x} \, d\zeta$$

for $f(x,y) = f(z)$ analytic. Deduce that

$$u(x,0) = \frac{1}{\pi} \text{ P.V. } \int_{-\infty}^{\infty} \frac{v(\zeta,0)}{\zeta - x} \, d\zeta$$

and that

$$v(x,0) = -\frac{1}{\pi} \text{ P.V. } \int_{-\infty}^{\infty} \frac{u(\zeta,0)}{\zeta - x} \, d\zeta.$$

Note. u and v are called *Hilbert transforms* of one another.

29. Explain what is wrong with the following reasoning.

$$\int_{0}^{\infty} \frac{\sin x}{x} \, dx = \frac{1}{2} \int_{-\infty}^{\infty} \frac{\sin x}{x} \, dx = \frac{1}{2} \lim_{R \to \infty} \int_{\gamma_R} \frac{\sin z}{z} \, dz,$$

where γ_R is the x-axis from $-R$ to R plus the circumference $\{z = Re^{i\theta} | 0 \leq \theta \leq \pi\}$. But $\sin z/z$ is analytic everywhere, including zero, so by Cauchy's Theorem,

$$\int_{\gamma_R} \frac{\sin z}{z} \, dz = 0.$$

†An exact answer to this seemingly simple problem is not known. An approximate value is 1.202. The sum is a special value of the Riemann zeta function ζ; see example 1, Section 3.1. This function ζ is very important in analysis and number theory and is the source of several famous open problems in mathematics.

Hence

$$\int_0^\infty \frac{\sin x}{x}\, dx = 0.$$

30. Prove that

$$\sum_{-\infty}^{\infty} \frac{(-1)^n}{(a+n)^2} = \pi^2 \mathrm{cosec}(\pi a)\, \cot(\pi a).$$

Chapter 5

Conformal Mappings

Chapter 1 included a brief investigation of some geometric aspects of analytic functions. Chapter 5 will return to this topic to develop some further techniques. We want to be able to map one given region to another given region by a one-to-one, onto, analytic function. Section 5.2 discusses several concrete cases for which such mappings can be written explicitly. That such mappings always exist in theory is the statement of the famous Riemann Mapping Theorem, which is discussed but not proved in Section 5.1.

The theory of conformal mappings has several important applications to the Dirichlet Problem and to harmonic functions. These applications are basic to heat conduction, electrostatics, and hydrodynamics, which will be discussed in Section 5.3. The basic idea of such applications is that an analytic mapping can be used to map a given region to a simpler region on which the problem can be solved by inspection. By transforming back to the original region, the desired answer is obtained.

5.1   Basic Theory of Conformal Mappings

The following definition was presented in Section 1.4: A mapping $f: A \rightarrow B$ is called *conformal* if, for each $z_0 \in A$, f rotates tangent vectors to curves through z_0 by a definite angle θ and stretches them by a definite factor r.

Let us recall the following theorem proved in Section 1.4.

Theorem 1. *Let $f: A \rightarrow B$ be analytic and let $f'(z_0) \neq 0$ for each $z_0 \in A$. Then f is conformal.*

Actually, if f merely preserves angles and if certain conditions of regularity hold, then f must be analytic and $f'(z_0) \neq 0$ (see exercise 9). Therefore, we can say that "conformal" means *analytic with a nonzero derivative*. We shall find it convenient to assume this meaning in the remainder of this text.

Let $A = \{z \mid \operatorname{Re} z > 0 \text{ and } \operatorname{Im} z > 0\}$ and $B = \{z \mid \operatorname{Im} z > 0\}$. Then $f: A \rightarrow B$; $z \mapsto z^2$ is conformal. (The student can easily check this.) Figure 5.1 illustrates Theorem 1 by showing preservation of angles in this case.

Notice that if $f'(z) = 0$, angles need not be preserved. (In the preceding example in which $z \mapsto z^2$, the x- and y-axes intersect at an angle $\pi/2$ but the images intersect at an angle π.) Such a point where $f'(z) = 0$ for an analytic function f is called a *singular point*. Singular points are studied in greater detail in Chapter 6.

Theorem 2.
  i. *If $f: A \rightarrow B$ is conformal and bijective (that is, one-to-one and onto), then $f^{-1}: B \rightarrow A$ is also conformal.*
  ii. *If $f: A \rightarrow B$ and $g: B \rightarrow C$ are conformal and bijective, then $g \circ f: A \rightarrow C$ is conformal and bijective.*

Proof.
  i. Since f is bijective, the mapping f^{-1} certainly exists. By the Inverse Function Theorem (Theorem 20, Chapter 1), f^{-1} is analytic with $df^{-1}(w)/dw = 1/(df(z)/dz)$ where $w = f(z)$. Thus $df^{-1}(w)/dw \neq 0$, so f^{-1} is conformal.
  ii. Certainly $g \circ f$ is bijective and analytic, since g and f are. (The inverse of $g \circ f$ is $f^{-1} \circ g^{-1}$.) The derivative of $g \circ f$ at z is $g'(f(z)) \cdot f'(z) \neq 0$. Therefore, $g \circ f$ is conformal by definition. ∎

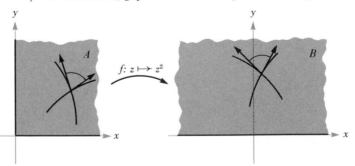

Figure 5.1 A conformal map.

Because of the two properties in Theorem 2 (and the obvious fact that the identity map $z \mapsto z$ is conformal), we can justifiably refer to the bijective conformal maps as a *group*.

Property (i) is important because we will be able to use it to solve various problems (such as the Dirichlet Problem) for a given region A. The method will be to find a bijective conformal map $f: A \to B$ where B is a simpler region on which the problem can be solved. To obtain the answer on A we then transform our answer from B to A by f^{-1}.

Thus we should check that harmonic functions remain harmonic when we compose them with a conformal map. To do so we prove the following result.

Theorem 3.   *Let u be harmonic on a region B and let $f: A \to B$ be analytic. Then $u \circ f$ is harmonic on A.*

A direct justification using coordinates is possible but quite laborious. The following proof is much shorter.

Proof.   Let $z \in A$ and $w = f(z)$. Let U be an open disk in B around w and let $V = f^{-1}(U)$. It suffices to show that $u \circ f$ is harmonic on V (Why?).

By Theorem 32, Chapter 2, there is an analytic function g on U such that $u = \text{Re } g$. Then $u \circ f = \text{Re}(g \circ f)$ (Why?), and we know that $g \circ f$ is analytic by the Chain Rule. Thus $\text{Re}(g \circ f)$ is harmonic. ∎

Riemann Mapping Theorem

There is a basic but more sophisticated theorem concerning conformal mappings that guarantees the existence of such mappings between given regions A and B. The applicability of this theorem in several special cases is verified in the next section. The general theorem is of little practical value because it does not tell us explicitly how to find conformal maps. Nevertheless, it is an important theorem that we should be aware of. The full proof is not given here since it requires more techniques than can be developed without more exposition. Uniqueness, however, can be proved fairly easily. (For the complete proof of this theorem, see E. Hille, *Analytic Function Theory*, Vol. II, p. 322; or L. Ahlfors, *Complex Analysis*, p. 222.)

Theorem 4 (Riemann Mapping Theorem).   *Let A be a simply connected region such that $A \neq \mathbb{C}$. Then there exists a bijective conformal map $f: A \to D$ where $D = \{z \mid |z| < 1\}$. Furthermore, we can specify that, for any fixed $z_0 \in A$, $f(z_0) = 0$ and $f'(z_0) > 0$. With such a specification, f is unique.*

From this result we see that if A and B are any two simply connected regions with $A \neq \mathbb{C}$, $B \neq \mathbb{C}$, then there is a bijective conformal map g: $A \longrightarrow B$. Indeed, if f: $A \longrightarrow D$ and h: $B \longrightarrow D$ are conformal, we can set $g = h^{-1} \circ f$ (see Figure 5.2). Two regions A and B are called *conformal* if there is a bijective conformal map from A to B. Thus the Riemann Mapping Theorem implies that two simply connected regions (unequal to $\mathbb{C}$) are conformal.

Proof of Uniqueness in Theorem 4.   Let f_1 and f_2 satisfy the conditions of Theorem 4; thus $f_i(z_0) = 0$ and $f_i'(z_0) > 0$ for some fixed z_0 and $i = 1, 2$. Let $g = f_1 \circ f_2^{-1}$. We shall show that $g(z) = z$. Similarly, we can show that $f_2 \circ f_1^{-1}(z) = z$, which will prove the result.

This mapping, g: $D \longrightarrow D$, has $g(0) = 0$ and $g'(0) > 0$; the inverse map g^{-1}: $D \longrightarrow D$ satisfies the same conditions. Let $w = g(z)$; that is, $z = g^{-1}(w)$. The Schwarz Lemma (see Theorem 31, Chapter 2) applies to g and gives $|g(z)| \leq |z|$ and $|g^{-1}(w)| \leq |w|$; that is, $|w| \leq |z|$ and $|z| \leq |w|$. Thus $|w| = |z|$, or $|g(z)| = |z|$. Therefore, $g(z) = e^{i\theta} z$ for θ a real constant (see exercise 16, Section 2.5). But if $g'(0) > 0$, we must have $e^{i\theta} = 1$. Hence $g(z) = z$. ∎

Notice that the condition $f'(z_0) > 0$ is equivalent to saying that $\arg f'(z_0) = 0$. Using the preceding argument, we can easily extend the uniqueness so that $f(z_0)$ and $\arg f'(z_0)$ are specified. The student is asked to prove this in exercise 7.

There is another useful fact we should know about conformal maps. Let A and B be two (connected) regions with boundaries bd(A) and bd(B). Suppose that f: $A \longrightarrow f(A)$ is conformal. If $f(A)$ has boundary bd(B) and if, for some $z_0 \in A$, we have $f(z_0) \in B$, then $f(A) = B$. In

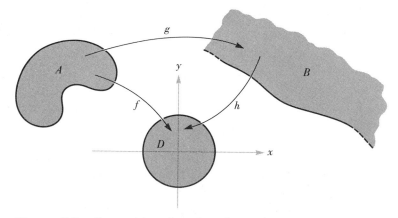

Figure 5.2 Composition of conformal mappings.

other words, to determine the image of a conformal map, we merely need to check the boundaries and a single point inside.

To prove this we argue as follows. Since B is open, $B \cap \text{bd}(B) = \varnothing$. The closure of B is just $B \cup \text{bd}(B)$, so we can decompose the plane as a disjoint union $\mathbb{C} = B \cup \text{bd}(B) \cup \text{ext } B$ where ext B is open. Since f' never vanishes on A, the inverse function theorem shows $f(A)$ is open. Thus $f(A) \cap \text{bd}(f(A)) = \varnothing$. But $\text{bd}(f(A)) = \text{bd}(B)$, so $f(A)$ is contained in the union of the disjoint open sets B and ext B. Since f is continuous on the connected set A, $f(A)$ is connected. So either $f(A) \subset B$ or $f(A) \subset \text{ext } B$. As $f(z_0) \in B$, we must have $f(A) \subset B$. Since $f(A)$ is open, it is open relative to B. Finally,

$$
\begin{aligned}
f(A) &= f(A) \cap B = (f(A) \cap B) \cup (\text{bd}(B) \cap B) \\
&= (f(A) \cap B) \cup (\text{bd}(f(A)) \cap B) \\
&= (f(A) \cup \text{bd}(f(A))) \cap B = \text{cl}(f(A)) \cap B
\end{aligned}
$$

so that $f(A)$ is closed relative to B. Since B is connected, we get $f(A) = B$ by the theorem on p. 149.

Theorem 5 outlines a simple procedure by which we can check one other case directly; our conditions are restrictive enough that we do not need to check the point $z_0 \in A$.

Theorem 5.   *Let A be a bounded region with $f: A \to \mathbb{C}$ a bijective conformal map onto its image $f(A)$. Suppose that f extends to be continuous on $\text{cl}(A)$ and that f maps the boundary of A onto a circle of radius R. Then $f(A)$ equals the inside of that circle. More generally, if B is a bounded region that, together with its boundary, can be mapped conformally onto the unit disk and its boundary and if f maps $\text{bd}(A)$ onto $\text{bd}(B)$, then $f(A) = B$.*

Proof.   By composing f with the conformal map h that takes B to the unit disk it is sufficient to consider the special case in which B equals $D = \{z \mid |z| < 1\}$. On $\text{bd}(A)$, $|f(z)| = 1$, so by the Maximum Modulus Theorem (Theorem 30, Section 2.5), $|f(z)| \leq 1$ on A. Thus since f cannot be constant, $f(A) \subset D$. (In other words, at no $z \in A$ is the maximum $|f(z)| = 1$ reached.) We have assumed that $f(\text{bd}(A)) = \text{bd}(D)$, but this is also equal to $\text{bd}(f(A))$. (Exercise. *Hint:* Use compactness of $\text{cl}(A)$, continuity of f, and $D \cap \text{bd}(D) = f(A) \cap \text{bd}(f(A)) = \varnothing$.) Thus our earlier argument applies to show that $f(A) = D$. ∎

Simple connectivity is essential in Theorem 4. It is easy to show (see example 1) that only a simply connected region can be mapped bijectively by an analytic map to D. A related result that can be shown is that the annuli $0 < |z| < 1$ and $1 < |z| < 2$ are not conformal.

Riemann sphere

For some purposes it is convenient to introduce a point "∞" in addition to the points $z \in \mathbb{C}$. This has not been done systematically in this text because such a device often leads to confusion and abuse of the symbol ∞, which, in fact, is merely a convenience.

In contrast to the real line to which $+\infty$ and $-\infty$ are added, we have only one ∞ for $\mathbb{C}$; the reason is that $\mathbb{C}$ has no natural ordering as $\mathbb{R}$ does.

Formally we add a symbol "∞" to $\mathbb{C}$ to obtain the *extended complex plane* $\overline{\mathbb{C}}$ and define operations with ∞ by the "rules"

$$z + \infty = \infty$$
$$z \cdot \infty = \infty$$
$$\infty + \infty = \infty$$
$$\infty \cdot \infty = \infty.$$

We also define, for example, $\underset{z \to \infty}{\text{limit }} f(z) = z_0$ to mean that *for any $\epsilon > 0$ there is an R such that $|z| \geq R$ implies that $|f(z) - z_0| < \epsilon$.*

Thus a point $z \in \mathbb{C}$ is "close to ∞" when it lies outside a large circle. This type of closeness can be pictured geometrically by means of the Riemann sphere. The Riemann sphere S is shown in Figure 5.3. By "stereographic projection," illustrated in this figure, a point z' on the sphere is associated with each point z in $\mathbb{C}$. Exactly one point on the sphere S has been omitted—the "north" pole. We assign ∞ in $\overline{\mathbb{C}}$ to the north pole. Then we see geometrically that z is close to ∞ iff, on the Riemann sphere, these points are close in the usual sense of closeness in $\mathbb{R}^3$. Proof of this assertion is requested in exercise 11.

In summary, then, the Riemann sphere represents a convenient geometric picture of the extended plane $\overline{\mathbb{C}} = \mathbb{C} \cup \{\infty\}$. This picture can also be used to make precise the ideas of analytic at ∞ and residue at ∞ (see exercise 10, Section 4.2). It can also be used conveniently to observe points that correspond to ∞ in conformal mappings. We shall have occasion to use it for this purpose in the next section.

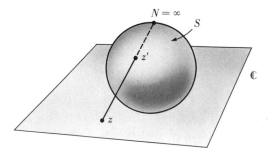

Figure 5.3   Riemann sphere.

Worked Examples

1. Find a bijective conformal map that takes a bounded region to an unbounded region. Can you find one that takes a simply connected region to
a region that is not simply connected?

> *Solution.*   Consider $f(z) = 1/z$ on $A = \{z| \ 0 < |z| < 1\}$. Clearly, A is
> bounded. Also, $B = f(A) = \{z| \ |z| > 1\}$; f is conformal from A to B and has
> an inverse $g^{-1}(w) = 1/w$. But B is unbounded.
>
> The answer to the second part of the question is no. If A is simply con-
> nected and $f: A \rightarrow B$ is a bijective conformal map, then B must be simply
> connected. To show this, let γ be a closed curve in B and let $\tilde{\gamma} = f^{-1} \circ \gamma$.
> Then if $H(t,s)$ is a homotopy shrinking $\tilde{\gamma}$ to a point, $f \circ H(t,s)$ is a homotopy
> shrinking γ to a point.

2. Consider the harmonic function $u(x,y) = x + y$ on the region $A = \{z|0 <$
$\text{Im } z < 2\pi\}$. What is the corresponding harmonic function on $B = \mathbb{C}\backslash(\text{posi-}$
tive real axis) when A is transformed by $z \mapsto e^z$?

> *Solution.*   Let $f(z) = e^z$. We know from Chapter 1 that f is one-to-one,
> onto B and that $f'(z) = e^z \neq 0$. Thus f is conformal from A to B, and there-
> fore, by Theorem 3, the corresponding function on B is harmonic. This
> function is:
>
> $$\begin{aligned} v(x,y) &= u(f^{-1}(x,y)) \\ &= u(\log(x + iy)) \\ &= u\left(\log \sqrt{x^2 + y^2} + i\tan^{-1}\left(\frac{y}{x}\right)\right) \\ &= \log \sqrt{x^2 + y^2} + \tan^{-1}\left(\frac{y}{x}\right), \end{aligned}$$
>
> where $\tan^{-1}(y/x) = \arg(x + iy)$ lies in $]0,2\pi[$. Note that to check directly that
> v is harmonic would be slightly tedious, but we know it must be so by
> Theorem 3.

3. What is the image of the region $A = \{z|(\text{Re } z)(\text{Im } z) > 1$, and $\text{Re } z > 0$,
$\text{Im } z > 0\}$ under the transformation $z \mapsto z^2$?

> *Solution.*   Figure 5.4 depicts the region A.
>
> On the right half plane $\{z|\text{Re } z > 0\}$, we know that $f(z) = z^2$ is conformal
> (Why?). To find the image of A we first find the image of the curve $xy = 1$.
> Let $w = z^2 = u + iv$. Then $u = x^2 - y^2$, $v = 2xy$. Thus the image of $xy = 1$ is
> the curve $v = 2$. We must check the location of the image of a point in A,
> say, $z = 2 + 2i$. Here $z^2 = 8i$, and therefore the image is the shaded region
> in Figure 5.5.

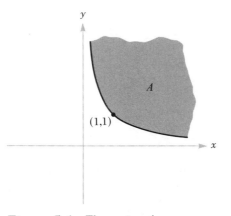

Figure 5.4 The region A.

Exercises

1. Suppose that f is a conformal map from a bounded region A onto an unbounded region B. Show that f cannot be extended in such a way as to be continuous on $A \cup \text{bd}(A)$. (Note: The full force of conformality is not needed in this problem.)

2. Consider $f = u + iv$ where $u(x,y) = 2x^2 + y^2$ and $v = y^2/x$. Show that the curves $u = $ constant and $v = $ constant intersect orthogonally but that f is not analytic.

3. Let A and B be regions whose boundaries are smooth arcs. Let f be conformal on a region including $A \cup \text{bd}(A)$; map A onto B and $\text{bd}(A)$ onto $\text{bd}(B)$. Let u be harmonic on B and $u = h(z)$ on the boundary of B. Let $v = u \circ f$ so that v equals $h \circ f$ on the boundary of A. Prove that $\partial v/\partial n = 0$ at z_0 iff $\partial u/\partial n = 0$ at $f(z_0)$ where $z_0 \in \text{bd}(A)$ and $\partial/\partial n$ denotes the derivative in the normal direction to the boundary.

4. Near what points are the following maps conformal?
   a. $f(z) = z^3 + z^2$.
   b. $f(z) = z/(1 + 5z)$.
   c. $f(z) = \bar{z}$.
   d. $f(z) = \sin z/\cos z$.

5. Consider the harmonic function $u(x,y) = 1 - y + x/(x^2 + y^2)$ on the upper half plane $y > 0$. What is the corresponding harmonic function on the first quadrant $x > 0$, $y > 0$, under the transformation $z \mapsto z^2$?

6. What is the image of the first quadrant under the mapping $z \mapsto z^3$?

7. Let A and B be regions as in Theorem 4. Given $z_0 \in A$, $w_0 \in B$, and θ, and by assuming Theorem 4, show that there exists a conformal map f: $A \to B$ with $f(z_0) = w_0$ and $\arg f'(z_0) = \theta$; also show that such an f is unique.

8. If $f: A \to B$ is bijective and analytic with analytic inverse, prove that f is conformal.

9. Let $f: A \rightarrow B$ be a function such that $\partial f/\partial x$ and $\partial f/\partial y$ exist and are continuous. Suppose that f is one-to-one and onto and preserves angles; prove that f is analytic and conformal. Can the map in exercise 2 preserve all angles?

   Hint.   Let $c(t)$ be a curve with $c(0) = z_0$ and let $d(t) = f(c(t))$. Prove

   $$d'(t) = \frac{1}{2}\left(\frac{\partial f}{\partial x} - i\frac{\partial f}{\partial y}\right)c'(t) + \frac{1}{2}\left(\frac{\partial f}{\partial x} + i\frac{\partial f}{\partial y}\right)\overline{c'(t)}$$

   and examine the statement that $d'(0)/c'(0)$ has constant argument in order to establish the Cauchy-Riemann equations for f.

10. Let $f: \mathbb{C} \rightarrow \mathbb{C}$, $z \mapsto az + b$. Show that f can be written as a rotation followed by a magnification followed by a translation. Does every bijective conformal transformation from $\mathbb{C}$ onto $\mathbb{C}$ have this form?

11. Introduce the *chordal metric* ρ on $\overline{\mathbb{C}}$ by setting

    $$\rho(z_1, z_2) = d(z_1', z_2')$$

    where z_1', z_2' are the corresponding points on the Riemann sphere and d is the usual distance between points in $\mathbb{R}^3$.
    a. Show that $z_n \rightarrow z$ in $\mathbb{C}$ iff $\rho(z_n, z) \rightarrow 0$.
    b. Show that $z_n \rightarrow \infty$ iff $\rho(z_n, \infty) \rightarrow 0$.
    c. If $f(z) = (az + b)/(cz + d)$, $ad - bc \neq 0$, show that f is continuous at ∞.

5.2 Fractional Linear and Schwarz-Christoffel Transformations

This section will investigate some ways of obtaining specific conformal maps between two given regions. No general prescription can be given for obtaining these maps; however, after a little practice the student will be able to combine bilinear transformations (studied in this section) with other transformations with which he is already acquainted (like

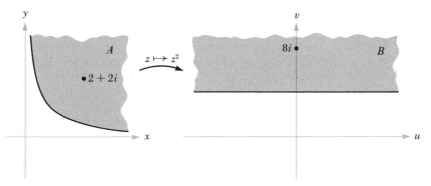

Figure 5.5   Image of the set A under the conformal map $z \mapsto z^2$.

z^2, e^z, or sin z) and thus be able to handle many useful situations. To aid him in this effort, some common transformations are illustrated in Figure 5.10 at the end of the section.

In addition, the Schwarz-Christoffel Formula will be studied briefly, although it yields answers that usually can only be given in terms of integrals and is somewhat awkward to handle.

The simplest and one of the most useful kinds of conformal mappings will be discussed first.

Fractional linear transformations

A fractional linear transformation (also called a bilinear transformation or Möbius transformation) is a mapping of the form

$$T(z) = \frac{az + b}{cz + d} \tag{1}$$

where a, b, c, d are fixed complex numbers. We shall assume that $ad - bc \neq 0$, because otherwise T would be a constant (Why?) and we want to omit that case.

The properties of these transformations will be developed in the next four theorems.

Theorem 6. *The map T defined by equation (1) is bijective and conformal from*

$$A = \left\{ z \middle| cz + d \neq 0, \text{ that is, } z \neq \frac{-d}{c} \right\} \text{ onto } B = \left\{ w \middle| w \neq \frac{a}{c} \right\}.$$

In fact, the inverse of T is also fractional linear and is given by

$$T^{-1}(w) = \frac{-dw + b}{cw - a}. \tag{2}$$

Proof.   Certainly T is analytic on A and $S(w) = (-dw + b)/(cw - a)$ is analytic on B. The map T will be bijective if we can show that $T \circ S$ and $S \circ T$ are the identities since this means that T has S as its inverse. Indeed, this is seen by straightforward computation:

$$T(S(w)) = \frac{a\left(\dfrac{-dw + b}{cw - a}\right) + b}{c\left(\dfrac{-dw + b}{cw - a}\right) + d}$$

$$= \frac{-adw + ab + bcw - ab}{-cdw + bc + dcw - da}$$

$$= \frac{(bc - ad)w}{bc - ad} = w.$$

We can cancel because $cw - a \neq 0$ and $bc - ad \neq 0$. Similarly, $ST(z) = z$. Finally, $T'(z) \neq 0$ because

$$\frac{d}{dz}\,S(T(z)) = \frac{d}{dz}\,z = 1$$

so

$$S'(T(z)) \cdot T'(z) = 1.$$

Therefore, $T'(z) \neq 0.\blacksquare$

It is sometimes convenient to say that $T(-d/c) = \infty$ (although we must, as always, be careful to avoid the erroneous answers that we would obtain if we canceled ∞/∞ or $0/0$). In fact, we can show that all fractional linear transformations are conformal maps of the extended plane $\overline{\mathbb{C}}$ to itself. Some special cases should be noted. For example, if $a = 1$, $c = 0$, and $d = 1$, we get $T(z) = z + b$, which is a translation or "shift" that merely translates by the vector b (see Figure 5.6).

Also, in the case in which $b = c = 0$, $d = 1$, T becomes $T(z) = az$. This map, multiplication by a, is a rotation by arg a and magnification by $|a|$. The student should review the geometric meaning in this case.

Finally, $T(z) = 1/z$ is an inversion. It is geometrically pictured in Figure 5.7.

The fractional linear transformations are useful because of their geometric properties. This advantage is illustrated by the following theorem.

Theorem 7. *Any conformal map of $D = \{z|\ |z| < 1\}$ onto itself is a fractional linear transformation of the form*

$$T(z) = e^{i\theta}\,\frac{(z - z_0)}{1 - \bar{z}_0 z} \tag{3}$$

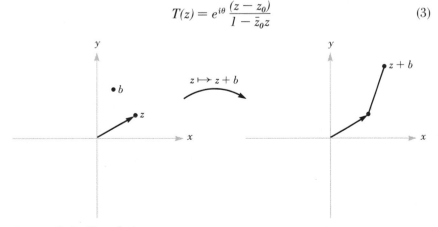

Figure 5.6 Translation.

276]

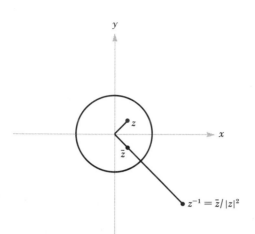

Figure 5.7 Inversion.

for some fixed $z_0 \in D$, $\theta \in [0,2\pi[$, and any T of this form is a conformal map of D onto D.

Proof. First we check that for T of this form, $|z| = 1$ implies that $|T(z)| = 1$. Indeed,

$$|T(z)| = \left|\frac{z - z_0}{1 - \bar{z}_0 z}\right| = \frac{|z - z_0|}{|z|\,|z^{-1} - \bar{z}_0|}.$$

But $|z| = 1$ and so $z^{-1} = \bar{z}$. Hence we get

$$|T(z)| = \frac{|z - z_0|}{|\bar{z} - \bar{z}_0|} = 1$$

since $|w| = |\bar{w}|$. The only singularity of T is at $z = \bar{z}_0^{-1}$, which lies outside the unit circle.

Thus by the Maximum Modulus Theorem, T maps D to D. But by Theorem 6,

$$T^{-1}(w) = e^{-i\theta}\left\{\frac{w - (-e^{i\theta}z_0)}{1 - (-e^{-i\theta}\bar{z}_0)w}\right\},$$

which, since it has the same form as T, is also a map from D to D. Thus T is conformal from D onto D.

Let $R: D \rightarrow D$ be any conformal map. Let $z_0 = R^{-1}(0)$ and let $\theta = \arg R'(z_0)$. Now T defined by equation (3) also has $T(z_0) = 0$ and $\theta = \arg T'(z_0)$. (We see that

$$T'(z) = e^{i\theta}\left\{\frac{1 - |z_0|^2}{(1 - \bar{z}_0 z)^2}\right\},$$

which, at $z = z_0$, equals

$$e^{i\theta}\left\{\frac{1}{1 - |z_0|^2}\right\},$$

a real constant times $e^{i\theta}$.) Thus, by uniqueness of conformal maps (see Theorem 4 and exercise 7, Section 5.1), $R = T$. ∎

Therefore, the only way to map a disk onto itself conformally is by means of a fractional linear transformation. But these transformations have two additional properties, as will be shown in the two fundamental theorems that follow.

Theorem 8.   *Let T be a fractional linear transformation. If $L \subset \mathbb{C}$ is a straight line and $S \subset \mathbb{C}$ is a circle, then $T(L)$ is either a straight line or a circle and $T(S)$ is either a straight line or a circle.*

We should be aware that a line can map either to a circle or to a line. If we regard lines as circles of infinite radius, then Theorem 8 can be summarized by saying that circles transform into circles.

Proof.   This theorem can be proved in several ways but perhaps the easiest is the following. We can write $T = T_4 \circ T_3 \circ T_2 \circ T_1$, where $T_1(z) = z + d/c$, $T_2(z) = 1/z$, $T_3(z) = (bc - ad)z/c^2$, and $T_4(z) = z + a/c$. (If $c = 0$, we merely write $T(z) = (a/d)z + b/d$.) This is easily verified (see exercise 1). It is obvious that T_1, T_3, and T_4 map lines to lines and circles to circles. Thus if we can verify the conclusion for $T(z) = 1/z$, the proof will be complete. We know from analytic geometry that a line or circle is determined by the equation

$$Ax + By + C(x^2 + y^2) = D$$

for constants A, B, C, D, with not all A, B, C zero. Let $z = x + iy$, suppose that $z \neq 0$, and let $1/z = u + iv$ so that $u = x/(x^2 + y^2)$, and $v = -y/(x^2 + y^2)$. Thus the preceding equation is equivalent to

$$Au - Bv - D(u^2 + v^2) = -C,$$

which is also a line or circle. ∎

The second property of fractional linear transformations is described in Theorem 9.

Theorem 9.   *Given two sets of distinct points z_1, z_2, z_3, and w_1, w_2, w_3 (that is, $z_1 \neq z_2$, $z_1 \neq z_3$, $z_2 \neq z_3$, and $w_1 \neq w_2$, $w_2 \neq w_3$, $w_1 \neq w_3$, but we* *could have $z_1 = w_2$, and so on), there is a unique fractional linear transformation T taking $z_i \mapsto w_i$, $i = 1, 2, 3$. In fact, if $T(z) = w$, then*

$$\frac{w - w_1}{w - w_2} \cdot \frac{w_3 - w_2}{w_3 - w_1} = \frac{z - z_1}{z - z_2} \cdot \frac{z_3 - z_2}{z_3 - z_1}. \tag{4}$$

Remark.   The student will find that instead of trying to remember formula (4) it is often easier to proceed directly (see example 2).

Proof.   Formula (4) defines a fractional linear transformation $w = T(z)$ (Why?). By direct substitution we see that it has the desired properties $T(z_i) = w_i$, $i = 1, 2, 3$. Let us show that it is unique. Define

$$S(z) = \frac{(z - z_1)}{(z - z_2)} \cdot \frac{(z_3 - z_2)}{(z_3 - z_1)}.$$

Then S is a fractional linear transformation taking z_1 to 0, z_3 to 1, and z_2 to ∞. (z_2 is the singularity of S.) Let R be any other fractional linear transformation $R(z) = (az + b)/(cz + d)$ with $R(z_1) = 0$, $R(z_3) = 1$, and $R(z_2) = \infty$ (that is, $cz_2 + d = 0$). Then $az_1 + b = 0$, $cz_2 + d = 0$, and $(az_3 + b)/(cz_3 + d) = 1$. Thus we get $a = -b/z_1$ and $c = -d/z_2$, so the last condition gives $b(z_1 - z_3)/z_1 = d(z_2 - z_3)/z_2$. Substituting in R we see, after simplification (which the student should do), that $R = S$.

We use this result to prove that T is unique as follows. Let T be any fractional linear transformation taking z_i to w_i, $i = 1, 2, 3$. The fractional linear transformation ST^{-1} takes $w_1 = Tz_1$ to 0, $w_3 = Tz_3$ to 1, and $w_2 = Tz_2$ to ∞. Therefore, ST^{-1} is uniquely determined by the preceding computation. Hence T is uniquely determined since $T = (ST^{-1})^{-1}S.$ ∎

We can therefore map any three points to any other three points. Three points lie on either a (uniquely determined) line or circle. Thus, by Theorem 8, the transformation maps the line (or circle) through z_1, z_2, z_3 to the line (or circle) through w_1, w_2, w_3. For example, we could have the situation depicted in Figure 5.8. We know from Section 5.1 that the inside of the circle in Figure 5.8 maps onto one of the two half planes determined by the line through w_1, w_2, w_3. To determine which half, we can check to see where the center of the circle goes. The half that is the image can be altered by interchanging z_1 and z_2. If z_1, z_2, z_3 and w_1, w_2, w_3 each determine a circle, the inside of the circle through z_1, z_2, z_3 must map to the inside of the circle through w_1, w_2, w_3 by Theorem 5.

As mentioned earlier, fractional linear transformations can be combined with other transformations to obtain a fairly large class of conformal maps. This is illustrated in example 1.

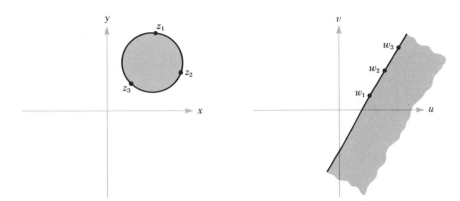

Figure 5.8 Effect of a fractional linear transformation.

The Schwarz-Christoffel Formula

The Schwarz-Christoffel Formula gives a method, in terms of integrals, for mapping the upper half plane or unit circle to the interior of a given polygon. The case of the upper half plane will be discussed here; the case of a circle is left as an exercise.

Theorem 10 (Schwarz-Christoffel Formula). *Suppose that P is a polygon in the w-plane with vertices at w_1, w_2, $\cdots$, w_n and with exterior angles $\pi\alpha_i$, where $-1 < \alpha_i < 1$ (see* Figure 5.9*). Then conformal maps from $A = \{z \mid \operatorname{Im} z > 0\}$ onto B, the interior of P, have the form*

$$f(z) = a\left(\int_{z_0}^{z} (\zeta - x_1)^{-\alpha_1} \cdot \ \cdot \ \cdot \ \cdot (\zeta - x_{n-1})^{-\alpha_{n-1}} \, d\zeta\right) + b, \qquad (5)$$

where a and b are constants and the integration is along any path in A joining $z_0 \in A$ to z; the principal branch is used for the powers in the integrand. Furthermore,

i. *Two of the points x_1, $\cdots$, x_n may be chosen arbitrarily;*
ii. *a and b determine the size and position of P;*
iii. *$f(x_i) = w_{i+1}$, $i = 1$, $\cdots$, $n-1$;*
iv. *f takes the point at infinity to w_1.*

The geometric meaning of the constants a and b in Theorem 10 is explained in more detail in the following proof. It can be shown that the function f can be extended to be continuous on the x-axis and that it maps the x-axis to the polygon P. However, the function f is not analytic on the x-axis because it does not preserve angles at x_i. But f will be analytic on A itself. Only the main ideas of the proof of Theorem 10 will be given here because to make the proof absolutely precise would be rather tedious.

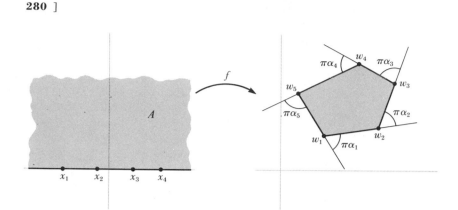

Figure 5.9 Schwarz-Christoffel Formula.

Sketch of the proof of Theorem 10.   The first step is to show that if $x_1, \cdots, x_{n-1}$ have already been chosen, then f maps the real axis to a polygon having the correct angles. Let

$$g(z) = a(z - x_1)^{-\alpha_1} \cdot \cdots \cdot (z - x_{n-1})^{-\alpha_{n-1}}$$

so that on A, $f'(z) = g(z)$. Then

$$\arg f'(z) = \arg g(z)$$
$$= \arg a - \alpha_1 \arg(z - x_1) - \cdots - \alpha_{n-1} \arg(z - x_{n-1}).$$

At a point where $f'(z)$ exists, $\arg f'(z)$ represents the amount f rotates tangent vectors. Thus, as z moves along the real axis, $f(z)$ moves along a straight line for z on each of the segments $]-\infty, x_1[, \cdots,]x_i, x_{i+1}[, \cdots,]x_{n-1}, \infty[$. As z crosses x_i, $\arg f(z)$ jumps by an amount $\alpha_i \pi$. (If $z - x_i < 0$, $\arg(z - x_i) = \pi$; if $z - x_i > 0$, $\arg(z - x_i) = 0$.) Thus the real axis is mapped to a polygon with the correct angles. The last angle of the polygon is determined since we must have $\Sigma_{i=1}^n \alpha_i \pi = 2\pi$.

Next, we adjust this polygon to obtain P. Equality of angles forces similarity of polygons only for triangles. (For example, not all rectangles are squares.) This is the basic reason why two of the points x_i may be chosen arbitrarily (three if we count the point at infinity). The positions of the other points relative to these points control the ratios of the lengths of the sides of the image polygon. By choosing the x_i correctly we thus obtain a polygon similar to P. Another way to understand this problem is to consider the similar problem of mapping the upper half plane to a disk. We know that this can be accomplished by a linear fractional transformation and that this transformation is completely determined by its value at three of the boundary points. (Two of the finite points and the value at infinity are specified.) Choosing a and b properly means performing a scaling, a rotation, and a translation to bring this polygon to P. ■

Although the Schwarz-Christoffel Formula expresses the conformal map f only as an integral, numerical techniques can be used to obtain approximate answers. Often this is the only available recourse.

Worked Examples

1. Find a conformal map taking $A = \{z \mid 0 < \arg z < \pi/2,\ 0 < |z| < 1\}$ to $D = \{z \mid |z| < 1\}$.

   Solution.   The answer is not given by $z \longmapsto z^4$ because this map does not map A onto D; its image omits the positive real axis.
   First consider $z \longmapsto z^2$. This maps A to B where B is the intersection of D and the upper half plane (Figure 5.11). Next (consult Figure 5.10(iv)) map B to the first quadrant by $z \longmapsto (1 + z)/(1 - z)$ and then square to get the upper half plane; then map $z \longmapsto (z - i)/(z + i)$ to give the unit circle.
   Thus we obtain our transformation by successive substitution:

   $$w_1 = z^2,\ w_2 = \frac{1 + w_1}{1 - w_1} = \frac{1 + z^2}{1 - z^2},\ w_3 = w_2^2 = \left(\frac{1 + z^2}{1 - z^2}\right)^2,$$

   $$w_4 = \frac{w_3 - i}{w_3 + i} = \frac{\left(\dfrac{1 + z^2}{1 - z^2}\right)^2 - i}{\left(\dfrac{1 + z^2}{1 - z^2}\right)^2 + i}.$$

   Therefore,

   $$f(z) = \frac{(1 + z^2)^2 - i(1 - z^2)^2}{(1 + z^2)^2 + i(1 - z^2)^2}$$

   is the required transformation. Note that it is not a fractional linear transformation but is more complicated.

2. Verify Figure 5.10(vi).

   Solution.   We seek a fractional linear transformation $T(z) = (az + b)/(cz + d)$ such that $T(-1) = i$, $T(0) = -1$, $T(1) = -i$. Thus $(-a + b)/(-c + d) = i$, $b = -d$, and $(a + b)/(c + d) = -i$. Solving gives $-a - d = i(-c + d)$, $b = -d$, $a - d = -i(c + d)$. Adding the first and last equations, we get $-2d = i(-2c)$ or $d = ic$, and subtracting gives us $a = -id$. We can set, say, $b = 1$ (because numerator and denominator can be multiplied by a constant), so $d = -1$, $a = i$, $c = i$, and thus

   $$T(z) = \frac{iz + 1}{iz - 1} = \frac{z - i}{z + i}.$$

   We must check that $T(i)$ lies inside the unit circle. This is true because $T(i) = 0$. (If it lay outside, we would interchange $A = -1$ and $B = 0$.)

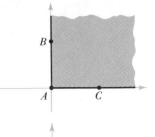

$$z \mapsto z^2$$
$$\sqrt{z} \longmapsfrom z$$

(i)

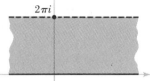

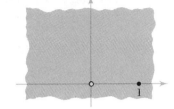

$$z \mapsto e^z$$
$$\log z \longmapsfrom z$$

(ii)

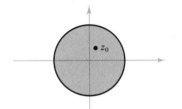
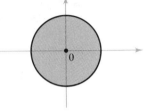

$$z \mapsto e^{i\theta}\,\frac{z - z_0}{1 - \bar{z}_0 z}$$

(iii)

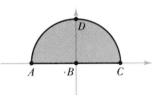

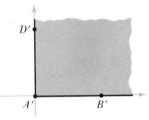

$$z \mapsto \frac{1 + z}{1 - z}$$
$$-\frac{1 - z}{1 + z} \longmapsfrom z$$

(iv)

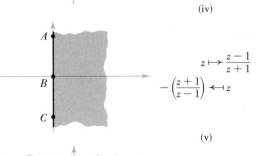
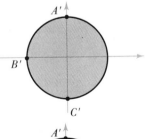

$$z \mapsto \frac{z - 1}{z + 1}$$
$$-\left(\frac{z + 1}{z - 1}\right) \longmapsfrom z$$

(v)

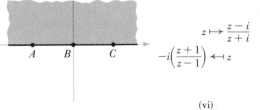
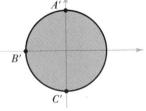

$$z \mapsto \frac{z - i}{z + i}$$
$$-i\left(\frac{z + 1}{z - 1}\right) \longmapsfrom z$$

(vi)

Figure 5.10   Some common transformations.

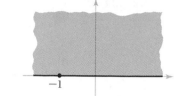

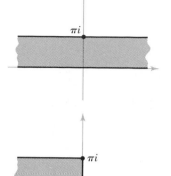

$$z \mapsto e^z$$
$$\log z \longleftarrow z$$

(vii)

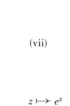

$$z \mapsto e^z$$
$$\log z \longleftarrow z$$

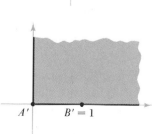

(viii)

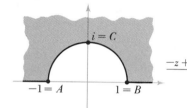

$$z \mapsto z + \frac{1}{z}$$
$$\frac{-z + \sqrt{z^2 - 4}}{2} \longleftarrow z$$

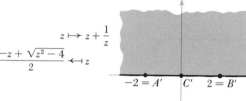

(ix)

$$z \mapsto \frac{1}{z}$$
$$\frac{1}{z} \longleftarrow z$$

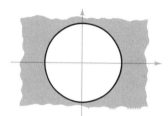

(x)

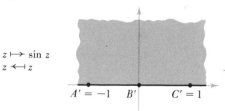

$$z \mapsto \sin z$$
$$\sin^{-1} z \longleftarrow z$$

(xi)

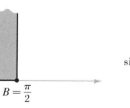

$$z \mapsto \sin z$$
$$\sin^{-1} z \longleftarrow z$$

(xii)

Figure 5.10 (*continued*)

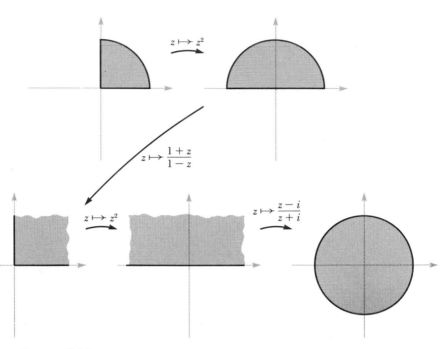

Figure 5.11

3. Find a conformal map that takes the half plane shown in Figure 5.12 onto the unit circle.

    Solution.   Consider $S(z) = e^{-i\alpha}z$. This maps the region A to the upper half plane (Why?). Then, using Figure 5.10(vi), we get

    $$T(z) = \frac{e^{-i\alpha}z - i}{e^{-i\alpha}z + i}$$

    as the required transformation.

Exercises

1. Prove: Any fractional linear transformation with $c \neq 0$ can be written $T = T_4 \circ T_3 \circ T_2 \circ T_1$, where $T_1(z) = z + d/c$, $T_2(z) = 1/z$, $T_3(z) = [(bc - ad)/c^2]z$, and $T_4(z) = z + a/c$. Interpret T geometrically.
2. Prove that if both T and R are fractional linear transformations, then so is $T \circ R$.
3. Find a conformal map of the unit disk onto itself that maps $1/2$ to 0.
4. Find all conformal maps that take the disk of radius R onto the unit disk.
5. Suppose that a, b, c, d are real and that $ad > bc$; show that $T(z) = (az + b)/(cz + d)$ leaves the upper half plane invariant. Show that any conformal map of the upper half plane onto itself is of this form.

6. Prove: The most general conformal transformation that takes the upper half plane onto the unit circle is

$$T(z) = e^{i\theta}\left(\frac{z - \lambda}{z - \bar\lambda}\right),$$

where $\operatorname{Im} \lambda > 0$.

7. Establish (iii), (iv), and (v) of Figure 5.10.

8. The *cross ratio* of four distinct points z_1, z_2, z_3, z_4 is defined by

$$[z_1, z_2, z_3, z_4] = \left(\frac{z_4 - z_1}{z_4 - z_2}\right) \cdot \left(\frac{z_3 - z_2}{z_3 - z_1}\right).$$

Show that any fractional linear transformation has the property that $[Tz_1, Tz_2, Tz_3, Tz_4] = [z_1, z_2, z_3, z_4]$.

Hint.   Use exercise 1.

9. Show that $[z_1, z_2, z_3, z_4]$ is real iff z_1, z_2, z_3, z_4 lie on a line or circle. Use exercise 8 to give another proof of Theorem 8.

10. Find a conformal map that takes $\{z \mid 0 < \arg z < \pi/8\}$ onto the unit disk.

11. Show that a fractional linear transformation T that is not the identity map has at most two fixed points (that is, points z for which $T(z) = z$). Give an example to show that T need not have any fixed points. Find the fixed points of $T(z) = z/(z + 1)$.

12. Let γ_1 and γ_2 be two circles that intersect orthogonally. Let T be a fractional linear transformation. What can be said about $T(\gamma_1)$ and $T(\gamma_2)$?

13. Conformally map $A = \{z \mid \operatorname{Re} z < 0, 0 < \operatorname{Im} z < \pi\}$ onto the first quadrant.

14. Conformally map $\mathbb{C}\backslash\{\text{nonpositive real axis}\}$ onto the region $A = \{z \mid -\pi < \operatorname{Im} z < \pi\}$.

15. Conformally map $A = \{a \mid |z - 1| < 1\}$ onto $B = \{z \mid \operatorname{Re} z > 1\}$.

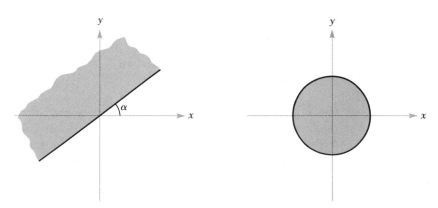

Figure 5.12   Mapping a sector to the disk.

16. Argue that the conformal maps that take $|z| < 1$ to the interior of a polygon with vertices $w_1, \cdots, w_n$ and points $z_1, \cdots, z_n$ on the unit circle $|z| = 1$ to the points $w_1, \cdots, w_n$ are given by

$$f(z) = a\left(\int_0^z (\zeta - z_1)^{-\alpha_1} \cdots (\zeta - z_n)^{-\alpha_n} \, d\zeta\right) + b$$

where α_i are as in Theorem 10.

17. Show that

$$f(z) = \int_0^z \frac{d\zeta}{\sqrt{\zeta(\zeta - 1)(\zeta - c)}},$$

where $c > 0$, maps the upper half plane to a rectangle. (The integrand is called an elliptic integral and generally cannot be computed explicitly.)

18. Is it possible to map the upper half plane conformally to a triangle using fractional linear transformations? Devise a formula that is based on the Schwarz-Christoffel Formula.

19. Verify from the Schwarz-Christoffel Formula that the conformal map from the upper half plane to $\{z \mid \operatorname{Im} z > 0 \text{ and } -\pi/2 < \operatorname{Re} z < \pi/2\}$ is $z \mapsto \sin^{-1} z$.

20. Consider a fractional linear transformation of the form

$$f(z) = a\left(\frac{z - b}{z - d}\right).$$

Show that

a. Circles through the points b and d are mapped to lines through the origin.

b. The *circles of Apollonius* with equation $|(z - b)/(z - d)| = r/|a|$ are mapped to circles with center 0, radius r.

c. The circles in (a) and (b), when located in the z-plane, are called *Steiner circles*. Sketch them and verify that both these circles and their images meet in right angles.

5.3 Applications of Conformal Mapping to Laplace's Equation, Heat Conduction, Electrostatics, and Hydrodynamics

We are now in a position to apply the theory of conformal maps developed in Sections 5.1 and 5.2 to some physical problems. In doing so we will solve the Dirichlet Problem* and related problems for several types of two-dimensional regions. We will then apply these results to the three

---

*The problem of finding a harmonic function on a region A whose values are specified on the boundary of A is called the *Dirichlet Problem*. This problem was discussed in Section 2.5.

classes of physical problems mentioned in the title of this section. Only a very meager knowledge of elementary physics is needed to understand these examples. The student is cautioned that the variety of problems that can explicitly be solved in this way is somewhat limited and that the methods discussed apply only to two-dimensional problems.

The Dirichlet and Neumann problems

Let us recall that $u(x,y)$ is said to satisfy *Laplace's equation* (or be *harmonic*) on a region A when

$$\nabla^2 u = \frac{\partial^2 u}{\partial x^2} + \frac{\partial^2 u}{\partial y^2} = 0.$$

In addition to this condition, some boundary behavior determined by the physical problem to be solved is generally specified. This boundary behavior (or boundary conditions) usually determines u uniquely. For example, Theorem 37, Section 2.5, indicated that a harmonic function $u(x,y)$, whose value on the boundary of A is specified, is uniquely determined.

We shall also have occasion to consider the boundary condition in which $\partial u/\partial n = \operatorname{grad} u \cdot \mathbf{n}$ is specified on the boundary. ($\partial u/\partial/n$ equals the derivative in the direction normal to $\operatorname{bd}(A)$). (For $\partial u/\partial n$ to be well defined, the boundary of A should be at least piecewise smooth, so that it has a well-defined normal direction. The *outward* normal direction of $\mathbf{n}$ can be defined precisely, but since the general emphasis of this section is on computational methods, an absolutely precise treatment of such a question will not be given here. Thus we accept as clear what is meant by the outward unit normal, $\mathbf{n}$ (see Figure 5.13).

The problem of finding a harmonic function u with $\partial u/\partial n$ specified on the boundary is called the *Neumann Problem*. We cannot specify $\varphi = \partial u/\partial n$ arbitrarily because if such a u exists, then we must have

$$\int_{\gamma} \frac{\partial u}{\partial n} = 0,$$

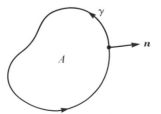

Figure 5.13   Neumann Problem.

where γ is the boundary of A. To prove this, we apply Green's Theorem (see Section 2.2), which can be written

$$\int_\gamma \mathbf{X} \cdot \mathbf{n} = \int_A \operatorname{div} \mathbf{X} dxdy;$$

where

$$\operatorname{div} \mathbf{X} = \frac{\partial X^1}{\partial x} + \frac{\partial X^2}{\partial y} \tag{1}$$

and where $\mathbf{X}$ is a given vector function with components (X^1, X^2). (In this form the theorem is called *Gauss' Theorem.*) Applying formula (1) to $\mathbf{X} = \operatorname{grad} u$ gives

$$\int_\gamma \frac{\partial u}{\partial n} = \int_\gamma (\operatorname{grad} u) \cdot \mathbf{n} = \int_A \operatorname{div} \operatorname{grad} u$$

$$= \int_A \nabla^2 u = 0$$

because div grad $u = \nabla^2 u = 0$.

If we are given a boundary condition φ on γ with $\int_\gamma \varphi = 0$, then it can be shown (although with difficulty) that the Neumann Problem has a solution.

Let us indicate the proof of the following fact: *The solution of the Neumann Problem on a bounded simply connected region is unique up to the addition of a constant.* Let u_1 and u_2 be two solutions with $\partial u_1/\partial n = \partial u_2/\partial n$ on $\gamma = \operatorname{bd}(A)$. Let v_1 and v_2 be harmonic conjugates of u_1 and u_2 and set $u = u_1 - u_2$, $v = v_1 - v_2$. Now $\partial u/\partial n = 0$, so, by Theorem 33, Section 2.5, v is constant along γ. Thus by uniqueness of the solution to the Dirichlet Problem, v is constant on A. Therefore, since u is the harmonic conjugate of v, u is constant on A as well. This proves our claim.

If the boundary values specified in the Dirichlet and Neumann problems are not continuous, the uniqueness results are still valid, in a sense, but are more difficult to obtain. However, the student is cautioned that on an unbounded region we do not have uniqueness. For example, let A be the upper half plane. Then $u_1(x,y) = x$ and $u_2 = x + y$ have the same boundary values (at $y = 0$) and are harmonic but are not equal.

To recover uniqueness for unbounded regions a "condition at ∞" must also be specified; "u bounded on all of A" is such a condition. Some of these conditions will be illustrated in the examples that are integrated into this section.

The Dirichlet and Neumann problems can also be combined; for example, u can be specified on one part of the boundary and $\partial u/\partial n$ can be specified on another.

Method of solution

The basic method for solving the Dirichlet and Neumann problems in a given region A is as follows. Take the given region A and transform it by a conformal map to a "simpler" region B on which the problem can be solved. This procedure is justified by the fact that under a conformal map f, harmonic functions are transformed again into harmonic functions (see Theorem 3). When we have solved the problem on B, we can transform the answer back to A.

For the Dirichlet Problem we are given the boundary values on $\mathrm{bd}(A)$. These values obviously get mapped to the corresponding boundary values on B. (We assume that the conformal map f is defined on the boundary.) The specification of $\partial u/\partial n$ is more complicated. However, a special case is easy to see. Let $u \circ f = u_0$ be the solution we seek; that is, $u_0(x,y) = u(f(x,y))$ (see Figure 5.14). Then we claim that $\partial u_0/\partial n = 0$ iff $\partial u/\partial n = 0$ *on corresponding portions*. This follows because $\partial u_0/\partial n = 0$ and $\partial u/\partial n = 0$ mean that the conjugates are constant on those portions, and if v is the conjugate of u, then $v \circ f = v_0$ is the conjugate of u_0. This proves the claim. These are the only types of boundary conditions for $\partial u/\partial n$ that will be dealt with in this text.

To use this method, then, we need to be able to solve the problem in some simple region B. We already saw in Section 2.5 that the unit disk is suitable for this purpose because in that case we have the Poisson Formula for the solution of the Dirichlet Problem. However, we can sometimes get more explicit solutions than those yielded by that formula.

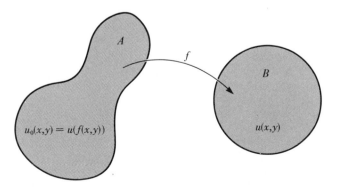

Figure 5.14   Transformation of harmonic functions.

The following situation is used to illustrate the method. It will be used in subsequent examples. We consider the upper half plane H and the problem of finding a harmonic function that takes the constant boundary values c_0 on $]-\infty, x_1[$, c_1 on $]x_1, x_2[$, $\cdots$, and c_n on $]x_n, \infty[$ where $x_1 < x_2 < \cdots < x_n$ are points on the real axis. We claim that a solution is given by

$$u(x,y) = c_n + \frac{1}{\pi}\{(c_{n-1} - c_n)\theta_n + \cdots + (c_0 - c_1)\theta_1\} \qquad (2)$$

where $\theta_1, \cdots, \theta_n$ are as indicated in Figure 5.15; $0 \le \theta_i \le \pi$.

This is easy to see. First, since u is the real part of

$$c_n + \frac{1}{\pi i}\{(c_{n-1} - c_n)\log(z - x_n) + \cdots + (c_0 - c_1)\log(z - x_1)\},$$

it is harmonic. Also, on $]x_i, x_{i+1}[$, u reduces to c_i. (The student should check this.)

As mentioned previously, the Dirichlet Problem does not have a unique solution, so the question arises: Why was this solution chosen? Another solution could have been obtained by adding $u(x,y) = y$ to the solution (2).

The answer is that the u that is given by formula (2) is bounded (Why?). The student will find this answer physically reasonable after studying the examples that follow.

Thus if a problem can be transformed to the one described by Figure 5.15, we can use formula (2) to find a solution. This will be done in the examples in this section.

Heat conduction

Physical laws tell us that if a two-dimensional region is maintained at a steady temperature T (that is, a temperature not changing in time,

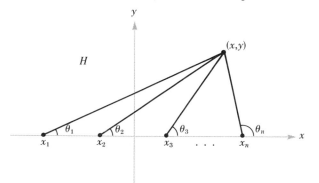

Figure 5.15   Dirichlet Problem in the upper half plane.

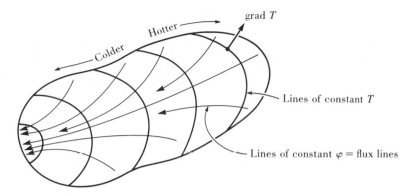

Figure 5.16 Heat conduction.

accomplished by fixing the temperature at the walls, or by insulating them), then T should be harmonic †

The negative of the gradient of T represents the direction in which heat flows. Thus we can, by using Theorem 33, Section 2.5, *interpret the level surfaces of the harmonic conjugate φ of T as the lines along which heat flows and the temperature is decreasing.* Lines of constant T are called *isotherms*; lines on which the conjugate φ are constant are called *flux lines* (Figure 5.16).

Thus to say that T is prescribed on a portion of the boundary means that the portion is maintained at a preassigned temperature (for example, by a heating device). The condition $\partial T/\partial n = 0$ means that the flux line (or $-\text{grad } T$) is parallel to the boundary; in other words, the boundary is *insulated*. (No heat flows across the boundary.)

Example 1.   Consider A, the first quadrant; the x-axis is maintained at $T = 0$ while the y-axis is maintained at $T = 100$. Find the temperature distribution in the first quadrant. (Physically, the region is approximated by a thin metal sheet.)

---

†This is a consequence of conservation of energy and Gauss' Theorem (see formula (1)). The heat flows in the direction of the vector field $= \kappa \nabla T$ (κ = conductivity) and the energy density is $c\rho T$ (c = specific heat, ρ = density). Then the law of conservation of energy states that the rate of change of energy in any region V equals the rate at which energy enters V; that is,

$$\frac{d}{dt} \int_V c\rho T \, dx = -\int_{\text{bd}(V)} -\kappa \nabla T \cdot dS.$$

By Gauss' Theorem, this condition is equivalent to

$$\frac{\partial}{\partial t} (c\rho T) = \kappa \nabla^2 T.$$

If c, ρ, T are independent of t, we conclude that T is harmonic: $\nabla^2 T = 0$.

Solution.   We map the first quadrant to the upper half plane by
$z \mapsto z^2$ (Figure 5.17).

It is physically reasonable that the temperature should be a
bounded function; otherwise we would obtain arbitrarily high (or low)
temperatures. Therefore, the solution in the upper half plane is given
by formula (2):

$$u(x,y) = \frac{1}{\pi}\,(100\ \arg z) = \frac{100}{\pi}\ \tan^{-1}\!\left(\frac{y}{x}\right).$$

Thus the solution we seek is

$$u_0(x,y) = u(f(x,y))$$

where $f(x,y) = z^2 = x^2 - y^2 + 2ixy = (x^2 - y^2, 2xy)$. Hence

$$u_0(x,y) = \frac{100}{\pi}\ \tan^{-1}\!\left(\frac{2xy}{x^2 - y^2}\right)$$

is the desired answer. It is understood that $\tan^{-1}$ is taken in the
interval $[0,\pi]$. Another form of the answer may be obtained as
follows:

$$u_0(x,y) = u(z^2) = \frac{100}{\pi}\ \arg(z^2) = \frac{200}{\pi}\ \arg z = \frac{200}{\pi}\ \tan^{-1}\!\left(\frac{y}{x}\right).$$

The isotherms and flux lines are indicated in Figure 5.18.

Example 2.   Let A be the upper half of the unit circle. Find the
temperature inside if the circular portion is insulated; $T = 0$ for $x > 0$
and $T = 10$ for $x < 0$ on the real axis.

Solution.   For this type of problem where there is a portion of the
boundary where $\partial T/\partial n = 0$ (insulated), it is convenient to map the
region to a half strip. This can be done for A by means of $\log z$ (using

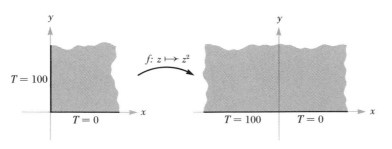

Figure 5.17

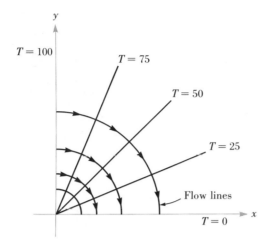

Figure 5.18

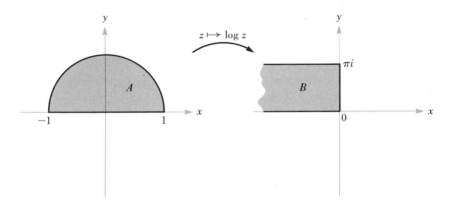

Figure 5.19

the principal branch); see Figures 5.19 and 5.10(viii). For strip B we obtain, by inspection, the solution

$$T_0(x,y) = \frac{10y}{\pi}.$$

(Note that $\partial T_0/\partial n = \partial T_0/\partial x = 0$.)

Thus our answer is

$$T(x,y) = T_0(\log(x + iy)) = \frac{10}{\pi} \tan^{-1}\left(\frac{y}{x}\right).$$

Electric potential

From physics we learn that if an electric potential φ is determined by static electric charges, φ must satisfy Laplace's Equation (that is, be harmonic). The conjugate function Φ of φ is interpreted as follows: Lines along which Φ is constant are lines along which a small test charge would travel. They are called *flux lines*. Tangent vectors to such lines are $-\text{grad } \varphi = E$, called the *electric field* (see Figure 5.20).

Thus *the flux lines and the equipotential lines (lines of constant φ) intersect orthogonally.*

The Dirichlet Problem usually arises in electrostatics; this is because the boundary is often maintained at a given potential (for example, by means of a battery or by grounding).

Example 3.   Consider the unit circle. The electric potential is maintained at $\varphi = 0$ on the lower semicircle and at $\varphi = 1$ on the upper semicircle. Find φ inside.

Solution.   We use the general procedure for solving the Dirichlet Problem by mapping our given region to the upper half plane. In the present case we may use a fractional linear transformation (see Figures 5.10(vi) and 5.21).

As with temperature, it is physically reasonable that the potential is bounded. Thus, by formula (2), the solution in the upper half plane is

$$\varphi_0(x,y) = 1 - \frac{1}{\pi} \tan^{-1}\left(\frac{y}{x}\right),$$

so the solution on the unit circle is

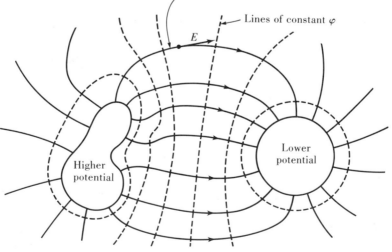

Figure 5.20   Electrical potential.

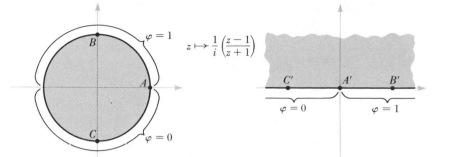

Figure 5.21

$$\varphi(x,y) = \varphi_0(f(x,y)),$$

where $f(z) = (z-1)/i(z+1)$. Now let $(z-1)/i(z+1) = u + iv$. Then

$$u = \frac{2y}{x^2 + y^2 + 2x + 1}$$

and

$$v = -\frac{x^2 + y^2 - 1}{x^2 + y^2 + 2x + 1}.$$

Thus the solution is

$$\varphi(x,y) = 1 - \frac{1}{\pi} \tan^{-1}\left(\frac{1 - x^2 - y^2}{2y}\right).$$

The equipotential and flux lines are shown in Figure 5.22.

This example could also be solved by using Poisson's Formula. The two answers would be equal although this would not be obvious from their form.

Example 4. The harmonic function $u(z) = (Q/2\pi) \log|z - z_0|$, which equals the real part of $(Q/2\pi) \log(z - z_0)$, represents the potential of a charge Q located at z_0.‡ Sketch the equipotential lines for two charges of like or opposite signs.

Solution. The potential of two charges is obtained by adding the respective potentials. Thus two charges $Q > 0$ located at z_1 and z_2 have the potential $(Q/2\pi) \log|z - z_1| |z - z_2|$; a charge $Q > 0$ at z_1 and a charge $-Q$ at z_2 have potential $(Q/2\pi) \log|z - z_1|/|z - z_2|$. The equipotential lines are sketched in Figure 5.23. The curves $\varphi = $ constant

‡The reason for this is that u is a radial field and has the property that if $\mathbf{E} = -\text{grad } \varphi$ is the electric field, the integral of $\mathbf{E} \cdot \mathbf{n}$ around a curve enclosing z_0 is Q (by Gauss' Theorem).

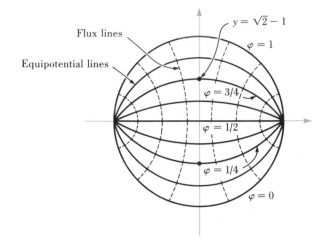

Figure 5.22

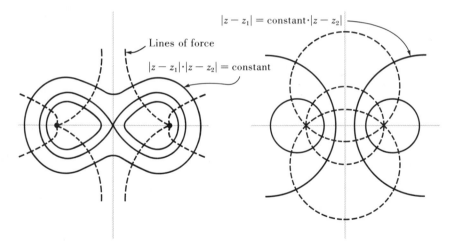

Figure 5.23   The field of like charges (*left*) and the field of opposite charges (*right*).

in the drawing on the left are called leminiscates; in the drawing on the right they are called circles of Apollonius.

Hydrodynamics

If we have a (steady state) incompressible, nonviscous fluid, we are interested in finding its velocity field, $V(x,y)$. From elementary vector analysis we know that "incompressible" means that the divergence div $V = 0$. (We say V is *divergence free*.) We shall assume that V is also circulation free; that is, $V = \text{grad } \varphi$ for some φ called the *velocity potential*. Thus φ must be harmonic because $\nabla^2\varphi = \text{div grad } \varphi = \text{div } V = 0$. Thus when we solve for φ we can obtain V by taking $V = \text{grad } \varphi$.

The conjugate ψ of the harmonic function φ (ψ will exist on any simply connected region) is called the *stream function*, and $F = \varphi + i\psi$ is called the *complex potential*. Lines of constant ψ have V as their tangents (Why?), so *lines of constant ψ may be interpreted as the lines along which particles of fluid move*, hence the name stream function (see Figure 5.24).

The natural boundary condition is that V should be parallel to the boundary. (The fluid flows parallel to the walls.) This means that $\partial\varphi/\partial n = 0$, so we have the Neumann Problem for φ.

Let us again consider the upper half plane. A physically acceptable motion is obtained by setting $V(x,y) = \alpha = (\alpha,0)$ or $\varphi(x,y) = \alpha x = \text{Re}(\alpha z)$; where α is real. The flow corresponding to V is parallel to the x-axis, with velocity α. Notice that now φ is not bounded; thus the behavior at ∞ for fluids, for temperature, and for electric potential is different because of the different physical circumstances (see Figure 5.25).

Thus to find the flow in a region we should map the region to the upper half plane and use the solution $\varphi(x,y) = \alpha x$. α may be specified as the velocity at infinity. It should be clear that if f is the conformal map from the given region to the upper half plane, the required complex potential is given by $F(z) = \alpha f(z)$.

Example 5.   Find the flow around the upper half of the unit circle if the velocity is parallel to the x-axis and is α at infinity.

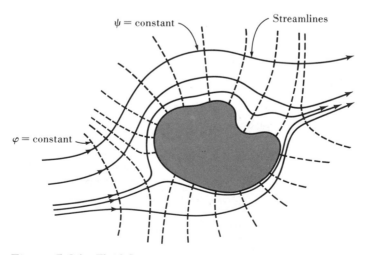

Figure 5.24   Fluid flow.

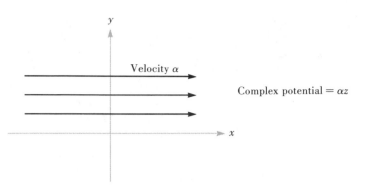

Figure 5.25   Flow in the upper half plane.

Solution.   We must map the exterior of the given region to the upper half plane. Such a conformal map is $z \mapsto z + 1/z$ (Figure 5.26). Thus $F_0(z) = \alpha z$ is the complex potential in the upper half plane, so the required complex potential is

$$F(z) = \alpha\left(z + \frac{1}{z}\right).$$

It is convenient to use the polar coordinates r,θ to express φ and ψ. Then we get

$$\varphi(r,\theta) = \alpha\left(r + \frac{1}{r}\right) \cos\,\theta$$

and

$$\psi(r,\theta) = \alpha\left(r - \frac{1}{r}\right) \sin\,\theta.$$

A few streamlines are shown in Figure 5.27.

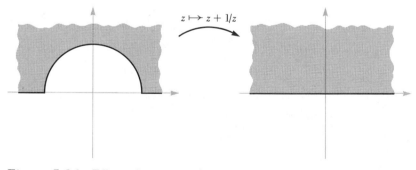

Figure 5.26   Effect of $z \mapsto z + 1/z$.

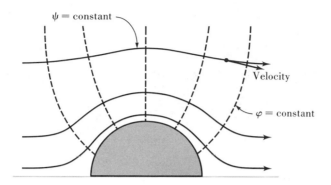

Figure 5.27   Streamlines for flow around a cylinder.

Note.   By slightly modifying the transformation $z \mapsto z + 1/z$ by the addition of higher-order terms, the half circle can be replaced by something more closely resembling an airplane wing; these are called *Joukowski transformations.*

Exercises

1.   Find a formula for determining the temperature in the region with the indicated boundary values shown in Figure 5.28.

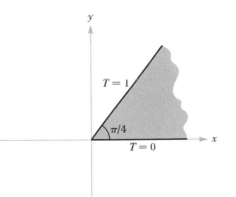

Figure 5.28

2.   Find a formula for determining the temperature in the region illustrated in Figure 5.29.

   Hint.   Consider the map $z \mapsto \sin z$.

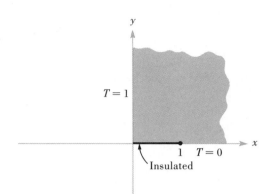

Figure 5.29

3. Find the electric potential in the region illustrated in Figure 5.30.

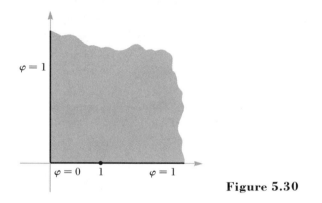

Figure 5.30

4. Find the electric potential in the region illustrated in Figure 5.31. Sketch a few equipotential curves.

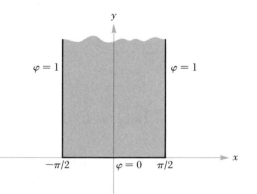

Figure 5.31

5. Find the flow around a circular disk if the flow is at an angle θ to the x-axis with velocity α at infinity (Figure 5.32).

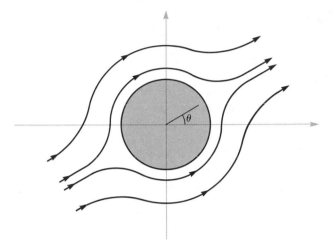

Figure 5.32 Flow around a disk.

6. Obtain a formula for determining the flow of a fluid in the region illustrated in Figure 5.33. (The velocity at ∞ is α.)

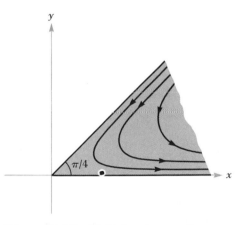

Figure 5.33 Fluid flow in a wedge (exercise 6).

Review Exercises for Chapter 5

1. Consider the map $z \mapsto z^3$. On what sets $A \subset \mathbb{C}$ is this map conformal (onto its image)?

2. Verify directly that the map $z \mapsto z^n$ preserves the orthogonality between rays from 0 and circles around 0.

3. Find a conformal map that takes the unit circle onto itself and maps $i/2$ to 0.

4. Find a conformal map that takes the region $A = \{z \mid |z - 1| > 1$ and $|z - 2| < 2\}$ onto $B = \{z \mid 0 < \operatorname{Re} z < 1\}$.

5. Let $z_1, z_2 \in \mathbb{C}$ and $a \in \mathbb{R}$, $a > 0$. Show that

$$\left| \frac{z - z_1}{z - z_2} \right| = a$$

defines a circle and z_1, z_2 are inverse points in that circle (that is, they are collinear with the center z_0 and $|z_1 - z_0| \cdot |z_2 - z_0| = \rho^2$ where ρ is the radius of the circle).

6. Examine the image of the set $\{z \in \mathbb{C} \mid \operatorname{Im} z \geq 0, 0 \leq \operatorname{Re} z \leq \pi/2\}$ under the map $z \mapsto \sin z$ by considering it to be the composition of the maps $z \mapsto e^{iz}$, $z \mapsto z - 1/z$, $z \mapsto z/2i$.

7. Let $f: A \to B$ be a conformal map, let γ be a curve in A, and let $\tilde{\gamma} = f \circ \gamma$. Show that

$$l(\tilde{\gamma}) = \int_a^b |f'(\gamma(t))| \cdot |\gamma'(t)| \, dt.$$

If f preserves the lengths of all curves, argue that $f(z) = e^{i\theta} z + a$ for some $a \in \mathbb{C}$, and $\theta \in [0, 2\pi[$.

8. Find a conformal map that takes $A = \{z \mid |z - i| < 1\}$ onto $B = \{z \mid |z - 1| < 1\}$.

9. From the Riemann Mapping Theorem, we know there is a conformal map from $A = \{z \mid 1 < |z| < 2\}$ onto $B = \{z \mid 0 < \operatorname{Re} z < 1\}$. A is bounded by two concentric circles; B is bounded by two parallel lines. Can this mapping be accomplished by a linear fractional transformation that takes these circles to these lines? If so, display the function; if not, why not?

10. Let $T(z) = (az + b)/(cz + d)$. Show that $T(T(z)) = z$; that is, $T \circ T =$ identity if and only if $a = -d$.

11. Find the temperature on the region illustrated in Figure 5.34.

Hint.   Use $z \mapsto \sin^{-1} z$.

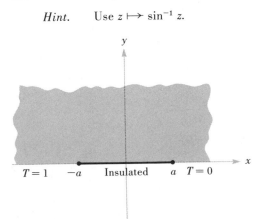

Figure 5.34 Boundary data for exercise 11.

12. Find the electric potential in the region shown in Figure 5.35.

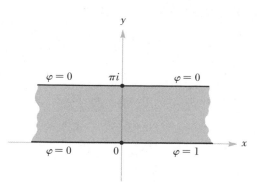

Figure 5.35 Boundary data for exercise 12.

13. Find the flow of a fluid in the region shown in Figure 5.36.

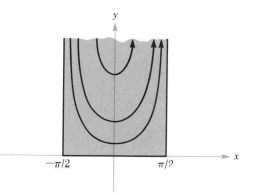

Figure 5.36 Region for exercise 13.

14. Is it possible to find a conformal map of the interior of the unit circle onto its exterior? Is $f(z) = 1/z$ such a map?

15. Let F_1 and F_2 be conformal maps of the unit disk onto itself and let $F_1(z_0) = F_2(z_0) = 0$ for some fixed z_0, $|z_0| < 1$. Show that there is a $\theta \in [0, 2\pi[$ such that $F_1(z) = e^{i\theta} F_2(z)$.

16. Give a complete list of all conformal maps of the first quadrant $A = \{z | \operatorname{Re} z > 0 \text{ and } \operatorname{Im} z > 0\}$ onto itself.

17. Let g_n be a sequence of analytic functions defined on a region A. Suppose that $\sum_{n=1}^{\infty} |g_n(z)|$ converges uniformly on A. Prove that $\sum_{n=1}^{\infty} |g_n'(z)|$ converges uniformly on closed disks in A.

18. Evaluate by residues:

$$\int_0^\infty \frac{\cos x}{x^2 + 3} \, dx.$$

19. Let f be analytic on $\mathbb{C}\setminus\{0\}$. Suppose that $f(z) \to \infty$ as $z \to 0$ and $f(z) \to \infty$ as $z \to \infty$. Prove that f can be written in the form

$$f(z) = \frac{c_k}{z^k} + \cdots + \frac{c_1}{z} + c_0 + d_1 z + \cdots + d_l z^l$$

for constants c_i and d_j.

20. Find a conformal map that takes the region in Figure 5.37 to the upper half plane. Use this map to find the electric potential φ with the stated boundary conditions.

Hint.   Consider a branch of $z \mapsto \sqrt{z^2 - 1}$ after rotating the figure by 90°.

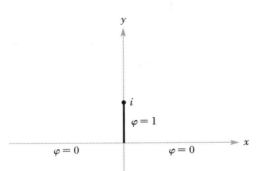

Figure 5.37   Boundary data for exercise 20.

21. Use exercise 20 to find the fluid flow over the obstacle in Figure 5.38. Plot a few streamlines.

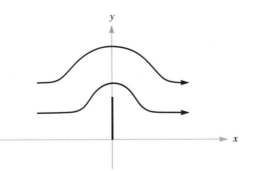

Figure 5.38   Flow over an obstacle.

22. If $\sum_0^\infty a_n z^n$ has radius of convergence ρ, what is the radius of convergence of $\sum_{n=0}^\infty a_n z^{2n}$, $\sum_{n=0}^\infty a_n^2 z^n$?

23. Find the Laurent expansion of $f(z) = z^4/(1 - z^2)$ that is valid on the annulus $1 < |z| < \infty$.

24. Use the Schwarz-Christoffel Formula to find a conformal map between the two regions shown in Figure 5.39. ($A = -1$, $B = 1$, $B' = 0$.)

25. Use exercise 24 to find the flow lines over the step in the bed of the deep channel shown in Figure 5.39.

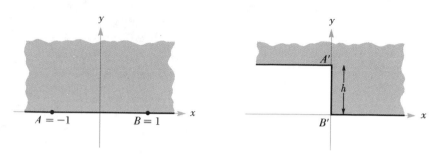

Figure 5.39

Chapter 6

Further Development of the Theory

This chapter continues the development of the theory of analytic functions that was begun in Chapters 3 and 4. The main tools used in this development are the Taylor series and the Residue Theorem.

The first topic that is dealt with in this chapter is analytic continuation, or the attempt to make the domain of an analytic function as large as possible. Further investigation of analytic continuation leads naturally to the concept of a Riemann surface, which is briefly discussed. Additional properties concerning analytic functions are developed in subsequent sections. Some of these properties deal with such topics as counting zeros of an analytic function; others are generalizations of the Inverse Function Theorem.

6.1 Analytic Continuation and Elementary Riemann Surfaces

The first theorem to be proved in this section is the Principle of Analytic Continuation, which is also referred to as the Identity Theorem. This theorem and its proof lead conveniently to a discussion of Riemann surfaces, which facilitates a more satisfactory treatment of what were previously referred to as "multiple-valued functions," such as $\log z$ and $\sqrt{z}$. The discussion is heuristic, being intended only to motivate more advanced work.

Analytic continuation

If two analytic functions agree on a small portion of a region, then they agree on the whole region on which they are both analytic. This is stated precisely in Theorem 1.

Theorem 1 (Principle of Analytic Continuation or Identity Theorem). *Let f and g be analytic in a region A. Suppose that there is a sequence z_1, z_2, $\cdot$ $\cdot$ $\cdot$ of distinct points of A converging to $z_0 \in A$, such that $f(z_n) = g(z_n)$ for all $n = 1, 2, 3, \cdot$ $\cdot$ $\cdot$. Then $f = g$ on all of A (see Figure 6.1). The conclusion is valid, in particular, if $f = g$ on some neighborhood of some point in A.*

Proof.   We must show that $f - g = 0$ on A. We shall do this by proving that for a given analytic function h on A, the following four assertions are equivalent:

   i. For some z_0, the nth derivative $h^{(n)}(z_0) = 0$, $n = 0, 1, 2, \cdot$ $\cdot$ $\cdot$.
   ii. $h = 0$ on some neighborhood of z_0.
   iii. $h(z_k) = 0$ where z_k is some sequence of distinct points converging to z_0.
   iv. $h = 0$ on all of A.

After these equivalences are proved, we prove the theorem by taking $f - g = h$ and applying (iv).

First, (i) $\Longleftrightarrow$ (ii) by Taylor's Theorem. (The student should write out the details if this assertion is not clear.) Next we show that (iii) $\Rightarrow$ (ii) by showing that the zeros of an analytic function h are isolated, unless $h = 0$ on a neighborhood of x_0. Indeed, if h is not identically zero on a disk around z_0, we can write $h(z) = (z - z_0)^k \varphi(z)$ where $\varphi(z_0) \neq 0$ and k is an integer ≥ 1 (Why?). Now $\varphi(z) \neq 0$ in a whole neighborhood of z_0 by continuity, so in that neighborhood $h(z) \neq 0$ except at $z = z_0$. This statement contradicts $h(z_n) = 0$ since, for n large enough, z_n lies in the neighborhood, and we can suppose that $z_n \neq z_0$.

Clearly, (iv) $\Rightarrow$ (iii). The proof will be complete when we show that $\{$(i) and (ii)$\} \Rightarrow$ (iv). Let $B = \{z \in A \mid h$ is zero in a neighborhood of $z\}$. By its definition, B is open and is nonempty by hypothesis. We shall show that B is closed in A as well by showing that if $z_k \rightarrow z$, $z_k \in B$, then $z \in B$ (see p. 137). It suffices (by (i)) to show that $h^{(n)}(z) = 0$ for all n. But $h^{(n)}(z) = \underset{k \to \infty}{\text{limit}}\, h^{(n)}(z_k) = 0$.

Thus B is closed, open, and nonempty, so $B = A$ because A is assumed to be connected (a region is open and connected by definition); see p. 149 ∎

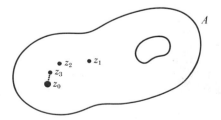

Figure 6.1   Identity Theorem: $\{f = g$ on z_1, z_2, $\cdot$ $\cdot$ $\cdot\}$ $\Rightarrow \{f(z) = g(z)$ for all $z \in A\}$.

An interesting application of this theorem is the following. There is exactly one analytic function on $\mathbb{C}$ that agrees with e^x on the x-axis. This is an immediate consequence of Theorem 1 because the x-axis contains a convergent sequence of distinct points (for example, $1/n$). Thus there was only one possible way for us to define e^z and still obtain an analytic function.

We have the following important consequence of Theorem 1.

Theorem 2.   *Let $f: A \to \mathbb{C}$ and $g: B \to \mathbb{C}$ be analytic on regions A and B. Suppose that $A \cap B \neq \phi$ and $f = g$ on $A \cap B$. Define*

$$h(z) = \begin{cases} f(z) \text{ if } z \in A; \\ g(z) \text{ if } z \in B. \end{cases}$$

Then h is analytic on $A \cup B$ and is the only analytic function on $A \cup B$ equaling f (or g on B). We say that h is an analytic continuation of f (or g) (see Figure 6.2).

Proof.   That h is analytic is obvious because f and g are. Uniqueness of h results at once from Theorem 1 and from the facts that $A \cup B$ is a region and that $A \cap B$ is open. ∎

Analytic continuation is important because it provides a method for making the domain of an analytic function as large as possible. However, the following phenomenon can occur. Let f on A be continued to a region A_1 and let A_2 be as pictured in Figure 6.3. If we continue f to be analytic on A_1, then continue this new function from A_1 to A_2, the result need not agree with the original function f on A. An example should clarify this point. Consider $\log z$, the principal branch ($-\pi < \arg z < \pi$) on the region A consisting of the right half plane union the lower half plane. The log function may be continued uniquely so as to include $A_1 =$ the upper half plane in its domain. Similarly, we can continue the log again from the upper half plane so as to include $A_2 =$ the left half plane in its domain by choosing the branch $0 < \arg z < 2\pi$. But these

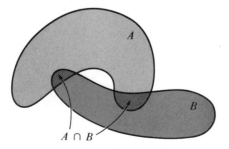

Figure 6.2   Analytic continuation.

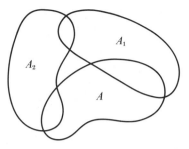

Figure 6.3 Continuation of a function from A to A_1 and from A_1 to A_2.

branches do not agree on the third quadrant; they differ by $2\pi i$ (see Figure 6.4). Therefore, in continuing a function we must be sure that the function on the extended region B agrees with the original on the whole intersection $A \cap B$ and not merely on part of it.

Schwarz Reflection Principle

There is a special case of analytic continuation that can be dealt with directly as follows.

Theorem 3. *Let A be a region in the upper half plane whose boundary bd(A) intersects the real axis in an interval $[a,b]$ (or finite union of disjoint intervals). Let f be analytic on A and continuous on $A \cup \,]a,b[$. Let $\bar{A}=\{z \mid \bar{z} \in A\}$, the reflection of A (see Figure 6.5), and define g on $\bar{A}$ by $g(z) = \overline{f(\bar{z})}$. Assume that f is real on $]a,b[$. Then g is analytic and is the unique analytic continuation of f to $A \cup \,]a,b[\, \cup \, \bar{A}$.*

Proof. Uniqueness is implied by Theorem 1 because f equals g on a set (namely, $]a,b[$) containing a convergent sequence of distinct points; note that $f = g$ on the real axis because there, $z = \bar{z}$ and $\bar{f} = f$. (f is assumed real on the real axis.) First we show that g is analytic at $z \in \bar{A}$. Indeed, if $f = u + iv$ and $g = U + iV$, then

$$U(x,y) = u(x,-y)$$

and

$$V(x,y) = -v(x,-y).$$

We know that U and V satisfy the Cauchy-Riemann equations because u and v do. Hence g is analytic. It remains to be proved that the extended function, call it h, is analytic at points of $]a,b[$. By construction, h is continuous. We shall use Morera's Theorem to show that it is analytic. It suffices to prove (see the proof of Theorem 28, Chapter 2) that $\int_\gamma h = 0$ where γ is any rectangle. Let $z_0 \in \,]a,b[$ and

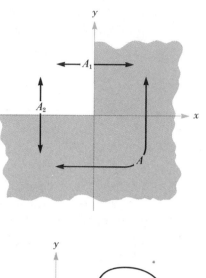

Figure 6.4 Continuing the log.

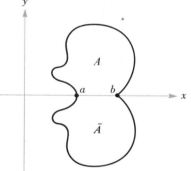

Figure 6.5 $\bar{A}$ is the reflection of A.

suppose first that γ has one edge along the real axis (Figure 6.6). Let $\eta > 0$ and consider the curve γ_η shown in Figure 6.6.

We know that h is continuous, hence uniformly continuous on the compact set γ. Therefore, for any $\epsilon > 0$, there is a δ with $\epsilon > \delta > 0$ such that $|z_1 - z_2| < 2\delta$ implies that $|h(z_1) - h(z_2)| < \epsilon$. Since h is analytic in A and $\bar{A}$, $\int_{\gamma_\delta} h(z)\,dz = 0$. Hence

$$\left| \int_\gamma h(z)\,dz \right| = \left| \int_\gamma h(z)\,dz - \int_{\gamma_\delta} h(z)\,dz \right|.$$

To estimate this, let $M = \sup\limits_{z \in \gamma} |h(z)|$. M is finite since h is continuous and γ is compact. Let x_1 and x_2 be the corners of γ on the real axis. The last equation becomes

$$\left| \int_{\gamma} h(z)\,dz \right| = \left| \int_{x_1}^{x_2} h(z)\,dz + \int_{x_2}^{x_2+i\delta} h(z)\,dz + \int_{x_2+i\delta}^{x_1+i\delta} h(z)\,dz + \int_{x_1+i\delta}^{x_1} h(z)\,dz \right|$$

$$\leq \left| \int_{x_1}^{x_2} h(z)\,dx + \int_{x_2}^{x_1} h(x+i\delta))\,dx \right| + 2\delta M$$

$$\leq \left| \int_{x_1}^{x_2} (h(x) - h(x+i\delta))\,dx \right| + 2\epsilon M$$

$$\leq \epsilon |x_2 - x_1| + 2\epsilon M$$

$$\leq \epsilon |b - a| + 2\epsilon M.$$

Since ϵ was arbitrary, we must have $\int_{\gamma} h(z)\,dz = 0$. Finally, if z_0 should happen to lie inside the rectangle γ, we break up $\int_{\gamma} h = \int_{\gamma_1} h + \int_{\gamma_2} h$ where γ_1 and γ_2 are as shown in Figure 6.6.

The preceding argument shows that $\int_{\gamma_1} h = 0$; an obvious slight modification shows that $\int_{\gamma_2} h = 0$. ■

This result is remarkable in that we only required f to be continuous and real on $]a,b[$. It followed automatically that f is analytic on $]a,b[$ when continued across the real axis.

A visually intuitive way to see that g (and thus h) is analytic on $\bar{A}$ is to consider the map in three steps:

$$z \mapsto \bar{z}; \qquad \bar{z} \mapsto f(\bar{z}); \qquad f(\bar{z}) \mapsto \overline{f(\bar{z})}.$$

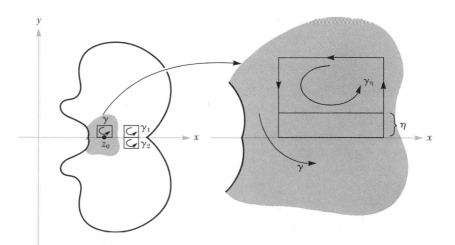

Figure 6.6 Detail for the proof of Theorem 3.

The middle map is conformal; the first and last are anticonformal in the sense that they reverse angles. Since angles are reversed twice, the net result is to preserve angles. The whole map is thus conformal.

Students who have worked the problems on inversion in a circle should note that a similar reflection principle can be formulated for the case in which A lies in the interior (or exterior) of a circle γ having part of its boundary on γ. If f is analytic on A and continuous on $A \cup \gamma$ and if $f(\gamma)$ is an arc of another circle Γ, the process just described may be carried out with the role of complex conjugation taken by inversion in the circles γ and Γ.

An argument very similar to that used in the preceding paragraphs to show that h is analytic will establish another analytic continuation theorem:

Let A and B be disjoint simply connected regions whose boundaries intersect in a simple smooth curve γ. Let $C = A \cup (interior \ \gamma) \cup B$ (where interior γ means the image of γ without its end points) and suppose that

i. Each point in interior γ has a neighborhood in C.

ii. f is analytic in A and continuous on $A \cup \gamma$.

iii. g is analytic in B and continuous on $B \cup \gamma$.

iv. For $t \in \gamma$, $\displaystyle\lim_{\substack{z \to t \\ z \in A}} f(z) = \lim_{\substack{z \to t \\ z \in B}} g(z)$.

Then there is a function h analytic on C that agrees with f on A and g on B.

Continuation by power series along curves

Suppose that f is analytic in a neighborhood U of z_0 and that γ is a curve joining z_0 to another point z' (as in Figure 6.7). If we want to continue f to z' we can proceed as follows. For z_1 on γ in U consider the Taylor series of f expanded around z_1:

$$\sum_{n=0}^{\infty} \frac{f^{(n)}(z_1)}{n!} (z - z_1)^n.$$

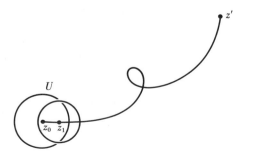

Figure 6.7 Continuation by power series.

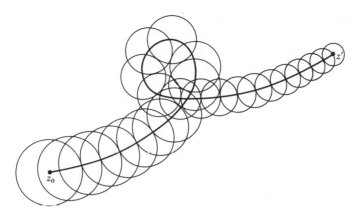

Figure 6.8 Continuation can lead to self-intersections.

This power series may have a radius of convergence such that the power series is analytic farther along γ than the portion of γ in U. The power series so obtained then defines an analytic continuation of f. We can continue this way along γ in hopes of reaching z', which will be possible if the successive radii of convergence do not shrink to 0 before we reach z'. If we succeed, we say f can be *analytically continued* along γ. However, we must be careful because the analytic continuation of f so defined might not be single-valued if γ intersects itself (as in Figure 6.8).

Thus we may merely obtain local functions defined on disks but with agreement not necessarily holding on overlaps. This construction is one basic way in which multiple-valued functions arise. If f can be analytically continued along γ, it should be fairly clear that we can always analytically continue f by power series in a finite number of steps (see exercise 7).

The following basic result tells us that multiple-valued functions never arise on a simply connected region.

Theorem 4 (Monodromy Principle).   *Let A be simply connected and let $z_0 \in A$. Let f be analytic in a neighborhood of z_0. Suppose that f can be analytically continued along any arc joining z_0 to another point $z \in A$. Then this continuation defines a (single-valued) analytic continuation of f on A.*

Sketch of Proof.   The machinery needed for the proof of Theorem 4 was established in Section 2.3. Only the main steps will be given here; the details will be left to the student. Consider the proof of Theorem 15, Section 2.3. It suffices to show that if γ_1 and γ_2 are any two curves joining z_0 to z', then the continued values of f along γ_1 and γ_2 are the same at z'. We can show (a task that is left to the student) by simple connectivity of A (see Section 2.3) that γ_1 and γ_2 are homotopic in A with fixed end points. The function f then plays the role of λ in Theorem 15, Section 2.3, and that argument shows (see also exercise 1, Section 2.3) that $f_2(z') - f_2(z_0) = f_1(z') - f_1(z_0)$, where subscripts mean "f continued along the respective arcs γ_1 and γ_2." Because $f_1(z_0) = f_2(z_0)$, we have $f_2(z') = f_1(z')$ as required. ■

However, for regions that are not simply connected, we can get different values for the continuation of f when we traverse two different paths. This fact was already mentioned at the beginning of this section in connection with $\log z$. For example in Figure 6.9, starting with $\log$ defined near 1 and continuing along γ_1, we get $\log(-1) = \pi i$, whereas along γ_2, we get $\log(-1) = -\pi i$. The reason for this is that the region we are considering, $\mathbb{C}\setminus\{0\}$, is not simply connected.

Riemann surfaces of some elementary functions

The phenomenon just described may naturally lead the student to ask if there is a definition of $\log$ that does not introduce any artificial branch lines (which, after all, can be chosen arbitrarily). The answer is given by a brilliant idea of Georg Riemann in his doctoral thesis in 1851 that is briefly described here.

For the logarithm, if $\log z$ is to be single-valued, we should merely regard γ_1 and γ_2 in Figure 6.9 as ending up in different places. This can

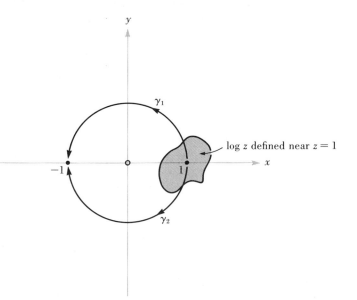

Figure 6.9   Continuation of $\log z$ along two different arcs from 1 to -1.

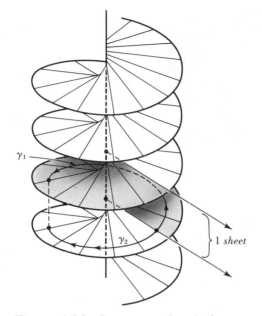

Figure 6.10   Riemann surface for log z.

be pictured as in Figure 6.10. Only a core of spiral staircase M is shown along the origin, but it should have infinite extent laterally. If we cut from 0 outward at any level and the one directly below it, we get a part of the surface of a *sheet* (the shaded portion in Figure 6.10). This can be identified with the domain for a branch of log. Thus we have stacked up infinitely many copies of the complex plane $\mathbb{C}$ joined through 0 and glued together as shown in Figure 6.10. The arcs γ_1 and γ_2 now go to different points so we can assign different values of log z to each without ambiguity.

The main property of this surface that enables us to define $\log z = \log |z| + i \arg z$ as a single-valued function is that on this surface arg z is well defined, and the different sheets correspond to different intervals of length 2π in which arg z takes its values.

Thus we can take care of multiple-valued functions by introducing an enlarged domain on which the function becomes single-valued.

Let us briefly consider another example, the square root function: $z \mapsto \sqrt{z} = \sqrt{r}\, e^{i\theta/2}$. Here the situation is slightly different from that for the log function. If we go around the origin once, $\sqrt{z}$ takes on a different value, but if we go around twice (increase θ by 4π), we arrive back at the same value, so we want to be at the same point on the Riemann surface. The surface is illustrated in Figure 6.11 (although it really cannot be visualized in three-space).

Even though the two sheets appear to intersect, they are not supposed to; this is a fault of our attempt at visualization in $\mathbb{R}^3$. Rather we

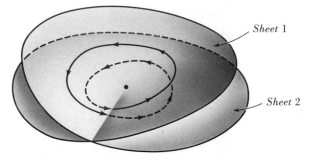

Sheet 1

Sheet 2

Figure 6.11   Riemann surface for $\sqrt{z}$.

must construct the surface abstractly; as we go around 0 once we pass from one sheet to another, and if we go around 0 twice we come back to the same sheet.

For more complicated functions like $\cos^{-1}(z)$ the Riemann surface can be constructed as follows. On certain regions of $\mathbb{C}$, $\cos z$ is one-to-one and we define $\cos^{-1}(z)$ to be the inverse function. The period strips defined in Section 1.3 are examples of such regions for e^{z} and $\log z$. Such a region for $\cos z$ is shown in Figure 6.12.

We can show that each such strip is mapped conformally onto $\mathbb{C}$ minus the intervals $]-\infty, -1]$ and $[1, \infty[$ on the real axis (Figure 6.13). The image region will be one sheet of the final Riemann surface. Two adjacent sheets should be "glued" along any line that is the image of two different lines under $\cos z$. Using $\cos(x + iy) = \cos x \cosh y - \sin x \sinh y$, we easily see that these lines are $]-\infty, -1]$ or $[1, \infty[$ as indicated in Figure 6.13.

If we look at a cross section along the line Re $z = 2$, for example, we see that this cross section of the Riemann surface looks something like that shown in Figure 6.14. The full Riemann surface is slightly difficult to visualize but the figure gives us a general idea. It shows that if we follow a suitably chosen curve possibly going from one branch line to the other, we may pass from any one sheet to any other.

Worked Examples

1.   Let f be an entire function equal to a polynomial in x on the x-axis, for $x \in [0,1]$. Show that f is a polynomial.

Solution.   Let $f(x) = a_0 + a_1 x + \cdots + a_n x^n$ on $[0,1]$. Then $f(z)$ and $a_0 + a_1 z + \cdots + a_n z^n$ agree for $z \in [0,1]$, and both are analytic on $\mathbb{C}$ (that is, both are entire). Then by Theorem 1 they are equal on all of $\mathbb{C}$ since $[0,1]$ contains a convergent sequence of distinct points (for example, $z_n = 1/n$). This proves the assertion.

2. Prove that if $z_n \to 0$ and $z_n \neq 0$, and if f is defined in a deleted neighborhood of 0, with $f(z_n) = 0$, then f has a nonremovable singularity at $z = 0$ unless f is identically zero. Illustrate with $\sin(1/z)$.

Solution. If the singularity of f at 0 were removable, then (by definition) we could define $f(0)$ so that f was analytic at 0. Thus if $f(z_n) = 0$, Theorem 1 would imply that f is identically zero. (The z_n have infinitely many distinct values — Why?) This is true with $\sin(1/z) = f(z)$ because for $z_n = 1/\pi n$, $z_n \to 0$ but $f(z_n) = 0$. Thus the singularity is not removable.

We can go one step further. For such an f the singularity must be essential, because if f had a pole at $0, f(z) \to \infty$ as $z \to 0$ (see exercise 2, Section 3.3).

3. Let $f(z) = \Sigma_{n=0}^{\infty} a_n(z - z_0)^n$ have a radius of convergence $R > 0$.
   a. Is there always a sequence z_n with $|z_n - z_0| < R$ for $n = 1, 2, 3, \cdots$, and $|z_n - z_0| \to R$ such that $f(z_n) \to \infty$?
   b. Can f be continued analytically to a circle $|z - z_0| < R + \epsilon$ for some $\epsilon > 0$?

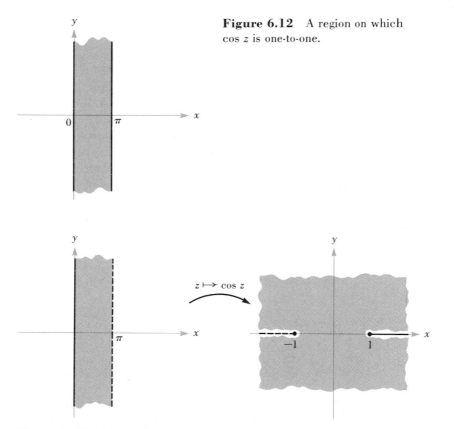

Figure 6.12 A region on which $\cos z$ is one-to-one.

$z \mapsto \cos z$

Figure 6.13 Image of $\cos z$.

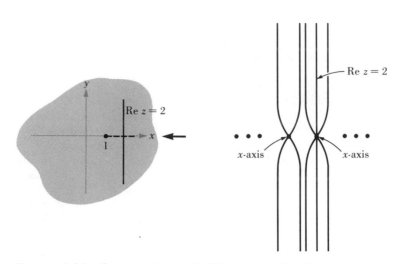

Figure 6.14 Cross section of the Riemann surface for $\cos^{-1} z$ at Re $z = 2$.

Solution.

a. Such a sequence does not necessarily exist. Consider the series $\sum_0^\infty z_n/n^2$. By the ratio test (see Theorem 4, Section 3.2), the radius of convergence is

$$\underset{n \to \infty}{\text{limit}} \left| \frac{a_n}{a_{n+1}} \right| = \underset{n \to \infty}{\text{limit}} \frac{(n+1)^2}{n^2} = 1.$$

But for $|z| \leq 1$ we have

$$\left| \sum_0^\infty \frac{z^n}{n^2} \right| \leq \sum_0^\infty \left| \frac{z^n}{n^2} \right| = \sum_0^\infty \frac{|z|^n}{n^2} \leq \sum_0^\infty \frac{1}{n^2} < \infty.$$

Thus $|f(z)|$ is bounded by $\sum_0^\infty 1/n^2$ on $\{z \mid |z| < 1\}$, so $f(z_n) \to \infty$ is impossible.

b. No. Suppose that there is an analytic function g on $|z - z_0| < R + \epsilon$ with $g(z) = f(z)$ for $|z - z_0| < R$. Since f and g are analytic and agree on $|z - z_0| < R$, the Taylor series of g, $\sum_0^\infty a_n (z - z_0)^n$, is valid (by Theorem 12, Section 3.2) for $|z - z_0| < R + \epsilon$. Hence the radius of convergence of the given series is greater than R, which is impossible (since it equals R).

Exercises

1. Let $h(x)$ be a function of a real variable $x \in \mathbb{R}$. Suppose that $h(x) = \sum_{n=0}^\infty a_n x^n$, which converges for x in some interval $]-\eta, \eta[$ around 0. Prove that h is the restriction of some analytic function defined in a neighborhood of 0.

2. a. Let $f(z) = e^{1/z} - 1$. If $z_n = 1/2\pi n i$, then $z_n \to 0$ and $f(z_n) = 0$, yet f is not identically zero. Does this contradict Theorem 1? Why or why not?

   b. Is the Identity Theorem true for harmonic functions?

3. Let f be analytic and not identically zero on A. Show that if $f(z_0) = 0$, there is an integer k such that $f(z_0) = 0 = \cdots = f^{(k-1)}(z_0)$, and $f^{(k)}(z_0) \neq 0$.

4. Let f be analytic in a region A and let z_1, $z_2 \in A$. Let $f'(z_1) \neq 0$. Show that f is not constant on a neighborhood of z_2.

5. Prove the following result of Karl Weierstrass. Let $f(z) = \sum_{n=0}^{\infty} z^{n!}$. Then f cannot be analytically continued to *any* open set containing $A = \{z \mid |z| < 1\}$.

    Hint.   Show that the series diverges for each $|z| = 1$. First consider $z = e^{2\pi i p/q}$ where p and q are integers.

6. Formulate a Schwarz Reflection Principle for harmonic functions.

7. Suppose that f can be continued analytically along a curve γ in the manner shown in Figure 6.15. Show that f can be continued by power series (in a finite number of steps).

8. Discuss the Riemann surface for $\sqrt[3]{z}$.

9. Discuss the Riemann surface for $\sqrt{z^2 - 1}$.

10. Discuss the relationship between Theorem 10, Section 2.2, and the Monodromy Theorem.

11. Consider the power series $\sum_{0}^{\infty} (-1)^n z^n$ defined in $|z| < 1$. To what domain in $\mathbb{C}$ can you analytically continue this function?

6.2   Rouché's Theorem and Principle of the Argument

Some further properties of analytic functions will now be developed, the main ones being those mentioned in the section title. Our main tool will be the Residue Theorem (see Theorem 8, Section 4.2). It will be convenient to begin with the following result.

Theorem 5.   *Let f be analytic on a region A except for poles at $b_1, \cdots, b_m$ and zeros at $a_1, \cdots, a_n$, counted with their multiplicities (that is, if b_1 is a pole of order k, b_1 is to be repeated k times in the list,*

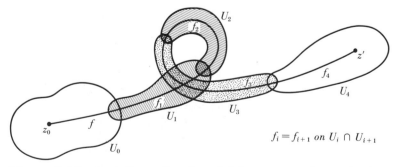

$$f_i = f_{i+1} \text{ on } U_i \cap U_{i+1}$$

Figure 6.15   Analytic continuation of f along a curve from z_0 to z'.

and similarly for the zeros a_j). Let γ be a closed curve homotopic to a point in A and passing through none of the points a_i or b_j. Then

$$\int_\gamma \frac{f'(z)}{f(z)}\, dz = 2\pi i \left\{\sum_{j=1}^{n} I(\gamma, a_j) - \sum_{l=1}^{m} I(\gamma, b_l)\right\}. \tag{1}$$

Note.   Formula (1) applies in particular to *meromorphic functions*; that is, functions defined on $\mathbb{C}$ except for poles (see Section 3.3). There can be only a finite number of poles in any bounded region since poles are, by definition, isolated.

Proof.   First, it is clear that $f'(z)/f(z) = g(z)$ is analytic except at $a_1, \cdots, a_n, b_1, \cdots, b_m$. If f has a zero of order k at a_j, f' has a zero of order $k-1$, so $f'/f = g$ has a simple pole at a_j and the residue there is k. This is because we can write $f(z) = (z - a_j)^k \varphi(z)$, as shown in Section 3.2, where φ is analytic and $\varphi(a_j) \neq 0$; therefore,

$$g(z) = \frac{k(z - a_j)^{k-1}\varphi(z)}{(z - a_j)^k \varphi(z)} + \frac{(z - a_j)^k \varphi'(z)}{(z - a_j)^k \varphi(z)}$$

$$= \frac{k}{z - a_j} + \frac{\varphi'(z)}{\varphi(z)}.$$

Thus the residue at a_j is clearly k. Similarly, if b_l is a pole of order k, we can write, near b_l,

$$f(z) = \frac{\varphi(z)}{(z - b_l)^k},$$

where φ is analytic and $\varphi(b_l) \neq 0$ (see Theorem 15(iv), Section 3.3). Then, as above, we see that near b_l,

$$g(z) = \frac{-k}{z - b_l} + \frac{\varphi'(z)}{\varphi(z)},$$

so the residue is $-k$.

By the Residue Theorem,

$$\int_\gamma g(z)\, dz = 2\pi i \left\{\sum_{j}' (\mathrm{Res}\ (g, a_j))\, I(\gamma, a_j) + \sum_{l}' (\mathrm{Res}\ (g, b_l))\, I(\gamma, b_l)\right\},$$

where Σ' means the sum over the distinct points. But since the residue equals the number of times a_j occurs and minus that number for the b_l, this expression becomes

$$2\pi i \left\{\sum_{j=1}^{n} I(\gamma, a_j) - \sum_{l=1}^{n} I(\gamma, b_l)\right\},$$

as required.∎

Theorem 5 is best understood in the following special case. Let γ be a simple closed curve, so $I(\gamma,\alpha_j) = 1$ if a_j lies inside γ and 0 if a_j lies outside γ. Then formula (1) becomes

$$\int_\gamma \frac{f'(z)}{f(z)}\, dz = 2\pi i\ (Z_f - P_f), \qquad (2)$$

where Z_f = number of zeros inside γ and P_f = number of poles inside γ counted with their multiplicities.

Principle of the argument

We now consider a useful consequence of Theorem 5. For a closed curve γ and z_0 not on γ we can legitimately say that the change in argument of $z - z_0$ as z traverses γ is $2\pi \cdot I(\gamma,z_0)$. This is actually the intuitive basis on which the index was developed; it is written $\Delta_\gamma \arg (z - z_0) = 2\pi\, I(\gamma,z_0)$ (see Figure 6.16).

Next we want to define $\Delta_\gamma \arg f$, that is, the change in $\arg f(z)$ as z goes around γ. Intuitively, and for practical computations, the meaning is clear; we merely compute $\arg f(\gamma(t))$ and let t run from a to b if $\gamma\colon [a,b] \to \mathbb{C}$, then look at the difference $\arg f(\gamma(b)) - \arg f(\gamma(a))$. We choose a branch of the argument so that $\arg f(\gamma(t))$ varies continuously with t. Equivalently, by changing variables, we can let $\tilde\gamma = f\circ\gamma$ and compute $\Delta_{\tilde\gamma} \arg z$. This leads to formulation of the following definition.

Definition 1.   *Let f be analytic on a region A and let γ be a closed curve in A homotopic to a point and passing through no zero of f. Then set*

$$\Delta_\gamma \arg f = 2\pi \cdot I(f\circ\gamma,0).$$

(The index makes sense because 0 does not lie on $f\circ\gamma$.)

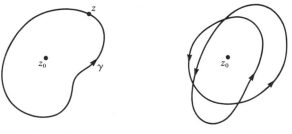

$\Delta_\gamma \arg (z - z_0) = 2\pi$ $\qquad\qquad$ $\Delta_\gamma \arg (z - z_0) = 4\pi$

Figure 6.16   Change in the argument of $z - z_0$ when the two curves are traversed.

In examples, we can make use of our previous intuition about the index to compute $\Delta_\gamma \arg f$. The argument principle is as follows.

Theorem 6 (Principle of the Argument). *Let f be analytic on a region A except for poles at $b_1, \cdots, b_m$ and zeros at $a_1, \cdots, a_n$ counted according to their multiplicity. Let γ be a closed curve homotopic to a point and passing through no a_j or b_l. Then*

$$\Delta_\gamma \arg f = 2\pi \left\{ \sum_{j=1}^{n} I(\gamma, a_j) - \sum_{l=1}^{m} I(\gamma, b_l) \right\}. \tag{3}$$

Proof.   By Theorem 5, it suffices to show that

$$i\Delta_\gamma \arg f = \int_\gamma \frac{f'(z)}{f(z)} \, dz$$

since f has no zeros or poles on γ. Indeed,

$$i\Delta_\gamma \arg f = 2\pi i \cdot I(f \circ \gamma, 0) = \int_{f \circ \gamma} \frac{dz}{z}$$

by the formula for the index (see Section 2.4). Letting $\gamma : [a,b] \to \mathbb{C}$, we have

$$\int_{f \circ \gamma} \frac{dz}{z} = \int_a^b \frac{\frac{d}{dt} f(\gamma(t))}{f(\gamma(t))} \, dt$$

$$= \int_a^b \frac{f'(\gamma(t))}{f(\gamma(t))} \gamma'(t) \, dt$$

by the definition of the integral and the Chain Rule. The latter integral equals $\int_\gamma [f'(z)/f(z)] \, dz$ by definition. (If γ is only piecewise C^1, this holds only on each interval where γ' exists, but we get the result by addition.) ∎

Formula (3) is usually applied in the case where γ is a *simple* closed curve. Then we may conclude that the change in $\arg f(z)$ as we go once around γ (in a counterclockwise direction) is $2\pi(Z_f - P_f)$ where Z_f (or P_f) is the number of zeros (or poles) inside γ counted with their multiplicities. It is somewhat surprising, a priori, that $Z_f - P_f$ and the argument change of f are even related.

Rouché's Theorem

Theorem 6 can be used to prove a very useful theorem that has many applications, some of which will be given throughout the remainder of this chapter.

Theorem 7 (Rouché's Theorem). *Let f and g be analytic on a region A except for a finite number of zeros and poles in A. Let γ be a closed curve in A homotopic to a point and passing through no zero or pole of f or g. Suppose that on γ,*

$$|f(z) - g(z)| < |f(z)|.$$

Then (i) Δ_γ arg $f = \Delta_\gamma$ arg g; *and* (ii) $Z_f - P_f = Z_g - P_g$ *where* $Z_f = \Sigma_{j=1}^n I(\gamma, a_j)$, a_j *being the zeros of f, counted with multiplicities, and* P_f, Z_g, P_g *being defined similarly.*

Proof. Since f and g have no zeros on γ, we can write our assumption as

$$\left|\frac{g(z)}{f(z)} - 1\right| < 1 \text{ on } \gamma.$$

Thus $g(z)/f(z) = h(z)$ maps γ into the unit disk centered at 1 (see Figure 6.17). Therefore, $I(h \circ \gamma, 0) = 0$, since $h \circ \gamma$ is homotopic to 1 in that disk (which does not contain 0). Therefore,

$$\int_\gamma \frac{h'(z)}{h(z)} \, dz = 0$$

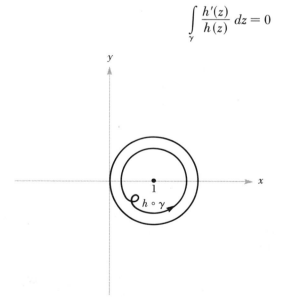

Figure 6.17 The image of γ under h.

by Theorems 5 and 6 and the definition of $\Delta_\gamma \arg h$. But we can easily compute that

$$\frac{h'(z)}{h(z)} = \frac{g'(z)}{g(z)} - \frac{f'(z)}{f(z)},$$

and thus

$$\int_\gamma \frac{g'(z)}{g(z)}\, dz = \int_\gamma \frac{f'(z)}{f(z)}\, dz.$$

Hence the result follows from Theorems 5 and 6. ∎

The important special case of Theorem 7 is the following. *Let γ be a simple closed curve and let f and g be analytic inside and on γ with γ passing through no zeros of f or g; suppose that $|f(z) - g(z)| < |f(z)|$ on γ. Then f and g have the same number of zeros inside γ.* For example, γ could be a circle. Note that if $|f(z) - g(z)| < |f(z)|$ on γ, then γ automatically can pass through no zeros of f or g (Why?).

Theorem 7 can be used to locate the zeros of a polynomial. An illustration is given in example 1. Rouché's Theorem can also be used to give a simple proof of the Fundamental Theorem of Algebra, including the fact that an nth degree polynomial has exactly n roots (see exercise 3).

Hurwitz' Theorem

One of the theoretical applications of Rouché's Theorem is the following result of Hurwitz.

Theorem 8 (Hurwitz' Theorem). *Let f_n be a sequence of analytic functions on a region A converging uniformly on every closed disk in A to f (thus f is analytic by Theorem 6, Section 3.1). Assume that f is not identically zero, and let $z_0 \in A$. Then $f(z_0) = 0$ iff there is a sequence $z_n \to z_0$ where $f_n(z_n) = 0$ (that is, a zero of f is a limit of zeros of the functions f_n).*

Proof.   Choose γ to be a small circle centered at z_0 and lying in A. Thus, by assumption, $f_n \to f$ uniformly on and inside γ. We may suppose that inside and on γ, z_0 is the only possible zero of f, since the zeros of f are isolated (by Theorem 1, Section 6.1).

Let m be the minimum of $|f(z)|$ on γ, that is, $|f(z)| \geq m > 0$ on γ. Choose N large enough so that $|f_n(z) - f(z)| < m$ on γ for $n \geq N$. Thus Rouché's Theorem applies and we conclude that $f_n(z)$ and $f(z)$ have the same number of zeros inside γ. The result follows immediately from this conclusion. ∎

We must assume that f is not identically zero. Consider, for example, the function $f_n(z) = e^z/n$, which approaches zero uniformly on closed disks (Why?) but f_n has no zeros.

Another illustration of the use of Rouché's Theorem is given in example 2.

One-to-one functions

Analytic functions that are one-to-one find many useful applications. (The term *schlicht* (simple) function is often used.) A class of such functions was studied in Chapter 5, where proof of the following result was referred to. Again Rouché's Theorem is the appropriate tool.

Theorem 9. *If $f: A \to \mathbb{C}$ is analytic and one-to-one, then $f'(z_0) \neq 0$ for all $z_0 \in A$. (It follows from Theorem 20, Section 1.4, that $f(A)$ is open and that f^{-1} is analytic from $f(A)$ to A.)*

Proof.   Suppose that, to the contrary, for some point z_0 we have $f'(z_0) = 0$. Then $f(z) - f(z_0)$ has a zero of order $k \geq 2$ at z_0. Now f is not constant and thus the zeros of f' are isolated. Thus there is a $\delta > 0$ and an $m > 0$ such that on the circle $|z - z_0| = \delta$, $|f(z) - f(z_0)| \geq m > 0$ and $f'(z) \neq 0$ for $0 < |z - z_0| \leq \delta$. For $0 < \eta < m$, we conclude that $f(z) - f(z_0) - \eta$ has k zeros inside $|z - z_0| = \delta$, by Rouché's Theorem. A zero cannot be a double zero since $f'(z) \neq 0$ for $|z - z_0| \leq \delta$, $z \neq z_0$. Thus $f(z) = f(z_0) + \eta$ for two distinct points z and therefore is not one-to-one. This contradiction means that $f'(z_0) \neq 0$, as was to be shown. ∎

Another basic property of one-to-one functions is the following.

Theorem 10.   *Let f be analytic on a region A and let γ be a closed curve homotopic to a point in A. Suppose that $I(\gamma, z) = 0$ or 1. Let $B = \{z \in A \mid I(\gamma, z) \neq 0\}$ (the "inside" of γ). If f is such that each point of $f(B)$ has index 1 with respect to the curve $\tilde{\gamma} = f \circ \gamma$, then f is one-to-one on B.*

Proof.   Consider, for $z_0 \in B$ and $w_0 = f(z_0)$,

$$N = \frac{1}{2\pi i} \int_\gamma \frac{f'(z)}{f(z) - w_0} \, dz.$$

By Theorem 5, N equals the number of times that $f(z) = w_0$ on B. We therefore must show that it equals 1. But letting $\tilde{\gamma} = f \circ \gamma$, we see (compare Theorem 6) that

$$N = \frac{1}{2\pi i} \int\limits_{\tilde{\gamma}} \frac{1}{z - w_0} \, dz,$$

which is the index of w_0 with respect to $\tilde{\gamma}$. Thus $N = 1$ and therefore $f(z) = w_0$ has exactly one solution, $z = z_0$. This means that f is one-to-one. ■

Theorem 10 becomes more intuitive if we use the Jordan Curve Theorem. Let γ be a simple closed curve and let B be its interior. Suppose that the set $f(B)$ is bounded by the curve $\tilde{\gamma} = f \circ \gamma$. The hypothesis of Theorem 10 will be fulfilled if $\tilde{\gamma}$ is a simple closed curve (since this means that f should be one-to-one on γ). Therefore, Theorem 10 may be rephrased as follows. To see if an analytic function is one-to-one on a region, it is sufficient to check that it is one-to-one on the boundary.

Worked Examples

1. Use Rouché's Theorem to determine the quadrants in which the zeros of $z^4 + iz^2 + 2$ lie and the number of zeros that lie inside circles of varying radii.

    Solution.   Let $g(z) = z^4$, $f(z) = z^4 + iz^2 + 2$, and note that

    $$|f(z) - g(z)| = |iz^2 + 2| \le |z|^2 + 2$$

    and that

    $$|g(z)| = |z|^4.$$

    Hence if $r = |z| > \sqrt{2}$, we have

    $$|f(z) - g(z)| < |g(z)|.$$

    Since g does not vanish on any circle of positive radius, the preceding inequality shows that f does not vanish on circles with radius $> \sqrt{2}$. Rouché's Theorem then shows that all four roots of f lie inside these circles, that is, inside the closed disk $|z| \le \sqrt{2}$.

    Next, let $h(z) = z^4 + 2iz^2 = z^2(z^2 + 2i)$. Clearly, h has a double root at 0 and two additional roots on the circle $|z| = \sqrt{2}$. Furthermore,

    $$|f(z) - h(z)| = |-iz^2 + 2| = |z^2 + 2i| = \frac{|h(z)|}{|z|^2}.$$

    For any choice of r with $1 < r < \sqrt{2}$, h and hence f does not vanish on the circle $|z| = r$ and $|f(z) - h(z)| < |h(z)|$. Rouché's Theorem shows that f has precisely two zeros in $|z| < r$ for any of these values of r. Letting r approach 1 and $\sqrt{2}$, we get that f has two roots in the closed disk $|z| \le 1$ and two on the circle $|z| = \sqrt{2}$.

Finally, let $k(z) = 2$. Then

$$|f(z) - k(z)| = |z^4 + iz^2| \le |z|^4 + |z|^2 < 2 = |k(z)|$$

whenever $|z| < 1$. Arguing as before, for any r with $0 < r < 1$, k and hence f does not vanish in $|z| < r$. Combining these three results we find that f has two zeros on $|z| = 1$ and two on $|z| = \sqrt{2}$.

Now we turn to an analysis of the quadrants in which the roots lie. For z either real or purely imaginary, $f(z) = z^4 + iz^2 + 2$ has a nonzero imaginary part unless $z = 0$. Thus f has no roots on the axes. Consider a large quarter circle as shown in Figure 6.18. We shall compute $\Delta_\gamma \arg(z^4 + iz^2 + 2)$ and use the Principle of the Argument. Along the x-axis z is real and $f(z)$ lies in the first quadrant. Also $f(0) = 2$, and as $R \to \infty$, $\arg f(R) \to 0$ since

$$\arg f(R) = \arg R^4 \left(1 + \frac{i}{R^2} + \frac{2}{R^4} \right)$$

$$= \arg \left(1 + \frac{i}{R^2} + \frac{2}{R^4} \right) \to 0 \quad \text{as} \quad R \to \infty.$$

Since f takes its values in the first quadrant, we can conclude that the change in the argument is zero as z moves from 0 to ∞. Along the curved portion of γ, z^4 clearly changes argument by 2π ($= 4 \times \pi/2$). As $R \to \infty$, 2π is the limiting change in argument for $f(z)$ as well, as we see by writing

$$f(z) = z^4 \left(1 + \frac{i}{z^2} + \frac{2}{z^4} \right)$$

as before.

Similarly, we see that coming down the imaginary axis there is, in the limit of large R, no change in the argument of f. (If $f(0)$ were not real, this device would still give the limiting behavior of the argument at infinity and the value at zero, so the change in the argument, at least up to multiples of 2π, can be inferred.) We conclude from our analysis that the change in argument as we traverse γ is 2π. From the Principle of the Argument, there is exactly one zero in the first quadrant. By inspection, $f(z) = f(-z)$,

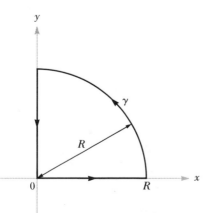

Figure 6.18   The curve γ used to locate the quadrants in which the zeros of the polynomial $z^4 + iz^2 + 2$ lie.

so $-z$ is a root when z is. Thus there must be a root in each quadrant. Therefore, we must have one of the two possibilities shown in Figure 6.19. The methods used here do not enable us to tell which of these possibilities actually occurs without more detailed analysis.

We can check this example by finding the roots directly; however, in other examples a direct computation may be impossible or impractical whereas the methods described here can nevertheless be used.

2.   If $a > e$, show that the equation $e^z = az^n$ has n solutions inside the unit circle.

Solution.   Let $f(z) = e^z - az^n$ and let $g(z) = -az^n$. Then g has exactly n roots. We shall show that f and g have the same number of roots inside the unit circle $|z| = 1$. To do so we must show that

$$|f(z) - g(z)| < |g(z)|$$

for $|z| = 1$. But

$$|f(z) - g(z)| = |e^z| = e^x \le e,$$

since $|x| \le 1$. Also $|g(z)| = |az^n| = a > e$, so the result follows by Rouché's Theorem.

3.   Let $f(z) = \sum_{n=0}^{\infty} a_n z^n$. Assume that $a_0 = 0$, $a_1 = 1$. Prove that f is one-to-one on the unit disk $\{z \mid |z| < 1\}$ if $\sum_{l=2}^{\infty} l|a_l| \le 1$.

Solution.   Certainly, f converges for $|z| < 1$ since, as a consequence of $\sum_{l=2}^{\infty} l|a_l| \le 1$, we get $|a_n| \le 1$, and thus $|a_n z^n| \le |z|^n$; we know that $\sum |z^n|$ converges for $|z| < 1$. Thus f is analytic on $\{z \mid |z| < 1\}$.

Let $|z_0| < 1$. We want to show that $f(z) = f(z_0)$ has exactly one solution, z_0. Let $g(z) = z - z_0$, which has exactly one zero, and thus if we set $h(z) = f(z) - f(z_0)$, then

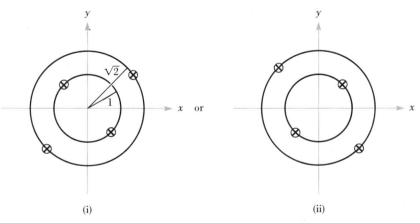

(i)   (ii)

Figure 6.19   Locating the roots of the polynomial $z^4 + iz^2 + 2$.

$$h(z) - g(z) = \sum_{n=2}^{\infty} a_n z^n - \sum_{n=2}^{\infty} a_n z_0^n.$$

To estimate this we can use the following trick. Let $\varphi(z) = \sum_{n=2}^{\infty} a_n z^n$. Then

$$|\varphi(z) - \varphi(z_0)| \le \{\max |\varphi'(\zeta)|\} \cdot |z - z_0|,$$

where the maximum is over those ζ on the line joining z_0 to z (Why?). But $|\varphi'(\zeta)| = |\sum_{n=2}^{\infty} n a_n \zeta^{n-1}| < \sum_{n=2}^{\infty} n|a_n| \le 1$ since $|\zeta| < 1$. Hence

$$|h(z) - g(z)| = |\varphi(z) - \varphi(z_0)| < |z - z_0| = |g(z)|.$$

Thus, by Rouché's Theorem, $h(z) = f(z) - f(z_0)$ has exactly one solution, $z = z_0$; this proves the assertion.

Exercises

1. Let f be analytic inside and on the unit circle $|z| = 1$. Suppose that $|f(z)| < 1$ if $|z| = 1$. Show that f has exactly one fixed point (a point z_0 such that $f(z_0) = z_0$) inside the unit circle.

2. Let $g_n = \sum_{k=0}^{n} z^k/k!$. Let $D(0,R)$ be the disk of radius $R > 0$. Show that for n large enough, g_n has no zeros in $D(0,R)$.

3. Use Rouché's Theorem to prove that if $f(z) = a_0 + a_1 z + \cdots + a_n z^n$, $n \ge 1$, and $a_n \ne 0$, then f has exactly n roots.

    Hint.   Let $g(z) = a_n z^n$ and estimate $f(z) - g(z)$, as was done in Theorem 27, Section 2.4.

4. Extend Theorem 5 to include the following result. If f is analytic on A except for zeros at $a_1, \cdots, a_n$ and poles at $b_1, \cdots, b_m$ (each repeated according to its multiplicity), if φ is analytic on A, and if γ is a closed curve homotopic to a point in A, passing through none of $a_1, \cdots, a_n$, $b_1, \cdots, b_m$, then

    $$\int_{\gamma} \frac{f'(z)}{f(z)} \varphi(z) dz = 2\pi i \left\{ \sum_{i=1}^{n} \varphi(a_i) I(\gamma, a_i) - \sum_{k=1}^{m} \varphi(b_k) I(\gamma, b_k) \right\}.$$

    Exercises 5 and 6 are based on the result of exercise 4.

5. a. Let $f: A \to B$ be analytic, one-to-one, and onto. Let $w \in B$ and let γ be a small circle centered at z_0 in A. Prove that

    $$f^{-1}(w) = \frac{1}{2\pi i} \int_{\gamma} \frac{f'(z)z}{f(z) - w} dz$$

    for w sufficiently close to $f(z_0)$.

    b. Explain the meaning of

    $$\frac{1}{2\pi i} \int_{\gamma} \frac{f'(z)}{f(z) - w} dz.$$

6. If $f(z)$ is a polynomial, prove that

$$\frac{1}{2\pi i} \int_{\gamma} \frac{f'(z)}{f(z)} z \, dz$$

is the sum of the zeros of f if the circle γ is large enough.

7. Let $f(z)$ be a polynomial of degree n, $n \geq 1$. Show that f maps $\mathbb{C}$ onto $\mathbb{C}$. In fact, show that $f(z) = w$ has n roots z for every $w \in \mathbb{C}$.

8. Locate the zeros (as was done in example 1) for the polynomial $z^4 - z + 5 = 0$.

9. If $f(z)$ is analytic and has n zeros inside the simple closed curve γ, must it follows that $f'(z)$ has $n - 1$ zeros inside γ?

10. Let f be analytic inside and on $|z| = R$ and let $f(0) \neq 0$. Let $M = \max |f(z)|$ on $|z| = R$. Show that the number of zeros of f inside $|z| = R/3$ does not exceed

$$\frac{1}{\log 2} \cdot \log\left(\frac{M}{|f(0)|}\right).$$

Hint.   Let $h(z) = f(z)/\{(z - z_1) \cdots (z - z_n)\}$ where z_n are the zeros of f inside $|z| = R/3$ and apply the Maximum Modulus Theorem to h.

11. Let $f_n \to f$ uniformly on closed disks in A. Suppose that the functions f_n are analytic and one-to-one. Show that if f is not identically constant, then f is also one-to-one.

12. Show that $z \mapsto z^2 + 3z$ is one-to-one on $\{z \mid |z| < 1\}$.

13. Find an $r > 0$ such that the polynomial $z^3 - 4z^2 + z - 4$ has exactly two roots inside the circle $|z| = r$.

14. Prove that the following statement is false: For every function f analytic on the annulus $1/2 < |z| < 3/2$, there is a polynomial p such that $|f(z) - p(z)| < 1/2$ for $|z| = 1$.

15. Let f be analytic on $\mathbb{C}$ and let $|f(z)| \leq 5\sqrt{|z|}$ for all $|z| \geq 1$. Prove that f is constant.

16. Use Rouché's Theorem to show that the roots of an nth degree polynomial (the roots being counted with multiplicities) depend continuously on the coefficients of the polynomial.

6.3   Mapping Properties of Analytic Functions

Some further properties of analytic functions that are of a local nature (that is, they depend only on the values of $f(z)$ for z in a neighborhood of a given point z_0) will be proved in this section. Additional proofs will be given here of the Inverse Function Theorem (Theorem 20, Section 1.4), the Maximum Modulus Theorem (Theorem 30, Section 2.5), and the Open Mapping Theorem (stated formally for the first time in this chapter but previously mentioned in exercise 4, Section 1.4). We can prove thes theorems and also obtain information concerning the behavior of a function near a point by using the Root Counting Formula:

$$\frac{1}{2\pi i} \int_{\gamma} \frac{f'(z)}{f(z) - w} \, dz = \text{the number of roots of } f(z) = w \text{ inside } \gamma,$$ counting multiplicities.

This formula was established in the preceding section. One result of the formula is Rouché's Theorem, from which, in turn, we get the following basic tool.

Theorem 11. *Let f be analytic and not constant on a region A and let $z_0 \in A$. Suppose that $f(z) - w_0$ has a zero of order $k \geq 1$ at z_0. Then there is an $\eta > 0$ such that for any $\epsilon \in]0,\eta]$, there is a $\delta > 0$ such that if $|w - w_0| < \delta$, then $f(z) - w$ has exactly k roots (counted with their multiplicities) in the disk $|z - z_0| < \epsilon$ (see Figure 6.20). In fact, there is a $\lambda > 0$ (probably smaller than η) such that for any $\epsilon \in]0,\lambda]$, there is a $\delta > 0$ such that if $0 < |w - w_0| < \delta$, then $f(z) - w$ has exactly k distinct roots in the disk $0 < |z - z_0| < \epsilon$.*

In other words, if f takes on the value w_0 at z_0 with multiplicity k, then for all w sufficiently near w_0, f takes on the value w k times near z_0 (counting multiplicities). For all w still nearer w_0, the k roots of $f(z) = w$ near z_0 are distinct.

As an example, let us consider the special case in which $f(z) = z^k$. This function has a zero of order k at $z_0 = 0$ (here $w_0 = 0$). Theorem 11 says that for all w near 0, $z^k = w$ has exactly k solutions near 0. The theorem tells us that this behavior of the power function, of which we already are aware, occurs more generally. We might have guessed this from the power series expansion of f around z_0:

$$f(z) - w_0 = \sum_{n=k}^{\infty} a_n (z - z_0)^n.$$

For $|z - z_0|$ that are small enough, we might guess (correctly) that the behavior of the lowest-degree nonvanishing term, $a_k(z - z_0)^k$, will dominate.

Proof. Since f is not constant, the zeros of $f(z) - w_0$ are isolated. Thus there is an $\eta > 0$ such that for $|z - z_0| \leq \eta$, $f(z) - w_0$ has no zeros other than z_0. On the compact set $\{z \mid |z - z_0| = \epsilon\}$ (the circle γ in Figure 6.20), $f(z) - w_0$ is continuous and never zero. Hence there is a $\delta > 0$ such that $|f(z) - w_0| \geq \delta > 0$ for $|z - z_0| = \epsilon$. Thus if w satisfies $|w - w_0| < \delta$, then for $|z - z_0| = \epsilon$, the following hold:

i. $f(z) - w_0 \neq 0$;

ii. $f(z) - w \neq 0$, (since $f(z) = w$ would mean that $|w - w_0| \geq \delta$);

iii. $|(f(z) - w) - (f(z) - w_0)| = |w - w_0| < \delta \leq |f(z) - w_0|$.

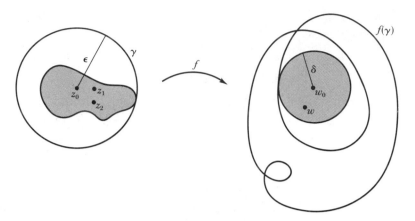

Figure 6.20 f is two-to-one near z_0.

By applying Rouché's Theorem, we see that $f(z) - w$ has the same number of zeros, counting multiplicities, as $f(z) - w_0$ inside the circle $|z - z_0| = \epsilon$. Thus we have proved the first part of the theorem. To prove the second part, we notice that f' is not identically zero on A. The zeros of f' are thus isolated. Therefore, for some $\lambda \le \eta$, neither $f(z) - w_0$ nor $f'(z)$ is zero in $|z - z_0| \le \lambda$ except at z_0. We observe that $f(z) - w$ still has the same number of roots as $f(z) - w_0$ for any w near enough to w_0, but now the roots must be first order, hence distinct, since f' is nonzero. ∎

Theorem 11 tells us that on some disk centered at z_0, f is exactly k-to-one. The theorem may not be directly helpful in finding the size of this disk (see the examples and exercises at the end of this section), but often knowledge of its existence can lead to interesting results.

A function $f: A \to \mathbb{C}$ is called *open* iff, for every open set $U \subset A$, $f(U)$ is open. By the definition of an open set, this statement is equivalent to: For every $\epsilon > 0$ sufficiently small, there is a $\delta > 0$ such that $|w - w_0| < \delta$ implies that there is a z, $|z - z_0| < \epsilon$ with $w = f(z)$. In other words, if f hits w_0, it hits every w sufficiently near w_0. Careful reading of the definition of open set and examination of Figure 6.20 show that Theorem 11 implies Theorem 12:

Theorem 12 (Open Mapping Theorem). *Let $f: A \to \mathbb{C}$ be non-constant and analytic. Then f is an open mapping.*

Using Theorem 11, we can also get an alternative proof of Theorem 20, Section 1.4:

Theorem 13 (Inverse Function Theorem). *Let $f: A \to \mathbb{C}$ be analytic, let $z_0 \in A$, and let $f'(z_0) \neq 0$. Then there is a neighborhood U of*

z_0 and a neighborhood V of $w_0 = f(z_0)$ such that $f: U \to V$ is one-to-one and onto and $f^{-1}: V \to U$ is analytic.

Proof. $f(z) - w_0$ has a simple zero at z_0 since $f'(z_0) \neq 0$. We can use Theorem 11 to find $\epsilon > 0$ and $\delta > 0$ such that each w with $|w - w_0| < \delta$ has exactly one pre-image x with $|z - z_0| < \epsilon$. Let $V = \{w \mid |w - w_0| < \delta\}$ and let U be the inverse image of V under the map f restricted to $\{z \mid |z - z_0| < \epsilon\}$ (the shaded region of Figure 6.20). By Theorem 11, f maps U one-to-one onto V. Since f is continuous, U is a neighborhood of z_0. By Theorem 12, $f = (f^{-1})^{-1}$ is an open map, so f^{-1} is continuous from V to U. To show that it is analytic, use

$$f^{-1}(w) = \frac{1}{2\pi i} \int_{|z - z_0| = \epsilon} \frac{f'(w)}{f(z) - w} z \, dz$$

(see exercise 5 at the end of the preceding section). That this is analytic in w follows from example 4, Section 2.4. ■

These ideas can be used as the basis for another proof of the Maximum Modulus Theorem (see Section 2.5), as follows.

Theorem 14 (Maximum Modulus Theorem).   *Let f be analytic on a region A. If $|f|$ has a relative maximum at $z_0 \in A$, then f is constant.*

Proof.   Suppose that f is not constant and that $z_0 \in A$. Since f is an open map, for $|w - f(z_0)|$ sufficiently small there is a z near z_0 with $w = f(z)$. Choose w with $|w| > |f(z_0)|$. Specifically, choose $w = (1 + \delta/2)f(z_0)$ if $f(z_0) \neq 0$ and $w = \delta/2$ if $f(z_0) = 0$ for δ small. Then it is clear that f does not have a relative maximum at z_0. ■

A similar proof shows that if $f(z_0) \neq 0$, then f has no minimum at z_0 unless f is constant. Here, Theorem 30, Section 2.5, again follows from Theorem 13.

Worked Examples

1.   Determine the largest disk around $z_0 = 0$ on which $f(z) = 1 + z + z^2$ is one-to-one.

   Solution.   Since $f'(0) = 1$, $f(z) - 1$ has a simple zero at 0, and Theorem 11 shows that f is one-to-one on some disk around $z_0 = 0$. Because $f(z) - 1 = z + z^2 = z(1 + z)$, which has roots at 0 and -1, we know that $f(z) - 1$ has only one root in the disk $\{z \mid |z| < 1\}$. This disk is the disk in the first part of Theorem 11, but Theorem 11 does not guarantee that f is one-to-one on the disk; in fact, it is not. Theorem 11 shows that f is one-to-one only on the subregion of the disk shaded in Figure 6.20, the pre-image of $\{w \mid |w - w_0| < \delta\}$. We can find out what causes this phenomenon by plotting the image of the unit circle. In this case $f(z) = 1 + z + z^2$, $z_0 = 0$, and $w_0 = 1$. Thus

$$f(0) = 1 \qquad\qquad f(e^{2\pi i/3}) = 0$$

$$f(1) = 3 \qquad\qquad f(e^{4\pi i/3}) = 0$$

$$f(i) = i \qquad\qquad f\left(\frac{1}{\sqrt 2} + \frac{1}{\sqrt 2}i\right) = \left(1 + \frac{1}{\sqrt 2}\right) + \left(1 + \frac{1}{\sqrt 2}\right)i$$

$$f(-1) = 1 \qquad\qquad f\left(-\frac{1}{\sqrt 2} - \frac{1}{\sqrt 2}i\right) = \left(1 - \frac{1}{\sqrt 2}\right) + \left(1 - \frac{1}{\sqrt 2}\right)i$$

$$f(-i) = -i$$

$$f\left(\frac{1}{\sqrt 2} - \frac{1}{\sqrt 2}i\right) = \left(1 + \frac{1}{\sqrt 2}\right) \qquad f\left(-\frac{1}{\sqrt 2} + \frac{1}{\sqrt 2}i\right) = \left(1 - \frac{1}{\sqrt 2}\right) - \left(1 - \frac{1}{\sqrt 2}\right)i$$

$$-\left(1 + \frac{1}{\sqrt 2}\right)i$$

By plotting these points, we find that the image of the unit circle is as shown in Figure 6.21. The index of the image curve with respect to the small shaded region is 2. Therefore, each point here is hit twice by points in the unit disk; for example, $f'(-1/2) = 0$ and $f(-1/2) = 3/4$. Theorem 11 shows that f is two-to-one on small neighborhoods of $-1/2$. Thus f will not be one-to-one on any disk containing a neighborhood of $-1/2$.

Consider the disk $D(0,r) = \{z \mid |z| < r\}$. The boundary curve is the circle $\gamma_r = \{z \mid |z| = r\}$. As r gets smaller, the troublemaking loop in the image curve shrinks. For some critical r_0 it disappears. For $r > r_0$, f is not one-to-one on γ_r. For $r < r_0$, f is one-to-one on γ_r. By Theorem 10, f is thus one-to-one on $D(0,r)$, and the desired disk is $D(0,r_0)$. To find r_0, suppose that $re^{i\theta}$ and $re^{i\psi}$ lie on γ_r and that $f(re^{i\theta}) = f(re^{i\psi})$. Then

$$1 + re^{i\theta} + r^2 e^{i2\theta} = 1 + re^{i\psi} + r^2 e^{i2\psi}.$$

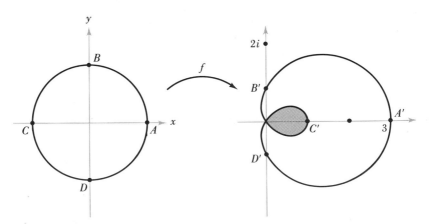

Figure 6.21   Image of the unit circle under $f(z) = 1 + z + z^2$.

Hence $e^{i\theta} + re^{i2\theta} = e^{i\psi} + re^{i2\psi}$, and so

$$re^{i(\theta+\psi)} \left[e^{i(\theta-\psi)} - e^{i(\psi-\theta)} \right] = e^{i(\theta+\psi)/2} \left[e^{i(\psi-\theta)/2} - e^{i(\theta-\psi)/2} \right].$$

Thus

$$re^{i(\theta-\psi)/2} \sin(\theta - \psi) = -\sin\left(\frac{\theta - \psi}{2}\right).$$

In other words,

$$2re^{i(\theta+\psi)/2} \sin\left(\frac{\theta - \psi}{2}\right) \cos\left(\frac{\theta - \psi}{2}\right) = -\sin\left(\frac{\theta - \psi}{2}\right).$$

Now one of two things must happen: either $\sin[(\theta - \psi)/2] = 0$, in which case $\theta - \psi = 2\pi n$ for some integer n, and thus $re^{i\theta} = re^{i\psi}$, or $\cos[(\theta-\psi)/2] = -(1/2r)e^{-i(\theta+\psi)/2}$. If $r > 1/2$, the latter can happen for $\psi = -\theta$; for example, at $r = 1$, it occurs at the points $e^{2\pi i/3}$ and $e^{4\pi i/3}$. If $r < 1/2$, this same condition cannot hold since $|\cos[(\theta - \psi)/2]| \le 1$. If $r = 1/2$, it can happen only for $\theta = \psi = \pi$. The critical radius is therefore $r_0 = 1/2$. Hence f is one-to-one on the disk $D(0,1/2) = \{z \mid |z| < 1/2\}$ but not on any larger open disk.

Remark. $D(0,1/2)$ is the largest disk around $z_0 = 0$ on which $f'(z)$ is never zero. It is not generally true that this will also be the disk on which f is one-to-one (see exercise 3).

2. Prove the following: If f is analytic near $z_0 \in A$ and if $f(z) - f(z_0)$ has a zero of order k at z_0, $1 \le k < \infty$, then there is an analytic function $h(z)$ such that $f(z) = f(z_0) + [h(z)]^k$ for z near z_0, and h is locally one-to-one.

Solution.   Since $k < \infty$, f is not constant. We can write $f(z) - f(z_0) = (z - z_0)^k \varphi(z)$ where $\varphi(z_0) \ne 0$ and φ is analytic. For z near z_0, $\varphi(z)$ lies in a small disk around $\varphi(z_0)$ not containing 0, by continuity. On such a disk we can define $\sqrt[k]{\varphi(z)}$ and let $h(z) = (z - z_0) \sqrt[k]{\varphi(z)}$. Then $h'(z_0) \ne 0$, so by Theorem 13, h is locally one-to-one. In exercise 10 the student is asked to diagram the results of this example.

Exercises

1. What is the largest disk around $z_0 = 1$ on which $f(z) = z + z^2$ is one-to-one?

2. What is the largest disk around $z_0 = 1$ on which $f(z) = e^z$ is one-to-one?

3. Let f be analytic on $D = \{z \mid |z - z_0| < r\}$. Let $f(z_0) = w_0$ and suppose that $f(z) - w_0$ has no roots in D other than z_0 and that $f'(z)$ is never zero in D. Show that it is not necessarily true that f is one-to-one on D.

Hint.   Consider z^3.

4. Let $f(z) = \Sigma_0^\infty a_n z^n$ have a radius of convergence R. Suppose that $|a_1| \ge \Sigma_{n=2}^\infty n |a_n| r^{n-1}$ for some $0 < r \le R$. Show that f is one-to-one on $\{z \mid |z| < r\}$ unless f is constant. Compare your method with that used to solve example 3, Section 6.2.

5. If f is analytic on A, $0 \in A$, and $f'(0) \neq 0$, then prove that near 0 we can write $f(z^n) = f(0) + h(z)^n$ for some analytic function h that is one-to-one near 0.

Hint.   Use example 2.

6. Let $u: A \to \mathbb{R}$ be harmonic and nonconstant on a region A. Prove that u is an open mapping.

7. Use exercise 6 to prove the Maximum and Minimum Principles for Harmonic Functions (see Section 2.5).

8. Let f be entire and have the property that if $B \subset \mathbb{C}$ is any bounded set, then $f^{-1}(B)$ is bounded (or perhaps empty). Show that for any $w \in \mathbb{C}$, there exists $z \in \mathbb{C}$ such that $f(z) = w$.

Hint.   Show that $f(\mathbb{C})$ is both open and closed and deduce that $f(\mathbb{C}) = \mathbb{C}$.

Apply this result to polynomials to deduce yet another proof of the Fundamental Theorem of Algebra.

9. Show that the equation $z = e^{z-a}$, $a > 1$ has exactly one solution inside the unit circle.

10. Consider example 2 and take the case where $k = 3$. Visualize the local mapping in three steps as follows:

$$z \longmapsto t = (z - z_0) \sqrt[3]{\varphi(z)}$$
$$t \longmapsto s = t^3$$
$$s \longmapsto w = s + f(z_0)$$

Sketch this mapping.

Review Exercises for Chapter 6

1. Let $f: A \to B$ be analytic and onto; assume that $z_1, z_2 \in A$, $z_1 \neq z_2$, implies that $f(z_1) \neq f(z_2)$. Prove that f^{-1} is analytic.

2. Let f be analytic and bounded on $|z + i| > 1/2$ and real on $]-1,1[$. Show that f is constant.

Hint.   Use the Schwarz Reflection Principle from Section 6.1.

3. Let f be analytic and bounded on $A = \{z \mid |z| < 1\}$. Show that if f is one-to-one on $\{z \mid 0 < |z| < 1\}$, then f is one-to-one on A.

4. a. Prove *Vitali's Convergence Theorem*: Let f_n be analytic on a region A such that
  i. For each closed disk B in A there is a constant M_B such that $|f_n(z)| \leq M_B$ for all $z \in B$ and $n = 1, 2, 3, \cdots$.
  ii. There is a sequence of distinct points z_k of A converging to $z_0 \in A$ such that $\lim_{n \to \infty} f_n(z_k)$ exists for $k = 1, 2, \cdots$.

Then f_n converges uniformly on every closed disk in A; the limit is an analytic function.

Hint.   First take the case of a disk B with radius R; $z_k \longrightarrow z_0 =$ the center of B. Use the Schwarz Lemma to show that $|f_n(z) - f_n(z_0)| \leq 2M|z - z_0|/R$. Then show that

$$|f_n(z_0) - f_{n+p}(z_0)| \leq \frac{4M\,|z - z_0|}{R} + |f_n(z) - f_{n+p}(z)|$$

and deduce that $f_n(z_0)$ converges. Let

$$g_n(z) = \frac{f_n(z) - f_n(z_0)}{z - z_0}$$

and conclude that $g_n(z_0)$ converges. Show that in general, if

$$f_n(z) = \sum_{k=0}^{\infty} a_{n,k}(z - z_0)^k,$$

then $a_{n,k} \longrightarrow a_k$ as $n \longrightarrow \infty$. Deduce that $f_n(z)$ converges uniformly in $|z - z_0| < R - \epsilon$. Then use connectedness of A to deduce uniform convergence on any closed disk.

b. Show that if condition (i) is omitted, the conclusion is false. (Let $f_n(z) = z^n$.)

c. Show that if f_n satisfy condition (i), there is a subsequence of the f_n that converges uniformly on every closed disk. (This result is called *Montel's Theorem* and is used in the proof of the Riemann Mapping Theorem.)

5. Let f be a polynomial. Show that the integral of f'/f around every sufficiently large circle centered at the origin is $2\pi i$ times the degree of f.

6. Let f be analytic on a region A and let γ be a closed curve in A homotopic to a point. Show that

$$\mathrm{Re}\left(\int_{\gamma} \frac{f'}{f}\right) = 0.$$

7. Show that for $n > 2$, all the roots of $z^n - (z^2 + z + 1)/4 = 0$ lie inside the unit circle.

8. Show that $f(z) = (z^2 + 1)/(z^2 - 1)$ is one-to-one on $\{z \mid \mathrm{Im}\, z > 0\}$. Is it one-to-one on any larger set?

9. Suppose that f is analytic in $\mathbb{C}$ except for poles at $n \pm i$, $n = 0, \pm 1, \pm 2,$ $\cdots$. What is the length of the longest interval $]x_0 - R, x_0 + R[$ in $\mathbb{R}$ on which $f(x_0) + f'(x_0)(x - x_0) + f''(x_0)(x - x_0)^2/2 + \cdots + f^{(k)}(x_0)(x - x_0)^k/k!$ $+ \cdots$ converges?

10. Let f be analytic and bounded on $\{z \mid \mathrm{Im}\, z < 1\}$ and suppose that f is real on the real axis. Show that f is constant.

11. Let f be entire and suppose that for $z = x$ real, $f(x + 1) = f(x)$. Show that $f(z + 1) = f(z)$ for all $z \in \mathbb{C}$.

12. Let f be analytic on $\{z \mid |z| < 1\}$ and let $f(1/n) = 0$, $n = 1, 2, \cdots$. What can be said about f?

13. If f is analytic on the set $\{z \mid |z| < 1\}$ and $f(1 - 1/n) = 0$, $n = 1, 2, 3, \cdots$, does it follow that $f = 0$?

14. Let $f(z)$ be analytic on $\{z \mid 0 < |z| < 2\}$ and suppose that

$$\int_{|z| = 1} z^n f(z) dz = 0, \quad n = 0, 1, 2, \cdots .$$

Show that f has a removable singularity at $z = 0$.

15. Let $|f(z)| \le 1$ when $|z| = 1$ and let $f(0) = 1/2$ with f analytic. Prove that

$$|f(z)| \le \begin{cases} \dfrac{(3|z| + 1)}{2} & \text{for all } |z| \le \dfrac{1}{3}; \\ \\ 1 & \text{for } \dfrac{1}{3} \le |z| \le 1. \end{cases}$$

16. Let f and g be continuous for $|z| \le 1$ and analytic for $|z| < 1$. Suppose that $f = g$ on the unit circle. Prove that $f = g$.

17. Prove the following theorem (*Phragmen-Lindelof Theorem*):
   i. Suppose that f is analytic in a domain that includes the strip $G = \{z \in \mathbb{C} \mid 0 \le \operatorname{Re} z \le 1\}$. If $\lim\limits_{\substack{z \to \infty \\ z \in G}} f(z) = 0$ and if $|f(it)| \le 1$ and $|f(1 + it)| \le 1$ for all real t, then $|f(z)| \le 1$ for all $z \in G$.
   ii. If g is analytic in a domain containing G, if $\lim\limits_{\substack{z \to \infty \\ z \in G}} g(z) = 0$, and if $|g(it)| \le M$ and $|g(1 + it)| \le N$ for all real t, then $|g(z)| \le M^{1 - \operatorname{Re} z} N^{\operatorname{Re} z}$.

   Hint.   Apply the result of (i) to $f(z) = g(z)/M^{1 - z}N^z$.

18. If $f(z)$ is analytic for $|z| < 1$ and if $|f(z)| \le 1/(1 - |z|)$, show that the coefficients of the expansion $f(z) = \sum_{n=0}^{\infty} a_n z^n$ are subject to the inequality

$$|a_n| \le (n + 1)\left(1 + \frac{1}{n}\right)^n < e(n + 1).$$

19. Which of the following statements is/are true?
   (1) The radius of convergence of $\sum_{n=0}^{\infty} 2^n z^{2n}$ is $1/\sqrt{2}$.
   (2) An entire function that is constant in the unit circle is a constant.
   (3) The residue of $1/[z^{10}(z - 2)]$ at the origin is $-(2)^{-10}$.
   (4) If f_n is a sequence of entire functions converging to a function f and if the convergence is uniform on the unit circle, then f is analytic in the open unit disk.
   (5) $\displaystyle\int_0^{\pi} \frac{d\theta}{a + \cos \theta} = \frac{2\pi}{a^2 - 1}$.
   (6) For sufficiently large r, $\sin z$ maps the exterior of the disk of radius r ($\{z \mid |z| > r\}$) into any preassigned neighborhood of ∞.

(7) Let $f: \mathbb{C} \rightarrow \mathbb{C}$ be analytic in the open unit disk and let f have a non-removable singularity at i. Then the radius of convergence of the Taylor series of f at 0 is 1.

(8) Let $f: \mathbb{C} \rightarrow \mathbb{C}$ be analytic and nonconstant and let D be a domain in $\mathbb{C}$. Then f maps the boundary of D into the boundary of $f(D)$.

(9) Let f be analytic on $\{z \mid 0 < |z| < 1\}$ and suppose that $|f(z)| \leq \log (1/|z|)$. Then f has a removable singularity at 0.

(10) Suppose that $f: \mathbb{C} \rightarrow \mathbb{C}$ is entire and that f has exactly k zeros in the open unit disk but none on the unit circle. Then there exists an $\epsilon > 0$ such that any entire function g that satisfies $|f(z) - g(z)| < \epsilon$ for $|z| = 1$ must also have exactly k zeros in the open unit disk.

20. Is it correct to say that $1/\sqrt{z}$ has a pole at $z = 0$?

21. Prove that for the principal value of the logarithm, $|\log z| \leq r/(1 - r)$ if $|1 - z| \leq r < 1$.

22. a. Let $f: \mathbb{C} \rightarrow \mathbb{C}$ be continuous on $\mathbb{C}$ and analytic on $\mathbb{C} \backslash \mathbb{R}$. Is f actually entire?

b. Let $f = \mathbb{C} \rightarrow \mathbb{C}$ be analytic on $\mathbb{C} \backslash \mathbb{R}$. Is f entire?

23. Let $P(z)$ be a polynomial. Prove that

$$\int_{|z|=1} P(z)d\bar{z} = -2\pi i P'(0).$$

24. Find the radius of convergence of the series $\Sigma_0^\infty 2^n z^n$.

Chapter 7

Asymptotic Methods

In this chapter the student will be given an introduction to the theory of asymptotic methods that will enable him to study functions $f(z)$ as $z \longrightarrow \infty$. The chapter begins with a study of infinite products and the gamma function. These topics are of interest in their own right but they also provide the student with a motivation to study asymptotic expansions, which are analyzed in Section 7.2. One of the main techniques used in this analysis, the method of steepest descent, and its variant, the method of stationary phase, are also studied and are applied to Stirling's Formula and to Bessel functions in Section 7.3.

7.1 Infinite Products and the Gamma Function

To be able to study the gamma function and subsequent topics, we shall first need to develop some of the basic properties of infinite products. They are somewhat analogous to the infinite sums considered in Section 3.1. For orientation and motivation, the student should note that any polynomial $p(z)$ can be written in the form

$$p(z) = a_n(z - \alpha_1) \ \cdots \ (z - \alpha_n) = a_n \prod_{j=1}^{n} (z - \alpha_j)$$

where $\alpha_1, \ \cdots \ , \alpha_n$ are the roots of $p(z) = a_n z^n + \ \cdots \ + a_1 z + a_0$, and Π stands for "take the product of" in the same way as Σ stands for "take the sum of." It is natural to attempt to generalize this expression to entire functions and in doing so we encounter the concept of an infinite product.

Infinite products

Let $z_1, z_2, \cdots$ be a given sequence of complex numbers. We want to consider

$$\prod_{n=1}^{\infty} (1 + z_n) = (1 + z_1)(1 + z_2) \cdot \cdot \cdot \cdot$$

We write $1 + z_n$ because if the product is to converge, the general term should approach 1; that is, $z_n \rightarrow 0$. Some technicalities are involved when $z_n = -1$. We want to allow the product to be zero yet be able to impose some convergence condition. The following definition fits our needs.

Definition 1.   *The product $\prod_{n=1}^{\infty} (1 + z_n)$ is said to converge iff only a finite number of z_n equal -1 and if $\prod_{k=m}^{n} (1+z_k) = (1+z_m) \cdots (1+z_n)$ (where $z_k \neq -1$ for $k \geq m$) converges as $n \rightarrow \infty$, to a nonzero number. We set*

$$\prod_{n=1}^{\infty} (1 + z_n) = \underset{n \rightarrow \infty}{\text{limit}} \prod_{k=1}^{n} (1 + z_k).$$

(This product will be zero if some $z_k = -1$ and nonzero otherwise.)

For example, consider

$$\prod_{n=2}^{\infty} \left(1 - \frac{1}{n}\right) = \frac{1}{2} \cdot \frac{2}{3} \cdot \frac{3}{4} \cdot \cdot \cdot \cdot$$

The nth partial product is

$$\frac{1}{2} \cdot \frac{2}{3} \cdot \quad \cdots \quad \cdot \frac{n-1}{n} = \frac{1}{n} \rightarrow 0.$$

Thus the product does not converge because we demanded convergence to a nonzero number. (We say the product *diverges to zero*.) If we started at $n = 1$, the product would still diverge. The reason for this rather strange condition is a technical one that is explained by the following theorem.

By starting a given product beyond the point where some $z_n = -1$, we can assume that $z_n \neq -1$ for all n. Such an assumption imposes no real restrictions when testing for convergence in Theorem 1.

Theorem 1.

i. *If $\prod_{n=1}^{\infty} (1 + z_n)$ converges, then $z_n \rightarrow 0$.*

ii. *Suppose that $|z_n| < 1$ for all $n = 1, 2, \cdots$ so that $z_n \neq -1$. Then $\prod_{n=1}^{\infty} (1 + z_n)$ converges if and only if $\sum_{n=1}^{\infty} \log (1 + z_n)$ converges.*

(log *is the principal branch;* $|z_n| < 1$ *implies that* $\log(1 + z_n)$ *is defined.)*

iii. $\Pi_{n=1}^{\infty} (1 + |z_n|)$ *converges iff* $\Sigma_{n=1}^{\infty} |z_n|$ *converges. (We say that* $\Pi_{n=1}^{\infty}$ $(1 + z_n)$ *converges absolutely in this case.)*

iv. *If* $\Pi_{n=1}^{\infty} (1 + |z_n|)$ *converges, then* $\Pi_{n=1}^{\infty} (1 + z_n)$ *converges.*

This theorem summarizes the main convergence properties of infinite products. Criteria (iii) and (iv) are particularly important and are easy to apply.

Because it is technical, the proof of Theorem 1 appears at the end of this section. Only the plausibility of the theorem will be discussed here. Criterion (i) was explained at the beginning of this section. To explain (ii), note that if we let $S_n = \Sigma_1^n \log(1 + z_k)$ and $P_n = \Pi_1^n (1 + z_k)$, then

$$P_n = e^{S_n}.$$

That (ii) is plausible easily follows from this equation. Indeed, if $S_n \to S$ it is clear that $P_n \to e^S$.

Once (ii) is shown, (iii) and (iv) follow easily.

The following corollary requires no proof since it is implicit in the preceding discussion.

Corollary 1. *If* $|z_n| < 1$ *and* $\Sigma \log(1 + z_n)$ *converges to* S, *then* $\Pi(1 + z_n)$ *converges to* e^S.

This corollary is sometimes useful but when it is applied to concrete problems, the sum of logarithms is often difficult to handle.

Let $f_n(z)$ be functions defined on a set $B \subset \mathbb{C}$. The way to define the concept of the uniform convergence of $\Pi_1^{\infty} (1 + f_n)$ should be fairly clear.

Definition 2. *The sequence*

$$\prod_{n=1}^{\infty} (1 + f_n(z))$$

converges uniformly on B *iff, for some* m, $f_n(z) \neq -1$ *for* $n \geq m$ *and all* $z \in B$, *the sequence* $P_n(z) = \Pi_{k=m}^{n} (1 + f_k(z))$ *converges uniformly on* B *to some* $P(z)$, *and* $P(z) \neq 0$ *for all* $z \in B$ (see Section 3.1 for the definition of uniform convergence of a sequence of functions).

Theorem 2 follows from Theorem 6, Section 3.1.

Theorem 2. *Suppose that* $f_n(z)$ *are analytic functions on a region* A *and that* $\Pi_{n=1}^{\infty} (1 + f_n(z))$ *converges uniformly to* $f(z)$ *on every closed disk in* A. *Then* $f(z)$ *is analytic on* A. *Such uniform convergence holds if*

$|f_n(z)| < 1$ for $n \geq m$ and if either $\Sigma_{n=m}^{\infty} \log(1 + f_n(z))$ converges uniformly or $\Sigma_{n=1}^{\infty} |f_n(z)|$ converges uniformly (on closed disks in both cases).

To check the validity of last statement the student must check that the proof of Theorem 1 can be applied to uniform convergence; this is left as an exercise.

Canonical products

The following useful theorem is a special case of a theorem of Weierstrass that constructs the most general entire function with a given set of zeros. The special case described here is applicable to most problems, yet it illustrates the main ideas of the general case. (For a statement of the general case, see exercises 7 and 8 at the end of this section.)

Theorem 3. *Let a_1, a_2, $\cdots$, be a given sequence (possibly finite) of nonzero complex numbers such that*

$$\sum_{n=1}^{\infty} \frac{1}{|a_n|^2} < \infty.$$

Then if $g(z)$ is any entire function, the function

$$f(z) = e^{g(z)} z^k \left(\prod_{n=1}^{\infty} \left(1 - \frac{z}{a_n} \right) e^{z/a_n} \right) \tag{1}$$

is entire. Actually, the product converges uniformly on closed disks, has zeros at a_1, a_2, $\cdots$, has a zero of order k at $z = 0$, but has no other zeros. Furthermore, if f is any entire function having these properties, it can be written in the same form (equation (1)). In particular, f is entire with no zeros if and only if f has the form $f(z) = e^{g(z)}$ for some entire function g.
The product $\Pi_{n=1}^{\infty} (1 - z/a_n)e^{z/a_n}$ is called a canonical product.

The technical proof of Theorem 3 appears at the end of this section. The result is quite plausible if we note that the product vanishes exactly when z is equal to some a_n and that z^k has a zero of order k at 0. Also, $e^{g(z)}$ vanishes nowhere since $e^w \neq 0$ for all $w \in \mathbb{C}$.

Note. The points a_1, a_2, $\cdots$, need not be distinct; each may be repeated finitely many times. If a_n is repeated l times, f will have a zero of order l at a_n.

Theorem 3 will be applied to a considerable extent in the remainder of this section. An important application of the theorem is found in example 2, where it is proved that

$$\sin \pi z = \pi z \prod_{\substack{n=-\infty \\ n \neq 0}}^{\infty} \left(1 - \frac{z}{n}\right) e^{z/n}. \tag{2}$$

Our first application of the theorem will be in our study of the main properties of the gamma function.

The gamma function

The importance of this function was realized by Euler and Gauss as early as the eighteenth century. Two equivalent definitions of the gamma function are given here; the first will be in terms of infinite products, the second will be in terms of an integral formula. These two formulas are credited to Euler. Significant contributions were also made by Gauss and Legendre.

The main facts that are included in the following discussion and in the end-of-section exercises are summarized in Table 7.1 (p. 355).

For the first definition, let us begin with an associated function that is defined by the canonical product

$$G(z) = \prod_{n=1}^{\infty} \left(1 + \frac{z}{n}\right) e^{-z/n}. \tag{3}$$

By Theorem 3, this function is entire with simple zeros at the negative integers $-1, -2, -3, \cdots$. This function satisfies the identity

$$z G(z) G(-z) = \frac{\sin \pi z}{\pi} \tag{4}$$

because of formula (2).

Now we consider the function

$$H(z) = G(z - 1). \tag{5}$$

This function has zeros at $0, -1, -2, \cdots$. Thus by Theorem 3, we can write

$$H(z) = e^{g(z)} z \prod_{1}^{\infty} \left(1 + \frac{z}{n}\right) e^{-z/n} = z e^{g(z)} G(z) \tag{6}$$

for some entire function $g(z)$. It will now be shown that $g(z)$ is constant. Using Theorem 1, we get

$$\log H(z) = \log z + g(z) + \sum_{n=1}^{\infty} \left(\log \left(1 + \frac{z}{n}\right) - \frac{z}{n}\right).$$

Since the convergence is uniform on closed disks, we may differentiate term by term:

$$\frac{d}{dz} \log H(z) = \frac{1}{z} + g'(z) + \sum_{n=1}^{\infty} \left(\frac{1}{z+n} - \frac{1}{n} \right). \qquad (7)$$

Similarly, by equation (3),

$$\frac{d}{dz} \log G(z-1) = \sum_{n=1}^{\infty} \left(\frac{1}{z-1+n} - \frac{1}{n} \right)$$

$$= \frac{1}{z} - 1 + \sum_{n=1}^{\infty} \left(\frac{1}{z+n} - \frac{1}{n+1} \right)$$

$$= \frac{1}{z} - 1 + \sum_{n=1}^{\infty} \left(\frac{1}{z+n} - \frac{1}{n} \right) + \sum_{n=1}^{\infty} \left(\frac{1}{n} - \frac{1}{n+1} \right)$$

$$= \frac{1}{z} + \sum_{n=1}^{\infty} \left(\frac{1}{z+n} - \frac{1}{n} \right). \qquad (8)$$

Comparing equations (7) and (8) and then using equation (5), we see that $g'(z) = 0$, so $g(z)$ is constant.

Note. Theorem 1 (ii) is actually valid only for $|z| < 1$, but this region of validity suffices since two entire functions that agree on $|z| < 1$ are equal by Taylor's Theorem.

The constant value $g(z) = \gamma$ is called *Euler's Constant.* We can determine an expression for it as follows. We have, by formulas (3), (5), and (6),

$$G(z-1) = z \, e^{\gamma} G(z); \qquad (9)$$

so if we let $z - 1$, then $G(0) - 1 = e^{\gamma} G(1)$. Thus, by equation (3),

$$e^{-\gamma} = \prod_{1}^{\infty} \left(1 + \frac{1}{n} \right) e^{-1/n}$$

$$= \prod_{1}^{\infty} \left(\frac{n+1}{n} \right) e^{-1/n}.$$

Noting that

$$\prod_{k=1}^{n} \left(\frac{k+1}{k} \right) e^{-1/k} = \frac{2}{1} \cdot \frac{3}{2} \cdot \frac{4}{3} \cdot \quad \cdots \quad \cdot \frac{n+1}{n} e^{-1-1/2-1/3-\cdots-1/n}$$

$$= (n+1) \, e^{-1-1/2-\cdots-1/n}$$

$$= ne^{-1-1/2-\cdots-1/n} + e^{-1-1/2-\cdots-1/n},$$

we get $e^{-\gamma} = \lim_{n \to \infty} ne^{-1-1/2-\cdots-1/n}$. Taking logs, we find that

$$\gamma = \lim_{n \to \infty} \left\{ 1 + \frac{1}{2} + \cdots + \frac{1}{n} - \log n \right\}. \qquad (10)$$

We do not need to prove that the limit in equation (10) exists and is finite because this follows from what we have done. Numerically, we can compute from the limit in (10) that $\gamma \approx .57716 \cdots$.

Now we are ready to define the gamma function. We set

$$\Gamma(z) = [z \, e^{\gamma z} G(z)]^{-1} = \left[z \, e^{\gamma z} \prod_{n=1}^{\infty} \left(1 + \frac{z}{n} \right) e^{-z/n} \right]^{-1}. \tag{11}$$

From the entireness of G, we can conclude that $\Gamma(z)$ is meromorphic with simple poles at $0, -1, -2, \cdots$. Since by equation (9), $G(z-1) = ze^{\gamma}G(z)$, we find that

$$\Gamma(z+1) = z \, \Gamma(z); \quad z \neq 0, -1, -2, \cdots, \tag{12}$$

which is called the *functional equation for the gamma function* (see exercise 10).

Also, $\Gamma(1) = 1$ since $\Gamma(z) = [ze^{\gamma z}G(z)]^{-1}$ and $G(1) = e^{-\gamma}$ by our construction of γ. Thus from formula (12) we see that $\Gamma(2) = 1 \cdot 1$, $\Gamma(3) = 2 \cdot 1$, $\Gamma(4) = 3 \cdot 2 \cdot 1$, and generally that

$$\Gamma(n+1) = n!. \tag{13}$$

This property is an important one and is one of the main reasons for studying the gamma function. In addition, it enables us to obtain manageable formulas for approximating $n!$, which are derived in Section 7.3. (Figure 7.1 is a graph of $\Gamma(x)$ for x real.)

From the equation $zG(z)G(-z) = (\sin \pi z)/\pi$ (see equation (4)), we get the relation

$$\Gamma(z) \, \Gamma(1-z) = \frac{\pi}{\sin \pi z}. \tag{14}$$

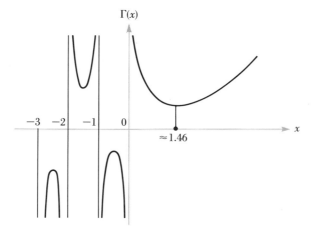

Figure 7.1   Graph of $\Gamma(x)$ for x real.

From this it follows that $\Gamma(z) \neq 0$ for all z. (Of course, $z \neq 0, -1, -2, \cdots$.) We know this to be true because if $\Gamma(z) = 0$, we would have $\pi = \Gamma(z) \, \Gamma(1-z) \sin \pi z = 0$ as long as $z \neq 0, \pm 1, \pm 2, \cdots$. (These are the points at which $\sin \pi z$ vanishes, so cross multiplication is invalid at those points.) But we also know that $\Gamma(z) \neq 0$ if $z = 1, 2, 3, \cdots$, since $\Gamma(n+1) = n!$, $n = 0, 1, 2, \cdots$. Thus we have proved that $\Gamma(z) \neq 0$ if $z \neq 0, -1, -2, \cdots$.

If we let $z = 1/2$ in formula (14), we get $[\Gamma(1/2)]^2 = \pi$. But $\Gamma(1/2) > 0$. To see this, note that Γ is real for real positive z; also, we have shown that Γ has no zeros and that $\Gamma(n+1) = n! > 0$. Therefore, since $\Gamma(x)$ is continuous for $x \in \,]0,\infty[$ (because Γ is analytic), it follows that $\Gamma(x) > 0$ for all $x \in \,]0,\infty[$ (see Figure 7.1). Thus $\Gamma(1/2) = \sqrt{\pi}$ (rather than the other possibility, $-\sqrt{\pi}$).

Euler's Formula for the gamma function is:

$$\Gamma(z) = \frac{1}{z} \prod_{n=1}^{\infty} \left\{ \left(1 + \frac{1}{n}\right)^z \left(1 + \frac{z}{n}\right)^{-1} \right\} = \lim_{n \to \infty} \frac{n! n^z}{z(z+1) \, \cdots \, (z+n)}. \tag{15}$$

This formula is proven as follows. By definition,

$$\frac{1}{\Gamma(z)} = z \left[\lim_{n \to \infty} e^{(1+1/2+\cdots+1/n - \log n)z} \right] \left[\lim_{n \to \infty} \prod_{k=1}^{n} \left(1 + \frac{z}{k}\right) e^{-z/k} \right]$$

$$= z \left[\lim_{n \to \infty} e^{(1+1/2+\cdots+1/n - \log n)z} \prod_{k=1}^{n} \left(1 + \frac{z}{k}\right) e^{-z/k} \right]$$

$$= z \lim_{n \to \infty} \left[n^{-z} \prod_{k=1}^{n} \left(1 + \frac{z}{k}\right) \right],$$

since $n^{-z} = e^{-\log n \cdot z}$. Thus we get

$$\frac{1}{\Gamma(z)} = z \lim_{n \to \infty} \left[\prod_{k=1}^{n-1} \left(1 + \frac{1}{k}\right)^{-z} \prod_{k=1}^{n} \left(1 + \frac{z}{k}\right) \right]$$

$$= z \lim_{n \to \infty} \left(1 + \frac{1}{n}\right)^z \left[\prod_{k=1}^{n} \left(1 + \frac{z}{k}\right) \left(1 + \frac{1}{k}\right)^{-z} \right].$$

The first equality in formula (15) now follows. The student is asked to prove the second in exercise 12.

Another important property of the gamma function is described by the *Gauss Formula*: For any fixed positive integer $n \geq 2$,

$$\Gamma(z) \, \Gamma\left(z + \frac{1}{n}\right) \, \cdots \, \Gamma\left(z + \frac{n-1}{n}\right) = (2\pi)^{(n-1)/2} n^{(1/2) - nz} \, \Gamma(nz). \tag{16}$$

To prove this formula we first note that we can write Euler's Formula (equation (15)) as

$$\Gamma(z) = \lim_{m \to \infty} \frac{m!\, m^z}{z(z+1) \, \cdots \, (z+m)}$$

$$= \lim_{m \to \infty} \frac{(m-1)!\, m^z}{z(z+1) \, \cdots \, (z+m-1)}$$

$$= \lim_{m \to \infty} \frac{(mn-1)!\, (mn)^z}{z(z+1) \, \cdots \, (z+mn-1)}.$$

We define $f(z)$ as follows:

$$f(z) = \frac{n^{nz}\, \Gamma(z)\, \Gamma\!\left(z+\frac{1}{n}\right) \cdots \Gamma\!\left(z+\frac{n-1}{n}\right)}{n\, \Gamma(nz)}$$

$$= \frac{n^{nz-1} \displaystyle\prod_{k=0}^{n-1} \lim_{m \to \infty} \dfrac{(m-1)!\,m}{\left(z+\dfrac{k}{n}\right)\!\left(z+\dfrac{k}{n}+1\right) \cdots \left(z+\dfrac{k}{n}+m-1\right)}}{\displaystyle\lim_{m \to \infty} \dfrac{(mn-1)!\,(mn)^{nz}}{nz(nz+1) \cdots (nz+nm-1)}}$$

$$= \lim_{m \to \infty} \frac{[(m-1)!]^n m^{(n-1)/2} n^{mn-1}(nz)(nz+1) \cdots (nz+mn-1)}{(mn-1)! \displaystyle\prod_{k=0}^{n-1} [(nz+k)(nz+k+n) \cdots (nz+k+mn-n)]}$$

$$= \lim_{m \to \infty} \frac{\{(m-1)!\}^n m^{(n-1)/2} n^{nm-1}}{(nm-1)!}.$$

Thus f is constant. Setting $z = 1/n$ gives

$$f(z) = \Gamma\!\left(\frac{1}{n}\right) \Gamma\!\left(\frac{2}{n}\right) \cdots \Gamma\!\left(\frac{n-1}{n}\right) > 0,$$

so

$$f(z)^2 = \frac{\pi^{n-1}}{\sin\!\left(\dfrac{\pi}{n}\right) \sin\!\left(\dfrac{2\pi}{n}\right) \cdots \sin\!\left(\dfrac{n-1}{n}\pi\right)}$$

using equation (14). From the fact that

$$\sin\!\left(\frac{\pi}{n}\right) \sin\!\left(\frac{2\pi}{n}\right) \cdots \sin\!\left(\frac{(n-1)\pi}{n}\right) = \frac{n}{2^{n-1}}; \quad n = 2, 3, \cdots$$

(see exercise 24, Section 1.2), we get

$$f(z)^2 = \frac{(2\pi)^{n-1}}{n}.$$

Since $f(z) > 0$,

$$f(z) = \frac{(2\pi)^{(n-1)/2}}{\sqrt{n}}.$$

The Gauss Formula therefore follows.

If we take the special case of equation (16), in which $n = 2$, we obtain the *Legendre Duplication Formula*:

$$2^{2z-1} \, \Gamma(z) \, \Gamma\left(z + \frac{1}{2}\right) = \sqrt{\pi} \, \Gamma(2z). \tag{17}$$

Let us next show that the residue of $\Gamma(z)$ at $z = -m, m = 0, 1, 2, \cdots$, is $(-1)^m/m!$. Indeed,

$$(z + m) \, \Gamma(z) = (z + m) \, \frac{\Gamma(z + 1)}{z} = (z + m) \, \frac{\Gamma(z + 2)}{z(z + 1)}.$$

More generally, we find that

$$(z + m) \, \Gamma(z) = \frac{\Gamma(z + m + 1)}{z(z + 1) \, \cdots \, (z + m - 1)}.$$

Letting $z \to -m$, we get

$$\frac{\Gamma(1)}{-m(-m + 1) \, \cdots \, (-1)} = \frac{(-1)^m}{m!}$$

as required.

There is an important expression for $\Gamma(z)$ as an integral. For Re $z > 0$, we shall establish the following formula:

$$\Gamma(z) = \int_0^\infty t^{z-1} \, e^{-t} \, dt. \tag{18}$$

The student might suspect that this expression can be evaluated by the methods of Section 4.3 (see Theorem 12 of that section). Unfortunately, the hypotheses of that theorem do not hold in this case, so another method is needed. Let us start by showing that

$$F_n(z) = \frac{n! \, n^z}{z(z + 1) \, \cdots \, (z + n)} \tag{19}$$

where F_n is defined by

$$F_n(z) = \int_0^n \left(1 - \frac{t}{n}\right)^n t^{z-1} \, dt.$$

By Euler's Formula (equation (15)), we will then have $F_n(z) \to \Gamma(z)$ as $n \to \infty$.

To prove formula (19), we note that by changing variables and letting $t = ns$,

$$F_n(z) = n^z \int_0^1 (1-s)^n s^{z-1} ds.$$

Now we integrate this expression successively by parts, the first step being

$$F_n(z) = n^z \left[\frac{1}{z} s^z (1-s)^n \Big|_0^1 + \frac{n}{z} \int_0^1 (1-s)^{n-1} s^z \, ds \right]$$

$$= n^z \frac{n}{z} \int_0^1 (1-s)^{n-1} s^z \, ds.$$

Repeating this procedure, we integrate by parts n times and get

$$F_n(z) = n^z \frac{n \cdot (n-1) \, \cdots \, 1}{z(z+1) \, \cdots \, (z+n-1)} \int_0^1 s^{z+n-1} \, ds$$

$$= \frac{n! \, n^z}{z(z+1) \, \cdots \, (z+n)},$$

which establishes equation (19).

From exercise 15 we obtain the formula that is well known in calculus:

$$\left(1 - \frac{t}{n} \right)^n \longrightarrow e^{-t} \quad \text{as} \quad n \longrightarrow \infty. \tag{20}$$

If we let $n \longrightarrow \infty$ in equation (19), the validity of equation (18) seems assured. However, such a conclusion is not easily justified. We proceed as follows.

From equations (19) and (15) we know that

$$\Gamma(z) = \lim_{n \to \infty} \int_0^n \left(1 - \frac{t}{n} \right)^n t^{z-1} \, dt. \tag{21}$$

Let $f(z) = \int_0^\infty e^{-t} t^{z-1} dt$. This integral converges since $|e^{-t} t^{z-1}| \leq e^{-t} t^{\operatorname{Re} z - 1}$ and $\operatorname{Re} z > 0$ (compare with $\int_1^\infty e^{-t} t^p dt$ and $\int_0^1 t^p dt$, $p > -1$).

We shall need to know "how fast" $[1 - (t/n)]^n \longrightarrow e^{-t}$. The following inequalities actually hold:

$$0 \leq e^{-t} - \left(1 - \frac{t}{n} \right)^n \leq \frac{t^2 e^{-t}}{n} \quad \text{for } 0 \leq t \leq n. \tag{22}$$

(These inequalities are a lemma from calculus whose proof is outlined in exercise 15.)

From equation (21) and the definition of f we have:

$$f(z) - \Gamma(z) = \lim_{n \to \infty} \left[\int_0^n \left\{ e^{-t} - \left(1 - \frac{t}{n}\right)^n \right\} t^{z-1} \, dt + \int_n^\infty e^{-t} t^{z-1} \, dt \right]. \qquad (23)$$

To show that the limit (equation (23)) is zero, note that $\int_n^\infty e^{-t} t^{z-1} \, dt \to 0$ as $n \to \infty$. Indeed if $t > 1$, then $|e^{-t} t^{z-1}| \le e^{-t} t^m$ where m is an integer $m \ge \operatorname{Re} z > 0$. But from calculus (or directly using integration by parts), we know that $\int_0^\infty e^{-t} t^m \, dt < \infty$, so $\int_n^\infty e^{-t} t^m \, dt \to 0$ as $n \to \infty$. It remains to be shown that

$$\int_0^n \left\{ e^{-t} - \left(1 - \frac{t}{n}\right)^n \right\} t^{z-1} dt \to 0 \text{ as } n \to \infty.$$

By inequality (22),

$$\left| \int_0^n \left\{ e^{-t} - \left(1 - \frac{t}{n}\right)^n \right\} t^{z-1} dt \right| \le \int_0^n \frac{e^{-t} t^{\operatorname{Re} z + 1}}{n} \, dt \le \frac{1}{n} \int_0^\infty e^{-t} t^{\operatorname{Re} z + 1} \, dt,$$

which approaches zero as $n \to \infty$ because the integral converges. This completes the proof of (18); that is, for $\operatorname{Re} z > 0$,

$$\Gamma(z) = \int_0^\infty e^{-t} t^{z-1} \, dt.$$

In fact, if we examine that proof, we see that for $0 < \epsilon < R, \epsilon \le |z| \le R$, and $(-\pi/2) + \delta \le \arg z \le (\pi/2) - \delta, \delta > 0$, the convergence is uniform in z (see exercise 19).

Technical proofs of Theorems 1 and 3

Theorem 1.

i. If $\prod_{n=1}^\infty (1 + z_n)$ converges, then $z_n \to 0$.

ii. Suppose that $|z_n| < 1$ for all $n = 1, 2, \cdots$ so that $z_n \ne -1$. Then $\prod_{n=1}^\infty (1 + z_n)$ converges if and only if $\sum_{n=1}^\infty \log(1 + z_n)$ converges. (log is the principal branch; $|z_n| < 1$ implies that $\log(1 + z_n)$ is defined.)

iii. $\prod_{n=1}^\infty (1 + |z_n|)$ converges iff $\sum_{n=1}^\infty |z_n|$ converges. (We say that $\prod_{n=1}^\infty (1 + z_n)$ converges absolutely in this case.)

iv. If $\prod_{n=1}^\infty (1 + |z_n|)$ converges, then $\prod_{n=1}^\infty (1 + z_n)$ converges.

Proof.

i. We can assume that $z_n \ne -1$ for all n. Let $P_n = \prod_{k=1}^n (1 + z_k)$; therefore, by assumption, $P_n \to P$ for some $P \ne 0$. Thus $P_n/P_{n-1} \to 1$ by the Quotient Theorem for Limits (see Section 1.4). But $P_n/P_{n-1} = 1 + z_n$. Thus $z_n \to 0$.

ii. Let $S_n = \Sigma_{k=1}^{n} \log(1 + z_k)$ and let $P_n = \Pi_{k=1}^{n}(1 + z_k)$. Then $P_n = e^{S_n}$. It is clear that if S_n converges, then P_n also converges because e^z is continuous.

Conversely, suppose that $P_n \longrightarrow P \neq 0$. To show that S_n converges, it suffices to show that for n sufficiently large, all S_n lie in a period strip (on which e^z has a continuous inverse).

We cannot write $\log \Sigma_{k=1}^{n} (1 + z_k) = \log P_n$, because P_n could be on the negative real axis. Instead, for purposes of this proof, let us choose the branch of log such that P lies in the domain A of the log chosen. Now $P_n \longrightarrow P$, so $P_n \in A$ if n is large and we can write $S_n = \log P_n + k_n \cdot 2\pi i$ for an integer k_n. Thus

$$(k_{n+1} - k_n) \cdot 2\pi i = \log(1 + z_{n+1}) - \log(P_{n+1} - \log P_n).$$

Since the left side of the equation is purely imaginary,

$$(k_{n+1} - k_n) \cdot 2\pi i = i\{\arg(1 + z_{n+1}) - \arg P_{n+1} + \arg P_n\}.$$

By (i), $z_{n+1} \longrightarrow 0$, so $\arg(1 + z_{n+1}) \longrightarrow 0$. Also, $\arg P_n \longrightarrow \arg P$, and so $k_{n+1} - k_n \longrightarrow 0$ as $n \longrightarrow \infty$. Since k_n are integers, they must equal a fixed integer k for n large. Thus $S_n = \log P_n + k \cdot 2\pi i$. Therefore, as $n \longrightarrow \infty$, $S_n \longrightarrow S = \log P + k \cdot 2\pi i$.

iii. By (ii) it suffices to show that for $x_n \geq 0$, Σx_n converges iff $\Sigma \log(1 + x_n)$ converges. Indeed, since

$$\log(1 + z) = z - \frac{z^2}{2} + \frac{z^3}{3} - \cdots , \; |z| < 1,$$

we see that

$$\frac{\log(1 + z)}{z} = 1 - \frac{z}{2} + \frac{z^2}{3} - \cdots$$

has a removable singularity at $z = 0$ and that

$$\lim_{z \to \infty} \frac{\log(1 + z)}{z} = 1.$$

Suppose that Σx_n converges. Then $x_n \longrightarrow 0$. Thus, given $\epsilon > 0$, $0 \leq \log(1 + x_n) \leq (1 + \epsilon)x_n$ for sufficiently large n. By the comparison test, $\Sigma \log(1 + x_n)$ converges. If we use $(1 - \epsilon)x_n \leq \log(1 + x_n)$, we obtain the converse.

iv. Suppose that $\Pi (1 + |z_n|)$ converges. Then by (ii), $\Sigma \log(1 + |z_n|)$ converges. (We must begin with terms such that the conditions in (ii) hold.) In fact, the argument in (iii) shows that $\Sigma \log(1 + z_n)$ converges absolutely and hence converges. Thus by (ii), $\Pi (1 + z_n)$ converges. ∎

Theorem 3. *Let $a_1, a_2, \cdots$, be a given sequence (possibly finite) of nonzero complex numbers such that*

$$\sum_{n=1}^{\infty} \frac{1}{|a_n|^2} < \infty.$$

Then if g(z) is any entire function, the function

$$f(z) = e^{g(z)} z^k \left(\prod_{n=1}^{\infty} \left(1 - \frac{z}{a_n}\right) e^{z/a_n} \right) \tag{1}$$

is entire. Actually, the product converges uniformly on closed disks, has zeros at $a_1, a_2, \cdots$, has a zero of order k at $z = 0$, but has no other zeros. Furthermore, if f is any entire function having these properties, it can be written in the same form (equation (1)). In particular, f is entire with no zeros if and only if f has the form $f(z) = e^{g(z)}$ for some entire function g.

Proof.   First we show that $\Pi\ (1 - z/a_n)e^{z/a_n}$ is entire. For each $R > 0$, let $A_R = \{z \mid |z| \le R\}$. Since $a_n \to \infty$, only a finite number of a_n lie in A_R, say, $a_1, \cdots, a_{n-1}$. Therefore, for $z \in A_R$, only a finite number of terms $(1 - z/a_n)$ vanish.

To effect the proof of Theorem 3, we need to prove the following lemma:

Lemma 1.   *If $1 + w = (1 - a)e^a$ and $|a| < 1$, then*

$$|w| \le \frac{|a|^2}{1 - |a|}.$$

Proof.   We have

$$(1 - a)e^a = 1 - \frac{a^2}{2} - \cdots - \left(1 - \frac{1}{n}\right) \frac{a^n}{(n-1)!} - \cdots$$

Thus

$$|(1 - a)e^a - 1| = |w|$$
$$\le \frac{|a|^2}{2} + \cdots + \frac{(n-1)}{n!}|a|^n + \cdots$$
$$\le |a|^2 + |a|^3 + \cdots$$
$$= \frac{|a|^2}{1 - |a|}, \text{ since } |a| < 1. \blacksquare$$

The next step in the proof of Theorem 3 is to show that the series

$$\sum_{n=1}^{\infty} w_n(z) = \sum_{n=1}^{\infty} \left\{ \left(1 - \frac{z}{a_n}\right) e^{z/a_n} - 1 \right\}$$

converges uniformly and absolutely on $A_{R/2}$. This will show that

$$\prod_{n=1}^{\infty} \left(1 - \frac{z}{a_n}\right) e^{z/a_n}$$

is entire (by Theorem 2).

Indeed, from the preceding lemma, for $n \geq N$, $|z/a_n| < 1$ if $|z| \leq R/2$, so

$$|w_n(z)| \leq \frac{\left|\dfrac{z}{a_n}\right|^2}{1 - \left|\dfrac{z}{a_n}\right|} \leq \frac{\left(\dfrac{R}{2}\right)^2}{1 - \dfrac{1}{2}} \cdot \frac{1}{|a_n|^2},$$

since $|z| \leq R/2$ and $|a_n| \geq R$ for $n \geq N$. Thus

$$|w_n(z)| \leq \frac{R^2}{2} \cdot \frac{1}{|a_n|^2} = M_n.$$

By assumption, $\Sigma\, M_n$ converges, so by the Weierstrass M Test, $\Sigma\, w_n(z)$ converges uniformly and absolutely.

Thus $f_1(z) = z^k\, \Pi_{n=1}^{\infty}\, [1 - (z/a_n)]e^{z/a_n}$ is entire. From the definition of the product it is clear that f_1 has exactly the required number of zeros. Thus so does $e^g f_1$. If f has the given number of zeros, then f/f_1 will be entire and have no zeros (by Theorem 1 of Section 4.1). Therefore, we have only to prove the following lemma.

Lemma 2.   *Let $h(z)$ be entire with no zeros. Then there is an entire function $g(z)$ such that $h = e^g$.*

Proof.   It is not quite correct to set $g(z) = \log h(z)$ because it is not obvious that $h(\mathbb{C})$ is simply connected. Therefore, we do not know whether log has an analytic branch on $h(\mathbb{C})$. However, we can circumvent this problem as follows.

Since h'/h is entire, we can write

$$\frac{h'(z)}{h(z)} = a_0 + a_1 z + \cdots$$

and let $g_1(z) = a_0 z + a_1 z^2/2 + a_2 z^3/3 + \cdots$; that is, $g_1' = h'/h$. (The series for g_1 has an infinite radius of convergence since $\Sigma\, a_n z^n$ does.) Let $f_2(z) = e^{g_1(z)}$. Then

$$\frac{f_2'(z)}{f_2(z)} = g'(z) = \frac{h'(z)}{h(z)},$$

and therefore $d/dz(f_2/h) = 0$, so $h = K \cdot f$ for a constant K. If we let $z = 0$, we see that $K = h(0) \neq 0$. Since $K \neq 0$, we can let $K = e^c$ for some c, so $h = e^c \cdot e^{g_1} = e^g$ where $g = g_1 + c$. ∎

With this lemma our proof of Theorem 3 is complete. ■

Table 7.1   Summary of Properties of the Gamma Function

Definition.

$$\Gamma(z) = \frac{1}{ze^{\gamma z}\left[\prod\limits_{n=1}^{\infty}\left(1+\dfrac{z}{n}\right)e^{-z/n}\right]}; \qquad \gamma = \lim_{n\to\infty}\left(1+\frac{1}{2}+\cdots+\frac{1}{n}-\log n\right)\approx .577.$$

1. Γ is meromorphic with simple poles at $0, -1, -2, \cdots$.

2. $\Gamma(z+1) = z\,\Gamma(z)$; $z\neq 0, -1, -2, \cdots$.

3. $\Gamma(n+1) = n!$; $n = 0, 1, 2, \cdots$.

4. $\Gamma(z)\,\Gamma(1-z) = \dfrac{\pi}{\sin \pi z}$.

5. $\Gamma(z)\neq 0$ for all z.

6. $\Gamma\left(\dfrac{1}{2}\right) = \sqrt{\pi}$; $\Gamma\left(n+\dfrac{1}{2}\right) = \dfrac{1\cdot3\cdot5\ \cdots\ (2n-1)}{2^n}\sqrt{\pi}$.

7. $\Gamma(z) = \dfrac{1}{z}\prod\limits_{n=1}^{\infty}\left\{\left(1+\dfrac{1}{n}\right)^{z}\left(1+\dfrac{z}{n}\right)^{-1}\right\}$.

8. $\Gamma(z) = \lim\limits_{n\to\infty}\dfrac{n!n^z}{z(z+1)\ \cdots\ (z+n)}$.

9. $\Gamma(z)\,\Gamma\left(z+\dfrac{1}{n}\right)\ \cdots\ \Gamma\left(z+\dfrac{n-1}{n}\right) = (2\pi)^{(n-1)/2}\,n^{(1/2)-nz}\,\Gamma(nz)$.

10. $2^{2z-1}\,\Gamma(z)\,\Gamma\left(z+\dfrac{1}{2}\right) = \sqrt{\pi}\,\Gamma(2z)$.

11. Residue at $-m$ equals $(-1)^m/m!$.

12. For Re $z > 0$, $\Gamma(z) = \int_0^\infty t^{z-1}e^{-t}dt$. The convergence is uniform and absolute for $-\pi/2 + \delta \leq \arg z \leq \pi/2 - \delta$, $\delta > 0$, and $\epsilon \leq |z| \leq R$, where $0 < \epsilon < R$.

13. $\dfrac{\Gamma'(z)}{\Gamma(z)} = -\gamma - \dfrac{1}{z} + \sum\limits_{1}^{\infty}\left(\dfrac{1}{n}-\dfrac{1}{z+n}\right) = \int\limits_{0}^{\infty}\left(\dfrac{e^{-t}}{t}-\dfrac{e^{-zt}}{1-e^{-t}}\right)dt$.

14. $\Gamma(z) = -\dfrac{1}{2i\sin\pi z}\int\limits_{C}(-t)^{z-1}e^{-t}dt$ (C as in Figure 7.2).

15. $\dfrac{1}{\Gamma(z)} = \dfrac{i}{2\pi}\int\limits_{C}(-t)^{-z}e^{-t}dt$ (C as in Figure 7.2).

16. $\Gamma(z+1) \approx \sqrt{2\pi}\,z^{z+1/2}\,e^{-z}$ for $|z|$ large, Re $z > 0$. (This is Stirling's Formula, which will be proved in Section 7.3.)

Worked Examples

1. For what z does $(1+z) \prod_{n=1}^{\infty} (1+z^{2^n})$ converge absolutely? Show that the product is $1/(1-z)$.

    Solution.   By Theorem 1(iii), we have absolute convergence iff $\sum_{n=1}^{\infty} z^{2^n}$ converges absolutely. This occurs for $|z| < 1$, since the radius of convergence of the series is one. Thus the product converges absolutely for $|z| < 1$.

    Evaluation of the product requires a trick, which, in this case, is quite simple. Our product is $(1 + z)(1 + z^2)(1 + z^4)(1 + z^8) \cdots$. Now $(1 + z)(1 + z^2) = 1 + z + z^2 + z^3$, so $(1+z)(1+z^2)(1+z^4) = 1 + z + z^2 + z^3 \cdots + z^7$. Generally, $\prod_{k=1}^{n} (1+z^{2^k}) = 1 + z + z^2 + \cdots + z^{2^{n+1}-1}$. We know that this series converges to $1/(1-z)$ as $n \rightarrow \infty$ since it is the power series around $z = 0$ for $1/(1-z)$.

2. Prove that

$$\sin z = z \prod_{n=1}^{\infty} \left(1 - \frac{z^2}{n^2\pi^2}\right).$$

    Solution.   The zeros of $\sin z$ occur at 0 and $\pm n\pi$; say, $a_1 = \pi$, $a_2 = -\pi$, $a_3 = 2\pi$, $a_4 = -2\pi$, $\cdots$. All the zeros are simple, and $\sum 1/|a_n|^2$ converges. Therefore, by Theorem 3, we can write

$$\sin z = e^{g(z)} z \prod_{n=1}^{\infty} \left(1 - \frac{z}{a_n}\right) e^{z/a_n}$$

$$= e^{g(z)} z \cdot \left[\left(1 - \frac{z}{\pi}\right) e^{z/\pi}\right] \left[\left(1 + \frac{z}{\pi}\right) e^{-z/\pi}\right] \left[\left(1 - \frac{z}{2\pi}\right) e^{z/2\pi}\right] \left[\left(1 + \frac{z}{2\pi}\right) e^{-z/2\pi}\right] \cdots$$

$$= e^{g(z)} z \prod_{n=1}^{\infty} \left(1 - \frac{z^2}{n^2\pi^2}\right)$$

    (gathering the terms in pairs). Thus it remains to be shown that $e^{g(z)} = 1$.

    This is not so simple, and the student would not be expected routinely to carry out the following argument; it is, admittedly, a trick. Let

$$P_n(z) = e^{g(z)} z \prod_{k=1}^{n} \left(1 - \frac{z^2}{k^2\pi^2}\right).$$

    We know that $P_n(z) \rightarrow \sin z$ (uniformly on disks), so $P_n'(z) \rightarrow \cos z$. Thus

$$\frac{P_n'(z)}{P_n(z)} \rightarrow \cot z \text{ for } z \neq 0, \pm\pi, \pm2\pi, \cdots.$$

    But

$$\frac{P_n'(z)}{P_n(z)} = \frac{d}{dz} \log P_n(z)$$

$$= \frac{d}{dz} \left\{ g(z) + \log z + \sum_{k=1}^{n} \log\left(1 - \frac{z^2}{k^2\pi^2}\right) \right\}$$

$$= g'(z) + \frac{1}{z} + \sum_{k=1}^{n} \left(\frac{2z}{z^2 - k^2\pi^2}\right).$$

However, we know from Section 4.3 (see the example following Theorem 15) that

$$\cot z = \frac{1}{z} + \sum_{n=1}^{\infty} \frac{2z}{z^2 - n^2\pi^2}, \ z \neq n\pi.$$

Thus $g'(z) = 0$, so $g(z)$ is a constant, say, c. Therefore,

$$\frac{\sin z}{z} = e^c \prod_{n=1}^{\infty} \left(1 - \frac{z^2}{n^2\pi^2}\right).$$

Let $z \to 0$. The left side approaches 1 while the right side approaches e^c (Why?). Thus $e^c = 1$ and we have our formula.

3. Prove that

$$\frac{\Gamma'(z)}{\Gamma(z)} = -\gamma - \frac{1}{z} + \lim_{n \to \infty} \sum_{k=1}^{n} \left(\frac{1}{k} - \frac{1}{z+k}\right).$$

Solution.   We have

$$\frac{1}{\Gamma(z)} = ze^{\gamma z} \prod_{n=1}^{\infty} \left(1 + \frac{z}{n}\right)e^{-z/n}.$$

Thus, taking logs (which we know we can do near each z for which we have convergence), we obtain

$$-\log \Gamma(z) = \gamma z + \log z + \sum_{1}^{\infty} \left\{\log\left(1 + \frac{z}{n}\right) - \frac{z}{n}\right\}.$$

Differentiating, we get

$$-\frac{\Gamma'(z)}{\Gamma(z)} = \gamma + \frac{1}{z} + \sum_{1}^{\infty} \left(\frac{\frac{1}{n}}{1 + \frac{z}{n}} - \frac{1}{n}\right)$$

$$= \gamma + \frac{1}{z} + \sum_{1}^{\infty} \left(\frac{1}{n+z} - \frac{1}{n}\right).$$

This is the result claimed.

Exercises

1. Show that $\prod_{n=2}^{\infty} \left(1 - \frac{1}{n^2}\right) = \frac{1}{2}.$

2. Show that $\prod_{n=1}^{\infty} (1 + z_n)$ converges absolutely iff $\sum_{n=1}^{\infty} \log(1 + z_n)$ converges absolutely.

   Hint.   Study the proof of Theorem 1.

3. Complete the proof of Theorem 2.

4. Show that $\displaystyle\prod_{n=2}^{\infty}\left(1-\frac{2}{n(n+1)}\right)=\frac{1}{3}$.

5. Show that $\displaystyle\prod_{n=2}^{\infty}\left(1-\frac{2}{n^3+1}\right)=\frac{2}{3}$.

6. Show that $\Pi_{n=1}^{\infty}(1+z_n)$ converges (assuming that $z_n \ne -1$) iff, for any $\epsilon > 0$, there is an N such that $n \ge N$ implies that

$$|(1+z_n) \cdots (1+z_{n+p})-1| < \epsilon \text{ for all } p = 0, 1, 2, \cdots.$$

Hint.   Use the Cauchy criterion for sequences.

7. Let $a_1, a_2, \cdots$ be points in $\mathbb{C}$, let $a_i \ne 0$, and let

$$\sum_{1}^{\infty} \frac{1}{|a_n|^{1+h}} < \infty$$

for a fixed integer $h \ge 0$. Show that the most general entire function having zeros at $a_1, a_2, \cdots$ and a zero of order k at 0 is

$$f(z) = e^{g(z)}z^k \prod_{1}^{\infty}\left\{\left(1-\frac{z}{a_n}\right)e^{[z/a_n+(z/a_n)^2/2+\cdots+(z/a_n)^h/h]}\right\}.$$

(Each of the points a_i may be repeated finitely often.)

Hint.   Prove the following lemma: If $1+w=(1-a)e^{a+a^2/2+\cdots+a^h/h}$ for $|a| < 1$, then $|w| \le |a|^{h+1}/(1-|a|)$.

8. i. Let $E(z,h)=(1-z)e^{z+z^2/2+\cdots+z^h/h}$. Show that the most general entire function having zeros at $a_1, a_2, \cdots$, each repeated according to its multiplicity, where $a_n \to \infty$, and having a zero of order k at 0 is

$$f(z) = e^{g(z)}z^k \prod_{n=1}^{\infty} E\left(\frac{z}{a_n}, n\right).$$

   (This is the *Weierstrass Factorization Theorem*.)
   ii. Conclude that every meromorphic function is the quotient of two entire functions.

9. Use example 2 to show that

$$\frac{\pi}{2}=\frac{2}{1}\cdot\frac{2}{3}\cdot\frac{4}{3}\cdot\frac{4}{5}\cdot\frac{6}{5}\cdot\frac{6}{7}\cdot \quad \cdots \quad .$$

   (This is *Wallis' Formula*.)

10. Prove formula 2 of Table 7.1.

11. Prove formula 4 of Table 7.1.

12. Prove formula 8 of Table 7.1.

13. Using Euler's Formula (formula 7 of Table 7.1), prove that $\Gamma(z+1)= z\,\Gamma(z)$.

14. Show that in the neighborhood $|z + m| < 1$ for m a fixed positive integer,

$$\Gamma(z) - \frac{(-1)^m}{m!(z + m)}$$

is analytic (that is, has a removable singularity at $z = -m$).

15. Prove that, for $0 \le t \le n$,

$$0 \le e^{-t} - \left(1 - \frac{t}{n}\right)^n \le \frac{t^2 e^{-t}}{n}$$

as follows. First, show that $1 + y \le e^y \le (1 - y)^{-1}$ for $0 \le y \le 1$ by using power series. Set $y = t/n$ to get $0 \le e^{-t} - (1 - t/n)^n$ and show that $e^{-t} - (1 - t/n)^n \le e^{-t}\{1 - (1 - t^2/n^2)^n\}$. Use the inequality $(1 - k)^n \ge 1 - nk$ for $0 \le k \le 1$ to get

$$1 - \left(1 - \frac{t^2}{n^2}\right)^n \le \frac{t^2}{n}$$

for $0 \le t \le n$.

16. Prove that, for $\mathrm{Re}\ z \ge 0$,

$$\frac{\Gamma'(z)}{\Gamma(z)} = \int\limits_{0}^{\infty} \left(\frac{e^{-t}}{t} - \frac{e^{-zt}}{1 - e^{-t}}\right) dt.$$

Hint. If $\mathrm{Re}\ z \ge 0$, then $1/(z + n) = \int_0^\infty e^{-t(z+n)}\, dt$. Use $\gamma = \lim\limits_{n \to \infty} (1 + 1/2 + \cdots + 1/n - \log n)$ and example 3.

17. Let γ be a circle of radius $1/2$ around $z_0 = 0$. Show that $\int_\gamma \Gamma(z)dz = 2\pi i$.

18. Prove the following (*Hankel's Formula*):

$$\Gamma(z) = \frac{-1}{2i\sin \pi z} \int\limits_{C} (-t)^{z-1} e^{-t} dt$$

where C is the contour illustrated in Figure 7.2. For what z is this formula valid? Using $\Gamma(z)\, \Gamma(1 - z) = \pi/\sin \pi z$, conclude that

$$\frac{1}{\Gamma(z)} = \frac{i}{2\pi} \int\limits_{C} (-t)^{-z}\, e^{-t}\, dt.$$

19. Establish the uniform convergence in formula 12 of Table 7.1.

20. In answering this question, refer back to Section 6.1. Define $\Gamma(z) = \int_0^\infty t^{z-1} e^{-t}\, dt$ for $\mathrm{Re}\ z > 0$.
  i. Show directly that $\Gamma(z)$ is analytic on $\mathrm{Re}\ z > 0$ by showing that $\int_0^n t^{z-1} e^{-t}\, dt$ converges uniformly on closed disks as $n \to \infty$.
  ii. Show that $\Gamma(z + 1) = z\Gamma(z)$; $\mathrm{Re}\ z > 0$.
  iii. Use (ii) and analytic continuation to prove that $\Gamma(z)$ can be extended to a meromorphic function having simple poles at $0, -1, -2, \cdots$.

Hint. The procedure used is analogous to that used in proving the Schwarz Reflection Principle; see Section 6.1.

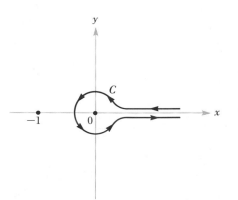

Figure 7.2 Contour for Hankel's Formula.

21. Show that $\int_{-\infty}^{\infty} e^{-y^2} \, dy = \sqrt{\pi}$ and that $\int_{-\infty}^{\infty} y^2 e^{-y} \, dy = \sqrt{\pi}/2$.

Hint. Relate these equations by integrating by parts and use $\Gamma(1/2) = \sqrt{\pi}$.

7.2 Asymptotic Expansions and the Method of Steepest Descent

The purpose of asymptotic expansions is to develop accurate formulas for determining the approximate value of a function $f(z)$ when z is large. Stirling's Formula says that $\Gamma(x) \approx \{e^{-x} \, x^{x-1/2}(2\pi)^{1/2}\}$ (recall that $\Gamma(n) = (n-1)!$), and the approximation becomes more accurate as x increases. (This will be explained in Definition 3.) The formula is useful because it is easier to handle than the Γ function itself. Moreover, it has important applications in general fields such as probability theory and statistical mechanics.

The basic theory of asymptotic expansions that is considered in this section will be applied in Section 7.3 when Stirling's Formula is proved and Bessel functions are analyzed.

There are other methods for studying the asymptotic behavior of functions $f(z)$. For example, if f satisfies a differential equation, then this equation frequently can be used to obtain an asymptotic formula. However, such methods will not be considered here. The student who wishes to delve more deeply into these topics should consult the references listed at the end of this text.

Asymptotic expansions

The following terminology will be useful. Let $[\alpha, \beta]$ be a given interval for arg z; $\alpha \le \beta$. A function $\varphi(z)$ is said to be $o(1/z^n)$ iff $z^n \varphi(z) \rightarrow 0$ as

$z \longrightarrow \infty$, with arg z lying in the specified range. (In other words, for given $\epsilon > 0$, there is an R such that $|z| \geq R$ and arg $z \in [\alpha,\beta]$ implies that $|z^n \varphi(z)| < \epsilon$. We sometimes write $\varphi(z) = o(1/z^n)$ but this is an abuse of notation. Informally, the assertion $\varphi(z) = o(1/z^n)$ means that $\varphi(z) \longrightarrow 0$ as $z \longrightarrow \infty$ faster than does $1/z^n$. A related notation, $O(1/z^n)$, means that $|z^n \varphi(z)|$ is bounded as $z \longrightarrow \infty$; that is, there are constants R, M such that $|z| \geq R$ and arg $z \in [\alpha,\beta]$ imply that $|z^n \varphi(z)| \leq M$.

Notice that if φ is $O(1/z^{n+1})$, then φ is $o(1/z^n)$, but in general, the converse is not true.

Consider the series

$$S = a_0 + \frac{a_1}{z} + \frac{a_2}{z^2} + \cdots$$

and let

$$S_n = a_0 + \frac{a_1}{z} + \cdots + \frac{a_n}{z^n}.$$

Thus S_n is well defined for $z \neq 0$ but we make no demand that S converge; in fact, as we shall see on p. 362, divergence is often more common. The correct way to say that S is asymptotic to a given function f is given in the following definition.

Definition 3.   *We say that $f \sim S$, or that f is asymptotic to S, or that S is an asymptotic expansion of f,* iff

$$f - S_n = o\left(\frac{1}{z^n}\right)$$

for arg z *lying in a specified range* $[\alpha,\beta]$ (see Figure 7.3).

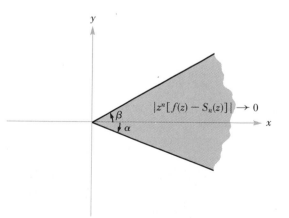

Figure 7.3   Asymptotic expansion.

Thus, although S may be divergent, the partial sums usually result in accurate approximations of f, the error being approximately $1/z^n$. This will be illustrated with an example in the following paragraphs.

If we allowed the full range $[-\pi,\pi]$ for arg z, we might expect $a_0 + a_1/z + a_2/z^2 + \cdot \cdot \cdot$ to converge if $f(z)$ were analytic outside a large circle, because f has a convergent Laurent series of that form. However, f usually has poles $z_n \longrightarrow \infty$ (such as $\Gamma(z)$, which has poles at $0, -1, -2, \cdot \cdot \cdot$), and therefore in many examples, we do not have a Laurent series that is valid on the exterior of any circle.

If f has poles $z_n \longrightarrow \infty$ in the sector arg $z \in [\alpha,\beta]$ and $f \sim S$, then S cannot converge at any z_0. If it did, then S would converge uniformly for all $|z| > |z_0| + 1$. (See Section 3.3.) Definition 3 and the uniform convergence of S_n to S would say that for large enough $|z|$ in that sector, we have $|f(z) - S(z)| < 1$. But this cannot hold near the poles of f.

The following example should help to clarify the concept of asymptotic expansion.

Example.   Let x be real, $x \geq 0$, and set

$$f(x) = \int_x^\infty t^{-1} e^{x-t} \, dt.$$

(This is not the gamma function.) Integration by parts gives us

$$f(x) = \frac{1}{x} - \frac{1}{x^2} + \frac{2!}{x^3} - \cdot \cdot \cdot + \frac{(-1)^{n-1}(n-1)!}{x^n} + (-1)^n n! \int_x^\infty \frac{e^{x-t}}{t^{n+1}} \, dt.$$

We claim that

$$f(x) \sim S(x) = \frac{1}{x} - \frac{1}{x^2} + \frac{2!}{x^3} - \frac{3!}{x^4} + \cdot \cdot \cdot .$$

Note that the series diverges. Here the sector is $\alpha=\beta=0$; that is, we are restricting z to the positive real axis.

Indeed, if

$$S_n = \frac{1}{x} - \frac{1}{x^2} + \cdot \cdot \cdot + \frac{(-1)^{n-1}(n-1)!}{x^n},$$

we have

$$|x^n(f(x) - S_n(x))| = x^n n! \int_x^\infty \frac{e^{x-t}}{t^{n+1}} \, dt = n! \int_x^\infty \left(\frac{x}{t}\right)^n \frac{e^{x-t}}{t} \, dt$$

$$\leq n! \int_x^\infty \frac{e^{x-t}}{t} \, dt \leq \frac{n!}{x} \int_x^\infty e^{x-t} \, dt = \frac{n!}{x},$$

which approaches zero as $x \longrightarrow \infty$. Thus $f(x) - S_n(x)$ is $o(1/x^n)$ and so $f \sim S$ as required. Note that even though $n!$ grows quickly, we still have an accurate approximation because

$$|f(x) - S_n(x)| \leq \frac{n!}{x^{n+1}} = O\left(\frac{1}{x^{n+1}}\right),$$

and if x is, say, greater than n, then $n!/x^{n+1}$ is very small.

Returning to the concept of asymptotic expansions, we shall also write

$$f(z) \sim g(z)\left\{a_0 + \frac{a_1}{z} + \frac{a_2}{z^2} + \cdots\right\}$$

to mean that

$$f(z) = g(z)\left\{a_0 + \cdots + \frac{a_n}{z^n} + o\left(\frac{1}{z^n}\right)\right\};$$

in other words, if $g(z) \neq 0$, then

$$\frac{f(z)}{g(z)} \sim a_0 + \frac{a_1}{z} + \cdots.$$

Some basic properties of asymptotic expansions are summarized in the next theorem.

Theorem 4.
  i. *If*

$$f(z) \sim S(z) = a_0 + \frac{a_1}{z} + \frac{a_2}{z^2} + \cdots,$$

  then

$$f(z) - S_n(z) = O\left(\frac{1}{z^{n+1}}\right)$$

  and conversely.
 ii. *If*

$$f \sim a_0 + \frac{a_1}{z} + \frac{a_2}{z^2} + \cdots$$

and

$$f \sim \tilde{a}_0 + \frac{\tilde{a}_1}{z} + \frac{\tilde{a}_2}{z^2} + \cdots,$$

then $a_i = \tilde{a}_i$. (Asymptotic expansions are unique.)

iii. *If*

$$f \sim a_0 + \frac{a_1}{z} + \frac{a_2}{z^2} + \cdots$$

and

$$g \sim b_0 + \frac{b_1}{z} + \frac{b_2}{z^2} + \cdots,$$

both being valid in the same range of arg *z, then in that range*

$$f + g \sim (a_0 + b_0) + \frac{(a_1 + b_1)}{z} + \frac{(a_2 + b_2)}{z^2} + \cdots$$

and

$$fg \sim c_0 + \frac{c_1}{z} + \frac{c_2}{z^2} + \cdots \qquad where \qquad c_n = \sum_{k=0}^{n} a_k b_{n-k}.$$

(Asymptotic series may be added and multiplied.)

iv. *Two different functions can have the same asymptotic expansion.*

v. *Let $\varphi: [a, \infty[\longrightarrow \mathbb{R}$ be continuous and suppose that $\varphi(x) = o(1/x^n)$, $n \geq 2$. Then*

$$\int_x^{\infty} \varphi(t)dt = o\left(\frac{1}{x^{n-1}}\right).$$

Proof.

i. Since $f \sim S$, we have, by definition, $f - S_{n+1} = o(1/z^{n+1})$. Therefore, $f - S_n = f - S_{n+1} + S_{n+1} - S_n = o(1/z^{n+1}) + a_{n+1}/z^{n+1} = O(1/z^{n+1})$. (If the student finds this o, O notation vague, he should write out these steps in detail.)

ii. It can be shown that $a_n = \tilde{a}_n$ by induction on n. First, by the definition of $f \sim S$, $f(z) - a_0 \to 0$ as $z \to \infty$, so $a_0 = \lim_{z \to \infty} f(z)$.

Thus $a_0 = \tilde{a}_0$. Suppose we have proven that $a_0 = \tilde{a}_0, \cdots, a_n = \tilde{a}_n$. We shall show that $a_{n+1} = \tilde{a}_{n+1}$. Given $\epsilon > 0$, there is an R such that if $|z| \geq R$, we have

$$\left| z^{n+1} \left\{ f(z) - \left(a_0 + \frac{a_1}{z} + \cdots + \frac{a_{n+1}}{z^{n+1}} \right) \right\} \right| < \epsilon$$

and

$$\left| z^{n+1} \left\{ f(z) - \left(\tilde{a}_0 + \frac{\tilde{a}_1}{z} + \cdots + \frac{\tilde{a}_{n+1}}{z^{n+1}} \right) \right\} \right| < \epsilon.$$

Therefore, by the triangle inequality,

$$|a_{n+1} - \tilde{a}_{n+1}| = |z^{n+1}| \frac{|a_{n+1} - \tilde{a}_{n+1}|}{|z^{n+1}|}$$

$$= |z^{n+1}| \left| \left[f(z) - \left(a_0 + \cdots + \frac{a_{n+1}}{z^{n+1}} \right) \right] \right.$$

$$\left. - \left[f(z) - \left(\tilde{a}_0 + \cdots + \frac{\tilde{a}_{n+1}}{z^{n+1}} \right) \right] \right| < \epsilon + \epsilon = 2\epsilon.$$

Thus $|a_{n+1} - \tilde{a}_{n+1}| < 2\epsilon$ for any $\epsilon > 0$. Hence $a_{n+1} = \tilde{a}_{n+1}$.

iii. Let $S_n(z) = a_0 + \cdots + a_n/z^n$ and $\tilde{S}_n(z) = b_0 + \cdots + b_n/z^n$. We must show that $f + g - (S_n + \tilde{S}_n) = o(1/z^n)$. To do this, write $f + g - (S_n + \tilde{S}_n) = (f - S_n) + (g - \tilde{S}_n) = o(1/z^n) + o(1/z^n) = o(1/z^n)$ (Why?).

To establish the formula for the product, note that $c_0 + c_1/z + \cdots + c_n/z^n = S_n \tilde{S} + o(1/z^n)$ since $S_n \tilde{S}_n = c_0 + c_1/z + \cdots + c_n/z^n$ plus higher-order terms. Thus $fg - (c_0 + c_1 + \cdots + c_n/z^n) = fg - S_n \tilde{S}_n + o(1/z^n)$. Now write $fg - S_n \tilde{S}_n = (f - S_n)g + S_n(g - \tilde{S}_n)$ and note that both terms are $o(1/z^n)$. (Remember that g and S_n are bounded as $z \to \infty$.)

iv. On $\mathbb{R}$, the function e^{-x} is $o(1/x^n)$ as $x \to \infty$ for any n. Thus if $f \sim a_0 + a_1/x + a_2/x^2 + \cdots$, then $f(x) + e^{-x} \sim a_0 + a_1/x + a_2/x^2 + \cdots$ as well.

v. Since $\varphi(t) = o(1/t^n)$, $\lim_{n \to \infty} t^n \varphi(t) = 0$. Given $\epsilon > 0$, there is an $x_0 > 0$ such that $t > x_0$ implies that $|t^n \varphi(t)| < \epsilon$. Thus for $x > x_0$,

$$\left| \int_x^\infty \varphi(t)dt \right| \le \int_x^\infty \frac{\epsilon}{t^n} \, dt = \frac{\epsilon}{n-1} \cdot \frac{1}{x^{n-1}};$$

and so, for $x > x_0$,

$$\left| x^{n-1} \int_x^\infty \varphi(t)dt \right| \le \epsilon.$$

Thus $\lim_{x \to \infty} x^{n-1} \int_x^\infty \varphi(t) \, dt = 0$, and so $\int_x^\infty \varphi(t) \, dt = o(1/x^{n-1})$. ∎

Method of steepest descent

We now turn to the problem of finding the asymptotic expansion of a given function. A fairly useful procedure is called the method of steepest descent (or the saddle point method). This method was discovered by P. Debye in approximately 1909.

We shall obtain an expansion of the form

$$f(z) \sim g(z)\left(1 + \frac{a_1}{z} + \frac{a_2}{z^2} + \cdots \right)$$

for a particularly simple function $g(z)$. We shall be mainly interested in obtaining the first term. Thus we seek to write $f(z) = g(z)(1 + O(1/z))$, which we can write $f \sim g$. We also want to be able to obtain the higher-order terms. The method described here works well only if f has the special form $f(z) = \int_\gamma e^{zh(\zeta)}\, d\zeta$. The precise conditions under which the method can be used are given in Theorem 5. We shall use contours $\gamma(t)$ defined for all $t \in \mathbb{R}$. We can integrate over such infinite contours in the same manner as we would integrate over ordinary ones, as long as we check convergence of the integrals.

The proofs of Theorems 5 and 7 omit a few of the technical details because complete proofs would take us too far afield. However, the student is asked to fill in some of the gaps in these proofs in the end-of-section exercises.

Theorem 5. *Let γ: $]-\infty,\infty[\to \mathbb{C}$ be a curve in $\mathbb{C}$. (γ may also be defined only on a finite interval.) Let $\zeta_0 = \gamma(t_0)$ be a point on γ and let $h(\zeta)$ be a function continuous along γ and analytic at ζ_0. Make the following hypotheses: For $|z| \geq R$ and arg z fixed,*
  i. *$f(z) = \int_\gamma e^{zh(\zeta)}\, d\zeta$ converges absolutely.*
 ii. *$h'(\zeta_0) = 0$; $h''(\zeta_0) \neq 0$.*
iii. *Im$(zh(\zeta))$ is constant for ζ on γ in some neighborhood of ζ_0.*
 iv. *Re$(zh(\zeta))$ has a strict maximum at ζ_0 along the entire curve γ.*
    Then

$$f(z) \sim \frac{e^{zh(\zeta_0)}\sqrt{2\pi}}{\sqrt{z}\sqrt{-h''(\zeta_0)}} \tag{1}$$

as $z \to \infty$, arg z fixed. The sign of the square root is chosen such that $\sqrt{z}\sqrt{-h''(\zeta_0)} \cdot \gamma'(t_0) < 0$.

Remarks.
    i. To achieve conditions (i)–(iv) in Theorem 5 it may be necessary to deform γ by applying Cauchy's Theorem. A path γ verifying these conditions is called a *path of steepest descent*.
   ii. The asymptotic expansion in the conclusion of Theorem 5 depends only on $h(\zeta_0)$ and $h''(\zeta_0)$ and not on the behavior of h elsewhere on γ (except, of course, that h must satisfy the hypotheses of the theorem). Higher-order derivatives would also occur if further terms in the expansion were needed.

iii. The origin of the term "steepest descent" can be traced to conditions (iii) and (iv) of Theorem 5 in the following way. Recall that $\text{Im}(zh(\zeta)) = v(\zeta)$ and $\text{Re}(zh(\zeta)) = u(\zeta)$ are harmonic conjugates, and the fact that v is constant on γ means that u is changing fastest in the direction of γ. Since ζ_0 is a maximum, $u(\zeta) = \text{Re}(zh(\zeta))$ is decreasing fastest when moving away from ζ_0 in the direction of γ. Hence the curve γ is called the path of steepest descent. The term "saddle point method" originated as follows. Consider again the function $u(\zeta) = \text{Re}(zh(\zeta))$. Along γ it has a maximum at ζ_0. But $h''(\zeta_0) \neq 0$ implies that u is not constant, so ζ_0 must be a saddle point of u. Recall (Theorem 35, Section 2.5) that harmonic functions never have local maxima or minima (see Figure 7.4).

iv. Often the correct sign for the square root may be determined by examining the sign of the integral defining $f(z)$.

Proof.   Break up γ into three portions, γ_1, C, and γ_2, as illustrated in Figure 7.5. Choose C so that it lies in a neighborhood of ζ_0 small enough that (1) $h(\zeta)$ is analytic and (2) condition (iii) of Theorem 5 holds. Clearly, $f(z) = I_1(z) + I_2(z) + J(z)$ where we use the notations $J(z) = \int_C e^{zh(\zeta)}\, d\zeta$ and $I_k(z) = \int_{\gamma_k} e^{zh(\zeta)}\, d\zeta$, $k = 1,2$.

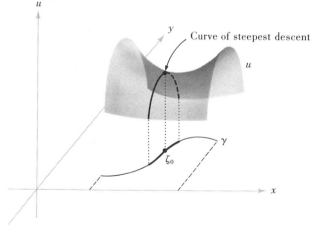

Figure 7.4 Saddle point.

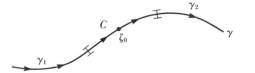

Figure 7.5 Method of steepest descent.

We first claim that $I_k(z)/J(z) = O(1/z)$ as $z \to \infty$. To prove this claim, note that

$$|I_k(z)| = \left| \int_{\gamma_k} e^{zh(\zeta)}\, d\zeta \right| \le \int_{\gamma_k} e^{\operatorname{Re} zh(\zeta)}\, |d\zeta|.$$

But

$$J(z) = \int_C e^{zh(\zeta)}\, d\zeta = \int_C e^{\operatorname{Re} zh(\zeta)}\, e^{i \operatorname{Im}(zh(\zeta))}\, d\zeta.$$

Since $\operatorname{Im}(zh(\zeta))$ is constant on C, we get

$$|J(z)| = \int_C e^{\operatorname{Re} zh(\zeta)}\, |d\zeta|.$$

Thus

$$\left| \frac{I_k(z)}{J(z)} \right| \le \frac{\displaystyle\int_{\gamma_k} e^{\operatorname{Re} zh(\zeta)}\, |d\zeta|}{\displaystyle\int_C e^{\operatorname{Re} zh(\zeta)}\, |d\zeta|}.$$

Now let $\tilde{C}$ be a strictly smaller subinterval of C, centered at ζ_0. Then

$$\left| \frac{I_k(z)}{J(z)} \right| \le \frac{\displaystyle\int_{\gamma_k} e^{\operatorname{Re} zh(\zeta)}\, |d\zeta|}{\displaystyle\int_{\tilde{C}} e^{\operatorname{Re} zh(\zeta)}\, |d\zeta|}.$$

Fix z_0 and let α be the minimum of $\operatorname{Re} zh(\zeta)$ on $\tilde{C}$. There is an $\epsilon > 0$ such that $\operatorname{Re} z_0 h(\zeta) \le \alpha - \epsilon$ for all $\zeta \in \gamma_k$. Thus, using the fact that z lies on the same ray as z_0,

$$\frac{\displaystyle\int_{\gamma_k} e^{\operatorname{Re} zh(\zeta)}\, |d\zeta|}{\displaystyle\int_{\tilde{C}} e^{\operatorname{Re} zh(\zeta)}\, |d\zeta|} = \frac{\displaystyle\int_{\gamma_k} e^{\operatorname{Re} z_0 h(\zeta)}\, e^{\operatorname{Re}(z - z_0)h(\zeta)}\, |d\zeta|}{\displaystyle\int_{\tilde{C}} e^{\operatorname{Re} z_0 h(\zeta)}\, e^{\operatorname{Re}(z - z_0)h(\zeta)}\, |d\zeta|}$$

$$\le \frac{\left(\displaystyle\int_{\gamma_k} e^{\operatorname{Re} z_0 h(\zeta)}\, |d\zeta| \right) e^{|z - z_0|(\alpha - \epsilon)/|z_0|}}{\left(\displaystyle\int_{\tilde{C}} e^{\operatorname{Re} z_0 h(\zeta)}\, |d\zeta| \right) e^{|z - z_0|\alpha/|z_0|}}.$$

This expression is a constant factor, say M, times $e^{-|z-z_0|\epsilon/|z_0|}$. The latter is certainly $O(1/z)$ (and in fact is $O(1/z^n)$ for all $n \geq 1$), so we have proved that $I_k(z)/J(z) = O(1/z)$.

Therefore, we have shown that $f(z) \sim J(z)$ as $z \to \infty$. This localizes the problem to a neighborhood around ζ_0 where the bulk of the contribution to the integral is made. Also, we can shrink the length of C without affecting the conclusion that $f(z) \sim J(z)$ as $z \to \infty$.

Next, we write

$$h(\zeta) = h(\zeta_0) - w(\zeta)^2$$

where $w(\zeta)$ is analytic and invertible (abusing notation, we denote the inverse by $\zeta(w)$), where $w(\zeta_0) = 0$, and where

$$[w'(\zeta_0)]^2 = \frac{h''(\zeta_0)}{2}.$$

That we can do this was proved in example 2, Section 6.3. The student who has not studied Chapter 6 can nevertheless do that example because it depends only on the Inverse Function Theorem (Theorem 20, Section 1.4, or Theorem 13, Section 6.3). Since $\text{Im}\,(zh(\zeta)) = \text{Im}\,(zh(\zeta_0))$ on C and $\text{Re}\,zh(\zeta) < \text{Re}\,zh(\zeta_0)$, we see that $zw(\zeta)^2$ is real and greater than zero on C; also, by our choice of branch for $\sqrt{\ }$, $\sqrt{z}w(\zeta)$ is real and, as a function of the curve parameter t, has positive derivative at t_0. Thus, by shrinking C if necessary, we can assume that $\sqrt{z}w(\zeta)$ is increasing along C.

Note that

$$J(z) = \int_C e^{zh(\zeta)}\,d\zeta = \int_C e^{zh(\zeta_0)} \cdot e^{-zw(\zeta)^2}\,d\zeta = e^{zh(\zeta_0)} \int_C e^{-zw(\zeta)^2}\,d\zeta.$$

We can change variables by setting $\sqrt{z}w(\zeta) = y$ and we get

$$J(z) = e^{zh(\zeta_0)} \int_{-\sqrt{|z|}\epsilon_1}^{\sqrt{|z|}\epsilon_2} e^{-y^2}\,\frac{d\zeta}{dw}\frac{dy}{\sqrt{z}}$$

$$= \frac{e^{zh(\zeta_0)}}{\sqrt{z}} \int_{-\sqrt{|z|}}^{\sqrt{|z|}\epsilon_2} e^{-y^2}\,\frac{d\zeta}{dw}\,dy,$$

since y is real on C; we choose $\epsilon_1, \epsilon_2 > 0$ so that $[-\sqrt{|z|}\,\epsilon_1, \sqrt{|z|}\,\epsilon_2]$ is the range of y corresponding to ζ on C.

Next we write

$$\zeta = \zeta_0 + a_1 w + a_2 w^2 + \cdots$$

where

$$a_1 = \frac{d\zeta}{dw}(0) = \frac{1}{\frac{dw}{d\zeta}(\zeta_0)} = \frac{1}{\sqrt{\frac{-h''(\zeta_0)}{2}}}.$$

Thus

$$\frac{d\zeta}{dw} = \frac{1}{\sqrt{\frac{-h''(\zeta_0)}{2}}} + O(w) = \sqrt{\frac{-2}{h''(\zeta_0)}} + O\left(\frac{y}{\sqrt{z}}\right).$$

Hence

$$J(z) = \frac{e^{zh(\zeta_0)}}{\sqrt{z}} \frac{\sqrt{2}}{\sqrt{-h''(\zeta_0)}} \int_{-\sqrt{|z|}\,\epsilon_1}^{\sqrt{|z|}\,\epsilon_2} e^{-y^2}\,dy + \frac{e^{zh(\zeta_0)}}{\sqrt{z}} \int_{-\sqrt{|z|}\,\epsilon_1}^{\sqrt{|z|}\,\epsilon_2} e^{-y^2}\,O\left(\frac{y}{\sqrt{z}}\right)\,dy.$$

By exercise 21, Section 7.1, $\int_{-\infty}^{\infty} e^{-y^2}\,dy = \sqrt{\pi}$. Since $e^{-y^2} = O(1/y^n)$ for any $n \geq 1$, we have, using Theorem 4(v), $\int_{-\sqrt{|z|}\,\epsilon_1}^{\sqrt{|z|}\,\epsilon_2} e^{-y^2}\,dy - \sqrt{\pi} = O(1/|z|^{(n-1)/2})$, and as a consequence this expression is $O(1/z)$. The preceding equation thus proves that

$$J(z) \sim \frac{e^{zh(\zeta_0)}}{\sqrt{z}} \cdot \frac{\sqrt{2\pi}}{\sqrt{-h''(\zeta_0)}},$$

which yields the theorem. ∎

To obtain higher-order terms in the expansion

$$f(z) \sim \frac{e^{zh(\zeta_0)}}{\sqrt{z}} \cdot \frac{\sqrt{2\pi}}{\sqrt{-h''(\zeta_0)}} \left(1 + \frac{A_1}{z} + \frac{A_2}{z^2} + \cdots\right), \qquad (2)$$

we must be able to compute more terms in the series $\zeta = \zeta_0 + a_1 w + a_2 w^2 + \cdots$ in the preceding proof. To do so, we use the following formulas.

$$\int_{-\infty}^{\infty} e^{-y^2/2} y^{2m}\,dy = \sqrt{2\pi} \cdot 1 \cdot 3 \cdots (2m-1); \qquad m = 1, 2, \cdots.$$

$$\int_{-\infty}^{\infty} e^{-y^2} y^{2m+1}\,dy = 0. \qquad (3)$$

These formulas are established by integration by parts from the case in which $m = 0$. In simple cases, these higher-order terms can be evaluated explicitly; see Theorem 8. The details of the method of obtaining higher-order terms will not be given here because such terms are needed only in very refined calculations. The leading term given in Theorem 5 is the important one.

The applications of Theorem 5 given in the next section deal primarily with the case in which z real and positive. Clearly, in that case conditions (iii) and (iv) of Theorem 5 can be written equivalently with or without the z.

A proof that is similar to that of Theorem 5 leads to the following generalization (see exercise 6).

Theorem 6. *Let the conditions of Theorem 5 hold but let f have the form*

$$f(z) = \int_\gamma e^{zh(\zeta)} \, g(\zeta) \, d\zeta$$

where $g(\zeta)$ is a continuous function on γ. Then

$$f(z) \sim e^{zh(\zeta_0)} \frac{\sqrt{2\pi}}{\sqrt{z}} \cdot \frac{g(\zeta_0)}{\sqrt{-h''(\zeta_0)}}. \tag{4}$$

Method of stationary phase

If we strengthen the hypotheses of Theorem 5 slightly we can obtain another useful result by using the same method that was used to prove Theorem 5. This variant is often referred to as the *method of stationary phase* because of the nature of the result. This technique will be applied to Bessel functions in the next section. The result is as follows.

Theorem 7. *Let γ be a segment on the real axis, say, $[a,b]$, let $h(\zeta)$ be analytic on a region containing $[a,b]$, and let $\zeta_0 \in \;]a,b[$. Suppose that $g(t)$ is continuous on $[a,b]$; define*

$$f(z) = \int_a^b e^{izh(t)} \, g(t) \, dt$$

and make the following hypotheses:
  i. *h is real on $[a,b]$.*
 ii. *h has a strict maximum at $\zeta_0 \in \;]a,b[$.*
iii. *$h'(\zeta_0) = 0$, $h''(\zeta_0) < 0$.*
 iv. *$h(\zeta) - h(\zeta_0) = -w(\zeta)^2$ where w is analytic and one-to-one on a region containing ζ_0 and $w(\zeta)$ is real and differentiable on $[a,b]$, is decreasing on $[a,\zeta_0]$, and is increasing on $[\zeta_0,b]$.*
  Then we have

$$f(z) \sim \frac{e^{izh(\zeta_0)} \sqrt{2\pi}}{\sqrt{z} \sqrt{-h''(\zeta_0)}} \cdot e^{-\pi i/4} \cdot g(\zeta_0)$$

as $z \longrightarrow \infty$ on the positive real axis.

If, instead, h has a minimum, $h''(\zeta_0) > 0$, and $h(\zeta) - h(\zeta_0) = w(\zeta)^2$, the formula becomes:

$$\frac{e^{izh(\zeta_0)}\sqrt{2\pi}}{\sqrt{z}\sqrt{h''(\zeta_0)}} \cdot e^{\pi i/4} \cdot g(\zeta_0) \tag{6}$$

as $z \longrightarrow \infty$ on the positive real axis.

Remarks.
   i. This is formally the same conclusion as that of Theorem 6, except that ih replaces h in equation (4).
   ii. The first part of condition (iv) will always hold inside a sufficiently small circle centered at ζ_0. This statement was proved in Section 6.3; the student who has not studied Chapter 6 should merely accept this remark.
   iii. As in Theorems 5 and 6, we can obtain a result if z is not real, and we can also get higher-order terms, but equations 5 and 6 cover the most important case.

Proof.   By hypothesis (iv), we have

$$f(z) = e^{izh(\zeta_0)} \int_a^b e^{-izw(t)^2}g(t)\ dt.$$

As in the proof of Theorem 5, let $\zeta(w)$ be the inverse function for $w(\zeta)$ and write $y = \sqrt{i}\ w(\zeta) = e^{\pi i/4}\ w(\zeta)$. We can use this as a valid change of variables because of (iv). Then

$$f(z) = e^{izh(\zeta_0)} \int_{-\epsilon_1 e^{\pi i/4}}^{\epsilon_2 e^{\pi i/4}} e^{-zy^2} \frac{d\zeta}{dw} \cdot \tilde{g}(y) \frac{dy}{e^{\pi i/4}}$$

where

$$\tilde{g}(y) = g\left(\zeta\left(\frac{y}{e^{\pi i/4}}\right)\right),$$

$\epsilon_1 = w(a)$, and $\epsilon_2 = w(b)$. The path of integration is indicated in Figure 7.6. Write $\zeta = \zeta_0 + a_1 w + \cdots$, so that, as in the proof of Theorem 5,

$$\frac{d\zeta}{dw} = \frac{1}{\sqrt{\dfrac{-h''(\zeta_0)}{2}}} + O(w).$$

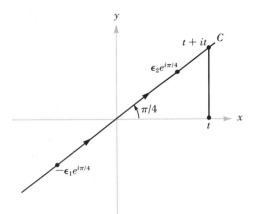

Figure 7.6 Contour for the method of stationary phase.

Substituting this expression into that for f, the first term is:

$$\frac{e^{-\pi i/4}e^{izh(\zeta_0)}}{\sqrt{\dfrac{-h''(\zeta_0)}{2}}} \int_{-\epsilon_1 e^{\pi i/4}}^{\epsilon_2 e^{\pi i/4}} e^{-zy^2}\tilde{g}(y)\, dy.$$

For z large, the graph of e^{-zy^2} is "peaked sharply" at $y = 0$; that is, e^{-zy^2} approaches zero rapidly as y moves away from 0. We then can argue that by continuity of g, except for a term that is $O(1/\sqrt{z})$ as $z \to \infty$, the preceding equation becomes

$$g(\zeta_0)\frac{e^{-\pi i/4}e^{izh(\zeta_0)}}{\sqrt{\dfrac{-h''(\zeta_0)}{2}}} \int_{-\epsilon_1 e^{\pi i/4}}^{\epsilon_2 e^{\pi i/4}} e^{-zy^2}\, dy. \tag{7}$$

Note that $\tilde{g}(0) = g(\zeta_0)$ and see exercise 7.

Now the integral equals

$$\frac{1}{\sqrt{z}} \int_{-\sqrt{z}\epsilon_1 e^{\pi i/4}}^{\sqrt{z}\epsilon_2 e^{\pi i/4}} e^{-y^2}\, dy. \tag{8}$$

Using Cauchy's Theorem, we can show that

$$\int_C e^{-y^2}\, dy = \int_{-\infty}^{\infty} e^{-y^2}\, dy$$

where C is the curve shown in Figure 7.6. This is because the integral along the vertical lines approaches zero as $t \to \infty$ (see exercise 8). Thus, using exercise 21, Section 7.1,

$$\int_C e^{-y^2} \, dy = \sqrt{\pi}.$$

Reinserting this equation in equations (7) and (8) gives us the desired result. (Compare a similar argument in the proof of Theorem 5.) The case in which h has a minimum is entirely analogous. ∎

Watson's Theorem

In the preceding discussion we encountered integrals of the form

$$\int_{-\infty}^{\infty} e^{-zy^2/2} g(y) \, dy.$$

The arguments used there can be applied to obtain another asymptotic formula that includes all the higher-order terms.

Theorem 8 (Watson's Theorem). *Let $g(z)$ be analytic and bounded on a domain containing the real axis. Set*

$$f(z) = \int_{-\infty}^{\infty} e^{-zy^2/2} g(y) \, dy$$

for z real. Then

$$f(z) \sim \frac{\sqrt{2\pi}}{\sqrt{z}} \left\{ a_0 + \frac{a_2}{z} + \frac{a_4 \cdot 1 \cdot 3}{z^2} + \frac{a_6 \cdot 1 \cdot 3 \cdot 5}{z^3} + \cdots \right\} \qquad (8)$$

as $z \to \infty$; $\arg z = 0$, where $g(z) = \sum_{n=0}^{\infty} a_n z^n$ near zero.

Proof.   We first observe that

$$h(z) = \frac{g(z) - (a_0 + a_1 z + \cdots + a_{2n-1} z^{2n-1})}{z^{2n}}$$

is bounded on the real axis since g is bounded and since $h(z) \to a_{2n}$ as $z \to 0$. Therefore, we obtain

$$\left| \int\limits_{-\infty}^{\infty} e^{-zy^2/2} [g(y) - (a_0 + a_1 y + \cdots + a_{2n-1} y^{2n-1})] \, dy \right|$$

$$\leq M \int\limits_{-\infty}^{\infty} e^{-zy^2/2} y^{2n} \, dy.$$

Now we use the fact that for $z > 0$,

$$\int\limits_{-\infty}^{\infty} e^{-zy^2/2} y^{2k} dy = \sqrt{2\pi} \cdot 1 \cdot 3 \cdots \frac{(2k-1)}{z^{k+1/2}}, \quad \int\limits_{-\infty}^{\infty} e^{-zy^2} y^{2k+1} dy = 0$$

(see equation (3)), to obtain

$$\left| \int\limits_{-\infty}^{\infty} e^{-zy^2} g(y) \, dy - \left(\frac{a_0 \sqrt{2\pi}}{z^{1/2}} + \frac{a_2 \sqrt{2\pi}}{z^{1+1/2}} + \frac{a_4 \sqrt{2\pi} \cdot 1 \cdot 3}{z^{2+1/2}} + \cdots \right. \right.$$

$$\left. \left. + \frac{a_{2n-2} \cdot \sqrt{2\pi} \cdot 1 \cdot 3 \cdots (2n-3)}{z^{n-1+1/2}} \right) \right|$$

$$\leq M \sqrt{2\pi} \frac{1 \cdot 3 \cdots (2n-1)}{z^{n+1/2}},$$

from which the theorem follows. ∎

Worked Examples

1.   Let $h(\zeta) = \zeta^2$, $\zeta_0 = 0$. Find a curve γ satisfying the hypotheses of Theorem 5. (In other words, find a path of steepest descent.) Take arg $z = 0$; that is, z is real, $z > 0$.

Solution.   Let $h(\zeta) = u + iv$, so if $\zeta = \xi + i\eta$, $u = \xi^2 - \eta^2$ and $v = 2\xi\eta$. The discussion following Theorem 5 indicated that the path of steepest descent is defined by $v = $ constant (since in our case z is real, $z > 0$). Thus the line of steepest descent through $\zeta_0 = 0$ is either $\xi = 0$ or $\eta = 0$. Since u must have a maximum at $\zeta_0 = 0$, the curve γ is defined by $\xi = 0$.

2.   Prove that

$$f(z) = \int\limits_{-\infty}^{\infty} e^{-zy^2/2} \cos y \, dy \sim \frac{\sqrt{2\pi}}{\sqrt{z}} \left(1 - \frac{1}{2z} + \frac{1}{2^2 2! z^2} - \cdots \right)$$

as $z \to \infty$; arg $z = 0$.

Solution.   We apply Watson's Theorem. Here

$$\cos y = 1 - \frac{y^2}{2!} + \frac{y^4}{5!} - \cdots ;$$

therefore,

$$a_0 = 1, \qquad a_2 = \frac{-1}{2!}, \qquad a_4 = \frac{1}{4!} \cdots ,$$

and thus

$$f(z) \sim \sqrt{\frac{2\pi}{z}} \left\{ 1 - \frac{1}{2z} + \frac{1 \cdot 3}{4!z^2} - \cdots \right\} = \sqrt{\frac{2\pi}{z}} \left\{ 1 - \frac{1}{2z} + \frac{1}{2^2 2! z^2} - \cdots \right\}.$$

Exercises

1. If $f(x) \sim a_2/x^2 + a_3/x^3 + \cdots$ for $x \in [0,\infty[$, show that

$$g(x) = \int_x^\infty f(t) \, dt \sim \frac{a_2}{x} + \frac{a_3}{2x^2} + \frac{a_4}{3x^3} + \cdots$$

2. Show that if $f(x) = \int_x^\infty e^{-t}/t \, dt$, then

$$f(x) \sim e^{-x}\left(\frac{1}{x} - \frac{1}{x^2} + \frac{1 \cdot 2}{x^3} - \cdots \right).$$

3. Show that

$$f(x) = \int_0^\infty \frac{e^{-xt}}{1 + t^2} \, dt \sim \frac{1}{x} - \frac{2!}{x^3} + \frac{4!}{x^5} - \cdots .$$

4. The expansion

$$\int_x^\infty t^{-1} e^{x-t} \, dt \sim \frac{1}{x} - \frac{1}{x^2} + \cdots$$

was discussed on p. 362. Compute $S_4(10)$ and $S_5(10)$ numerically and find an upper bound for the respective errors. Discuss how the errors change in $S_n(x)$ as n and x increase. For example, for a given x, are errors reduced if we take n very large?

5. Let $g(z)$ be analytic at z_0 and let $g'(z_0) = 0$ and $g''(z_0) \neq 0$, so that near z_0, $g(z) - g(z_0) = w(z)^2$ for w analytic, $w'(z_0) \neq 0$. Prove that there are exactly two perpendicular curves on which Re g (alternatively, Im g) are constant through z_0. (Recall that Theorem 33, Section 2.5, indicated that if $f'(z_0) \neq 0$, Re f has exactly one level curve through z_0.) Show also that lines of constant Re g and Im g intersect at $45° = \pi/4$.

6. Sketch the proof of Theorem 6, using the type of argument employed in the proof of Theorem 7.

7. Prove the statement made in the proof of Theorem 7 (see equation (7)) concerning bringing out the factor $g(\zeta)$.

Hint. Let $h_n(t) = \sqrt{n}\, e^{-nt^2}$. The area under the graph of $h_n(t)$ is $\sqrt{\pi}$ and for any $\epsilon > 0$, $h_n(t) \rightarrow 0$ uniformly outside $]-\epsilon,\epsilon[$. Show that $\int_{-\infty}^{\infty} g(t) h_n(t)\, dt \rightarrow \sqrt{\pi}\, g(0)$, if g is continuous. (Such a sequence h_n is called an *approximating δ sequence*; see Figure 7.7.)

8. Prove that $\int_C e^{-y^2}\, dy = \sqrt{\pi}$, where C is the curve shown in Figure 7.6, by showing that

$$\int_C e^{-y^2}\, dy = \int_{-\infty}^{\infty} e^{-y^2}\, dy.$$

Hint. Show that $\int_{\gamma_x} e^{-\zeta^2}\, d\zeta \rightarrow 0$ as $x \rightarrow \infty$ where γ_x is the vertical line joining x to $x + ix$.

9. Show that the first term in Watson's Theorem may be obtained as a special case of Theorem 6 if $g \geq 0$ on the real axis.

10. Find an asymptotic expansion for

$$f(z) = \int_{-\infty}^{\infty} e^{-zy^2/2} \sin(y^2)\, dy$$

(Assume that $z \rightarrow \infty$, $z > 0$.)

11. Find the path of steepest descent through $t_0 = 0$ if $h(t) = \cos t$. (Take z real, $z > 0$.)

12. Repeat exercise 11 but assume that z lies on the positive imaginary axis.

13. Use Theorem 5 to obtain the asymptotic formula for f using the path γ that was obtained in example 1.

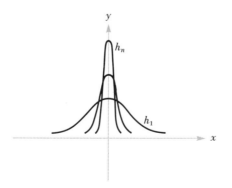

Figure 7.7   Approximating δ sequence.

14. Find the asymptotic formula for f when the path found in exercise 11 is used in Theorem 5.

15. Find the asymptotic formula for f when the path γ found in exercise 12 is used in Theorem 5.

7.3 Stirling's Formula and Bessel Functions

In this section the method of steepest descent (Theorem 5) will be applied to prove Stirling's Formula for the gamma function $\Gamma(z)$. Some properties of Bessel functions $J_n(z)$, $n = \cdots, -1, 0, 1, \cdots$, will also be defined and the method of stationary phase (Theorem 7) will be used to obtain an asymptotic formula for these functions.

Stirling's Formula

The formula may be stated as follows.

Theorem 9 (Stirling's Formula).

$$\Gamma(z+1) \sim \sqrt{2\pi}\, z^{z+1/2} e^{-z} \tag{1}$$

as $z \to \infty$ on the positive real axis.

Remark.   An extension of the proof given here shows that this result also holds for $-\pi/2 + \delta \leq \arg z \leq \pi/2 - \delta$, for any $\delta > 0$ (see Figure 7.8).

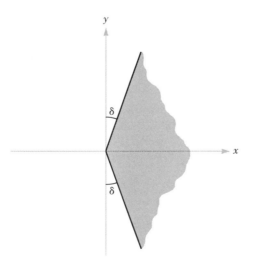

Figure 7.8   Region of validity for Stirling's Formula.

Proof.   Recall (Section 7.1) that for Re $z > 0$,

$$\Gamma(z) = \int_0^\infty e^{-t} t^{z-1} dt.$$

We are concerned with the case in which z is real and positive. We want to rewrite the integral so that Theorem 5 will apply. To do this, we make the change of variables $t = z\tau$. We get

$$\Gamma(z+1) = \int_0^\infty e^{-t} t^z dt = z^{z+1} \int_0^\infty e^{z(\log \tau - \tau)} d\tau.$$

Thus $\Gamma(z+1)/z^{z+1}$ has the form

$$\int_\gamma e^{zh(\zeta)} \, d\zeta$$

where $h(\zeta) = \log \zeta - \zeta$ and γ is the positive real axis, $[0,\infty[$. We must check the hypotheses of Theorem 5. Let $\zeta_0 = 1$. Clearly, $h(\zeta_0) = -1$, $h'(\zeta_0) = 0$, and $h''(\zeta_0) \neq 0$; thus (i) and (ii) of Theorem 5 hold. Also, $h(\zeta)$ is real on γ so (iii) is valid. To prove (iv), we know that Re $(zh(t)) = xh(t)$ has a maximum iff $h(t)$ does. But $h(t)$ has a maximum of -1 at $\zeta_0 = 1$ on γ (see Figure 7.9). Thus (iv) holds. Therefore,

$$\frac{\Gamma(z+1)}{z^{z+1}} \sim \frac{e^{zh(\zeta_0)}}{\sqrt{z}\sqrt{-h''(\zeta_0)}} \cdot \sqrt{2\pi} = \frac{e^{-z}}{\sqrt{z}} \cdot \sqrt{2\pi}.$$

Hence $\Gamma(z+1) \sim z^{z+1/2} e^{-z} \sqrt{2\pi}$, as required. ∎

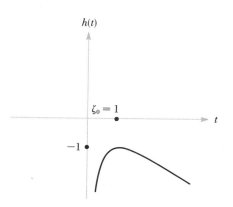

Figure 7.9   Graph of $h(t) = \log t - t$.

Remarks.   If $z = re^{i\theta}$ were not real, the x-axis would no longer be the path of steepest descent and the path of integration would have to be deformed into such a path.

If we analyze Theorem 5 more carefully, we find that the first few terms are

$$\Gamma(z+1) \sim \sqrt{2\pi}\, z^{z+1/2}e^{-z}\left\{1 + \frac{1}{12z} + \frac{1}{288z^2} + \cdots\right\}, \qquad (2)$$

that is,

$$\Gamma(z+1) = \sqrt{2\pi}\, z^{z+1/2}e^{-z}\left\{1 + \frac{1}{12z} + \frac{1}{288z^2} + O\left(\frac{1}{z^3}\right)\right\}.$$

However, when solving particular problems, we usually find that the first term is the most important one.

Since $\Gamma(z+1) = \Gamma(z)$, we obtain $\Gamma(z) \sim e^{-x}x^{x-1/2}(2\pi)^{1/2}$ mentioned on p. 360.

Bessel functions

The remainder of this section will include a brief discussion of some basic properties of Bessel functions and will describe a way in which the method of stationary phase (Theorem 7) can be applied to obtain an asymptotic formula. Bessel functions (the main properties of which are listed in Table 7.2, p. 384) are studied because they arise naturally in solutions to certain partial differential equations, such as Laplace's Equation, when these equations are expressed in terms of cylindrical coordinates.

Bessel functions can be defined in several different ways. We will find the following definition to be convenient.

Definition 4.   *Let $z \in \mathbb{C}$ be fixed and consider the function*

$$f(\zeta) = e^{z(\zeta - 1/\zeta)/2}.$$

Expand $f(\zeta)$ in a Laurent series around 0. The coefficient of ζ^n where n is positive or negative is denoted $J_n(z)$ and is called the Bessel function of order n. We call $e^{z(\zeta - 1/\zeta)/2}$ the generating function.

Thus, by definition,

$$e^{z(\zeta - 1/\zeta)/2} = \sum_{n=-\infty}^{\infty} J_n(z)\zeta^n. \qquad (3)$$

From the formula for the coefficients of a Laurent expansion (see Theorem 13, Section 3.3), we see that

$$J_n(z) = \frac{1}{2\pi i} \int_\gamma \zeta^{-n-1} \, e^{z(\zeta - 1/\zeta)/2} \, d\zeta$$

where γ is any circle around 0. If we use the unit circle $\zeta = e^{i\theta}$ and write out the integral explicitly, we get

$$J_n(z) = \frac{1}{2\pi} \int_0^{2\pi} e^{-(n+1)i\theta} \, e^{iz\sin\theta} \, e^{i\theta} \, d\theta$$

$$= \frac{1}{2\pi} \int_0^\pi e^{iz\sin\theta - ni\theta} \, d\theta + \frac{1}{2\pi} \int_0^\pi e^{-iz\sin\theta + ni\theta} \, d\theta$$

$$= \frac{1}{\pi} \int_0^\pi \cos(n\theta - z\sin\theta) \, d\theta. \tag{4}$$

Note. Although $J_n(z)$ will be defined for noninteger values of n in equation (10), equation (4) is valid only if n is an integer.

From equation (4), we see that $|J_n(z)| \le 1$ for z real. The graphs of $J_0(x)$ and $J_1(x)$ are shown in Figure 7.10.

Next it will be shown that $J_n(z)$ is entire and a method will be described for finding its power series. To carry out these tasks, it is convenient to change variables by $\zeta \mapsto 2\zeta/z$ and obtain, for each fixed z,

$$J_n(z) = \frac{1}{2\pi i} \left(\frac{z}{2}\right)^n \int_\gamma \zeta^{-n-1} \exp\left\{\zeta - \frac{z^2}{4\zeta}\right\} d\zeta. \tag{5}$$

Now

$$\exp\left(\frac{-z^2}{4\zeta}\right) = 1 - \frac{z^2}{4\zeta} + \frac{z^4}{2 \cdot (4\zeta)^2} - \cdots$$

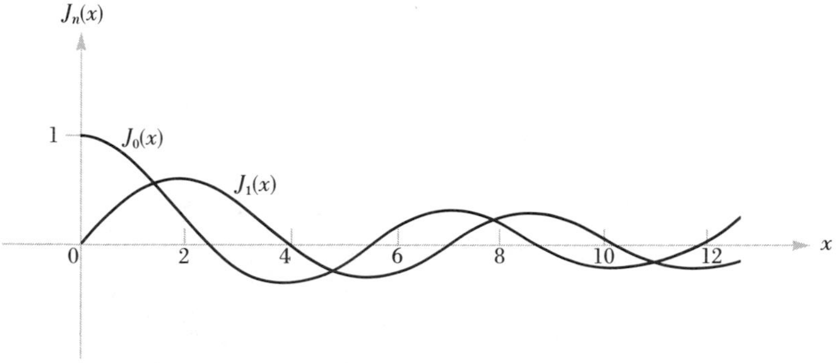

Figure 7.10 Bessel functions $J_0(x)$, $J_1(x)$.

converges uniformly in ζ on γ (Why?), so we can integrate equation (5) term by term (see Theorem 7, Section 3.1) and obtain

$$J_n(z) = \frac{1}{2\pi i} \sum_{k=0}^{\infty} \frac{(-1)^k}{k!} \left(\frac{z}{2}\right)^{n+2k} \int_{\gamma} \frac{e^{\zeta}}{\zeta^{n+k+1}} \, d\zeta.$$

If $n \geq 0$, the residue of e^{ζ}/ζ^{n+k+1} at $\zeta = 0$ is $1/(n+k)!$ (Why?), so

$$J_n(z) = \sum_{k=0}^{\infty} \frac{(-1)^k (z)^{n+2k}}{2^{n+2k} k! (n+k)!}$$

$$= \frac{z^n}{2^n n!} \left\{ 1 - \frac{z^2}{2^2 \cdot 1 (n+1)} + \frac{z^4}{2^4 \cdot 1 \cdot 2 (n+1)(n+2)} - \cdots \right\}. \qquad (6)$$

Thus $J_n(z)$ is entire for $n \geq 0$ and has zero of order n at $z = 0$.

Similarly, for $n \leq 0$,

$$J_n(z) = \sum_{k=0}^{\infty} \frac{(-1)^{k-n} z^{-n+2k}}{2^{-n+2k} (k-n)! k!}. \qquad (7)$$

It follows that if $n \leq 0$, then

$$J_n(z) = (-1)^n J_{-n}(z) \qquad (8)$$

(see exercise 11).

The relationship of Bessel functions to differential equations is as follows. $J_n(z)$ is a solution of *Bessel's Equation*:

$$\frac{d^2 J_n}{dz^2} + \frac{1}{z} \frac{dJ_n}{dz} + \left(1 - \frac{n^2}{z^2}\right) J_n = 0. \qquad (9)$$

Equation (9) is obtained by differentiating either equation (4) or equation (5) for $J_n(z)$ and inserting the result into Bessel's Equation (see exercise 2). Note that both J_n and J_{-n} satisfy equation (9).

Remark.   If $n \geq 0$ but n is not an integer, we set

$$J_n(z) = \sum_{k=0}^{\infty} \frac{(-1)^k z^{n+2k}}{2^{n+2k} k! \, \Gamma(n+k+1)}. \qquad (10)$$

For purposes of simplicity, we shall restrict ourselves to the case where n is an integer.

Some basic identities can be obtained from the following relations:

$$\frac{d}{dz} \left(z^{-n} J_n(z)\right) = -z^{-n} J_{n+1}(z); \qquad z \neq 0. \qquad (11)$$

This can be proven directly by differentiating the power series. (Term-by-term differentiation is, of course, valid — see Sections 3.1 and 3.2.) The student is requested to establish such a proof in exercise 3.

From identity (11) we differentiate out $z^{-n}J_n(z)$ and obtain

$$\frac{d}{dz}J_n(z) = \frac{n}{z}[J_n(z)] - J_{n+1}(z). \tag{12}$$

Let us write $-n$ for n in equation (11). We get

$$\frac{d}{dz}\{z^nJ_{-n}(z)\} = -z^nJ_{-n+1}(z).$$

But $J_{-n}(z) = (-1)^nJ_n(z)$, so

$$\frac{d}{dz}(z^nJ_n(z)) = z^nJ_{n-1}(z),$$

that is,

$$\frac{d}{dz}J_n(z) = J_{n-1}(z) - \frac{n}{z}J_n(z). \tag{13}$$

Combining equations (12) and (13), we get

$$\frac{d}{dz}J_n(z) = \frac{1}{2}\{J_{n-1}(z) - J_{n+1}(z)\}. \tag{14}$$

Relations (11), (12), and (14) are called the *recurrence relations* for Bessel functions; they are summarized in Table 7.2.

This brief study is concluded with a statement of the asymptotic formula for $J_n(z)$.

Theorem 10.   *The following formula holds:*

$$J_n(z) \sim \sqrt{\frac{2}{\pi z}}\left\{\cos\left(z - \frac{n\pi}{2} - \frac{\pi}{4}\right)\right\} \tag{15}$$

as $z \to \infty$, z real and greater than zero.

(Actually, this theorem is also valid for $|\arg z| < \pi$.)

Proof.   We use Theorem 7 and the representation

$$J_n(z) = \frac{1}{2\pi}\left\{\int_0^\pi e^{iz\sin\theta - ni\theta}\, d\theta + \int_0^\pi e^{-iz\sin\theta + ni\theta}\, d\theta\right\} \tag{16}$$

obtained in equation (4). First, let us consider

$$f(z) = \int_0^\pi e^{iz\sin\theta - ni\theta}\, d\theta.$$

Table 7.2   Properties of Bessel Functions

1. $J_n(z) = \dfrac{1}{\pi} \displaystyle\int_0^\pi \cos(n\theta - z\sin\theta)\, d\theta$, where n is an integer.

2. $|J_n(z)| \le 1$ for z real.

3. $J_n(z) = \displaystyle\sum_{k=0}^\infty \dfrac{(-1)^k z^{n+2k}}{2^{n+2k} k! (n+k)!};\ n \ge 0.$

4. J_n is entire and has a zero of order n at $z = 0$; $J_0(0) = 1$.

5. $J_n(z) = (-1)^n J_{-n}(z)$.

6. Bessel's Equation:

$$\frac{d^2 J_n}{dz^2} + \frac{1}{z}\frac{dJ_n}{dz} + \left(1 - \frac{n^2}{z^2}\right) J_n = 0.$$

7. $\dfrac{d}{dz}\{z^{-n} J_n(z)\} = -z^{-n} J_{n+1}(z).$

8. $\dfrac{d}{dz} J_n(z) = \dfrac{n}{z} J_n(z) - J_{n+1}(z).$

9. $\dfrac{d}{dz} J_n(z) = \dfrac{\{J_{n-1}(z) - J_{n+1}(z)\}}{2}.$

10. $J_n(z) \sim \sqrt{\dfrac{2}{\pi z}}\cos\left(z - \dfrac{n\pi}{2} - \dfrac{\pi}{4}\right)$ as $z \to \infty$; z real and greater than zero.

According to the notation of Theorem 7, $h(\zeta) = \sin\zeta$, $g(\zeta) = e^{-ni\zeta}$. We must check the hypotheses of that theorem. Condition (i) is clear. Condition (ii) is valid because the maximum of h on $[0,\pi]$ is at $\zeta_0 = \pi/2$ where $h(\zeta_0) = \sin\zeta_0 = 1$. Also, $h'(\zeta_0) = 0$, $h''(\zeta_0) = -1 < 0$, so (iii) holds. To check (iv) we proceed as follows. Because of remark (ii) following Theorem 7, it suffices to check that we can define a real and differentiable square root of $1 - \sin\zeta$ on $\zeta \in [0,\pi]$. This follows since $1 - \sin\zeta \ge 0$ and $\cos(\pi/2) = 0$. Moreover, $1 - \sin\zeta$ is decreasing on $[0,\pi/2]$ and is increasing on $[\pi/2,\pi]$. Thus (iv) holds. Therefore,

$$f(z) \sim \frac{e^{iz}\sqrt{2\pi}\,e^{-\pi i/4}}{\sqrt{z}} \cdot e^{-ni\pi/2} = \sqrt{\frac{2\pi}{z}}\, e^{i(z - n\pi/2 - \pi/4)}. \qquad (17)$$

Similarly, if we set $g(z) = \int_0^\pi e^{-iz\sin\theta + in\theta}\, d\theta$, we get

$$g(z) \sim \sqrt{\frac{2\pi}{z}}\, e^{-i(z - n\pi/2 - \pi/4)}. \qquad (18)$$

(Proof of this is requested in exercise 8.) Adding equations (17) and (18), we obtain, from equation (16),

$$J_n(z) \sim \left(\frac{1}{2\pi}\right) \sqrt{\frac{2\pi}{z}} \cdot 2 \cos\left(z - \frac{n\pi}{2} - \frac{\pi}{4}\right),$$

which is the result claimed. ∎

Thus, for large x, $J_n(x)$ behaves like $\sqrt{2/\pi x} \{\cos(x - \theta)\}$ where θ is a certain angle called the *phase shift*.

Exercises

1. Show that

$$\Gamma(z + 1) = \sqrt{2\pi} \, z^{z+1/2} \, e^{-z}\left\{1 + O\!\left(\frac{1}{z}\right)\right\}$$

and that

$$\Gamma(z + 1) = \sqrt{2\pi} \, z^{z+1/2} \, e^{-z}\left\{1 + \frac{1}{12z} + O\!\left(\frac{1}{z^2}\right)\right\}$$

as $z \to \infty$. (z is real and greater than zero.)

2. Prove that $J_n(z)$ satisfies Bessel's Equation (formula 6 of Table 7.2).
3. Prove that $d[z^{-n}J_n(z)]/dz = -z^{-n}J_{n+1}(z)$, for all n.
4. Prove that $J_0'(z) = -J_1(z)$.
5. Prove that $J_2(z) = J_0''(z) - J_0'(z)/z$.
6. Use the recurrence relations for Bessel functions and Rolle's Theorem from calculus to show that between two consecutive real positive zeros of $J_n(x)$, there is exactly one zero of $J_{n+1}(x)$. Show that $J_n(x)$ and $J_{n+1}(x)$ have no common roots.
7. Prove that $J_{1/2}(z) = \sqrt{2/\pi z} \{\sin z\}$, using the definition of $J_n(z)$ for non-integral n given by equation (10).
8. Complete the proof that

$$J_n(z) \sim \sqrt{\frac{2}{\pi z}} \cos\left(z - \frac{n\pi}{2} - \frac{\pi}{4}\right)$$

by showing that

$$\int_0^\pi e^{-iz\sin\theta + ni\theta} \, d\theta \sim \sqrt{\frac{2\pi}{z}} \, e^{-i(z - n\pi/2 - \pi/4)}.$$

(The function on the left side of the expression is called a *Hankel function*.)
9. Verify that the asymptotic expansion for $J_n(z)$ is consistent with Bessel's Equation.
10. Verify that $\varphi(x) = J_n(kx)$ is a solution of

$$\frac{d}{dx}(x\varphi') + \left(k^2 x - \frac{n^2}{x}\right)\varphi(x) = 0$$

with $\varphi(0) = 0$, $\varphi(a) = 0$, where ka is any of the zeros of J_n; $n \neq 0$.
11. Establish equations (7) and (8) on p. 382.

Review Exercises for Chapter 7

1. Show that $|d^k J_n(z)/dz^k| \leq 1$; $k = 0, 1, 2, \cdots$, for any n, z real.

2. Show that $\lim\limits_{x \to \infty} J_n(x) = 0$.

3. Let $f(z) = \int_0^\pi e^{iz\sin\theta} \cos^2\theta \, d\theta$. Using Theorem 7, find an asymptotic formula for $f(z)$. Assume that as $z \to \infty$, z is real and positive.

4. Let $H_m = \sum_{n=1}^\infty 1/n^m$. Prove that

$$\log \Gamma(1 + z) = -\gamma z + \sum_{n=2}^\infty \frac{(-1)^n}{n} H_n z^n; \qquad |z| < 1,$$

where γ is Euler's Constant.

5. Prove that

$$\Gamma(z) = \sum_{n=0}^\infty \frac{(-1)^n}{n!(z+n)} + Q(z)$$

where $Q(z)$ is entire. In addition, show that

$$Q(z) = \int_1^\infty t^{z-1} e^{-t} dt.$$

6. Prove that $x^n J_n(x) = \int_0^x t^n J_{n-1}(t) \, dt$; $n = 1, 2, \cdots$.

7. Prove that $\Gamma|(1/2) + iy| \to 0$ as $y \to \infty$.

8. Write out $\int_0^\infty e^{-t^3} \, dt$ in terms of the gamma function.

   Hint.   Change to the variable $y = t^3$.

9. Obtain an asymptotic expansion for $\int_{-\infty}^\infty e^{-zy^2/2} \cos y^2 \, dy$. (As $z \to \infty$, $z > 0$.)

10. Prove that $J_n(iy) \sim i^n e^y / \sqrt{2\pi y}$. (As $y \to \infty$, $y > 0$.)

11. In this exercise you are asked to develop some properties of the *Legendre functions* (see review exercise 34, Chapter 3). These functions are encountered in the study of differential equations (specifically Laplace's Equation in three dimensions, which describes a wide range of physical phenomena) when spherical coordinates are used.*
    a. For $-1 < x < 1$, set

$$P_n(x) = \frac{1}{2^{n+1}\pi i} \int_\gamma \frac{(t^2 - 1)^n}{(t - x)^{n+1}} \, dt$$

where γ is the contour as shown in Figure 7.11. By differentiating under the integral sign (see Theorem 4, Section 2.4), show that $P_n(x)$ solves *Legendre's Equation*:

$$(1 - x^2)y'' - 2xy' + n(n+1)y = 0.$$

*Consult, for example, G. F. D. Duff and D. Naylor, *Differential Equations of Applied Mathematics* (New York: Wiley, 1965).

Figure 7.11

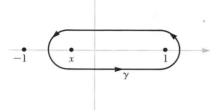

b. For an integer n, derive *Rodrigues' Formula*:

$$P_n(x) = \frac{1}{2^n n!} \frac{d^n}{dx^n} (x^2 - 1)^n.$$

Remark. This formula yields an obvious analytic extension $P_n(z)$.

c. Show that

$$P_n(x) = \frac{1}{n!} \frac{d^n}{dt^n} \frac{1}{(t^2 - 2tx + 1)^{1/2}} \bigg|_{t=0}$$

and deduce from Taylor's Theorem that

$$\frac{1}{(1 - 2tz + t^2)^{1/2}} = \sum_{n=0}^{\infty} P_n(z) t^n.$$

d. Develop recurrence relations for the coefficients of solutions of Legendre's Equation. Then use these relations to show that entire solutions must be of the form

$$\psi(x) = \sum_{k=0}^{\infty} a_k x^{2k}, \qquad n \text{ even;}$$

or of the form

$$\rho(x) = \sum_{k=0}^{\infty} b_k x^{2k+1}, \qquad n \text{ odd.}$$

Show that these are actually polynomials; that is, a_k and b_k vanish for large k.

e. Using (c), show that $nP_n(x) = (2n-1)xP_{n-1}(x) - (n-1)P_{n-2}(x)$.

f. Prove that $P_0(x) = 1$, $P_1(x) = x$, $P_2(x) = (3x^2 - 1)/2$.

g. Show that

$$\int_{-1}^{1} P_n(x) \, P_m(x) \, dx = \begin{cases} 0, & n \neq m; \\ \dfrac{2}{(2n+1)}, & n = m. \end{cases}$$

Hint. Use (b) to prove the case where $n \neq m$; use (c) to prove the case where $n = m$.

h. Obtain the asymptotic formula $P_n(z) \sim [(2n)!/2^n(n!)^2]z^n$ as $z \to \infty$, using (b).

Chapter 8

The Laplace Transform and Applications

This final chapter gives an introduction to and some elementary applications of the Laplace transform. The first section introduces the Laplace transform and develops some of its basic properties. There are two key properties that make the Laplace transform useful for differential equations. First, it behaves well with respect to differentiation, and second, a function can be recovered if its Laplace transform is known. The closely related Fourier transform also enjoys these properties but will not be discussed here in as much detail because it is not as closely connected with complex analysis as is the Laplace transform. However, there is some discussion of the Fourier transform and its applications in Section 8.4.

Section 8.2 develops techniques for inverting Laplace transforms; Section 8.3 considers some applications of Laplace transforms to ordinary differential equations. The final section shows how some techniques of complex analysis are used in theoretical physics.

The Laplace transform is a powerful tool that is used in both pure and applied mathematics. It is important, therefore, to have a good grasp of both its basic theory and its usefulness.

8.1 Basic Properties of Laplace Transforms

Consider a (real- or complex-valued) function $f(t)$ defined on $[0,\infty[$. The *Laplace transform* of f is defined to be the function $\tilde{f}$ of a complex variable z given by

$$\tilde{f}(z) = \int_0^\infty e^{-zt} f(t) \ dt; \tag{1}$$

$\tilde{f}$ will be defined for those $z \in \mathbb{C}$ for which the integral converges. Other common notations for $\tilde{f}$ are $\mathscr{L}(f)$ or simply F.

For technical reasons, it will be convenient to impose a mild restriction on the functions we consider. We will want $f\colon [0,\infty[\longrightarrow \mathbb{C}$ (or $\mathbb{R}$) to be of *exponential order*. This means that there should be constants $A > 0$, $B \in \mathbb{R}$, such that

$$|f(t)| \le A \ e^{tB} \tag{2}$$

for all $t \ge 0$. In other words, f should not grow too fast; for example, any polynomial satisfies this condition (Why?).

All functions considered in the remainder of this chapter will be assumed to be of exponential order. It will also be assumed that on any finite interval $[0,a]$, f is bounded and integrable. (If, for example, we assume that f is piecewise continuous, this last condition will hold.)

Abscissa of convergence

The first important result in this chapter concerns the nature of the set on which $\tilde{f}(z)$ is defined and is analytic.

Theorem 1.   *Let $f\colon [0,\infty[\longrightarrow \mathbb{C}$ (or $\mathbb{R}$) be of exponential order and let*

$$\tilde{f}(z) = \int_0^\infty e^{-zt} f(t)dt.$$

There exists a unique number $\sigma, -\infty \le \sigma < \infty$, such that

$$\int_0^\infty e^{-zt} f(t)dt \begin{cases} converges \ if \ \mathrm{Re} \ z > \sigma; \\ diverges \ if \ \mathrm{Re} \ z < \sigma. \end{cases} \tag{3}$$

Furthermore, $\tilde{f}$ is analytic on the set $A = \{z \mid \mathrm{Re} \ z > \sigma\}$ and we have

$$\frac{d}{dz} \tilde{f}(z) = -\int_0^\infty te^{-zt} f(t)dt \tag{4}$$

for $\mathrm{Re} \ z > \sigma$.

The number σ is called the abscissa of convergence and if we define the number ρ by

$$\rho = \inf\{B \in \mathbb{R} \mid \ there \ exists \ an \ A > 0 \ such \ that \ |f(t)| \le Ae^{Bt}\}, \tag{5}$$

then $\sigma \le \rho$.

The set $\{z \mid \mathrm{Re} \ z > \sigma\}$ is called the *half plane of convergence*. (If $\sigma = -\infty$, it is all of $\mathbb{C}$; see Figure 8.1.)

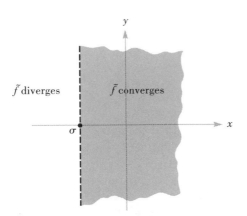

Figure 8.1 Half plane of convergence
of the Laplace transform.

In general, it is impossible to tell whether $\tilde{f}(z)$ will converge on the
vertical line Re $z = \sigma$.

The proof of Theorem 1 and more detailed convergence results
appear at the end of the section.

If there is any danger of confusion we can write $\sigma(f)$ for σ or $\rho(f)$
for ρ. A convenient way to compute $\sigma(f)$ is described in examples 2
and 3.

The map $f \mapsto \tilde{f}$ is clearly linear in the sense that $(af + bg)^{\sim} = a\tilde{f} + b\tilde{g}$,
valid for Re $z > \max\{\sigma(f), \sigma(g)\}$. It is also true that the map is
one-to-one; that is, $\tilde{f} = \tilde{g}$ implies that $f = g$; a function $\varphi(z)$ is the
Laplace transform of at most one function. Precisely:

Theorem 2. *Suppose that f and h are continuous and that $\tilde{f}(z) = \tilde{h}(z)$
for Re $z > y_0$ for some y_0. Then $f(t) = h(t)$ for all $t \in [0, \infty[$.*

This theorem is not as simple as it seems. There is not enough
machinery to give a complete proof, but the main ideas of a proof are
given at the end of the section. Using ideas from integration theory
we could extend the result of Theorem 2 to discontinuous functions as
well, but we would have to modify what we mean by "equality of functions." Clearly, if $f(t)$ is changed at a single value of t, $\tilde{f}$ is unchanged.

Theorem 2 enables us to give a meaningful answer to the problem,
"Given $g(z)$, find $f(t)$ such that $\tilde{f} = g$," because it makes clear that
there can be at most one such (continuous) f. We call f the *inverse
Laplace transform* of g; methods for finding f when g is given are considered in the next section.

Laplace transforms of derivatives

The main utility of Laplace transforms is that they enable us to transform differential problems into algebraic problems. When the latter are solved, the answers are obtained by using the inverse Laplace transform. The procedure is based on the following theorem.

Theorem 3. *Let $f(t)$ be continuous on $[0,\infty[$ and piecewise C^1; that is, piecewise continuously differentiable. Then for Re $z > \rho$ (as defined in Theorem 1),*

$$\left(\frac{\widetilde{df}}{dt}\right)(z) = z\tilde{f}(z) - f(0). \tag{6}$$

Proof.   By definition,

$$\left(\frac{\widetilde{df}}{dt}\right)(z) = \int_0^\infty e^{-zt}\frac{df}{dt}(t)\ dt.$$

Integrating by parts, we get

$$\lim_{t_0 \to \infty}\left(e^{-zt}f(t)\Big|_0^{t_0}\right) + \int_0^\infty ze^{-zt}f(t)\ dt.$$

(If the student has not encountered this method previously, he should check that integration by parts is valid for piecewise C^1 functions.)

By definition of ρ, $|e^{-Bt_0}\cdot f(t_0)| \le A$ for some $B < $ Re z. Thus $|e^{-zt_0}\cdot f(t_0)| = |e^{-(z-B)t_0}|\ |e^{-Bt_0}\cdot f(t_0)| \le e^{-(\text{Re }z - B)t_0}A$, which approaches zero as $t_0 \to \infty$. Therefore, we get $-f(0) + z\tilde{f}(z)$, as asserted. ∎

Note that we have proved that $(df/dt)^{\sim}$ exists for Re $z > \rho$, although its abscissa of convergence might be smaller than ρ.

If we apply equation (6) to d^2f/dt^2, we obtain

$$\left(\frac{\widetilde{d^2f}}{dt^2}\right)(z) = z^2\tilde{f}(z) - zf(0) - \frac{df}{dt}(0). \tag{7}$$

Equation (4) of Theorem 1 is the related formula $\tilde{g}(z) = d\tilde{f}(z)/dz$, where $g(t) = -tf(t)$. In exercise 3 the student is asked to prove Theorem 4, which contains a similar formula for integrals.

Theorem 4.   *Let $g(t) = \int_0^t f(\tau)d\tau$. Then for Re $z > \max\{0,\rho(f)\}$,*

$$\tilde{g}(z) = \frac{\tilde{f}(z)}{z}. \tag{8}$$

Shifting theorems

Table 8.1 (p. 400) lists some formulas that are useful for computing $\tilde{f}(z)$. The proofs of these formulas are straightforward and are included in the exercises and examples. However, three of the formulas are sufficiently important to be given separate explanation, which is done in the following three theorems. The first two of these are called the shifting theorems.

Theorem 5.   *Fix $a \in \mathbb{C}$ and let $g(t) = e^{-at}f(t)$. Then for $\operatorname{Re} z > \sigma(f)$ − $\operatorname{Re} a$,*

$$\tilde{g}(z) = \tilde{f}(z + a). \tag{9}$$

Proof.   By definition,

$$\tilde{g}(z) = \int_0^\infty e^{-zt} \, e^{-at} \, f(t) \, dt$$

$$= \int_0^\infty e^{-(z+a)t} \, f(t) \, dt$$

$$= \tilde{f}(z + a),$$

which is valid if $\operatorname{Re}(z + a) > \sigma.$ ■

Theorem 6.   *Let*

$$H(t) = \begin{cases} 0 & \text{if } \ t < 0; \\ 1 & \text{if } \ t \ge 0. \end{cases}$$

(This is called the Heaviside, or unit step function.) Also, let $a \ge 0$ and let $g(t) = f(t - a) \, H(t - a)$; that is,

$$g(t) = \begin{cases} 0 & \text{if } \ t < a; \\ f(t - a) & \text{if } \ t \ge a \end{cases}$$

(see Figure 8.2). *Then for $\operatorname{Re} z > \sigma$, we have*

$$\tilde{g}(z) = e^{-az} \, \tilde{f}(z). \tag{10}$$

Proof.   By definition and because $g = 0$ for $0 \le t < a$,

$$\tilde{g}(z) = \int_0^\infty e^{-zt} \, g(t) \, dt = \int_a^\infty e^{-zt} \, f(t - a) \, dt.$$

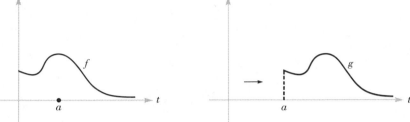

Figure 8.2 The shifting theorems.

Letting $\tau = t - a$, we get

$$\tilde{g}(z) = \int_0^\infty e^{-z(\tau+a)} f(\tau) d\tau = e^{-za} \tilde{f}(z). \blacksquare$$

From equation (10) we can deduce that if $a \geq 0$, and $g(t) = f(t) H(t-a)$, then $\tilde{g}(z) = e^{-az} \tilde{F}(z)$ where $F(t) = f(t + a)$, $t \geq 0$ (see Figure 8.3).

Convolutions

The *convolution* of functions $f(t)$ and $g(t)$ is defined by

$$(f * g)(t) = \int_0^\infty f(t - \tau) \cdot g(\tau) \ d\tau, \qquad t \geq 0, \tag{11}$$

where we set $f(t) = 0$ if $t < 0$. (Thus the integration is merely from 0 to t.)

The convolution operation is related to Laplace transforms in the following way.

Theorem 7. *We have $f * g = g * f$ and*

$$(f \widetilde{* g})(z) = \tilde{f}(z) \cdot \tilde{g}(z) \tag{12}$$

whenever Re $z > $ max $(\rho(f), \rho(g))$.

In brief, equation (12) states that the Laplace transform of a convolution of two functions is the product of their Laplace transforms. It is precisely this property that makes the convolution an operation of interest to us.

394]

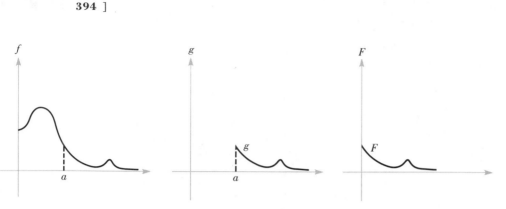

Figure 8.3 *g* is obtained from *f* by shifting and truncating.

Proof. We have

$$(f \widetilde{*} g)(z) = \int_0^\infty e^{-zt}\left(\int_0^\infty f(t-\tau)\cdot g(\tau)\, d\tau\right) dt$$

$$= \int_0^\infty e^{-z\tau} e^{-z(t-\tau)} \int_0^\infty f(t-\tau)g(\tau)\, d\tau dt.$$

For Re $z > \rho(f),\ \rho(g)$ the integrals for $\tilde{f}(z)$ and $\tilde{g}(z)$ converge absolutely, so we can interchange the order of integration* and thus obtain

$$\int_0^\infty e^{-z\tau}\left\{\int_0^\infty e^{-z(t-\tau)} f(t-\tau)\, dt\right\}g(\tau)\, d\tau.$$

Letting $x = t - \tau$ and remembering that $f(s) = 0$ if $s < 0$, we get

$$\int_0^\infty e^{-z\tau} \tilde{f}(z)g(\tau)\, d\tau = \tilde{f}(z)\cdot\tilde{g}(z).$$

By changing variables, it is not difficult to verify that $f * g = g * f$, but such verification also follows from what we have done if f and g are continuous. We have $(f * g)\widetilde{\ } = \tilde{f}\cdot\tilde{g} = \tilde{g}\cdot\tilde{f} = (g * f)\widetilde{\ }$. Thus $(f * g - g * f)\widetilde{\ } = 0$, so by Theorem 2, $f * g - g * f = 0$. ∎

*This is a theorem concerning integration theory from advanced calculus. See, for instance, J. Marsden, *Elementary Classical Analysis* (San Francisco: W. H. Freeman and Co., 1974). Chapter 9.

Technical proofs of theorems

Theorem 1. *Let $f: [0,\infty[\to \mathbb{C}$ (or $\mathbb{R}$) be of exponential order and let*

$$\tilde{f}(z) = \int\limits_{0}^{\infty} e^{-zt} f(t)\,dt.$$

There exists a unique number σ, $-\infty \leq \sigma < \infty$, such that

$$\int\limits_{0}^{\infty} e^{-zt} f(t)\,dt \begin{cases} converges\ if\ \mathrm{Re}\ z > \sigma; \\ diverges\ if\ \mathrm{Re}\ z < \sigma. \end{cases} \tag{3}$$

Furthermore, f is analytic on the set $A = \{z \mid \mathrm{Re}\ z > \sigma\}$ and we have

$$\frac{d}{dz}\tilde{f}(z) = -\int\limits_{0}^{\infty} t e^{-zt} f(t)\,dt \tag{4}$$

for $\mathrm{Re}\ z > \sigma$.

  The number σ is called the abscissa of convergence and if we define the number ρ by

$$\rho = \inf\{B \in \mathbb{R} \mid \text{there exists an } A > 0 \text{ such that } |f(t)| \leq Ae^{Bt}\}, \tag{5}$$

then $\sigma \leq \rho$.

To prove this theorem we shall use the following important result.

Lemma 1. *Suppose that $\tilde{f}(z) = \int_0^\infty e^{-zt}f(t)\,dt$ converges for $z = z_0$. Let $0 \leq \theta < \pi/2$ and define the set*

$$S_\theta = \{z \mid |\arg(z - z_0)| \leq \theta\}$$

(see Figure 8.4). *Then $\tilde{f}$ converges uniformly on S_θ.*

Proof of Lemma 1.   Let $h(x) = \int_0^x e^{-z_0 t}f(t)\,dt - \int_0^\infty e^{-z_0 t}f(t)\,dt$ so that $h \to 0$ as $x \to \infty$. We must show that for every $\epsilon > 0$, there is a t_0 such that $t_1, t_2 \geq t_0$ implies that

$$\left| \int\limits_{t_1}^{t_2} e^{-zt}f(t)\,dt \right| < \epsilon$$

for all $z \in S_\theta$. It follows that $\int_0^x e^{-zt}f(t)\,dt$ converges uniformly on S_θ as $x \to \infty$, by the Cauchy criterion. We will make use of the function $h(x)$ as follows. Write

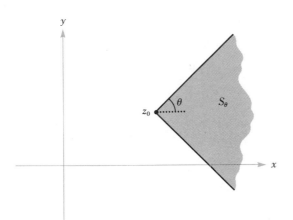

Figure 8.4 Sector of uniform convergence.

$$\int_{t_1}^{t_2} e^{-zt}f(t)\,dt = \int_{t_1}^{t_2} e^{-(z-z_0)t}[e^{-z_0 t}f(t)]\,dt.$$

If we integrate by parts, we get (as the student can easily check)

$$e^{-(z-z_0)t_2}h(t_2) - e^{-(z-z_0)t_1}h(t_1) + (z-z_0)\int_{t_1}^{t_2} e^{-(z-z_0)t}h(t)\,dt.$$

Given $\epsilon > 0$, choose t_0 such that $|h(t)| < \epsilon/3$ and $|h(t)| < \epsilon'$ $= \epsilon/(6\sec\theta)$ if $t \geq t_0$. Then for $t_2 > t_0$,

$$\left|e^{-(z-z_0)t_2}h(t_2)\right| \leq |h(t_2)| < \frac{\epsilon}{3},$$

since $\left|e^{-(z-z_0)t_2}\right| = e^{-(\text{Re}\,z - \text{Re}\,z_0)t_2} \leq 1$ because $\text{Re}\,z > \text{Re}\,z_0$. Similarly, for $t_1 > t_0$,

$$\left|e^{-(z-z_0)t_1}h(t_1)\right| < \frac{\epsilon}{3}.$$

We must still estimate the last term:

$$\left|(z-z_0)\int_{t_1}^{t_2} e^{-(z-z_0)t}h(t)\,dt\right| \leq |z-z_0|\,\epsilon'\int_{t_1}^{t_2} e^{-(x-x_0)t}\,dt$$

where $x = \text{Re}\,z$, and $x_0 = \text{Re}\,z_0$. If $z = z_0$, this term is zero. If $z \neq z_0$, then $x \neq x_0$, and we get

$$\epsilon' \frac{|z - z_0|}{x - x_0} \{e^{-(x-x_0)t_1} - e^{-(x-x_0)t_2}\} < 2\epsilon' \frac{|z - z_0|}{x - x_0} \leq 2\epsilon' \sec \theta = \frac{\epsilon}{3}$$

(see Figure 8.5). Note that the restriction $0 \leq \theta < \pi/2$ is necessary for $\sec \theta = 1/\cos \theta$ to be finite.

Combining the preceding inequalities, we get

$$\left| \int_{t_1}^{t_2} e^{-zt} f(t) \; dt \right| < \epsilon$$

if $t_1, t_2 \geq t_0$ for all $z \in S_\theta$, thus completing the proof of the lemma. ∎

Proof of Theorem 1.   Let $\sigma = \inf\{x \in \mathbb{R} \mid \int_0^\infty e^{-xt} f(t) \; dt$ converges$\}$, where inf stands for "greatest lower bound." We note from Lemma 1 that if $\tilde{f}(z_0)$ converges, then, more specifically, $\tilde{f}(z)$ converges if $\operatorname{Re} z > \operatorname{Re} z_0$ because z lies in some S_θ for z_0 (Why?).

Let $\operatorname{Re} z > \sigma$. By the definition of σ there is an $x_0 < \operatorname{Re} z$ such that $\int_0^\infty e^{-x_0 t} f(t) dt$ converges. Hence $\tilde{f}(z)$ converges by Lemma 1. Conversely, let $\operatorname{Re} z < \sigma$ and $\operatorname{Re} z < x < \sigma$. If $\tilde{f}(z)$ converges, then so does $\tilde{f}(x)$, and therefore to say that $\sigma \leq x$ is a contradiction. Thus $\tilde{f}(z)$ does not converge if $\operatorname{Re} z < \sigma$.

Next, using Theorem 6, Section 3.1, we prove that f is analytic on $\{z \mid \operatorname{Re} z > \sigma\}$. Let $g_n(z) = \int_0^n e^{-zt} f(t) dt$. Then $g_n(z) \longrightarrow \tilde{f}(z)$. By example 4, Section 2.4, g_n is analytic with $g_n'(z) = -\int_0^n te^{-zt} f(t) dt$. We must show that $g_n \longrightarrow \tilde{f}$ uniformly on closed disks in $\{z \mid \operatorname{Re} z > \sigma\}$.

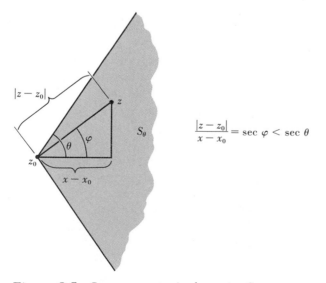

$$\frac{|z - z_0|}{x - x_0} = \sec \varphi < \sec \theta$$

Figure 8.5   Some geometry in the region S_θ.

But each disk lies in some S_θ relative to some $\dot{z}_0$ with Re $z_0 > \sigma$ (Figure 8.6). This statement is geometrically clear and can also be checked analytically.

Thus, by Theorem 6, Section 3.1, $\tilde{f}$ is analytic on $\{z \mid \text{Re } z > \sigma\}$ and

$$(\tilde{f})'(z) = -\int_0^\infty te^{-zt}f(t)\,dt.$$

We know automatically that this integral representation for the derivative of $\tilde{f}$ will converge for Re $z > \sigma$, as will all the iterated derivatives.

It remains to be shown that $\sigma \le \rho$. To prove this we merely need to show that $\sigma \le B$ if $|f(t)| \le Ae^{Bt}$. This will hold, by what we have proven, if $\tilde{f}(z)$ converges whenever Re $z > B$. Indeed we shall show absolute convergence.

Note that $|e^{-zt}f(t)| = |e^{-(z-B)t}e^{-Bt}f(t)| \le e^{-(\text{Re }z - B)t}A$. Since $\int_0^\infty e^{-\alpha t}dt = 1/\alpha$ converges for $\alpha > 0$, $\int_0^\infty e^{-zt}f(t)\,dt$ converges absolutely. ∎

To prove that $\tilde{f} = \tilde{h}$ implies that $f = h$ for continuous functions f and h, it suffices, by considering $f - h$, to prove the following special case of Theorem 2.

Theorem 2'.   *Suppose that f is continuous and that $\tilde{f}(z) = 0$ whenever* Re $z > y_0$ *for some y_0. Then $f(t) = 0$ for all $t \in [0,\infty[$.*

The crucial lemma we use to prove Theorem 2' is the following.

Lemma 2.   *Let f be continuous on $[0,1]$ and suppose that $\int_0^1 t^n f(t)\,dt = 0$ for all $n = 0, 1, 2, \cdots$. Then $f = 0$.*

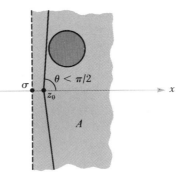

Figure 8.6   Each disk lies in S_θ for some θ, $0 \le \theta < \pi/2$.

This assertion is reasonable since it follows that $\int_0^1 P(t)f(t)\,dt = 0$ for any polynomial P.

Proof of Lemma 2.   The precise proof depends on the *Weierstrass Approximation Theorem*, which states that any continuous function is the uniform limit of polynomials; see, for example, J. Marsden, *Elementary Classical Analysis* (San Francisco: W. H. Freeman, 1974), Chapter 5. By this theorem we get $\int_0^1 g(t)f(t)\,dt = 0$ for any continuous g. The result follows by taking $g = f$ and applying the fact that if the integral of a nonnegative continuous function is zero, then the function is zero. ∎

Proof of Theorem 2′.   We now can easily prove Theorem 2′. Suppose that

$$f(z) = \int_0^\infty e^{-zt}f(t)\,dt = 0$$

whenever Re $z > \sigma$. Fix $x_0 > y_0$ real and let $s = e^{-t}$. By changing variables to express the integrals in terms of s, we get, at $z = x_0 + n$ ($n = 0, 1, 2, \cdots$),

$$\int_0^\infty e^{-nt}e^{-x_0 t}f(t)\,dt = \int_0^1 s^n h(s)\,ds = 0, \qquad n = 0, 1, 2, \cdots,$$

where $h(s) = e^{-x_0 t + t}f(t)$. Thus, by Lemma 2, h, and hence f, are zero. ∎

Remark.   It is useful to note that $\tilde{f}(z) \to 0$ as Re $z \to \infty$. This follows from the arguments used to prove Theorem 1 (see review exercise 5).

Worked Examples

1.   Prove formula 9 in Table 8.1 and find $\sigma(f)$ in that case.

Solution.   By definition,

$$\tilde{f}(z) = \int_0^\infty e^{-at}e^{-zt}dt$$

$$= \int_0^\infty e^{-(a+z)t}dt$$

$$= -\frac{e^{-(a+z)t}}{a+z}\bigg|_0^\infty = \frac{1}{z+a}.$$

Table 8.1   Some Common Laplace Transforms

Definition. $\tilde{f}(z) = \displaystyle\int_0^{\infty} e^{-zt} f(t)\, dt.$

1. $\tilde{g}(z) = -\dfrac{d}{dz}\tilde{f}(z)$ where $g(t) = tf(t)$.

2. $(af + bg)^{\sim} = a\tilde{f} + b\tilde{g}.$

3. $\left(\dfrac{df}{dt}\right)^{\sim}(z) = z\tilde{f}(z) - f(0)$. (Assume that f is piecewise C^1.)

4. $\tilde{g}(z) = \dfrac{1}{z}\tilde{f}(z)$ where $g(t) = \displaystyle\int_0^t f(\tau)\, d\tau.$

5. $\tilde{g}(z) = \tilde{f}(z + a)$ where $g(t) = e^{-at}f(t)$.

6. $\tilde{g}(z) = e^{-az}\tilde{f}(z)$ where $a > 0$ and $g(t) = \begin{cases} 0, & t < a; \\ f(t - a), & t \ge a. \end{cases}$

7. $\tilde{g}(z) = e^{-az}\tilde{F}(z)$, where $a \ge 0$, $F(t) = f(t + a)$, and $g(t) = \begin{cases} 0, & 0 \le t < a; \\ f(t), & t \ge a. \end{cases}$

8. $(f \overset{\sim}{*} g)(z) = \tilde{f}(z) \cdot \tilde{g}(z)$, where $f * g(t) = \displaystyle\int_0^{\infty} f(t - \tau)g(\tau)\, d\tau.$

9. If $f(t) = e^{-at}$, then $\tilde{f}(z) = \dfrac{1}{z + a}$, and $\sigma(f) = -\operatorname{Re} a$.

10. For $f(t) = \cos at$, $\tilde{f}(z) = \dfrac{z}{z^2 + a^2}$ and $\sigma(f) = |\operatorname{Im}(a)|$.

11. If $f(t) = \sin at$, $\tilde{f}(z) = \dfrac{a}{z^2 + a^2}$ and $\sigma(f) = |\operatorname{Im}(a)|$.

12. If $f(t) = t^a$, $a > -1$, $\tilde{f}(z) = \dfrac{\Gamma(a + 1)}{z^{a+1}}$ and $\sigma(f) = 0$.

13. If $f(t) = 1$, $\tilde{f}(z) = \dfrac{1}{z}$ and $\sigma(f) = 0$.

The evaluation at $t = \infty$ is justified by noting that $\displaystyle\lim_{t \to \infty} e^{-(a+z)t} = 0$ if $\operatorname{Re}(a + z) > 0$, since $|e^{-(a+z)t}| = e^{-\operatorname{Re}(a+z)t} \to 0$ as $t \to \infty$. Thus the formula is valid if $\operatorname{Re} z > -\operatorname{Re} a$.

Note.   The formula for $\tilde{f}$ is valid only for $\operatorname{Re} z > -\operatorname{Re} a$, although $\tilde{f}$ coincides there with a function that is analytic except at $z = -a$. This situation is similar to that for the gamma function (see formula 12 of Table 7.1).

Finally, we show that for $f(t) = e^{-at}$, $\sigma(f) = -\operatorname{Re} a$. We have already shown that $\sigma(f) \leq -\operatorname{Re} a$. But the integral diverges at $z = a$, so $\sigma(f) \geq -\operatorname{Re} a$, and thus $\sigma(f) = -\operatorname{Re} a$.

In the case where $a = 0$, this example specializes to formula 13 of Table 8.1.

2. Suppose that we have computed $\tilde{f}(z)$ and found it to converge for $\operatorname{Re} z > \gamma$. Suppose also that $\tilde{f}$ coincides with an analytic function that has a pole on the line $\operatorname{Re} z = \gamma$. Show that $\sigma(f) = \gamma$.

Solution. We know that $\sigma(f) \leq \gamma$ by the basic property of σ in Theorem 1. Also, since $\tilde{f}$ is analytic for $\operatorname{Re} z > \sigma$, there can be no poles in the region $\{z \mid \operatorname{Re} z > \sigma\}$. If $\sigma(f) < \gamma$, there would be a pole in this region. Hence $\sigma(f) = \gamma$ (see Figure 8.7).

3. Let $f(t) = \cosh t$. Compute $\tilde{f}$ and $\sigma(f)$.

Solution. $f(t) = \cosh t = (e^t + e^{-t})/2$. Thus, by formulas 2 and 9 of Table 8.1,

$$\tilde{f}(z) = \frac{1}{2}\left\{\frac{1}{z-1} + \frac{1}{z+1}\right\} = \frac{z}{z^2-1}.$$

Here $\sigma(f) = 1$ by example 2. ($\sigma(e^t) = 1$ and $\sigma(e^{-t}) = -1$, so $\sigma(f) \leq 1$ but it cannot be < 1.)

Exercises

1. Let $f(t) = e^{-e^t}$, $t \geq 0$. Show that $\sigma(f) = -\infty$.

2. Referring to Theorem 1, show that, in general, $\sigma \neq \rho$.

Hint. Consider $f(t) = e^t \sin(e^t)$ and show that $\sigma = 0$, $\rho = 1$.

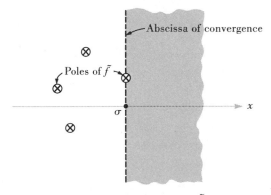

Figure 8.7 Location of poles of $\tilde{f}$.

3. Prove Theorem 4. First establish that $\rho(g) \leq \max\{0, \rho(f)\}$.

4. Give a direct proof that $f * g = g * f$ (see Theorem 7).

5. Prove formulas 10–13 of Table 8.1.

6. Compute the Laplace transform and abscissa of convergence for each of the following functions:
   a. $f(t) = t^2 + 2$.
   b. $f(t) = t + e^{-t} + \sin t$.
   c. $f(t) = (t + 1)^n$, n is a positive integer.

7. Use the shifting theorems to show the following:
   a. If $f(t) = e^{-at}\cos bt$, then

$$\tilde{f}(z) = \frac{z + a}{(z + a)^2 + b^2}.$$

   b. If $f(t) = e^{-at}t^n$, then

$$\tilde{f}(z) = \frac{\Gamma(n + 1)}{(z + a)^{n+1}}.$$

   What is $\sigma(f)$ in each case?

8. Let $f(t) = \begin{cases} 0, & 0 \leq t \leq 1; \\ 1, & 1 < t < 2; \\ 0, & t \geq 2. \end{cases}$
   Compute $\tilde{f}(z)$ and $\sigma(f)$.

9. Let $f(t) = \begin{cases} \sin t, & 0 \leq t \leq \pi; \\ 0, & t > \pi. \end{cases}$
   Compute $\tilde{f}(z)$.

10. Let $g(t) = \int_0^t e^{-s}\sin s \, ds$. Compute $\tilde{g}(z)$. Compute $\tilde{f}(z)$ if $f(t) = tg(t)$.

11. Suppose that f is periodic with period p (that is, $f(t + p) = f(t)$ for all $t \geq 0$). Prove that

$$\tilde{f}(z) = \frac{\int_0^p e^{-zt}f(t) \, dt}{1 - e^{-pz}}$$

   is valid if $\text{Re } z > 0$.

   Hint. Write out $\tilde{f}(z)$ as an infinite sum.

12. Use exercise 11 to prove that

$$\tilde{f}(z) = \frac{1}{z} \cdot \frac{1 - e^{-z}}{1 - e^{-2z}}$$

   where $f(t)$ is as illustrated in Figure 8.8.

13. Let $f(t) = t\cos at$. Show that

$$\tilde{f}(z) = \frac{z^2 - a^2}{(z^2 + a^2)^2}, \qquad \sigma(f) = |\text{Im } a|.$$

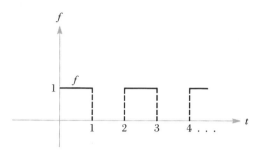

Figure 8.8   The unit pulse function.

14.   Let $f(t) = t\sin at$. Compute $\tilde{f}(z)$.
15.   Let $f(t) = (\sin at)/t$. Show that $\tilde{f}(z) = \tan^{-1}(a/z)$.

8.2   The Complex Inversion Formula

As previously mentioned, it is important to be able to compute $f(t)$ when $\tilde{f}(z)$ is known. The general formula for such a computation, called the Complex Inversion Formula, will be established in this section. In addition, by using the formulas of Table 8.1 in reverse order, we can also obtain a number of useful alternative techniques (see examples 1 and 2).

Complex Inversion Formula

The proof of the Complex Inversion Formula draws on many of the main points developed in the first four chapters of this book. It should be regarded as one of the key results of our analysis of the Laplace transform.

Theorem 8 (Complex Inversion Formula).   *Let $F(z)$ be analytic on $\mathbb{C}$, except that it has a finite number of poles. Suppose that $F(z)$ is analytic on a half plane $\{z \mid \text{Re } z > \sigma\}$ and that $|F(z)| \leq M/|z|$ for all $|z| \geq R$ for constants $M > 0$ and $R > 0$. (Actually, it is only necessary that $|F(z)| \leq M/|z|^{\beta}$ for some $\beta > 0$ and $|z| \geq R$.) In particular, this assumption holds if $F(z) = P(z)/Q(z)$ for polynomials P and Q with $\deg Q \geq \deg P + 1$.*
    For $t \geq 0$, let

$$f(t) = \sum \{Residues \ of \ e^{zt}F(z) \ at \ each \ of \ its \ poles\}. \qquad (1)$$

Then $\tilde{f}(z) = F(z)$ for $\text{Re } z > \sigma$.

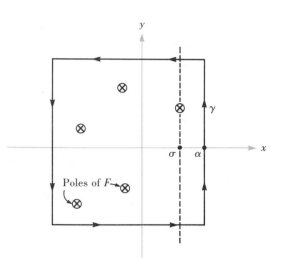

Figure 8.9 Contour for the Complex Inversion Formula.

Thus Theorem 8 gives us a formula for $f(t)$ when $F = \tilde{f}$ is known. By Theorem 2 there can be only one such (continuous) f. Recall that we call f the *inverse Laplace transform of F*, so equation (1) is a formula for computing f. Equation (1) is closely related to the Inversion Formula for Fourier Transforms and may be deduced from it (see Section 8.4 and review exercise 2).

Proof.   Let $\alpha > \sigma$ and consider the curve γ shown in Figure 8.9. Let the sides of γ be $\operatorname{Re} z = x_1$, $\operatorname{Re} z = \alpha$, $\operatorname{Im} z = -y_1 < 0$, and $\operatorname{Im} z = y_2 > 0$. Choose x_1, y_1, y_2 large enough so that all the poles of F lie within γ. By the Residue Theorem,

$$2\pi i \sum \{\text{Residues of } e^{zt} F(t)\} = \int_{\gamma} e^{zt}\, F(z)\, dz.$$

By exactly the same proof as that of Theorem 10, Section 4.3 (except that by that proof, Figure 8.9 would be rotated by $\pi/2$),

$$\int_{\gamma} e^{zt} F(z)\, dz \longrightarrow \int_{\gamma_{\alpha}} e^{zt} F(z)\, dz$$

as $x_1, y_1, y_2 \longrightarrow \infty$, where γ_{α} is the vertical curve $\operatorname{Re} z = \alpha$.
     By definition,

$$2\pi i \cdot f(t) = \int_{\gamma} e^{zt} F(z)\, dz$$

for Re $z > \alpha$. Note that

$$\int_0^r e^{-zt}\left(\int_\gamma e^{\zeta t}F(\zeta)\ d\zeta\right) dt = \int_\gamma\int_0^r e^{(\zeta-z)t}F(\zeta)\ dt\ d\zeta.$$

We can interchange the order of integration because these integrals are finite. This equation thus becomes

$$\int_\gamma \frac{1}{\zeta-z}\ (e^{(\zeta-z)r}-1)F(\zeta)\ d\zeta.$$

Since Re $z >$ Re ζ, the term $e^{(\zeta-z)r}$ approaches zero uniformly as $r \to \infty$. Therefore,

$$2\pi i\ \tilde{f}(z) = -\int_\gamma \frac{F(\zeta)}{\zeta-z}\ d\zeta.$$

As $x_1,\ y_1,\ y_2 \to \infty$, the preceding expression converges to

$$-\int_{\gamma_\alpha} \frac{F(\zeta)}{\zeta-z}\ d\zeta$$

since $|F(\zeta)/(\zeta-z)| \le M/|\zeta|^2$ for $|\zeta|$ large, as in Theorem 10, Section 4.3. We could also "turn over" our contour to $\tilde{\gamma}$ (see Figure 8.10) and get, as in Theorem 9, Section 4.3,

$$\int_{\tilde{\gamma}} \frac{F(\zeta)}{\zeta-z}\ d\zeta \to -\int_{\gamma_\alpha} \frac{F(\zeta)}{\zeta-z}\ d\zeta.$$

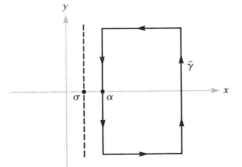

Figure 8.10   Reflection of the curve γ in the line Re $z = \alpha$.

But by the Cauchy Integral Formula,

$$\int_{\tilde{\gamma}} \frac{F(\zeta)}{\zeta - z}\, d\zeta = 2\pi i F(z).$$

Therefore, we have proven that $2\pi i F(z) = 2\pi i \tilde{f}(z)$ is valid for $\operatorname{Re} z > \alpha$. Since α is any number $> \sigma$, this conclusion holds for all z with $\operatorname{Re} z > \sigma$. ■

The following corollary of this theorem should be noted.

Corollary 1.

i. If $F(z)$ is analytic for $\operatorname{Re} z > \sigma$ and has a pole on the line $\operatorname{Re} z = \sigma$, then $\sigma = \sigma(f)$ (see Theorem 1 for the definition of $\sigma(f)$).

ii.

$$f(t) = \frac{1}{2\pi i} \int_{\alpha - i\infty}^{\alpha + i\infty} e^{zt} F(z)\, dz \tag{2}$$

for $\alpha > \sigma$. (The integral, which is convergent, is taken along the vertical line $\operatorname{Re} z = \alpha$.)

Proof.

i. Theorem 8 shows that $\sigma(f) \le \sigma$ since $\tilde{f}(z)$ converges for $\operatorname{Re} z > \sigma$. If $\sigma(f) < \sigma$, then $f(z)$ would be analytic for $\operatorname{Re} z > \sigma(f)$ by Theorem 1 and we would have $\tilde{f} = F$ for $\operatorname{Re} z > \sigma$. The fact that F has a pole at, say, z_0 on $\operatorname{Re} z_0 = \sigma$ means, specifically, that $\lim\limits_{\substack{z \to z_0 \\ \operatorname{Re} z > \operatorname{Re} z_0}} F(z) = \infty$, so $\tilde{f}(z) \to \infty$ as $z \to z_0$, which cannot be true since $\tilde{F}$ is analytic there. (Alternatively, by the Principle of Analytic Continuation stated in Theorem 1, Section 6.1, $\tilde{f} = F$ in a neighborhood of z_0.) Hence $\sigma(f) = \sigma$.

ii. This is obvious from the proof of Theorem 8. ■

The student is cautioned that in solving examples, all the hypotheses of Theorem 8 must be checked. If they do not hold, the formula for $f(t)$ may not be valid.

For computing inverse Laplace transforms, Theorem 8 is sometimes more convenient to use than Table 8.1 because it is systematic and requires no guesswork as to which formula is appropriate. However, the table is useful in some situations where the hypotheses of Theorem 8 fail.

Heaviside Expansion Theorem

When Theorem 8 is applied to the case in which $F(z) = P(z)/Q(z)$ where P and Q are polynomials, we get what are usually called the

Heaviside Expansion Theorems. This can be illustrated with the following
simple case.

Theorem 9.   *Let $P(z)$ and $Q(z)$ be polynomials with $\deg Q \ge \deg P + 1$.
Suppose that the zeros of Q are located at the points $z_1, \cdots, z_m$ and
are simple zeros. Then the inverse Laplace transform of $F(z) = P(z)/Q(z)$
is given by*

$$f(t) = \sum_{i=1}^{n} e^{z_i t}\, \frac{P(z_i)}{Q'(z_i)}. \tag{3}$$

Furthermore, $\sigma(f) = \max\ \{\operatorname{Re} z_i \mid i = 1, 2, \cdots, m\}$.

Proof.   Since $\deg Q \ge \deg P + 1$, the conditions of Theorem 8 are
met (compare Theorem 10, Section 4.3). Thus $f(t) = \Sigma$ {Residues of
$e^{zt}\,(P(z)/Q(z))$]}. But the poles are all simple and so, by formula 4
of Table 4.1, we have

$$\operatorname{Res}\left(e^{zt}\, \frac{P(z)}{Q(z)},\, z_i\right) = e^{z_i t}\, \frac{P(z_i)}{Q'(z_i)}.$$

The formula for $\sigma(f)$ is an immediate consequence of Corollary 1. ∎

The advantage of using Theorem 8 rather than more direct methods
is obvious from the preceding proof. A direct proof of Theorem 9 is
possible but is not as simple.

Worked Examples

1.   If $\tilde{f}(z) = 1/(z - 3)$, find $f(t)$.

   Solution.   Refer to formula 9 of Table 8.1. Let $a = -3$; then $f(t) = e^{3t}$.
   Alternatively, we could get the same result by using Theorem 9. In this
   example $\sigma(f) = 3$.

2.   If $\tilde{f}(z) = \log(z^2 + z)$, what is $f(t)$?

   Solution.   Note that if $g(t) = tf(t)$, then, by formula 1 of Table 8.1,

   $$\tilde{g}(z) = -\frac{d}{dz}\,\tilde{f}(z) = -\frac{d}{dz}\,\log(z^2 + z) = -\frac{2z + 1}{z^2 + z}.$$

   To find $g(t)$ we use partial fractions:

   $$\tilde{g}(z) = -\frac{2z + 1}{z^2 + z} = -\frac{1}{z} - \frac{1}{z + 1}.$$

   Therefore, $g(t) = -1 - e^{-t}$, and thus

   $$f(t) = -\frac{1}{t}\,(1 + e^{-t}).$$

Although this argument seems satisfactory, it is deceptive because there is in fact no $f(t)$ whose Laplace transform is $\log(z^2 + z)$. If there were, then this procedure would yield $f(t) = -(1 + e^{-t})/t$. For any x real,

$$\int_0^\infty e^{-xt}f(t)\ dt$$

cannot converge because near 0, e^{-xt} is $\geq 1/2$, and $f(t) \geq 1/t$ but $1/t$ is not integrable. Thus $\tilde{f}$ does not exist. The previous argument obtaining $f(t)$ is specious because it assumes the existence of an $f(t)$; see also the remark at the end of Section 8.1.

3. Compute the inverse Laplace transform of

$$F(z) = \frac{z}{(z+1)^2(z^2 + 3z - 10)}.$$

Then compute $\sigma(f)$, the abscissa of convergence of f.

Solution.   In this case Theorem 8 is convenient because its hypotheses clearly hold. Thus

$$f(t) = \sum \left\{\text{Residues of } \frac{e^{zt}z}{(z+1)^2(z^2 + 3z - 10)} = \frac{e^{zt}z}{(z+1)^2(z+5)(z-2)}\right\}.$$

The poles are at $z = -1$, $z = -5$, and $z = 2$. The pole at -1 is double whereas the others are simple. The residue at -1 is, by formula 7 of Table 4.1, $g'(-1)$ where $g(z) = (e^{zt}z)/(z^2 + 3z - 10)$. Thus we obtain

$$\frac{-te^{-t}}{-12} + \frac{e^{-t}}{-12} - \frac{(-e^{-t})\cdot(2\cdot(-1)+3)}{144} = \frac{1}{12}\left\{te^{-t} - e^{-t} + \frac{e^{-t}}{12}\right\}.$$

The residue at -5 is $e^{-5t}\cdot 5/16\cdot 7$; the residue at 2 is $e^{2t}\cdot 2/9\cdot 7$. Thus

$$f(t) = \frac{1}{12}\left\{te^{-t} - e^{-t} + \frac{e^{-t}}{12}\right\} + \frac{5e^{-5t}}{16\cdot 7} + \frac{2e^{2t}}{63}.$$

By Corollary 1, $\sigma(f) = 2$.

4. Check formula 9 of Table 8.1 using Theorem 8.

Solution.   We can find the inverse Laplace transform of $1/(z + a)$ by using Theorem 8. The only pole, which is simple, is at $z = -a$. The residue of $e^{zt}/(z + a)$ at $z = -a$ is clearly e^{-at}, which agrees with Table 8.1. Also, $\sigma(f) = -\text{Re } a$ because the pole of F lies on the line $\text{Re } z = -\text{Re } a$.

Exercises

1. Check formulas 10 and 11 of Table 8.1 by using Theorem 8.

2. Prove a Heaviside Expansion Theorem for P/Q when Q has double zeros.

3. Compute the inverse Laplace transform of each of the following:

    a. $F(z) = \dfrac{z}{z^2 + 1}.$

    b. $F(z) = \dfrac{1}{(z + 1)^2}.$

    c. $F(z) = \dfrac{z^2}{z^3 - 1}.$

4. Explain what is wrong with the following reasoning. Let

$$g(t) = \begin{cases} 0, & 0 \le t < 1; \\ 1, & t \ge 1. \end{cases}$$

Then, by formulas 6 and 13 of Table 8.1, $\tilde{g}(z) = e/z$. By Theorem 8, $g(t) = \text{Res } (e^{z(t-1)}/z; 0) = 1$. Therefore, $1 = 0$.

5. Use the shifting theorem (formula 7 of Table 8.1) to find a formula for the inverse Laplace transform of $e^{-z}/(z^2 + 1)$.

6. Compute the inverse Laplace transform of each of the following:

    a. $\dfrac{z}{(z + 1)(z + 2)}.$

    b. $\sinh z.$

    c. $\dfrac{1}{(z + 1)^3}.$

7. Use Theorem 7 to find a formula for the inverse Laplace transform of $\tilde{f}(z)/(z^2 + 1)$ for given $f(t)$.

8. Find the inverse Laplace transform of $1/\sqrt{z}$ by suitably modifying the proof of Theorem 8.

9. Let $f(t) = J_1(kt)/t$ where J_1 is the Bessel function. Show that

$$\tilde{f}(z) = \frac{(\sqrt{k^2 + z^2} - z)}{k}.$$

10. Find the inverse Laplace transform of $(z + 1)/[z(z + 3)^2]$.

8.3 Application of Laplace Transforms to Ordinary Differential Equations

This section will present a brief introduction to one of the many applications of Laplace transforms. Some examples of the application will be given but no general theorems will be formulated. The method is based on the formula

$$\left(\widetilde{\frac{df}{dt}}\right)(z) = z\tilde{f}(z) - f(0)$$

and the techniques for finding inverse Laplace transforms that were developed in the preceding section.

We shall assume that solutions exist and attempt to find formulas for them. When the formula is found, we may verify that it is indeed a solution. However, sometimes the solutions are not differentiable and thus do not strictly satisfy the equation. Such solutions should be regarded as *generalized solutions*. Further analysis of these solutions would lead to the subject of distribution theory. It should be remarked that even if f is not differentiable but is continuous at 0, $z\tilde{f}(z) - f(0)$ is still defined and can be regarded as the *generalized definition* of $(df/dt)^{\tilde{}}$. This is illustrated in example 2.

Example 1.   Solve the equation $y'' + 4y' + 3y = 0$ for $y(t)$, $t \geq 0$, subject to the conditions that $y(0) = 0$ and $y'(0) = 1$. Here y' stands for dy/dt.

Solution.   We take the Laplace transform on each side of the equation using the formulas

$$\left(\frac{\widetilde{dy}}{dt}\right)(z) = z\tilde{y}(z) - y(0) = z\tilde{y}(z)$$

and

$$\left(\frac{\widetilde{d^2y}}{dt^2}\right)(z) = z^2\tilde{y}(z) - zy(0) - y'(0) = z^2\tilde{y}(z) - 1.$$

Therefore, $z^2\tilde{y}(z) - 1 + 4z\tilde{y}(z) + 3\tilde{y}(z) = 0$, and so

$$\tilde{y}(z) = \frac{1}{z^2 + 4z + 3} = \frac{1}{(z+1)(z+3)}.$$

The inverse Laplace transform of this function is, by Theorem 8 or 9,

$$y(t) = \sum \left\{ \text{Residues of } \frac{e^{zt}}{(z+1)(z+3)} \text{ at } -1, -3 \right\}.$$

Thus

$$y(t) = \frac{e^{-t} - e^{-3t}}{2}.$$

This is the desired solution, as can be checked directly by substitution into the differential equation.

Example 2.   Solve the equation $y'(t) - y(t) = H(t - 1)$, $t \geq 0$, $y(0) = 0$, where H is the Heaviside function.

Solution.   Again we take the Laplace transforms of both sides of the equation. We get $z\tilde{y}(z) - y(0) - \tilde{y}(z) = e^{-z}/z$. Therefore,

$\tilde{y}(z) = e^{-z}/[z(z-1)]$. The inverse Laplace transform of $1/[z(z-1)]$ is $1 - e^{-t}$, so that of $e^{-z}/[z(z-1)]$ is, by formula 6 of Table 8.1,

$$y(t) = \begin{cases} 0, & 0 \le t < 1; \\ -1 + e^{t-1}, & t \ge 1. \end{cases}$$

(Note that Theorem 8 does not apply.)

This solution (see Figure 8.11) is not differentiable thus cannot be considered a solution in the strict sense. However, it is a solution in a generalized sense, as previously explained.

In Figure 8.11, the discontinuity in $H(t-1)$ causes the sudden jump in $y'(t)$. We say that $y(t)$ receives an "impulse" at $t = 1$.

Example 3.   Find a particular solution of $y''(t) + 2y'(t) + 2y(t) = f(t)$.

Solution.   Let us find the solution with $y(0) = 0$, $y'(0) = 0$. Taking Laplace transforms, $z^2\tilde{y}(z) + 2z\tilde{y}(z) + 2\tilde{y}(z) = \tilde{f}(z)$, and so $\tilde{y}(z) = \tilde{f}(z)/(z^2 + 2z + 2)$. The inverse Laplace transform of $1/(z^2 + 2z + 2)$ is

$$g(t) = \frac{e^{z_1 t}}{2(z_1 + 1)} + \frac{e^{z_2 t}}{2(z_2 + 1)},$$

where z_1, z_2 are the two roots of $z^2 + 2z + 2$, namely, $-1 \pm i$. By simplifying, we get $g(t) = e^{-t} \sin t$. Thus, by formula 8 of Table 8.1,

$$y(t) = (g * f)(t) = \int_0^t f(t - \tau) g(\tau) \, d\tau$$

$$= \int_0^t f(t - \tau) e^{-\tau} \sin \tau \, d\tau.$$

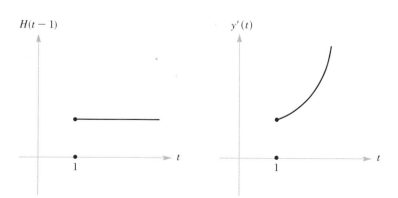

$H(t-1)$ $y'(t)$

Figure 8.11   At $t = 1$, y receives an impulse.

This is the particular solution we sought. Generally such particular solutions to equations of the form

$$a_n y^{(n)} + \cdots + a_1 y = f; \qquad a_1, \cdots, a_n \text{ are constants}$$

will be expressed in the form of a convolution. To obtain a solution with $y(0)$, $y'(0)$, $\cdots$, $y^{(n-1)}(0)$ prescribed, we can add a particular solution y_p satisfying $y_p(0) = 0$, $y_p'(0) = 0$, $\cdots$, $y_p^{(n-1)}(0) = 0$ to a solution y_c of the homogeneous equation in which f is set equal to zero and with $y_c(0)$, $y_c'(0)$, $\cdots$, $y_c^{(n-1)}(0)$ prescribed. The sum $y_p + y_c$ is the solution sought. (These statements are easily checked.)

The method of Laplace transforms, then, is a systematic method for handling constant coefficient differential equations. (Of course, these equations can be handled by other means as well.) If the coefficients are not constant the method fails, because transformation of a product then involves a convolution, and finding a solution for $\tilde{y}(z)$ becomes difficult.

Exercises

Solve the differential equations in exercises 1–7 by using Laplace transforms.

1. $y'' + 6y - 7 = 0$, $y(0) = 1$, $y'(0) = 0$.
2. $y'' - 4y = 0$, $y(0) = 2$, $y'(0) = 1$.
3. $y' + y = e^t$, $y(0) = 0$.
4. $y'' + 9y = H(t - 1)$, $y(0) = y'(0) = 0$.
5. $y'' + 9y = H(t)$, $y(0) = y'(0) = 0$.
6. $y' + y + \int_0^t y(\tau)\, d\tau = f(t)$ where $y(0) = 1$ and

$$f(t) = \begin{cases} 0, & 0 \le t < 1,\, t \ge 2; \\ 1, & 1 \le t < 2. \end{cases}$$

7. $y' + y = \cos t$, $y(0) = 1$.
8. Solve the following systems of equations for $y_1(t)$, $y_2(t)$ by using Laplace transforms.

   a. $\begin{cases} y_1' + y_2 = 0 \\ y_2' + y_1 = 0 \end{cases}$ $\quad y_1(0) = 1$, $y_2(0) = 0$

   b. $\begin{cases} y_1' + y_2' + y_1 = 0 \\ y_2' + y_1 = 3 \end{cases}$ $\quad y_1(0) = 0$, $y_2(0) = 0$

9. Solve $y'' + y = t \sin t$, $y(0) = 0$, $y'(0) = 1$.
10. Study the solution of $y'' + \omega_0^2 y = \sin \omega t$, $y(0) = y'(0) = 0$, and examine the behavior of solutions for various ω, especially those near $\omega = \omega_0$. Interpret these solutions in terms of forced oscillations.

8.4   Applications of Complex Analysis to Theoretical Physics

This final section gives a short exposition of some applications of the techniques used in complex analysis to problems in theoretical physics. The discussion focuses on the wave equation, which complex variables enable us to solve in a simple and elegant way. One-dimensional homogeneous, inhomogeneous, and scattering problems are the first topics to be discussed. The scattering amplitude T is defined and we shall see that the analytic properties of T can be concisely represented by using the Cauchy Integral Theorem. The dispersion relations for T are then described.

Subsequent paragraphs discuss the three-dimensional case for scattering problems that are spherically symmetric. A scattering amplitude is again defined, and some of the ways in which its analytic properties have been studied are enumerated.

The discussion is meant to be descriptive rather than rigorous, and several of the proofs are left as exercises. Indeed, detailed treatments of these topics have been a major objective of several theoretical physicists for the last two decades, and several monographs dealing with some of these questions are listed at the end of the book. It is hoped that this brief exposition will increase the student's awareness of the challenging and exciting problems that await those who have mastered the fundamentals of the complex variable theory.

The wave equation

The wave equation is the equation of motion that describes the development of a wave disturbance propagating in a medium. It describes, for example, the vertical displacement of a vibrating string (see Figure 8.12), the propagation of an electromagnetic wave through space and of a sound wave in a concert hall, and some types of water wave motion.

Let us first consider the homogenous problem, the simplest case of which is a wave traveling down a string of constant density ρ and under constant tension T. The vertical displacement $\varphi(x,t)$ at position x and time t satisfies the partial differential equation

$$\frac{1}{c^2} \cdot \frac{\partial^2 \varphi}{\partial t^2} = \frac{\partial^2 \varphi}{\partial x^2} \tag{1}$$

Figure 8.12 φ is the wave amplitude.

where $c = \sqrt{T/\rho}$ is the velocity of propagation, a constant. We accept this fact from elementary physics.

Note that if we were to have $c = \sqrt{-1}$ in equation (1), the equation of motion, we would recover the Laplace Equation (see Sections 2.5 and 5.3). Indeed, just as that equation admitted solutions of the form $f(x \pm \sqrt{-1}\, y)$, the solutions to wave equation take the form $f(x \pm ct)$. The fact that equation (1) is of second order in the t variable suggests that a solution is uniquely given when two pieces of initial data at $t = 0$ are specified. These data consist of $\varphi(x,0)$ and $\partial\varphi/\partial t$ at $(x,0)$; equation (1) then gives the development of $\varphi(x,t)$ for all subsequent t.

We will solve equation (1) in the same way that we solved the examples of Section 8.3: We perform a transform on the x variable to obtain a simpler equation involving the transform variable k. However, here x runs from $-\infty$ to $+\infty$, and so instead of using the Laplace transform we use the *Fourier transform*. Let $f: \mathbb{R} \to \mathbb{C}$; the Fourier transform function $\hat{f}$ of f is then defined by

$$\hat{f}(k) = \int_{-\infty}^{+\infty} e^{-ikx} f(x)\ dx.$$

There is an inversion formula that is analogous to the Laplace inversion formula (see Corollary 1, Section 7.1, and review exercise 2 at the end of this chapter). It is:

$$f(x) = \frac{1}{2\pi} \int_{-\infty}^{+\infty} e^{ikx} \hat{f}(k)\ dk$$

The Fourier transform of the function $\varphi(x,t)$ thus can be defined by

$$\hat{\varphi}(k,t) = \int_{-\infty}^{+\infty} e^{-ikx} \varphi(x,t)\ dx. \tag{2}$$

Here we perform the integral with respect to the x variable, regarding t as a fixed parameter. The Fourier Inversion Formula is

$$\varphi(x,t) = \frac{1}{2\pi} \int_{-\infty}^{+\infty} e^{ikx} \hat{\varphi}(k,t)\ dk. \tag{2'}$$

We are now ready to solve equation (1). By applying equation (1) to equation (2') and differentiating under the integral, we obtain*

$$\frac{1}{c^2} \cdot \frac{\partial^2 \hat{\varphi}}{\partial t^2}\ (k,t) + k^2 \hat{\varphi}(k,t) = 0. \tag{3}$$

*Equation (3) can also be obtained by taking the Fourier transform of equation (1) and using the fact that $\varphi(x,t)$ and $\partial\varphi(x,t)/\partial x$ converge to zero as $x \to \infty$ (cf. Section 8.3).

In other words, our transformation technique has replaced the partial differential equation for $\varphi(x,t)$ with a simple ordinary differential equation for $\hat{\varphi}(k,t)$. Equation (3) is easily solved. The solution is

$$\hat{\varphi}(k,t) = A(k)e^{ikct} + B(k)e^{-ikct} \qquad (4)$$

where $A(k)$, $B(k)$ are two constants of integration that may depend on the parameter k. Applying the inversion formula (2'), we have

$$\varphi(x,t) = \frac{1}{2\pi} \int\limits_{-\infty}^{+\infty} [A(k)e^{ik(x+ct)} + B(k)e^{ik(x-ct)}]\, dk. \qquad (4')$$

This is our solution to equation (1). The functions $A(k)$, $B(k)$ are determined by the initial data $\varphi(x,0)$ and $\partial\varphi(x,0)/\partial t$.

Also note that the first integral in (4') depends only on the variable $x + ct$, whereas the second depends only on $x - ct$; that is, $\varphi(x,t)$ takes the form

$$\varphi(x,t) = f(x + ct) + g(x - ct),$$

where the functions f, g are again determined by φ and $\partial\varphi/\partial t$ at $t = 0$. We can verify by substitution into equation (1) that this formula for φ does indeed give a solution of equation (1).

Some special solutions of equation (1) deserve separate attention. These are monochromatic (single frequency) waves and are of the form

$$\varphi(x,t) = e^{i(x/c-t)\omega}$$

where ω is the frequency. This φ represents a wave of frequency ω traveling to the right down the string. Generally, $f(x + ct)$ is a wave moving to the left whose shape is that of the graph of f, with velocity c. Similarly, $g(x - ct)$ is a wave moving to the right.

Next we shall deal with the inhomogeneous problem, which occurs when an external force is applied to the wave. For example, suppose that the string illustrated in Figure 8.12 were given a constant charge density q and then placed in an external electric field $E(x,t)$ pointing in the y-direction. This would result in the application of a force $F(x,t)$, proportional to $qE(x,t)$, to the string. We must then solve the following equation of motion for the displacement $\varphi(x,t)$:

$$\frac{1}{c^2} \cdot \frac{\partial^2 \varphi}{\partial t^2} = \frac{\partial^2 \varphi}{\partial x^2} + F(x,t). \qquad (5)$$

For simplicity, we take $F(x,t)$ in equation (5) to be periodic of frequency ω:

$$F = f(x,\omega)e^{i\omega t}.$$

This allows us to consider the simpler problem

$$\frac{1}{c^2} \cdot \frac{\partial^2 \varphi}{\partial t^2} = \frac{\partial^2 \varphi}{\partial x^2} + f(x,\omega) e^{i\omega t}. \tag{6}$$

We write the solution of equation (6) as $\varphi(x,t,\omega)$. Once we have solved equation (6), we can deal with the problem of a general force $F(x,t)$. First we "Fourier analyze" it; that is, we write F as follows:

$$F(x,t) = \frac{1}{2\pi} \int\limits_{-\infty}^{+\infty} f(x,\omega) e^{-i\omega t} \, d\omega,$$

where

$$f(x,\omega) = \int\limits_{-\infty}^{+\infty} F(x,t) e^{it\omega} \, dt.$$

We then superimpose the solutions of equation (6) to obtain

$$\varphi(x,t) = \frac{1}{2\pi} \int\limits_{-\infty}^{+\infty} \varphi(x,t,\omega) \, d\omega.$$

We solve (6) by again taking the Fourier transform with respect to the variable x, to get

$$\frac{1}{c^2} \cdot \frac{\partial^2 \hat{\varphi}}{\partial t^2} + k^2 \hat{\varphi} = \hat{f}(k,\omega) e^{i\omega t}, \tag{7}$$

where

$$\hat{f}(k,\omega) = \int\limits_{-\infty}^{+\infty} e^{ikx} f(x,\omega) \, dx.$$

Equation (7) is a simple, inhomogeneous, second-order differential equation. Adding the particular solution

$$\frac{\hat{f}(k,\omega) e^{i\omega t}}{k^2 - \left(\dfrac{\omega}{c}\right)^2}$$

of equation (7) to the homogeneous solutions in equation (4), we obtain the general solution to (7):

$$\hat{\varphi}(t,\omega) = A(k) e^{ikct} + B(k) e^{-ikct} + \frac{\hat{f}(k,\omega)}{k^2 - \left(\dfrac{\omega}{c}\right)^2} e^{i\omega t}. \tag{8}$$

The solution to equation (6) is thus

$$\varphi(x,t,\omega) = f(x + ct) + g(x - ct) + (G * f)e^{i\omega t}. \tag{8'}$$

The terms in equation (8') are explained as follows. The first two terms are solutions to the homogeneous equation (1), and again they are to be chosen so that the initial data at $t = 0$ are satisfied. The last term, a particular solution to the inhomogeneous equation (6), is given by taking the inverse Fourier transform of the last term in equation (8):

$$\frac{1}{2\pi} \int_{-\infty}^{\infty} e^{ikx} \left(\frac{\hat{f}(k,\omega)}{k^2 - \left(\frac{\omega}{c}\right)^2} \right) dk.$$

As with the Laplace transform (see p. 393), this term is actually the convolution of G and f where $\hat{G} = 1/(k^2 - (\omega/c)^2)$. This function $\hat{G}$ plays a central role in the theory of partial differential equations. Its transform,

$$G(x,\omega) = \frac{1}{2\pi} \int_{-\infty}^{+\infty} \left(\frac{e^{ikx}}{k^2 - \left(\frac{\omega}{c}\right)^2} \right) dk, \tag{9}$$

is called the "Green's function," and we can use contour integration to evaluate it in closed form as follows.

The integrand of equation (9) has simple poles at $k = \pm (\omega/c)$. In its present form, the integral in equation (9) is not convergent. To specify its value we must use the Cauchy Principal Value. Several possible values may be obtained depending on how we interpret our integrals. To select out the value we want, we shall evaluate (9) by closing the contour of integration in the upper half of the complex k-plane for $x > 0$ and in the lower half of the plane for $x < 0$. This is necessary if the integral over the semicircle is to approach zero as the radius approaches infinity. By Cauchy's Theorem, we pick up the residues of the enclosed poles. We still must specify how we are to go around the singularities at $k = \pm\omega/c$. Different choices will lead to different, but still mathematically acceptable, values of G. Our final choice is determined by the asymptotic behavior that we want G to have as $x \to \infty$. The homogeneous solutions to equation (1) in which we are interested behave like $\exp(\pm ikx)$ as a function of x, and we will require the same behavior of G. This can be specified by the so-called $i\epsilon$ prescription:

$$G(x,\omega) = \lim_{\substack{\epsilon \to 0 \\ \epsilon > 0}} \frac{1}{2\pi} \int_{-\infty}^{+\infty} \left(\frac{e^{ikx}\, dk}{k^2 - \left(\frac{\omega}{c} - i\epsilon\right)^2} \right), \tag{9'}$$

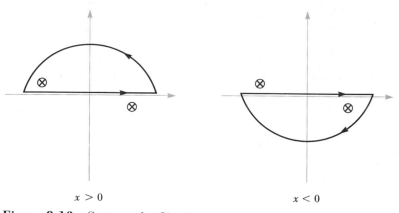

$$x > 0 \qquad\qquad x < 0$$

Figure 8.13   Contours for $G(x,\omega)$.

in which we still close the contour (as shown in Figure 8.13) according to the sign of x.

We can now evaluate $G(x,\omega)$. From equation (9′) and the Residue Theorem we obtain, for $x > 0$,

$$G(x,\omega) = \lim_{\substack{\epsilon \to 0 \\ \epsilon > 0}} \frac{1}{2\pi} \left[2\pi i \cdot \frac{e^{ix(i\epsilon + \omega/c)}}{-2\left(\dfrac{\omega}{c} - i\epsilon\right)} \right] = -\frac{ic}{2\omega} e^{i\omega x/c}.$$

Making a similar computation for $x < 0$ and using the right-hand contour in Figure 8.13, we obtain

$$G(x,\omega) = \begin{cases} \dfrac{-ic}{2\omega} e^{i\omega x/c}, & x > 0; \\[3mm] \dfrac{-ic}{2\omega} e^{-i\omega x/c}, & x < 0. \end{cases}$$

Equivalently,†

$$G(x,\omega) = \frac{c}{2i\omega} e^{i\omega |x|/c}. \tag{9″}$$

---

†In textbooks on the subject of differential equations, G is obtained as the solution to

$$\frac{d^2 G}{dx^2} + \omega^2 G = \delta(x - y),$$

where δ is the "Dirac δ function." The solution is found by the general formula

$$G(x \mid y,\omega) = \begin{cases} -u(x)v(y)/w, & x > y; \\ -u(y)v(x)/w, & x < y. \end{cases}$$

Here u, v are solutions of the corresponding homogeneous equation and w is their Wronskian; $w = uv' - vu'$. In this case, we have $u = e^{i\omega x}$ and $v = e^{-i\omega x}$. We recover equation (9″) by setting $y = 0$.

The scattering problem

When the medium through which the wave propagates is not homogeneous, we encounter the scattering problem. For example, suppose that the vibrating string of Figure 8.12 now consists of three pieces smoothly joined together, with one piece, of length a, having a density of ρ_2 (region II in Figure 8.14) and the other two pieces each having a density of ρ_1 (regions I, III in Figure 8.14). Let c_1 and c_2 denote the corresponding velocities of propagation. Assume that $\rho_2 > \rho_1$. Imagine that an incident wave $e^{i(x/c_1 - t)\omega}$ from the left travels down the string. As the wave moves onto the denser material at $x = 0$, part of it will be reflected backward, while part will be transmitted onward. At $x = a$, some of the wave will again be reflected backward while the rest travels forward (see Figure 8.15). We must solve the wave equation in each region:

$$\frac{\partial^2 \varphi}{\partial x^2} = \frac{1}{c_1^2} \cdot \frac{\partial^2 \varphi}{\partial t^2} \qquad \text{for regions I, III;}$$

$$\frac{\partial^2 \varphi}{\partial x^2} = \frac{1}{c_2^2} \cdot \frac{\partial^2 \varphi}{\partial t^2} \qquad \text{for region II.}$$

It is not unreasonable to expect solutions of the following forms:

$$\varphi_{\mathrm{I}}(x,t) = e^{i(x/c_1 - t)\omega} + R e^{i(-x/c_1 - t)\omega};$$

$$\varphi_{\mathrm{II}}(x,t) = A e^{i(x/c_2 - t)\omega} + B e^{i(-x/c_2 - t)\omega};$$

$$\varphi_{\mathrm{III}}(x,t) = T e^{i(x/c_1 - t)\omega}.$$

We shall require that, at $x = 0$ and $x = a$, the solutions join onto each other and have no sharp bend. Mathematically, this means that we impose the boundary conditions that φ and $\partial \varphi / \partial x$ be continuous at $x = 0, a$. In other words,

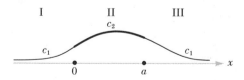

Figure 8.14 One-dimensional scattering.

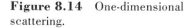

Figure 8.15 One-dimensional scattering.

$$\varphi_{\mathrm{I}}(0,t) = \varphi_{\mathrm{II}}(0,t); \qquad \varphi_{\mathrm{II}}(a,t) = \varphi_{\mathrm{III}}(a,t).$$

$$\frac{\partial \varphi_{\mathrm{I}}}{\partial x}(0,t) = \frac{\partial \varphi_{\mathrm{II}}}{\partial x}(0,t); \qquad \frac{\partial \varphi_{\mathrm{II}}}{\partial x}(a,t) = \frac{\partial \varphi_{\mathrm{III}}}{\partial x}(a,t).$$

These four equations allow us to solve for the coefficients R, A, B, T. We are particularly interested in T. After performing some algebraic manipulations we find that

$$T = \frac{4c_1 c_2 e^{it((1/c_2) - (1/c_1))\omega a}}{(c_1 + c_2)^2 - (c_1 - c_2)^2 e^{2ia\omega/c_2}} \qquad (10)$$

(see review exercise 8). T is called the *scattering amplitude*, and the square of its absolute value represents the intensity of the wave transmitted into region III.

We now allow ω in equation (10) to become a complex variable, and we see that $T(\omega)$ has the following properties:

1. T is meromorphic in ω and has poles in the lower half plane at $\omega = (c_2/a)(n\pi - i\rho)$, where ρ is determined by $e^{2\rho} = (c_1 + c_2)^2/(c_1 - c_2)^2$.

2. T has absolute value one at those values of ω for which $e^{2i\omega/c_2} = 1$.

3. As $\omega \to +i\infty$, $T \to 0$.

4. As $\omega \to -i\infty$, $T \to \infty$.

Dispersion relations

When a function of a complex variable $f(z)$ shares the same four properties as $T(\omega)$, the Cauchy Theorem can be used to obtain an interesting and useful representation for $f(z)$, as shown in Theorem 10.

Theorem 10.   *If $f(z)$ is analytic for $\mathrm{Im}\,(z) \ge 0$ and $f(z) \to 0$ uniformly as $z \to \infty$, $0 < \arg z < \pi$, then $f(z)$ satisfies the following integral relationships:*

$$f(z_0) = \frac{y_0}{\pi} \int_{-\infty}^{+\infty} \frac{f(x,0)}{(x_0 - x)^2 + y_0^2}\, dx, \qquad \mathrm{Im}\,(z_0) > 0; \qquad (11)$$

$$= \frac{1}{\pi i} \int_{-\infty}^{+\infty} \frac{f(x,0)(x - x_0)}{(x - x_0)^2 + y_0^2}\, dx, \qquad \mathrm{Im}\,(z_0) > 0; \qquad (12)$$

$$= \frac{1}{\pi i} \int_{-\infty}^{+\infty} \frac{f(x,0)}{x - x_0}\, dx, \qquad \mathrm{Im}\,(z_0) = 0. \qquad (13)$$

Proof.   Because of the assumptions of Theorem 10, we can apply Cauchy's Theorem, using a large semicircle in the upper half plane, to give:

$$f(z_0) = \frac{1}{2\pi i} \int\limits_{-\infty}^{+\infty} \frac{f(x,0)}{z - z_0}\, dx$$

(see Section 4.3). We also have

$$0 = \frac{1}{2\pi i} \int\limits_{-\infty}^{+\infty} \frac{f(x,0)}{z - \bar{z}_0}\, dx$$

where $\bar{z}_0$ lies in the lower half plane. If we subtract these last two equations we obtain equation (11); if we add them we get equation (12). Equation (13) follows from equation (6) on p. 254. ■

As an immediate corollary, by taking real and imaginary parts of each side of (11) we get:

$$u(x_0, y_0) = \frac{y_0}{\pi} \int\limits_{-\infty}^{+\infty} \frac{u(x,0)}{(x - x_0)^2 + y_0^2}\, dx \qquad (11')$$

where $f = u + iv$, and with a similar equation for $v(x,y)$. From equation (12) we have

$$u(x_0, y_0) = \frac{1}{\pi} \int\limits_{-\infty}^{+\infty} \frac{(x - x_0)v(x,0)}{(x - x_0)^2 + y_0^2}\, dx;$$

$$\qquad (12')$$

$$v(x_0, y_0) = \frac{-1}{\pi} \int\limits_{-\infty}^{+\infty} \frac{(x - x_0)u(x,0)}{(x - x_0)^2 + y_0^2}\, dx.$$

Finally, from equation (13), we obtain

$$u(x_0, 0) = \frac{1}{\pi}\, \text{P.V.} \int\limits_{-\infty}^{+\infty} \frac{v(x,0)}{x - x_0}\, dx;$$

$$\qquad (13')$$

$$v(x_0, y_0) = \frac{-1}{\pi}\, \text{P.V.} \int\limits_{-\infty}^{+\infty} \frac{u(x,0)}{x - x_0}\, dx.$$

Equation (11') gives us the values of a harmonic function in the upper half plane, in terms of its boundary values on the real axis, and thus provides a solution to the Laplace Equation in the upper half plane (see review exercise 9).

Note that if $f(z)$ satisfies the symmetry property $\bar{f}(-x) = f(x)$, then we can write

$$\text{Re}\, f(x_0) = \frac{2}{\pi}\, \text{P.V.} \int\limits_{0}^{\infty} \frac{\text{Im}\, f(x)}{x^2 - x_0^2}\, x\, dx,$$

whereas if $f(z)$ satisfies $\bar{f}(-x) = -f(x)$, then

$$\mathrm{Re}\, f(x_0) = \frac{2x_0}{\pi}\, \mathrm{P.V.} \int\limits_0^\infty \frac{\mathrm{Im}\, f(x)\; dx}{x^2 - x_0^2}.$$

Equations (12') and (13') can be regarded as integral versions of the Cauchy-Riemann equations; they simply tell us, for example, the values that the real part of an analytic function must take when the imaginary part is specified. When functions u,v satisfy the set of equations (13'), we say that u and v are "Hilbert transforms" of each other. Historically, the Hilbert transforms were the forerunners of a series of such relations called dispersion relations. They were first observed to hold for the complex dielectric constant as a function of incident frequency by H. A. Kramers and R. de L. Kronig in 1924. Since approximately 1950, they have been systematically studied and applied to the scattering amplitude $T(\omega)$ and to quite general classes of scattering problems for which this amplitude is defined. The extension of these relations to three-dimensional scattering problems will be considered later in this section.

The relations derived in Theorem 10 assumed that $f(z)$ is analytic only for $\mathrm{Im}\,(z) \geq 0$. However, there is a second class of dispersion relations for functions that are analytic in the z-plane except for a branch line along the real axis.

Theorem 11.   *If $f(z)$ is analytic in the z-plane with a branch line from $z = a$ to ∞, and if $|f(z)| = O(1/z)$, then*

$$f(z) = \lim_{\epsilon \to 0+} \frac{1}{2\pi i} \int\limits_a^\infty \frac{1}{x - z}\, [f(x + i\epsilon) - f(x - i\epsilon)]\; dx.$$

(The notation $O(1/z)$ is explained in Section 7.2; $\lim\limits_{\epsilon \to 0+}$ means the limit is taken through $\epsilon > 0$.)

Proof.   Take the contour of Figure 8.16 and apply the Cauchy Theorem to $f(\zeta)/(\zeta - z)$. ∎

If, in addition to the hypotheses of Theorem 11, $f(z)$ also satisfies the relation $\overline{f(\bar{z})} = f(z)$, that is, if $u(x,\epsilon) + iv(x,\epsilon) = u(x,-\epsilon) - iv(x,-\epsilon)$, so that the real part of f is continuous across the real axis, whereas the imaginary part is discontinuous, then from Theorem 11 we obtain

$$f(z) = \lim_{\epsilon \to 0+} \frac{1}{\pi} \int\limits_a^\infty \frac{1}{x - z}\, \mathrm{Im}\, f(x + i\epsilon)\; dx.$$

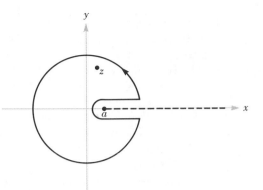

Figure 8.16 Contour used in the proof of Theorem 11.

When z actually moves onto the real axis, we can take real parts of each side of the preceding equation to obtain

$$\mathrm{Re}\, f(x_0) = \lim_{\epsilon \to 0+} \frac{\mathrm{P.V.}}{\pi} \int_a^\infty \frac{1}{x - x_0}\, \mathrm{Im}\, f(x + i\epsilon)\ dx.$$

The wave equation in three dimensions

The ideas that have been developed thus far in this section for wave motion in one dimension can easily be extended to higher-dimensional problems. In two dimensions the vibrating string is replaced by a vibrating membrane. In three dimensions we can think of sound waves propagating in air. The pressure $\varphi(\mathbf{r},t)$ then satisfies the equation of motion,

$$\frac{1}{c^2} \cdot \frac{\partial^2 \varphi}{\partial t^2}\, (\mathbf{r},t) = \nabla^2 \varphi(\mathbf{r},t) + F(\mathbf{r},t), \tag{14}$$

where F represents some external source of waves, $\mathbf{r} = (x,y,z)$, and

$$\nabla^2 = \frac{\partial^2}{\partial x^2} + \frac{\partial^2}{\partial y^2} + \frac{\partial^2}{\partial z^2}$$

is the Laplace operator. When equation (14) is expressed in terms of rectangular coordinates, the homogeneous and inhomogeneous solutions to equation (14) are obtained in much the same way as they were previously. The really new and exciting features of equation (14) not present in one dimension arise in the scattering problem, and these are most interesting and tractable when the scattering medium has spherical symmetry.

Consider an incident plane wave $e^{i(x/c-\omega)t}$ traveling from the left down the x-axis that impinges on a ball located at the origin (Figure 8.17). Part of the wave may penetrate the ball, part of the wave is scattered by the surface of the ball and then travels radially outward, and the remainder of the wave simply bypasses the ball. To solve for φ, we proceed as previously. We first obtain the solutions for φ in regions I and II separately and then require that φ and the radial derivative $\partial\varphi/\partial r$ should be continuous at the surface of the ball. This procedure eventually specifies the total wave in region I that results from the "impurity" of the medium in region II.

These calculations are not detailed here because such a task would take us too far afield into the subject of partial differential equations. However, the form of the final result is not too difficult to anticipate. The wave in region I will be a sum of the incident wave and the outgoing radial wave, and this will take the asymptotic form

$$\varphi_{\mathrm{I}}(\boldsymbol{r},t) \sim \left[e^{ix/c} + \frac{e^{ir/c}}{r} f(\omega,\theta) \right] e^{-i\omega t}$$

as $|\boldsymbol{r}| = r \rightarrow \infty$. Here $f(\omega,\theta)$ is the amount of scattering wave that is traveling outward at an angle θ to the axis of symmetry (see Figure 8.17); it is the three-dimensional analogue of the scattering amplitude $T(\omega)$ of the one-dimensional problem discussed earlier in the section.

Note that f is now a function of *two* complex variables, ω and θ. Physically observable scattering occurs, of course, only when ω and θ are real with $0 \le \theta \le \pi$. But by studying the properties of f for complex values of ω and θ, as we will do in the following paragraphs, we do gain a deeper understanding of the characteristics of f.

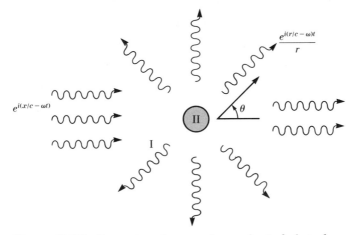

Figure 8.17   Scattering of a wave by a spherical obstacle.

Double dispersion relations

To write these dispersion relations, it is convenient to change variables as follows:

$$s = \frac{4\omega^2}{c^2};$$

$$t = -2\,\frac{\omega^2}{c^2}\,(1 - \cos\theta).$$

Now define the function A of the two complex variables s,t by $A(s,t) = f(\omega,\theta)$. For a large class of scattering problems, it can be shown that A has the following properties:

1. $A(s,t)$ is analytic in the two complex variables s and t with branch lines from $s = a$ to ∞ and $t = b$ to ∞.
2. $A(s,t) = O(1/s)$ as $s \to \infty$ for each t.
3. $A(\bar{s},t) = \bar{A}(s,t)$ for t real.

Applying Theorem 11, we get

$$A(s_0,t_0) = \lim_{\epsilon \to 0+} \frac{1}{2\pi i} \int_a^\infty \frac{1}{s - s_0} \left[A(s+i\epsilon,t_0) - A(s-i\epsilon,t_0)\right]\,ds \qquad (15)$$

$$= \lim_{\epsilon \to 0+} \frac{1}{\pi} \int_b^\infty \frac{1}{s - s_0}\,\text{Im}\,A(s+i\epsilon,t_0)\,ds. \qquad (16)$$

Equation (15) is valid for general complex s_0 and t_0; equation (16) assumes that t_0 is real so that property (3) of $A(s,t)$ can be used.

In equation (15) we are integrating $A(\zeta,t)$ for ζ slightly above and below the real axis. Now consider the integrand. By property (1) of $A(s,t)$, we can write a dispersion relation in the t variable as follows:

$$A(s+i\epsilon,t_0) = \lim_{\delta \to 0+} \frac{1}{2\pi i} \int_b^\infty \frac{1}{t - t_0} \left[A(s+i\epsilon,t+i\delta) - A(s+i\epsilon,t-i\delta)\right]\,dt.$$

Using a similar representation for the second term of equation (15), we finally obtain the *double dispersion relation*

$$A(s_0,t_0) = \frac{1}{\pi^2} \int_a^\infty \frac{1}{s - s_0} \left(\int_b^\infty \frac{1}{t - t_0}\,\rho(s,t)\,dt\right) ds \qquad (17)$$

where

$$\rho(s,t) = \lim_{\epsilon,\delta \to 0+} \left(\frac{1}{2i}\right)^2 [A(s+i\epsilon,t+i\delta) - A(s+i\epsilon,t-i\delta)$$

$$-A(s-i\epsilon,t+i\delta) + A(s-i\epsilon,t-i\delta)].$$

Equation (17) is the representation for $A(s,t)$ first obtained by S. Mandelstam in 1958.

Landau singularities and algebraic curves

The singularities in $A(s,t)$ that lead to equation (17) are "factorizable"; for example, we assumed that $A(s,t)$ has a branch point at $t = b$ for all complex values of s. For some types of scattering processes, it can be shown that A has further singularities that lie on curves of the form

$$g(s,t) = 0. \qquad (18)$$

This means that the singularities of the function A of two complex variables lie on a curve that defines a function of one complex variable given implicitly by equation (18). These singularities are called *Landau singularities* because they were first studied by L. D. Landau in 1959. Such singularity curves are often algebraic curves; that is, g often takes the form

$$g(s,t) = a_0 s^n + a_1 s^{n-1} t + \cdots + a_n t^n + b_0 s^{n-1} + \cdots + \cdots + d = 0.$$

Algebraic curves can themselves have singularities (which are called nodes), and the classical theory of algebraic curves can be invoked to relate the *degree* of the curve (g in the preceding paragraph has degree n) to the number and nature of these nodes. These ideas have been developed by S. Abhyankar and C. Risk (1968) and by T. Regge and collaborators (1968).

Hadamard-Gilbert Rules and pinching singularities

In many types of scattering processes, A cannot be evaluated in closed form; the best that can be achieved is an integral expression, often of the form

$$A(s,t) = \int F(\alpha,s,t)\ d\alpha, \qquad (19)$$

where F is a known function of α, s, and t, whose singularities are therefore known. The integration in the variable α represents a multidimensional integral that may be all but impossible to evaluate analytically. However, we can still obtain the singularities of A if we have some way to determine how the known singularities of F will be passed on to A. This question is answered by the *Hadamard-Gilbert Rules*, which were formulated by R. P. Gilbert in 1959 and which are contained in the following theorem.

Theorem 12. *Let $f(z) = \int_a^b g(z,w)\ dw$. If g is analytic in z and continuous in w, then f is analytic in z. If g has two singularities at $w_i = w_i(z)$,*

$i = 1, 2$, then f will be analytic at z as long as $w_i(z)$ does not pass through an end point ($w = a,b$) and as long as the two points $w_1(z)$ and $w_2(z)$ do not "pinch the contour of integration" (see the following proof and review exercises 13 and 14 for elucidation of this point).

Proof. As z ranges over complex values, the singularities $w_i(z)$ of g will wander over the w-plane. As long as they stay away from the contour of integration, g will be analytic in z and f will be analytic (see example 4, Section 2.4). If, for some value $z = z_0$, a singularity $w_1(z)$, say, hits the contour of integration (see Figure 8.18(i)), we can deform the contour to extend f analytically at z_0. But if $w_1(z)$ moves through an end point (Figure 8.18(ii)), or if the two singularities trap the contour of integration (Figure 8.18(iii)), then g has an irrevocable singularity, and this is passed on to f.■

Regge poles

A second way of studying analytic properties of the scattering amplitude $f(\omega,\theta)$ is to expand f in a Legendre series of the form

$$f(\omega,\theta) = \frac{1}{2i\omega} \sum_{l=0}^{\infty} (2l+1)a_l(\omega)P_l(\cos \theta), \tag{20}$$

where P_l is the Legendre polynomial of degree l (see review exercise 11, Chapter 7). The study of $f(\omega,\theta)$ in terms of ω and θ (or s and t) is now replaced by the study of the coefficients $a_l(\omega)$ in terms of l and ω. In equation (20), l is an integer, but we can show that $a_l(\omega)$ can be extended so that it is a function of *complex l and ω*, which we write as $a(l,\omega)$. The series (20) can then be replaced by a single integral in the complex l-plane, by means of the following theorem.

Theorem 13. *If $a(l,\omega)$ is analytic in the interior of the curve C of Figure 8.19, then*

$$f(\omega,\theta) = \frac{1}{2\omega} \int_C \frac{a(j,\omega)}{\sin \pi j} P_j(-\cos \theta)\left(j + \frac{1}{2}\right) dj. \tag{21}$$

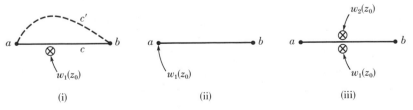

Figure 8.18 Hadamard-Gilbert Rules.

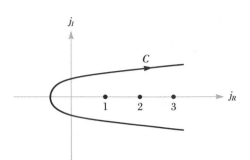

Figure 8.19 Contour used for the Regge Equation.

Proof. Replace C by small circles around the integers $j = l$ that enclose the poles of $1/\sin \pi j$; use the fact that $P_l(-\cos \theta) = (-1)^l P_l(\cos \theta)$ for integer l (cf. review exercise 34, Chapter 3). ∎

In many types of scattering problems, $a(j,\omega)$ will have a pole in the complex j-plane whose location depends on ω; denote this pole by $j = \alpha(\omega)$. Thus we have a function of the two variables j and ω having a singularity that will define a complex curve $j = \alpha(\omega)$. We now deform the contour of C of Figure 8.19 and write the integral in equation (21) symbolically as follows:

$$f(\omega,\theta) = \int_{C_\alpha} - \int_{C_1} - \int_{C_2}$$

(see Figure 8.20). The contribution from C_1 and C_2 can be shown to be small. By expressing the behavior of $a(j,\omega)$ near $\alpha(\omega)$ as

$$a(j,\omega) = \frac{\beta(\omega)}{j - \alpha(\omega)}$$

for j in a neighborhood of $\alpha(\omega)$, we obtain

$$f(\omega,\theta) \sim \frac{\pi i c \beta(\omega)\left(\alpha(\omega) + \frac{1}{2}\right)}{\omega \sin (\pi \alpha(\omega))} P_{\alpha(\omega)}(-\cos \theta).$$

If we now let the (complex) variable $\cos \theta$ become large and use the asymptotic formula for $P_j(\cos \theta)$ (see review exercise 11(h), Chapter 7), dropping multiplicative constants, we find that

$$f(\omega,\theta) \underset{\cos \theta \to \infty}{\sim} (\cos \theta)^{\alpha(\omega)},$$

that is,

$$A(s,t) \underset{t \to \infty}{\approx} t^{\alpha(s)}.$$

This is the rather profound result obtained by T. Regge in 1961: The asymptotic behavior of $A(s,t)$ as $t \to \infty$ is governed by the location of the singularities of $a(j,\omega)$ in the complex j-plane.

Review Exercises for Chapter 8

1. Let $f(t)$ be a bounded function of t. Show that $\sigma(f) \le 0$.

2. If $f(t) = 0$ for $t < 0$, then $\hat{f}(y) = \tilde{f}(iy)$ is called the *Fourier transform* of f. Using Corollary 1, show that, under suitable conditions,

$$f(x) = \frac{1}{2\pi} \int\limits_{-\infty}^{\infty} \hat{f}(y) e^{ixy} \, dy.$$

(This result is called the Inversion Theorem for Fourier Transforms.)

3. Find the Laplace transform of the following functions and compute the abscissa of convergence in each case:
a. $f(t) = \sinh t$.
b. $f(t) = H(t-1) + 3e^{-(t+6)}$.
c. $f(t) = H(t-1) \sin(t-1)$.

d. $f(t) = \begin{cases} t, & 0 \le t \le 1; \\ 1, & t > 1. \end{cases}$

e. $f(t) = \dfrac{e^t - 1}{t}$.

4. Find the inverse Laplace transform of each of the following and compute the abscissa of convergence in each case:

a. $F(z) = \dfrac{1}{(z+1)^2}$.

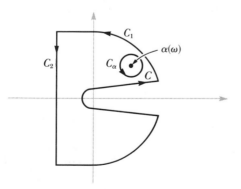

Figure 8.20 Deformation of the contour C.

    b. $F(z) = \dfrac{e^{-z}}{z^2 + 1}.$

    c. $F(z) = \dfrac{z}{(z+1)^2} + \dfrac{e^{-z}}{z}.$

5.   a. Let $\tilde{f}(z)$ be the Laplace transform of $f(t)$. Show that $\tilde{f}(z) \longrightarrow 0$ as Re $z \longrightarrow \infty$.

    b. Use (a) to show that, under suitable conditions, $z\tilde{f}(z) \longrightarrow f(0)$ as Re $z \longrightarrow \infty$.

    c. Can a nonzero polynomial be the Laplace transform of any $f(t)$?

    d. Can a nonzero entire function F be the Laplace transform of a function $f(t)$?

6. Solve the following differential equations using Laplace transforms:

    a. $y'' + 8y + 15 = 0$, $y(0) = 1$, $y'(0) = 0$.

    b. $y' + y = 3$, $y(0) = 0$.

    c. $y'' + y = H(t-1)$, $y(0) = 0$, $y'(0) = 0$.

    d. $y'' + 2y' + y = 0$, $y(0) = 1$, $y'(0) = 1$.

7. Suppose that $f(t) \geq 0$ and is infinitely differentiable. Prove that $(-1)^k \tilde{f}^{(k)}(z) \geq 0$, $k = 0, 1, 2, \cdots$, for $z \geq 0$. (The converse, called *Bernstein's Theorem*, is also true but is more difficult to prove.)

8. Apply the boundary conditions on p. 420 to obtain expressions for $T(\omega)$, $R(\omega)$, $A(\omega)$, $B(\omega)$. Verify that when ω is real,

$$|T(\omega)|^2 + |R(\omega)|^2 = 1.$$

(This last relation expresses "conservation of intensity": In a medium without dissipative loss, the unit intensity of the incident wave of Figure 8.15 is equal to the sum of the intensity of the wave R reflected back into region I and intensity of the wave T transmitted into region III.)

9. Since equation (11) solves the Dirichlet Problem for the upper half plane, apply it to the problem illustrated in the right side of Figure 5.17 and obtain the same solution. Thus we have two methods for solving Laplace's Equation: conformal mapping and integral relations of the type described by equation (11).

10. As a third way to solve the Laplace Equation, take Fourier transforms of

$$\frac{\partial^2 u}{\partial x^2}(x,y) + \frac{\partial^2 u}{\partial y^2}(x,y) = 0 \tag{22}$$

with respect to the variable x (use the method described on p. 414); show that this method gives the ordinary differential equation

$$\frac{\partial^2 \hat{u}}{\partial y^2}(k,y) - k^2 \hat{u}(k,y) = 0. \tag{23}$$

By solving, summing over all solutions, and keeping only those that decay exponentially as $y \longrightarrow +\infty$, obtain the formula

$$u(x,y) = \frac{1}{2\pi}\int_{-\infty}^{+\infty} A(k)e^{ikx-k|y|}\, dk. \tag{24}$$

With relative ease, $A(k)$ can now be evaluated in terms of the boundary data at $y=0$. Since at $y=0$ equation (23) reduces to

$$u(x,0) = \frac{1}{2\pi} \int_{-\infty}^{+\infty} A(k)e^{ikx}\, dk,$$

$A(k)$ must in fact be the Fourier transform of $u(x,0)$. Using this result, substituting in equation (23), and interchanging orders of integration, obtain

$$u(x,y) = \int_{-\infty}^{+\infty} u(z,0)\left(\frac{1}{2\pi} \int_{-\infty}^{+\infty} e^{ik(x-z)-|k|y}\, dk\right) dz.$$

Perform the integral on k to get

$$u(x,y) = \int_{-\infty}^{+\infty} u(z,0)\cdot\frac{1}{\pi}\cdot\frac{1}{(x-z)^2+y^2}\, dz. \tag{25}$$

11. Referring to equation (25), show that

$$\frac{y}{(x-z)^2+y^2} = \left.\frac{-\partial}{\partial w} G(x,y\mid z,w)\right|_{w=0}$$

where

$$G(x,y\mid z,w) = ln\,|\mathbf{r}-\mathbf{r}'|$$

with $\mathbf{r}=x+iy$, $\mathbf{r}'=z+iw$. (The function G is called the *Green's function* and is recognized as the potential at $\mathbf{r}$ caused by a unit charge at the point $\mathbf{r}'$ in the plane.)

12. As an application of Theorem 11, let

$$f(z) = \frac{1}{\sqrt{z}} = r^{-1/2}e^{-i\theta/2}, \qquad 0 < \theta < 2\pi$$

be defined on the z-plane with a branch line along the positive real axis. Show that application of Theorem 11 results in

$$\frac{1}{\sqrt{x_0+iy_0}} = -\frac{i}{\pi} \int_{0}^{\infty} \frac{1}{x-z_0}\frac{1}{\sqrt{x}}\, dx.$$

By making the change of variables $w=\sqrt{x}$, evaluate the right side of the last equation directly to verify the equality.

13. As an illustration of end point singularities in Theorem 12, consider

$$F(z) = \int_{0}^{1} \frac{dt}{t-z}.$$

As long as z does not lie on the contour of integration from $t=0$ to $t=1$, the integrand, and hence $F(z)$, are analytic in z. Now let z approach the point $z_0 = 1/2 + i\epsilon$ as $\epsilon \to 0$ from above and show that the denominator can be prevented from vanishing if the contour of integration is depressed, but that if z tends to either end point $z = 0$ or $z = 1$, then the denominator cannot avoid vanishing, and $F(z)$ becomes singular. By evaluating the integral in closed form, verify that $F(z)$ is indeed singular at $z = 0, 1$.

14.   As an example of pinching singularities in Theorem 12, consider the equation

$$F(z) = \int_{-1}^{1} \frac{dt}{(t - iz)(t + iz)}.$$

If z is nonzero (and not equal to $\pm i$), then $F(z)$ is analytic, but if $z \to 0$, then the two singularities trap the contour of integration and a singularity in $F(z)$ results. Verify this conclusion by direct evaluation.

15.   Let

$$F(z) = \int_{a}^{b} \frac{dw}{G(z,w)}.$$

Show that the following conditions are necessary for a singularity of $F(z)$:
1. $G(z,w_0) = 0$.
2. $w_0 = a$ or $w_0 = b$ or $\dfrac{\partial}{\partial w} G(z,w_0) = 0$.

16.   Many problems in applied mathematics include an infinite series

$$F(x) = \sum_{n=0}^{\infty} f_n(z),$$

in which the function $F(z)$ has singularities in addition to all those of each $f_n(z)$. These singularities are introduced by the failure of the series to converge.

The simplest example of this type of series is

$$\frac{1}{1 - z} = 1 + z + z^2 + \cdots.$$

The individual terms on the right of the equation are entire functions; the function their sum defines has a simple pole at $z = 1$. As a second example, consider

$$\frac{z}{z - 1} = 1 + \frac{1}{z} + \frac{1}{z^2} + \cdots.$$

The terms on the right of this equation have singularities at $z = 0$; their sum is singular at $z = 1$.

Consider

$$F(j) = \sum_{n=0}^{\infty} [g \, ln(j - \alpha)]^n,$$

where α, g are constants.

a. What are the analytic properties of the individual terms on the right side of the last equation?
b. By summing the series in closed form, verify that the sum in this equation has a pole at $j = \alpha + e^{1/g}$.

Suggestions for Further Study

Because of the enormous amount of material written on the subject of complex variables and its applications, this list will be extremely selective. Only the references that pertain to the recent work discussed in Section 8.4 are reasonably complete because they are fewer in number.

Despite the large number of basic texts written in recent years, some of the older classics remain the best; in fact, most modern references ultimately lean on them for their treatment of such topics as the gamma function and product expansions. Three that are well worth looking at are:

1. Knopp, K. 1947. *Theory of Functions*. New York: Dover.
2. Titchmarsh, E. C. 1939. *The Theory of Functions*. 2d ed. Oxford: Oxford Univ. Press.
3. Whittaker, E. T., and Watson, G. N. 1927. *A Course of Modern Analysis*. Cambridge: Cambridge Univ. Press.

Of the books that deal primarily with the theory of basic complex analysis, the following have been popular:

4. Ahlfors, L. V. 1966. *Complex Analysis*. 2d ed. New York: McGraw-Hill.
5. Caratheodory, C. 1954. *Theory of Functions*. 2 vols. New York: Chelsea.
6. Cartan, H. 1963. *Elementary Theory of Analytic Functions of One or Several Complex Variables*. Reading, Mass: Addison-Wesley.
7. Hille, E. 1959. *Analytic Function Theory*. 2 vols. Boston: Ginn.
8. Mackey, G. W. 1967. *Lectures on the Theory of Functions of a Complex Variable*. New York: Van Nostrand.
9. Markushevisch, A. I. 1965. *Theory of Functions of a Complex Variable*. 3 vols. Englewood Cliffs, N.J.: Prentice-Hall.

10. Phillips, E. G. 1957. *Functions of a Complex Variable*. 8th ed. London: Oliver and Boyd.

In addition to references 4, 6, 7, and 9, the following deal with advanced topics:

11. Hormander, L. 1966. *An Introduction to Complex Analysis in Several Variables*. New York: Van Nostrand.

12. Rudin, W. 1969. *Real and Complex Analysis*. New York: McGraw-Hill.

13. Veech, W. 1967. *A Second Course in Complex Analysis*. New York: Benjamin.

The theory of conformal mapping is found in most texts. Some that contain more detailed information include the following:

14. Bieberbach, L. 1953. *Conformal Mapping*. New York: Chelsea.

15. Caratheodory, C. 1941. *Conformal Representations*. Cambridge: Cambridge Univ. Press.

16. Kober, H. 1957 *Dictionary of Conformal Representations*. 2d ed. New York: Dover.

17. Nehari, Z. 1952. *Conformal Mapping*. New York: McGraw-Hill.

Some of the more popular general texts that include some applications of the theory of basic complex analysis are:

18. Carrier, G. F., Krook, F., and Pearson, C. E. 1966. *Functions of a Complex Variable*. New York: McGraw-Hill.

19. Churchill, R. V. 1960. *Complex Variables and Applications*. 2d ed. New York: McGraw-Hill.

20. Dettman, J. W. 1965. *Applied Complex Variables*. New York: Macmillan.

21. Kreyszig, E. 1962. *Advanced Engineering Mathematics*. New York: Wiley.

22. Kryala, E. 1972. *Applied Functions of a Complex Variable*. New York: Wiley.

23. Pennisi, L. L. 1963. *Elements of Complex Variables*. New York: Holt, Rinehart and Winston.

24. Wylie, C. R. 1960. *Advanced Engineering Mathematics*. New York: McGraw-Hill.

The following section includes a few references that discuss asymptotic methods. Reference 29 contains more references and information. Reference 25 explains how to determine the asymptotic behavior of solutions of differential equations. Reference 30 is the classical treatise on Bessel functions. These works are not light reading.

25. Cesari, L. 1963. *Asymptotic Behavior and Stability Problems in Ordinary Differential Equations*. New York: Academic Press.

26. DeBruijn, N. G. 1958. *Asymptotic Methods in Analysis.* Amsterdam, The Netherlands: North Holland.

27. Erdelyi, A. 1956. *Asymptotic Expansions.* New York: Dover.

28. Jeffreys, H. 1962. *Asymptotic Approximations.* Oxford: Oxford Univ. Press.

29. Jones, D. S. 1972. *Asymptotic Behavior of Integrals. SIAM Review* 14: 286–317.

30. Watson, G. 1944. *Theory of Bessel Functions.* Cambridge: Cambridge Univ. Press.

References 32 and 34 are the classical sources of information on the theory of transform methods, although readable and practical exposiions are also found in references 20, 21, and 24.

31. Churchill, R. V. 1958. *Modern Operational Mathematics in Engineering.* 2d ed. New York: McGraw-Hill.

32. Titchmarsh, E. C. 1950. *Introduction to the Theory of Fourier Integrals.* Oxford: Clarendon Press.

33. Tranter, C. J. 1962. *Integral Transforms in Mathematical Physics.* London: Methuen.

34. Widder, D. V. 1941. *The Laplace Transform.* Princeton, N.J.: Princeton Univ. Press.

The following is a general reference on the application of complex variable methods (as well as many other techniques) to the classical problems of theoretical physics.

35. Morse, P. M., and Feshbach, H. 1953. *Methods of Theoretical Physics.* New York: McGraw-Hill.

The discussion of the scattering problem in Section 8.4 dealt with the wave equation of classical physics. The Schroedinger Equation of quantum mechanics can be treated in an analogous manner, and this has been the subject of many of the published texts and articles written by theoretical physicists since approximately 1950. References 36 through 38 are general works on the subject. References 39, 40, 43, 44, 47, and 48 are readable and shorter introductions to the subject. Reference 45 reviews the application of homology methods to the scattering problem; applications of algebraic curve theory are studied in reference 49. References 42 and 47 give a systematic treatment of such applications as of 1967–1968.

This paragraph mentions some of the original papers. The idea of making the incident frequency of ω a complex variable was originally advanced in the work of Kramers and Kronig in 1926–1927 (reference 54); they worked with the index of refraction $N(\omega)$ of a medium. The application of this idea to the scattering amplitude $a_l(\omega)$ (see equation

(20), Section 8.4) was introduced by Jost in 1947 (reference 53). Single dispersion relations of the type in equations (11)–(13) were studied in the 1950's, particularly by Chew, *et al.* (reference 52). In 1958 Mandelstam introduced the double dispersion relation of equation (17) (reference 55). The validity of this representation depends partly on the asymptotic behavior of $A(s,t)$ as $s,t \rightarrow \infty$, and the analysis of this led Regge (reference 56) to return to the study of $a_l(\omega)$. He revived the idea of making l a complex variable, an idea that had been developed earlier by Sommerfeld and Watson. By making ω complex also, Regge now considered $a_l(\omega)$ to be a function of the two complex variables l and ω. The singularities of $a_l(\omega)$ in l are referred to as "Regge poles" and "Regge cuts."

The approach to the scattering problem in this text has been first to solve the wave equation, then obtain $A(s,t)$, and then study its analytic properties. A different approach to the high-energy scattering theory of elementary particles holds that the wave equation can be entirely dispensed with. Instead, several assumptions are made about $A(s,t)$: that it has certain analytic properties (such as those allowing the mathematician to write single and double dispersion relations), that it satisfies unitarity (see review exercise 9, Chapter 8 for the one-dimensional version), and that it has a few other properties. All these conditions together result in an integral equation that $A(s,t)$ must satisfy, and the assertion is made that the solution to this equation is indeed the one observed in nature.

The theoretical background for this approach is contained in reference 40. The integral equations obtained are difficult to solve but are solvable under several approximations. Examples of integral equations are developed in reference 50. Whether this approach is indeed valid can ultimately be determined only by actual experiments. Such experiments are currently under way at the National Accelerator at Batavia, New York, and the CERN accelerators in Geneva, Switzerland. It is hoped that the sought-for answers will soon be found.

General texts on scattering theory include the following:

36. Goldberger, M. L., and Watson, K. M. 1964. *Collision Theory.* New York: Wiley.

37. Newton, R. G. 1966. *Scattering Theory of Waves and Particles.* New York: McGraw-Hill.

38. Wu, T. Y., and Ohmura, T. 1962. *Quantum Theory of Scattering.* Englewood Cliffs, N.J.: Prentice-Hall.

References 39 through 48 include some advanced monographs on analyticity and scattering theory.

39. Alfaro, V., and Regge, T. 1965. *Potential Scattering*. Amsterdam, The Netherlands: North Holland.

40. Chew, G. 1966. *The Analytic S-Matrix: A Basis for Nuclear Democracy*. New York: Benjamin.

41. Collins, P. D. B., and Squires, E. 1968. *Regge Poles in Particle Physics*. New York: Springer-Verlag.

42. Eden, R. J. 1967. *High Energy Collisions of Elementary Particles*. Cambridge: Cambridge Univ. Press.

43. Eden, R. J., Landshoff, P. V., Olive, D. I., and Polkingorne, J. C. 1966. *The Analytic S-Matrix*. Cambridge: Cambridge Univ. Press.

44. Frautschi, S. C. 1963. *Regge Poles and S-Matrix Theory*. New York: Benjamin.

45. Hwa, R. C., and Teplitz, V. L. 1966. *Homology and Feyman Integrals*. New York: Benjamin.

46. Newton, R. G. 1964. *The Complex J-Plane*. New York: Benjamin.

47. Omnes, R., and Froissart, M. 1963. *Mandelstam Theory and Regge Poles*. New York: Benjamin.

48. Squires, E. 1963. *Complex Angular Momenta and Particle Physics*. New York: Benjamin.

The remaining references are research papers on scattering theory.

49. Abhyankar, S., and Risk, C. "Algebraic Curve Theory and the Envelope Diagrams." In *Analytic Methods in Mathematical Physics*, ed. R. P. Gilbert and R. G. Newton. 1968. New York: Gordon and Breach. p. 15.

50. Chew, G. F., and Frazer, W. R. 1969. *Phys. Rev.* 181:1914.

51. Chew, G. F., Goldberger, M., Low, F., and Nambu, Y. 1957. *Phys. Rev.* 106:1337.

52. Chew, G., and Pignotti, A. 1968. *Phys. Rev.* 176:2112.

53. Jost, R. 1947. *Helv. Phys. Acta.* 20:256.

54. Kronig, R. de L. 1926. *Jour. Opt. Soc. Am.* 12:547. Kramers, H. A. 1927. *Estratto dagli Atti de Congresso Internationale de Fisici Como, Bologna*.

55. Mandelstam, S. 1958, 1959. *Phys. Rev.* 112:1344; 1741, 1652.

56. Regge, T. 1959. *Nuovo Cimento* 14:951.

The modern treatment of complex analysis did not evolve rapidly or smoothly. The creators of this area of mathematics traveled over cobblestones and encountered numerous blind alleys before experiencing their brilliant insights. An appreciation of this history is important to the student's education in mathematics. A recommended text is

57. Klein, M. 1972. *Mathematical Thought from Ancient to Modern Times*. New York: Oxford Univ. Press.

Answers to Selected Exercises

1.1 Introduction to Complex Numbers

1. a. $6 + 4i$ b. $-5 + 14i$ c. $\dfrac{11}{17} + i\dfrac{10}{17}$ d. $28 + i96$

   e. $\dfrac{3}{2} - i\dfrac{5}{2}$ f. $4 - i\dfrac{15}{2}$

2. a. $x = \pm(2 - i)$

3. $\operatorname{Re}\left(\dfrac{1}{z^2}\right) = \dfrac{x^2 - y^2}{(x^2 + y^2)^2}$ $\operatorname{Im}\left(\dfrac{1}{z^2}\right) = -\dfrac{2xy}{(x^2 + y^2)^2}$

   $\operatorname{Re}\left(\dfrac{1}{3z + 2}\right) = \dfrac{3x + 2}{(3x + 2)^2 + 9y^2}$ $\operatorname{Im}\left(\dfrac{1}{3z + 2}\right) = \dfrac{-3y}{(3x + 2)^2 + 9y^2}$

   $\operatorname{Re}\left(\dfrac{z + 1}{2z - 5}\right) = \dfrac{2x^2 - 3x + 2y - 5}{(2x - 5)^2 + 4y^2}$ $\operatorname{Im}\left(\dfrac{z + 1}{2z - 5}\right) = \dfrac{-7y}{(2x - 5)^2 + 4y^2}$

   $\operatorname{Re}(z^3) = x^3 - 3xy^2$ $\operatorname{Im}(z^3) = 3x^2y - y^3$

6. No; let $z = w = i$.

14. a. -4 b. i c. $\dfrac{1}{2} + i\dfrac{1}{2}$ d. i

15. a. $\sqrt{1 + \sqrt{i}}$

$$= \pm\left(\frac{\sqrt{1 + \sqrt{2} + \sqrt{4 + 2\sqrt{2}}}}{2^{3/4}} + i\,\frac{\sqrt{-1 - \sqrt{2} + \sqrt{4 + 2\sqrt{2}}}}{2^{3/4}}\right)$$

or $\sqrt{1 + \sqrt{i}}$

$$= \pm\left(\sqrt{\frac{\sqrt{2} - 1 + \sqrt{4 - 2\sqrt{2}}}{2^{3/4}}} - i\frac{\sqrt{1 - \sqrt{2} + \sqrt{4 - 2\sqrt{2}}}}{2^{3/4}} \right)$$

b. $\sqrt{1 + i} = \pm\left(\sqrt{\frac{1 + \sqrt{2}}{2}} + i\sqrt{\frac{\sqrt{2} - 1}{2}} \right)$

1.2 Properties of Complex Numbers

1. a. $z = \sqrt[5]{2}\left(\cos\frac{2\pi k}{5} + i\sin\frac{2\pi k}{5} \right);$ $\quad k = 0, 1, 2, 3, 4.$

b. $z = \cos\left(\frac{3\pi}{8} + \frac{2\pi k}{4} \right) + i\sin\left(\frac{3\pi}{8} + \frac{2\pi k}{4} \right);$ $\quad k = 0, 1, 2, 3.$

3. $(3 - 8i)^4/(1 - i)^{10}$

5. $\cos 5x = \cos^5 x - 10\cos^3 x \cdot \sin^2 x + 5\cos x \cdot \sin^4 x$
$\sin 5x = \sin^5 x - 10\cos^2 x \cdot \sin^3 x + 5\cos^4 x \cdot \sin x$

8. $\sqrt{\frac{377}{5}}$

11. No; take $z = i.$ $\quad z^2 = |z|^2 \Longleftrightarrow z$ is real.

15. $|z - (8 + 5i)| = 3$

16. the real axis

1.3 Some Elementary Functions

1. a. $e^2(\cos(1) + i\sin(1))$

b. $\frac{1}{2}\sin(1)\left(\frac{1}{e} + e \right) + i\frac{1}{2}\left(\cos(1) \right) \cdot \left(e - \frac{1}{e} \right)$

c. $\frac{1}{2}(\cos 2)(e^3 + e^{-3}) + i\frac{1}{2}(\sin 2) \cdot (e^{-3} - e^3)$

2. a. $z = \pm\left(\frac{\pi}{4} + 2\pi n - i\frac{1}{2}\log 2 \right)$

b. $z = \pm(2\pi n - i\log(4 + \sqrt{15}))$

4. a. $\log(1) = 2\pi ni$

b. $\log(i) = \frac{\pi i}{2} + 2\pi ni$

c. $\log(-i) = -\dfrac{\pi i}{2} + 2\pi ni$

d. $\log(1+i) = \dfrac{1}{2}\log 2 + \dfrac{\pi i}{4} + 2\pi ni$

5. a. $e^{\pi/2}e^{-2\pi n}$ b. $e^{-\pi(2n+1)}$

c. $e^{-2\pi n}(\cos \log 2 + i\sin \log 2)$

d. $e^{1/2 \log 2 - 2\pi n - \pi/4}\left(\cos\left(\dfrac{1}{2}\log 2 + \dfrac{\pi}{4}\right) + i\sin\left(\dfrac{1}{2}\log 2 + \dfrac{\pi}{4}\right)\right)$

8. a. $e^{x^2-y^2}(\cos 2xy + i\sin 2xy)$ b. $e^{-y}(\cos x + i\sin x)$

c. $e^{x/(x^2+y^2)}\left(\cos \dfrac{y}{x^2+y^2} - i\sin \dfrac{y}{x^2+y^2}\right)$

9. $z = n\pi$ for any integer n

21. This holds iff $b \log a$ has imaginary part in $[-\pi, \pi[$. Otherwise, the formula is $\log a^b = b \log a + 2\pi ik$.

22. No, not even for real a and b. Let $a = 2$, $b = -1$. Then $|a^b| = |2^{-1}| = 1/2$, but $|a|^{|b|} = 2$.

27. a. $\approx 24 - i4.5$ b. $\approx (1.17) - i(1.19) + 2\pi ni$

c. $\approx 95.5 - i1644$

28. The maximum is $\cosh(2\pi) \approx 267$ attained at $z = 2\pi i$, $\pi + 2\pi i$, $2\pi + 2\pi i$.

1.4 Analytic Functions

2. Let $\{a_1, a_2, \cdots, a_n\}$ be a finite set of points and let z_0 be in its complement. Let $\delta_k = |z_0 - a_k|$ and let $\delta = \min\{\delta_1/2, \cdots, \delta_n/2\}$. Then no a_k can lie in $D(z_0, \delta)$ since $\delta < |z_0 - a_k|$.

5. Since f is continuous, there is a $\delta > 0$ such that $|z - z_0| < \delta \Rightarrow |f(z) - f(z_0)| < |f(z_0)|/2$. Thus $f(z) \neq 0$, for if $f(z) = 0$, then $|f(z_0)| < |f(z_0)|/2$, which is absurd.

6. a. Locally, f rotates by $\theta = 0$ and multiplies lengths by 1.

b. Locally, f rotates by an angle $\theta = 0$ and stretches lengths by a factor of 3.

c. Locally, f rotates by an angle π and stretches lengths by a factor of 2.

8. a. Analytic on all of $\mathbb{C}$. The derivative is $3(z+1)^2$.

b. Analytic on $\mathbb{C}\backslash\{0\}$. The derivative is $1 - 1/z^2$.

c. Analytic on $\mathbb{C}\backslash\{1\}$. The derivative is $-10[1/(z-1)]^{11}$.

d. Analytic on $\mathbb{C}\backslash\{e^{2\pi i/3}, e^{4\pi i/3}, 1, i\sqrt{2}, -i\sqrt{2}\}$. The derivative is $-[1/z^3 - 1)^2(z^2+2)^2][(z^3-1)2z + 3z^2(z^2+2)]$.

13. $\dfrac{\partial f}{\partial z} = \dfrac{1}{2}\left(\dfrac{\partial f}{\partial x} - \dfrac{1}{i}\dfrac{\partial f}{\partial y}\right) = \dfrac{1}{2}\left(\dfrac{\partial u}{\partial x} + i\dfrac{\partial v}{\partial x} - \dfrac{1}{i}\left(\dfrac{\partial u}{\partial y} + i\dfrac{\partial v}{\partial y}\right)\right)$

$= \dfrac{1}{2}\left(\dfrac{\partial u}{\partial x} - \dfrac{\partial v}{\partial y}\right) + i\dfrac{1}{2}\left(\dfrac{\partial v}{\partial x} + \dfrac{\partial u}{\partial y}\right).$

Thus the Cauchy-Riemann equations $\partial u/\partial x = \partial v/\partial y$ and $\partial u/\partial y = -\partial v/\partial x$ are equivalent to saying that this complex quantity $\partial f/\partial \bar{z}$ is zero.

15. a. If $n \ge 0$, it is analytic everywhere. If $n < 0$, it is analytic everywhere except at 0. The derivative is nz^{n-1}.

b. Analytic on $\mathbb{C}\backslash\{0,i,-i\}$. The derivative is given by

$$-2\,\dfrac{1}{\left(z+\dfrac{1}{z}\right)^3}\left(1-\dfrac{1}{z^2}\right).$$

c. Analytic except at the nth roots of 2, $\sqrt[n]{2}\,e^{2\pi i k/n}$. The derivative is given by

$$\dfrac{[(1-n)z^n - 2]}{(z^n - 2)^2}.$$

17. $\mathbb{C}\backslash\{1, e^{2\pi i/3}, e^{-2\pi i/3}\}$

19. $(f^{-1}\circ f)'(z) = (f^{-1})'(f(z))f'(z)$. But $(f^{-1}\circ f)(z) = z$, so $(f^{-1}\circ f)'(z) = 1$. Hence $(f^{-1})'(f(z))\cdot f'(z) = 1$.

21. $|z| < 1$

24. a. $\mathbb{C}\backslash\{1\}$ b. Yes   c. x-axis$\backslash\{1\}$, unit circle$\backslash\{1\}$ d. $90°$

25. Write

$$U = \dfrac{\partial u}{\partial x}, \qquad V = -\dfrac{\partial u}{\partial y}.$$

Then $f = U + iV$. By assumption, U and V have continuous partial derivatives. By assumption of continuous second partials for u and v, we get

$$\dfrac{\partial^2 u}{\partial x \partial y} = \dfrac{\partial^2 u}{\partial y \partial x};$$

that is,

$$\dfrac{\partial U}{\partial y} = -\dfrac{\partial V}{\partial x},$$

one of the Cauchy-Riemann equations for f. The other equation comes from

$$\frac{\partial U}{\partial x} = \frac{\partial^2 u}{\partial x^2} = -\frac{\partial^2 u}{\partial y^2} = \frac{\partial}{\partial y}\left(-\frac{\partial u}{\partial y}\right) = \frac{\partial V}{\partial y}.$$

f is thus analytic by Theorem 19.

26. $z = e^{\pi i(3+4k)/2m}$; $\quad k = 0, 1, 2, \cdots, m-1.$

1.5 Differentiation of the Elementary Functions

1. a. Analytic for all z; the derivative is $2z + 1$.
 b. Analytic on $\mathbb{C}\backslash\{0\}$; the derivative is $-1/z^2$.
 c. Analytic on $\mathbb{C}\backslash\{z = (2k+1)\pi/2 \mid k = 0, \pm1, \pm2, \pm3, \cdots\}$; the derivative is $1/\cos^2 z$.
 d. Analytic on $\mathbb{C}\backslash\{1\}$. The derivative is

$$\frac{2z^3 - 3z^2 - 1}{(z-1)^2} \exp\left(\frac{z^3+1}{z-1}\right).$$

2. a. 1 b. The limit does not exist.

3. No

7. Yes

10. a. Analytic on $\mathbb{C}\backslash\{\pm1\}$. The derivative is $\quad -\dfrac{(z^2+1)}{(z^2-1)^2}.$

 b. Analytic on $\mathbb{C}\backslash\{0\}$. The derivative is $\quad \left(1 - \dfrac{1}{z^2}\right)e^{z+1/z}.$

12. $z = 2\pi n \pm i \log(\sqrt{3} + \sqrt{2})$

14. The minimum is $1/e$, taken on at $z = \pm i$.

Review Exercises for Chapter 1

1. $e^i = \cos(1) + i\sin(1)$

$\log(1+i) = \dfrac{1}{2}\log 2 + \dfrac{\pi i}{4} + 2\pi n i,\ n$ an integer

$\sin i = i\dfrac{1}{2}\left(e - \dfrac{1}{e}\right)$

$\exp(2\log(-1)) = 1$

2. $e^{\pi i/16}1, e^{\pi i/16}\left(\dfrac{1}{\sqrt{2}} + \dfrac{i}{\sqrt{2}}\right),\ e^{\pi i/16}i,\ e^{\pi i/16}\left(\dfrac{-1}{\sqrt{2}} + \dfrac{i}{\sqrt{2}}\right),$

$e^{\pi i/16}(-1),\ e^{\pi i/16}\left(\dfrac{-1}{\sqrt{2}} - \dfrac{i}{\sqrt{2}}\right),\ e^{\pi i/16}(-i),\ e^{\pi i/16}\left(\dfrac{1}{\sqrt{2}} - \dfrac{i}{\sqrt{2}}\right)$

4. $(z^3 + 8)' = 3z^2$ on all of $\mathbb{C}$.

$$\left(\frac{1}{z^3 + 1}\right)' = \frac{-3z^2}{(z^3 + 1)^2} \text{ on } \mathbb{C}\backslash\{-1, e^{\pi i/3}, e^{-\pi i/3}\}.$$

$[\exp(z^4 - 1)]' = 4z^3 \exp(z^4 - 1)$ on all of $\mathbb{C}$.

$[\sin(\log z^2)]' = \dfrac{2}{z} \cos(\log z^2)$, on all $\mathbb{C}$ except the entire imaginary axis.

5. a. the real axis

b. a circle, centered at $(17/8, 0)$ of radius $3/8$.

8. if and only if f is constant

9. a. Analytic on $\mathbb{C}\backslash\{0\}$; $(e^{1/z})' = -e^{1/z}/z^2$.

b. Analytic on $\mathbb{C}\backslash\{z = (\pi/2) + 2\pi n \mid n = 0, \pm 1, \pm 2, \pm 3, \cdots\}$; $(1/(1 - \sin z)^2)' = 2 \cos z/(1 - \sin z)^3$.

c. $\mathbb{C}\backslash\{\pm a\}$; $(e^{az}/(a^2 + z^2))' = ((a^2 + z^2) a e^{az} - 2z e^{az})/(a^2 + z^2)^2$.

10. a. No   b. Yes   c. Yes

13. By hypothesis, $d/dz(f(z) - \log z) = 0$. Now use Theorem 21.

15. $\displaystyle\lim_{h \to 0} \frac{(z_0 + h)^n - z_0^n}{h} = f'(z_0)$ where $f(z) = z^n$

17. $x = y = -1$

18. Note that $v(x, y) = 0$ on A. Therefore,

$$\frac{\partial v}{\partial x} = \frac{\partial v}{\partial y} = 0$$

and the Cauchy-Riemann equations give

$$\frac{\partial u}{\partial x} = \frac{\partial u}{\partial y} = 0.$$

Thus f' is identically 0 on the connected set A. Hence by Theorem 21, f is constant on A.

20. Fix a branch on log, say, the principal branch, which is analytic on $\mathbb{C}\backslash\{\text{nonpositive real axis}\}$. Then we can write $z^z = e^{z \log z}$, which is analytic on the region of analyticity of logarithm chosen. The derivative is $z^z(1 + \log z)$.

22. $z = 2e^{i\pi/2}, 2e^{7\pi i/6}, 2e^{11\pi i/6}$

23. They are the real and imaginary parts of z^3.

27. a. 0   b. differentiable at all points $z \in \mathbb{C}$

28. a. $\dfrac{\partial^2 u}{\partial x^2} = -6y, \dfrac{\partial^2 u}{\partial y^2} = 6y \Rightarrow \dfrac{\partial^2 u}{\partial x^2} + \dfrac{\partial^2 u}{\partial y^2} = 0$

b. $v(x, y) = x^3 - 3xy^2$

30. $z = i, -1, -i, 1$

2.1 Contour Integrals

2. No; for example, let $f(z) = z$, $\gamma(t) = it$ for $t \in [0,1]$. Then

$$\int_\gamma \operatorname{Re} f = \int_0^1 0 \cdot i \, dt = 0,$$

but

$$\operatorname{Re} \int_\gamma f = \operatorname{Re} \int_0^1 it\, i \, dt = \operatorname{Re}\left(-\frac{t^2}{2}\Big|_0^1\right) = -\frac{1}{2}.$$

3. a. $2 + i\frac{1}{2}$ b. $\frac{1}{2}[\cos(2 + 2i) - \cos(2i)]$ c. 0

4. γ does not pass through or encircle 0.

7. a. $2\pi i$ b. $-\dfrac{i}{3}$

9. For $|z| = 1$ we have

$$\left|\frac{1}{2 + z^2}\right| = \frac{1}{|2 + z^2|} \le \frac{1}{2 - |z|^2} \le 1$$

since $|z_1 + z_2| \ge |z_1| - |z_2|$. Hence

$$\left|\int_\gamma \frac{dz}{2 + z^2}\right| \le 1 \cdot l(\gamma) = \pi.$$

11. a. $\displaystyle\int_{|z|=1} \frac{dz}{z} = 2\pi i, \quad \int_{|z|=1} \frac{dz}{|z|} = 0, \quad \int_{|z|=1} \frac{|dz|}{z} = 0, \quad \int_{|z|=1} \left|\frac{dz}{z}\right| = 2\pi.$

   b. $-\dfrac{i}{3}$

13. 0

2.2 Cauchy's Theorem: Intuitive Version

1. a. -6 b. 0 c. 0 `d. 0

2. 0, by Cauchy's Theorem

5. The integral will be zero iff γ encircles neither or both of the roots

$$-\frac{1}{2} \pm \frac{\sqrt{3}}{2}i \text{ of } z^2 + z + 1 = 0.$$

6. No; let $f(z) = z$, $\gamma(t) = e^{it}$, $t \in [0,2\pi]$ (the unit circle). Then $\int_\gamma \text{Re } f(z) \; dz = \pi i$, $\int_\gamma \text{Im } f(z) \; dz = -\pi$.

8. $-\dfrac{2}{3} - \dfrac{2}{3}i$

10. $4\pi i$

2.3   Cauchy's Theorem: Precise Version

2. If it were, then the circle $|z| = 1$ would be homotopic to a point in $\mathbb{C}\backslash\{0\}$ so that $\int_{|z|=1} dz/z = 0$. But $\int_{|z|=1} dz/z = 2\pi i$. This contradiction shows that $\mathbb{C}\backslash\{0\}$ is not simply connected.

3. Let $\gamma: [a,b] \to A$ be a closed curve in A. Define a homotopy $H: [a,b] \times [0,1] \to A$ by $H(t,s) = s\gamma(t) + (1-s)z_0$. H is clearly continuous and is into A since A is star-like at z_0. Thus γ is homotopic by H to the constant map at z_0.

6. a. $2\pi i$ b. 0   c. 0   d. πi

7. $\dfrac{\pi i}{2}$

10. a. $2\pi i$ b. $2\pi i$

2.4   Cauchy's Integral Formula

3. a. $2\pi i$ b. $2\pi i$ c. 0

8. The Cauchy inequalities (Theorem 25) show that $f^{(k)}(z)$ is identically 0 for $k > n$. The conclusion follows from exercise 14 of Section 1.4.

9. Let $\tilde{\gamma}$ be the circle $\tilde{\gamma}(t) = z_1 + re^{it}$, $0 \le t \le 2\pi$, $|z_0 - z_1| < r$. γ is homotopic to $\tilde{\gamma}$ by $H(t,s) = s(z_1 + re^{it}) + (1-s)(z_0 + re^{it})$. z_1 is not in the image of the homotopy, so

$$I(\gamma,z_1) = \frac{1}{2\pi i} \int_\gamma \frac{1}{z - z_1} = \frac{1}{2\pi i} \int_{\tilde{\gamma}} \frac{1}{z - z_1} = 1.$$

11. By Theorem 16 of Section 1.4, f is analytic on $A\backslash\{z_0\}$, so it is certainly also continuous there. Since

$$f(z_0) = F'(z_0) = \lim_{z \to z_0} \frac{F(z) - F(z_0)}{z - z_0} = \lim_{z \to z_0} f(z),$$

f is also continuous at z_0. By the corollary to Morera's Theorem, f is analytic on A.

12. a. 0 b. $-\dfrac{\pi i}{3}$ c. $2\pi i \sin(1) = \pi(e^i - e^{-i})$

13. a. 0 b. 0 c. 0 d. $\dfrac{\pi i}{2}$

14. Using the Cauchy inequalities, $|f'(0)| \leq 1/R$ for every $R < 1$. Hence $|f'(0)| \leq 1$. This is the best possible bound, as is clear from the example $f(z) = z$.

16. $4\pi i$

18. $1/f$ is entire and $|1/f(z)| \leq 1$ on $\mathbb{C}$. Therefore, $1/f$ is constant by Liouville's Theorem. Hence f is constant.

19. a. $\pi/2 + i(\pi/2)$ b. 0

2.5 Maximum Modulus Theorem and Harmonic Functions

2. Since $|f|$ is constant on $D = \{z \mid |z - z_0| < r\}$, $|f|$ attains its maximum on D at z_0, which is in the interior of D. By the Maximum Modulus Theorem, f is constant on D.

4. $f - g$ is continuous on cl (A) and analytic on A. $(f - g)(z) = 0$ for $z \in$ bd (A), so by the Maximum Modulus Theorem, $(f-g)(z) = 0$ for all $z \in A$. In other words, $f(z) = g(z)$ for all $z \in A$. Hence $f = g$ on all of cl $(A) = A \cup$ bd (A).

6. Locally in B, $w =$ Re g where g is analytic. Then $w \circ f =$ Re $(g \circ f)$. But $g \circ f$ is analytic, so $w \circ f$ is harmonic.

7. $|e^{z^2}|$ attains a maximum of e at ± 1.

9. a. $v(x,y) = -\cosh x \cos y$
 b. $v = \arctan(y/x)$ or $-\arctan(x/y)$ c. $v(x,y) = e^x \sin y$

11. If $z = x + iy$, Re $(e^z) = e^x \cos y$ and Im $(e^x) = e^x \sin y$. The normal vector to level curves of these functions is given by the gradient vectors

$$(e^x \cos y, -e^x \sin y) \quad \text{and} \quad (e^x \sin y, e^x \cos y).$$

Since these are orthogonal, the curves are also orthogonal.

16. $|g(0)| = |0| = 0$, so $g(0) = 0$. Also, $|g(z)| = |z| < 1$ for all $z \in \{z \mid |z| < 1\}$. Thus Schwarz's Lemma applies, and so $g(z) = cz$ with $|c| = 1$.

17. 0

19. a. No. Counterexample: $u(x,y) = x^2 - y^2$, $v(x,y) = x$ are harmonic, but $u(v(x,y),0) = x^2$ is not harmonic.
 b. No. Counterexample: $u(z) = v(z) = x$. Then $u(z) \cdot v(z) = x^2$ is not harmonic.
 c. Yes

Review Exercises for Chapter 2

1. Since f is analytic and nonconstant on A, z_0 cannot be a relative maximum. Thus in every neighborhood of z_0, *particularly in* $\{z \mid |z - z_0| < \epsilon\}$, there is a point z with $|f(z)| > |f(z_0)|$. If $f(z_0) \neq 0$, by continuity, $f(z) \neq 0$ in some small disk D centered at z_0. Thus $1/f(z)$ is analytic on D. By this last argument, there is a ζ close to z_0 such that

    $$\left| \frac{1}{f(\zeta)} \right| > \left| \frac{1}{f(z_0)} \right|;$$

    therefore,

    $$|f(\zeta)| < |f(z_0)|.$$

3. If $r_1 > r_2 > 1$, then γ_{r_1} is homotopic in $\{z \mid |z| > 1\}$ to γ_{r_2}, so the integrals are the same.

5. $-2/3$

6. a. No b. Yes c. Yes

7. The Maximum Modulus Theorem shows that f is constant on any bounded (connected) subregion of A containing z_0. Taking an increasing sequence of such subregions whose union is A gives the result.

8. a. 0 b. 0 c. $2\pi i$ d. $2\pi i \cos(1)$

9. $v(x,y) = \dfrac{-y}{(x-1)^2 + y^2}$ on $\mathbb{C} \backslash \{1\}$

11. Consider $\displaystyle\int_{|z|=1} \frac{e^z}{z^2}\, dz;$ 2π

13. a. i b. $\log i = i(\pi/2) + 2\pi i n$ c. $\log(-i) = -i(\pi/2) + 2\pi i n$

15. By the Cauchy integral formulas, f' is analytic on A. Since f is nonzero in A, f'/f is analytic on A and the integral is 0 by the Cauchy Integral Theorem.

17. No; let γ be the unit circle. Then $\int_\gamma x\,dx + x\,dy = \pi$.

18. $2e^{i\pi/6}$, $2e^{i5\pi/6}$, or $2e^{i3\pi/2}$.

20. By Theorem 34,

    $$u(0) = \frac{1}{2\pi} \int_0^{2\pi} u(Re^{i\theta})\, d\theta.$$

By Theorem 38,

$$u(re^{i\varphi}) = \frac{R^2 - r^2}{2\pi} \int\limits_0^{2\pi} \frac{u(Re^{i\theta})}{R^2 - 2Rr\cos(\theta - \varphi) + r^2} \, d\theta.$$

Using these equalities we get

$$\frac{R^2 - r^2}{2\pi} \int\limits_0^{2\pi} \frac{u(Re^{i\theta})}{R^2 + 2Rr + r^2} \, d\theta \le u(z) \le \frac{R^2 - r^2}{2\pi} \int\limits_0^{2\pi} \frac{u(Re^{i\theta})}{R^2 - 2Rr + r^2} \, d\theta;$$

that is,

$$\frac{(R+r)(R-r)}{(R+r)(R+r)} \frac{1}{2\pi} \int\limits_0^{2\pi} u(Re^{i\theta}) \, d\theta \le u(z)$$

$$\le \frac{(R-r)(R+r)}{(R-r)(R-r)} \frac{1}{2\pi} \int\limits_0^{2\pi} u(Re^{i\theta}) \, d\theta.$$

Therefore,

$$\frac{R - |z|}{R + |z|} u(0) \le u(z) \le \frac{R + |z|}{R - |z|} u(0).$$

3.1   Convergent Series of Analytic Functions

2. The sequence of partial sums converges uniformly, thus it converges to a continuous function and the assertion follows.

4. By example 1, $\zeta(z) = \Sigma_{n=1}^{\infty} n^{-z}$ converges uniformly on closed disks in A and thus it is analytic on A with $\zeta'(z) = \Sigma_{n=1}^{\infty} (-\log n)n^{-z}$, which also converges uniformly on closed disks in A and thus is analytic. By induction, $\zeta^{(k)}(z) = \Sigma_{n=1}^{\infty} (-\log n)^k n^{-z}$ converges uniformly on closed disks in A and thus is analytic. Therefore, $(-1)^k \zeta^k(z) = \Sigma_{n=1}^{\infty} (\log n)^k n^{-z}$ is also analytic.

6. Let D be a closed disk in A and let δ be its distance from the boundary Im $z = \pm 1$. Then for $z = x + iy \in D$, prove that $|e^{-n} \sin(nz)| \le e^{-n\delta}$.

7. $\{z \mid |z - 1| < 1\}$

10. No; let $f_n(z) = \Sigma_{k=1}^n z^k/k^2$.

11. Let D be any closed disk in A. Then there exists $r > 1$ such that $|z| > r$ for all $z \in D$. Hence $|1/z^n| < (1/r)^n$ for all $z \in D$. But $\Sigma_{n=1}^{\infty} (1/r)^n$ converges, since $1/r < 1$. Thus $\Sigma_{n=1}^{\infty} 1/z^n$ converges absolutely and uniformly on D. Since $1/z^n$ is analytic on A, Theorem 6 shows that $\Sigma_{n=1}^{\infty} 1/z^n$ is analytic on A.

13. The limit is one and the convergence is not uniform.

14. a. does not converge   b. 0

15. Neither of these series converges absolutely. However, both the real and imaginary parts of each are alternating series whose terms decrease in absolute value monotonically to zero and thus are convergent.

3.2   Power Series and Taylor's Theorem

2. a. 1   b. e c. e d. 1

4. a. $e^z = \sum_{n=0}^{\infty}(e/n!)\,(z-1)^n$, which converges everywhere.

b. $1/z = \sum_{n=0}^{\infty}(-1)^n(z-1)^n$; the series converges for $|z-1| < 1$.

6. a.
$$\frac{\sin z}{z} = \sin(1) + [\cos(1) - \sin(1)](z-1)$$
$$+ \left[\frac{\sin(1)}{2} - \cos(1)\right](z-1)^2$$
$$+ \left[\frac{5}{6}\cos(1) - \frac{1}{2}\sin(1)\right](z-1)^3 + \cdots$$

b. $z^2 e^z = \displaystyle\sum_{n=0}^{\infty}\frac{1}{n!}\,z^{n+2}$ c. $e^z \sin z = z + z^2 + \dfrac{1}{3}z^3 - \dfrac{1}{30}z^5 + \cdots$

7. $\sqrt{z^2 - 1} = i - \dfrac{i}{2}z^2 - \dfrac{i}{8}z^4 + \cdots$

8. The series is not in the form of a Taylor series. In fact, the series does not even converge at $z = 0$.

11. For $\sinh z$, the odd derivatives are 1 and the even derivatives are 0 at $z = 0$. Thus by Taylor's Theorem,
$$\sinh z = \sum_{n=1}^{\infty}\frac{z^{2n-1}}{(2n-1)!}.$$

12. If $|z| < R$, then $\sum a_n z^n$ converges absolutely; that is, $\sum |a_n|\,|z|^n$ converges. But $|(\mathrm{Re}\ a_n)z^n| \le |a_n|\,|z^n|$, so $\sum |(\mathrm{Re}\ a_n)z^n|$ and hence $\sum (\mathrm{Re}\ a_n)z^n$ converges. Thus $\sum (\mathrm{Re}\ a_n)z^n$ converges for any $|z| < R$. Thus the radius of convergence must be $\ge R$.

14. $\dfrac{1}{(1-z)^2} = \displaystyle\sum_{n=1}^{\infty} nz^{n-1};$ $\dfrac{1}{(1-z)^3} = \dfrac{1}{2}\displaystyle\sum_{n=2}^{\infty} n(n-1)z^{n-2}.$

17. The region for the first series is $A = \{z \mid |\mathrm{Im}\,(z)| < \log 2\}$. The second series is not analytic anywhere.

3.3   Laurent's Series and Classification of Singularities

2. Use the fact that there is an analytic function φ defined on a neighborhood of z_0 such that $\varphi(z_0) \neq 0$ and $f(z) = \varphi(z)/(z - z_0)^k$.

3.   c. $z - z^2 + z^3 - z^4 + z^5 - \cdots$, $|z| < 1$.

   d. $\dfrac{1}{z^2} + \dfrac{1}{z} + \dfrac{1}{2!} + \dfrac{1}{3!}z + \dfrac{1}{4!}z^2 + \cdots$, $0 < |z| < \infty$.

5.   a. No   b. No   c. Yes

9. $\dfrac{1}{e^z - 1} = \dfrac{1}{z} - \dfrac{1}{2} + \dfrac{1}{12}z - \dfrac{1}{720}z^3 + \cdots$

10. $\cot z = \dfrac{1}{z} - \dfrac{1}{3}z - \dfrac{1}{45}z^3 - \dfrac{2}{945}z^5 + \cdots$

15.   c. 1   d. 0

16. $\cos\left(\dfrac{1}{z}\right)$ has a zero of order 1 at

$$\frac{1}{z} = \frac{(2n + 1)\pi}{2}, \qquad n = 0, \pm 1, \pm 2, \cdots,$$

that is, at

$$z = \frac{2}{(2n + 1)\pi};$$

$1/\cos(1/z)$ has simple poles at these points. $z = 0$ is not an isolated singularity.

Review Exercises for Chapter 3

1. $\displaystyle\sum_{n=1}^{\infty} \frac{(-1)^{n-1}}{n!} (z - 1)^n$

3. $\dfrac{1}{z^2} - \dfrac{1}{z} + 1 - z + z^2 - z^3 + \cdots$

4. $\displaystyle\sum_{n=1}^{\infty} \frac{(-1)^{n+1}}{(2n - 1)!} z^{4n}$

6. Suppose that $w \in \mathbb{C}$, $w \neq 0$. Solve $e^{1/z} = w$ for z as follows.

$$\frac{1}{z} = \log w = \log |w| + i\,(\arg w + 2\pi n);$$

$$z = \frac{1}{\log|w| + i\,(\arg w + 2\pi n)}.$$

Infinitely many of these solutions lie in any deleted neighborhood of the origin.

12. Schwarz's Lemma applies to give $|f(z)| \le |z|$ for all $|z| < 1$. $|f(z)| = |z|$ for any $|z| < 1$ implies that $f(z) = cz$ for a constant c with $|c| = 1$.

14. $-\dfrac{2\pi i}{3}$

16. a. $= \dfrac{1}{z} - z + z^3 - z^5 + z^7 - \cdots$

   b. $= \dfrac{1}{z^3} - \dfrac{1}{z^5} + \dfrac{1}{z^7} - \dfrac{1}{z^9} + \dfrac{1}{z^{11}} - \cdots$

17. The coefficients for either series are given by

$$a_n = \frac{1}{2\pi i} \int_\gamma \frac{f(\zeta)}{\zeta^{n+1}}\, d\zeta,$$

where γ is the circle of radius 2 centered at the origin.

18. b. 1

19. Yes

21. Since f is bounded near z_0, z_0 is neither a pole nor an essential singularity.

23. 2π

24. $2\pi i$

25. a. poles of order 1 at $z = 1$ and $z = 5$
   b. removable singularity at $z = 0$
   c. pole of order 1 at $z = 0$
   d. pole of order 1 at $z = 1$

28. let $A = \mathbb{C}\backslash\{0\}$, $f(z) = e^z$

31. No

32. 1

33. a. 2π b. $B_n = \dfrac{n!}{2\pi} \int_0^{2\pi} \dfrac{e^{-i(n-1)\theta}}{e^{e^{i\theta}} - 1}\, d\theta$

36. $a_0 + z + z^2 + z^4 + \cdots$

4.1 Calculation of Residues

1. e. 0 f. 1 g. −1 h. 1/6 i. 0
2. Let $f(z) = 1/z$.
4. The correct residue is 2.
5. c. $\text{Res}(f,0) = 1/64$, $\text{Res}(f,-4) = -1/64$
   d. $\text{Res}(f,-1) = 0$
   e. $\text{Res}(f,\sqrt[3]{3}) = 3^{-5/3}$, $\text{Res}(f,\sqrt[3]{3}e^{2\pi i/3}) = 3^{-5/3}e^{-4\pi i/3}$,
   $\text{Res}(f,\sqrt[3]{3}e^{4\pi i/3}) = 3^{-5/3}e^{-2\pi i/3}$
6. 1/6
9. $\text{Res}(f_1f_2,z_0) = a_1\text{Res}(f_2,z_0) + a_2\text{Res}(f_1,z_0)$ where a_i is the constant term in the expansion of f_i.
11. a. 0 b. $-e/2$ c. $\text{Res}(f,0) = 1$, $\text{Res}(f,1) = -2$
    d. $\text{Res}(f,0) = 1$, $\text{Res}(f,1) = -e$

4.2 The Residue Theorem

2. a. 0 b. 0
3. 0
6. $-12\pi i$
7. a. 0 b. 0
9. $-\int_{\gamma} g\left(\frac{1}{z}\right)\frac{1}{z^2}\,dz = \int_{\tilde{\gamma}} g(w)\,dw$ where $\tilde{\gamma}$ is the curve $1/\gamma$
10. c. -8 e. $2\pi i$
11. $-\pi i$
12. e. $2\pi i$ f. $2\pi i$

4.3 Evaluation of Definite Integrals

1. $\pi/\sqrt{3}$

3. $\dfrac{\pi a}{(a^2 - b^2)^{3/2}}$

5. $\dfrac{\pi e^{-m/\sqrt{2}}}{2\sqrt{2}}\left[\cos\dfrac{m}{\sqrt{2}} + \sin\dfrac{m}{\sqrt{2}}\right]$ if $m > 0$

   $\dfrac{\pi e^{m/\sqrt{2}}}{2\sqrt{2}}\left[\cos\dfrac{m}{\sqrt{2}} - \sin\dfrac{m}{\sqrt{2}}\right]$ if $m < 0$

6. $\pi e^{-1}/2$

9. $-\pi i/(a-1)^2$

10. The function is even and line 3 of Table 4.2 applies.

14. 0

15. $-\pi i/2$ (or $\pi i/2$ if you use a different branch of $\sqrt{\ }$).

19. $\displaystyle\sum_{n=-\infty}^{\infty}(-1)^n f(n) = -\{$sum of residues of $\pi \csc(\pi z)f(z)$ at the poles of $f\}$.

    f should obey some conditions like: There is an $R > 0$ and $M > 0$ such that for $|z| > R$, $|f(z)| \le M/|z|^{\alpha}$ where $\alpha > 1$.

    If some of the poles of f should lie at the integers, the technique could still be used. After verifying that

    $$\int_{C_N} \frac{\pi f(z)}{\sin \pi z}\, dz \to 0 \quad \text{as } N \to \infty,$$

    we would have

    $$-\sum_{\substack{n=-\infty \\ n \text{ not a pole of } f}}^{\infty}(-1)^n f(n) = \text{the sum of the residues of } \pi \csc(\pi z)f(z) \text{ at the poles of } f.$$

21. Use exercise 20; $\mathrm{Res}\,((-z)^{a-1}f(z), 1) = (e^{\pi i})^{a-1}$.

Review Exercises for Chapter 4

1. $2\pi/\sqrt{3}$

2. $(\pi\sqrt{2})/2$

3. $\pi e^{-ab}/a$

4. $\mathrm{Res}\left(\dfrac{f''}{f'}, z_0\right) = \dfrac{2k^2}{k+1}\cdot\dfrac{f^{(k+1)}(z_0)}{f^{(k)}(z_0)}$

5. $\dfrac{\pi}{2}\left[2 - \dfrac{1}{e} - \cos(1)\right]$

6. a. $\mathrm{Res}(f,0) = -1$. The other residues are at z with $z^2 = 2\pi ni$, $n = \pm 1, \pm 2, \cdots$, and are equal to $-1/2$.
    b. $\mathrm{Res}(f,n\pi) = 2\pi n \cos(n^2\pi^2)$ c. $\cos(1)$

7. $\pi/\sqrt{5}$

8. $2\pi i \sin(1)$

9. 2π

13.   a. $\dfrac{1}{2} + \dfrac{3}{4}z + \dfrac{7}{8}z^2 + \cdots + \dfrac{2^{n+1}-1}{2^{n+1}}z^n + \cdots$

   b. $\dfrac{1}{z^2} + \dfrac{3}{z^3} + \dfrac{7}{z^4} + \dfrac{15}{z^5} + \cdots + \dfrac{2^n-1}{z^{n+1}} + \cdots$

14. $\pi/2$

19. $\dfrac{1}{1+(z-1)}$ has been expanded incorrectly.

20. $-\pi i$

21.   a. The radius of convergence is infinite.
   b. The radius of convergence is 1 (use the root test).
      Note. To use the ratio test, the following facts are used:

   1. $\displaystyle\lim_{n\to\infty}\left(1+\dfrac{z}{n}\right)^n = e^z$.

   2. In the power series $\Sigma\, a_n z^n$, if the coefficients a_n tend to a nonzero finite limit, then the radius of convergence must be 1.

22.   a. $1 + 3z + 6z^2 + 10z^3 + \cdots + \dfrac{(n+1)(n+2)}{2}z^n + \cdots$

   b. $-\dfrac{1}{z^3} - \dfrac{3}{z^4} - \dfrac{6}{z^5} - \cdots - \dfrac{(n+1)(n+2)}{2}\dfrac{1}{z^{n+3}} - \cdots$

   c. $\dfrac{1}{8} + \dfrac{3}{16}(z+1) + \dfrac{6}{32}(z+1)^2$

   $+ \cdots + \dfrac{(n+1)(n+2)}{2^{n+4}}(z+1)^n + \cdots$

   d. $\dfrac{-1}{(z-1)^3}$

23.   b. Use line 5 of Table 4.2.

28.   It will suffice for f to be analytic in a region containing the real axis and the upper half plane and to be such that the integral of $f(z)/(z-x)$ along the upper semicircle of $|z|=R \to 0$ as $R \to \infty$. These conditions will hold if $|f(z)| < M/R^\alpha$ for some $M > 0$ and $\alpha > 0$ for large enough R, and z lying in the upper half plane. Use exercise 27 for the last part.

29.   The last equality is wrong since the integral along the semicircle is omitted. We cannot conclude that the integral along the semicircle goes to 0 as $R \to \infty$ and thus must evaluate it more carefully.

30. Use exercise 19 of Section 4.3. $F(z) = (\pi \cosec \pi z)/(z+a)^2$ has a pole at $z = -a$ with residue $-\pi^2 \cosec(\pi a) \cot(\pi a)$.

5.1 Basic Theory of Conformal Mappings

5. $v(x,y) = 1 - 2xy + \dfrac{x^2 - y^2}{(x^2 + y^2)^2}$

6. the first three quadrants

8. From $f \circ f^{-1}(z) = z$ we get $f'(f^{-1}(z)) \cdot (f^{-1})'(z) = 1$. It follows that $f'(z_0) \neq 0$, so f is conformal by Theorem 1.

10. Suppose that $a = \rho e^{i\theta}$. Define maps $r(z) = e^{i\theta}z$, $m(z) = \rho z$, $t(z) = z + b$. Thus $f = t \circ m \circ r$. The answer to the second part of the question is yes.

5.2 Fractional Linear and Schwarz-Christoffel Transformations

1. T is the composition of a translation, inversion in the unit circle, reflection in the real axis, a rotation, a magnification, and another translation.

3. $\dfrac{2z - 1}{2 - z}$

4. $e^{i\theta} \dfrac{z - Rz_0}{R - \bar{z}_0 z}$

6. Suppose that T is such a map. Define W by

$$W(z) = T\left(\frac{1}{i} \cdot \frac{z+1}{z-1}\right).$$

W maps the unit disk conformally onto itself. Now use Theorem 6.

10. $\dfrac{z^8 - i}{z^8 + i}$

12. By Theorem 8, $T(\gamma_1)$ and $T(\gamma_2)$ are circles or straight lines and, by Theorem 1, they intersect orthogonally.

13. $f(z) = \dfrac{1 + e^z}{1 - e^z}$

14. Use a branch of log defined on $\mathbb{C} \setminus \{\text{nonpositive real axis}\}$ by $\log(re^{i\theta}) = \log r + i\theta$, $-\pi < \theta < \pi$.

17. By the Schwarz-Christoffel Formula, the image is a polygon with four sides, three of whose angles are 90°. Hence the image is a rectangle.

18. No.

$$f(z) = \int\limits_0^z (t-1)^{-\alpha_1} t^{-\alpha_2}\, dt$$

maps the upper half plane to a triangle with exterior angles $\pi\alpha_1,\ \pi\alpha_2,\ \pi\alpha_3$.

20. $f(b) = 0$ and $f(d) = \infty$. Thus a circle through b and d must map to a circle through 0 and ∞, that is, a line through the origin. Since

$$|f(z)| = \left|a\frac{z-b}{z-d}\right| = \left|\frac{z-b}{z-d}\right||a|,$$

$$\left|\frac{z-b}{z-d}\right| = \frac{r}{|a|} \Longleftrightarrow |f(z)| = r.$$

This establishes (a) and (b).

The easiest way to obtain the orthogonality is to notice that the images under the map f are trivially orthogonal. Since the inverse of a fractional linear transformation is of the same form, hence conformal, the same must have been true of the pre-image. To confirm this directly a straightforward but lengthy calculation is required.

5.3 Applications to Laplace's Equation, Heat Conduction, Electrostatics, and Hydrodynamics

1. $u(x,y) = \dfrac{1}{\pi}\arctan\left(\dfrac{4x^3y - 4xy^3}{x^4 - 6x^2y^2 + y^4}\right)$

$$= \frac{4}{\pi}\arctan\left(\frac{y}{x}\right)$$

4. $\varphi(x,y) = 1 - \dfrac{1}{\pi}\arctan\left(\dfrac{\cos x \cdot \sinh y}{\sin x \cdot \cosh y - 1}\right)$

$$+ \frac{1}{\pi}\arctan\left(\frac{\cos x \cdot \sinh y}{\sin x \cdot \cosh y + 1}\right)$$

5. $F(z) = |\alpha|\left(e^{-i\theta}z + \dfrac{1}{e^{-i\theta}z}\right)$

In polar coordinates,

$$\varphi(r,\Theta) = |\alpha|\left(r + \frac{1}{r}\cos(\Theta - \theta)\right)$$

$$\psi(r,\Theta) = |\alpha|\left(r - \frac{1}{r}\sin(\Theta - \theta)\right)$$

6. In polar coordinates,

$$\varphi(r,\theta) = \alpha r^4 \cos 4\theta$$

$$\psi(r,\theta) = \alpha r^4 \sin 4\theta$$

In rectangular coordinates,

$$\varphi(x,y) = (x^4 - 6x^2 y^2 + y^4)\alpha$$

$$\psi(x,y) = (4x^3 y - 4xy^3)\alpha$$

Review Exercises for Chapter 5

1. any region not containing zero

3. $\dfrac{2z - i}{2 + iz}$

4. $f(z) = \dfrac{2z - 4}{z}$

6. the first quadrant

8. $f(z) = z + 1 - i$

9. No

11. $T(x,y) = \dfrac{1}{2} - \dfrac{1}{\pi}\mathrm{Re}\left(\arcsin\left(\dfrac{z}{a}\right)\right)$

12. $\varphi(x,y) = 1 - \dfrac{1}{\pi}\arctan\left(\dfrac{e^x \sin y}{e^x \cos y - 1}\right)$

The arctan must be chosen between 0 and π.

14. No

15. Use Theorem 7 of Section 5.2.

16. $F(z) = \sqrt{\dfrac{az^2 + b}{cz^2 + d}}$ where a, b, c, d are real and $ad > bc$.

18. $\pi/6e^3$

20. $f(z) = \sqrt{z^2 + 1}$ where $\sqrt{}$ must be taken as the branch defined on $\mathbb{C}\backslash\{\text{positive real axis}\}$, which takes values in the upper half plane.

The desired potential is $\varphi(z) = \dfrac{1}{\pi}\arg\left(\dfrac{\sqrt{z^2+1}-1}{\sqrt{z^2+1}+1}\right).$

21. The desired complex potential must be $F(z) = \alpha\sqrt{z^2+1}.$

The formula

$$y^2 = \frac{K-1}{2x^2+K-1} + \frac{K-1}{2}$$

for $K > 1$ gives lines of constant ψ, which are streamlines.

22. $\sqrt{\rho}$; $\rho^2.$

23. $-z^2 - 1 - 1/z^2 - 1/z^4 - 1/z^6 - \cdots$

24. $f(z) = \dfrac{h}{\pi}\left(\sqrt{z^2-1} + \cosh^{-1}z\right)$

6.1 Analytic Continuation and Elementary Riemann Surfaces

2. a. No, it does not. An important condition of Theorem 1 is that the limit point z_0 must lie in A.

   b. No. Let A be the unit disk. Let $u_1(z) = \mathrm{Im}(z)$ and $u_2(z) = \mathrm{Im}(e^z)$. Both u_1 and u_2 are harmonic in A and are zero along the real axis, but they are not identically zero, nor do they agree with each other on A.

4. If f were constant on a neighborhood of z_2, it would be constant on all of A by Theorem 1. This would force $f'(z_1)$ to be 0, which is not true.

8. The situation is something like that existing for $\sqrt[3]{z}$, except that now there are three sheets, each a copy of the plane cut along, say, the negative real axis. They are joined along these cuts so that following a path that winds once around zero carries one from the first sheet to the second. When the path winds once more around zero one is carried to the third sheet; when the path winds a third time around zero one is carried back to the first sheet.

11. $\mathbb{C}\setminus\{z \mid z \text{ is real and } -\infty < z \le -1\}$

6.2 Rouché's Theorem and Principle of the Argument

6. Use exercise 4.

7. This follows from exercise 3.

9. No. Let $f(z) = e^z - 1$. f has three zeros inside a circle of radius 3π, center 0, but $f'(z)$ has no zeros.

13.   Any r such that $1 < r < 4$ will give the desired result. Using Rouché's Theorem we would get $r = 2$.

6.3   Mapping Properties of Analytic Functions

1. $\{z \mid |z| < 1/2\}$

3.   Let $f(z) = z^3$, $w_0 = z_0 = 1$, $r = 1$. The roots of $z^3 - 1$ lie at 1, $e^{2\pi i/3}$, and $e^{4\pi i/3}$. Of these only one lies in $D = \{z \mid |z - 1| < 1\}$. $f'(z) = 3z^2$ and is 0 only at 0, which does not lie in D. However, $f(re^{\pi i/3})$ $= f(re^{-\pi i/3}) = -r^3$, and for small enough r, these points lie in D.

7.   Let u be harmonic and nonconstant on a region A, $z_0 \in A$, and let U be any open neighborhood of z_0 lying in A. By exercise 6, u is an open mapping, so $u(U)$ is an open neighborhood of $u(z_0)$ in $\mathbb{R}$. This means that arbitrarily near z_0, u takes on values that are both larger and smaller than $u(z_0)$.

9.   Let $f(z) = e^{z-a} - z$ and $g(z) = -z$. Then, for $z = x + iy$ on the unit circle, $|f(z) - g(z)| = |e^{x+iy-a}| = e^{x-a} < 1 = |g(z)|$. Rouché's Theorem applies.

Review Exercises for Chapter 6

3.   Let $0 < r < 1$, $D(0,r) = \{z \mid |z| < r\}$, $\gamma_r = \{z \mid |z| = r\}$. By assumption, f is one-to-one on γ_r, so $f(\gamma_r)$ is a simple closed curve. Since f is bounded on A, f must map $D(0,r)$ to the interior of γ_r. Theorem 10 of Section 6.2 now shows that f is one-to-one on $D(0,r)$. Because this holds for any $r < 1$, f is one-to-one on A.

5.   Use the Root Counting Formula.

6.   Use $\displaystyle\int_\gamma \frac{f'}{f} = 2\pi i \sum_{j=1}^{n} I(\gamma, \alpha_j)$.

8.   No

9. $\sqrt{5}$

11.   Use the Identity Theorem.

12. f is identically zero on $\{z \mid |z| < 1\}$.

16.   Let $h(z) = f(z) - g(z)$ and use the Maximum Principle for Harmonic Functions.

19.   (1) true; (2) true; (3) true; (4) true; (5) false; (6) false; (7) true; (8) false; (9) true; (10) true.

20.   No

22. a. Yes   b. No

24. 1/2

7.1 Infinite Products and the Gamma Function

12. Start with the third line of the proof of Euler's Formula:

$$\frac{1}{\Gamma(z)} = z \lim_{n \to \infty} n^{-z} \prod_{k=1}^{n} \left(1 + \frac{z}{k}\right) = \lim_{n \to \infty} \frac{z}{n^z} \prod_{k=1}^{n} \left(\frac{k+z}{k}\right)$$

$$= \lim_{n \to \infty} \frac{z(z+1)(z+2) \cdot \quad \cdots \quad \cdot (z+n)}{n^1 \cdot 1 \cdot 2 \cdot 3 \cdot \quad \cdots \quad \cdot n}$$

$$= \lim_{n \to \infty} \frac{z(z+1) \cdot \quad \cdots \quad \cdot (z+n)}{n! n^z}$$

14. Use the Laurent expansion and $\text{Res}(\Gamma, -m) = (-1)^m / m!$.

17. Γ has simple poles at $0, -1, -2, \cdots$ and is analytic elsewhere. Therefore,

$$\int_\gamma \Gamma(z) \, dz = 2\pi i \, \text{Res}(\Gamma, 0) = 2\pi i \lim_{z \to 0} z\Gamma(z)$$

$$= 2\pi i \lim_{z \to 0} \Gamma(z+1) = 2\pi i \Gamma(1) = 2\pi i.$$

7.2 Asymptotic Expansions and the Method of Steepest Descent

4. For $S_4(10)$ the error is $\leq .00024$, and for $S_5(10)$ it is $.00012$. For fixed x, the error term decreases as n increases until n becomes larger than x, at which point it begins to increase again. In fact, for fixed x, our bound on the error term goes to infinity with n. For fixed n, $\lim_{x \to \infty} n!/x^{n+1} = 0$, so the error goes to zero as x increases.

10. $f(z) \sim \dfrac{\sqrt{2\pi}}{\sqrt{z}} \left\{ \dfrac{1}{z} - \dfrac{1 \cdot 3 \cdot 5}{3!} \dfrac{1}{z^3} + \dfrac{1 \cdot 3 \cdot 5 \cdot 7 \cdot 9}{5!} \dfrac{1}{z^5} - \cdots \right\}$

11. The path of steepest descent is the real axis.

12. $y = \log \left[\dfrac{1 - \sin x}{\cos x} \right]$

14. $f(z) \sim e^z \sqrt{\dfrac{2\pi}{z}}$

15. $f(z) \sim e^z (1 - i) \sqrt{\dfrac{\pi}{|z|}}$

7.3 Stirling's Formula and Bessel Functions

2. Differentiate any convenient formula for $J_n(z)$ twice and try it in the equation. For example:

$$J_n(z) = \frac{1}{2\pi i}\left(\frac{z}{2}\right)^n \int_\gamma t^{-n-1} e^{[t-(z^2/4t)]}\, dt.$$

4. This is the special case of exercise 3 for $n = 0$.

7. By definition,

$$J_{1/2}(z) = \sum_{k=0}^{\infty} \frac{(-1)^k z^{(1/2)+2k}}{2^{(1/2)+2k}\Gamma\left(\frac{1}{2}+k+1\right)k!}$$

$$= \sqrt{\frac{2}{z}}\sum_{k=0}^{\infty} \frac{(-1)^k z^{2k+1}}{\sqrt{\pi}\,\Gamma(2k+2)} \quad \begin{array}{l}\text{(by Legendre's}\\ \text{Duplication Formula)}\end{array}$$

$$= \sqrt{\frac{2}{\pi z}}\, \sin z.$$

Review Exercises for Chapter 7

3. $f(z) \sim 0$

8. $\Gamma(1/3)/3$

9. $\sqrt{\dfrac{2\pi}{z}}\left\{1 - \dfrac{1}{2\cdot 4}\cdot\dfrac{1}{z^2} + \dfrac{1}{2\cdot 4\cdot 6\cdot 8}\cdot\dfrac{1}{z^4} - \cdots\right\}$

10. Apply Theorem 6 after writing

$$J_n(iy) = \frac{1}{2\pi}\left\{\int_0^{2\pi} e^{-y\sin\theta}(\cos n\theta + 1)\,d\theta - \int_0^{2\pi} e^{-y\sin\theta}\,d\theta\right\}$$

$$-\frac{i}{2\pi}\left\{\int_0^{2\pi} e^{-y\sin\theta}(\sin n\theta + 1)\,d\theta - \int_0^{2\pi} e^{-y\sin\theta}\,d\theta\right\}$$

and using

$$h(\theta) = -\sin\theta, \qquad \theta_0 = 3\pi/2, \qquad \text{and} \qquad \gamma = [0,2\pi].$$

11. h. From part (b),

$$P_n(x) = \frac{1}{2^n n!}\cdot(2n)(2n-1)\,\cdots\,(n+1)\cdot x^n$$

$$+ \text{(lower-order terms)}$$

$$= \frac{(2n)!}{2^n(n!)^2}x^n + \text{(lower-order terms)}.$$

8.1   Basic Properties of Laplace Transforms

6. a. $\tilde{f}(z) = \dfrac{2}{z^3} + \dfrac{2}{z}$, $\sigma(f) = 0$.

b. $\tilde{f}(z) = \dfrac{1}{z^2} + \dfrac{1}{z+1} + \dfrac{1}{z^2+1}$, $\sigma(f) = 0$.

c. $\tilde{f}(z) = \dfrac{1}{z} + n \cdot \dfrac{1}{z^2} + n(n-1)\dfrac{1}{z^3} + \cdots + n!\dfrac{1}{z^{n+1}}$, $\sigma(f) = 0$.

10. $\tilde{g}(z) = \dfrac{1}{z} \cdot \dfrac{1}{(z+1)^2 + 1}$

$$\tilde{f}(z) = -\dfrac{3z^2 + 4z + 2}{(z^3 + 2z^2 + 2z)^2}$$

14. $\dfrac{2az}{(z^2 + a^2)^2}$

8.2   The Complex Inversion Formula

3. a. $\cos t$ b. $f(t) = te^{-t}$ c. $f(t) = [e^t + 2e^{-t/2}\cos\sqrt{3}t/2)]/3$

4. Theorem 8 cannot be applied since there are no constants M and R for which $|e^{-z}z| < M/|z|$ whenever $|z| > R$.

6. a. $f(t) = 2e^{-2t} - e^{-t}$
b. $\sinh z$ has no inverse Laplace transform.
c. $f(t) = t^2 e^{-t}/2$

7. $g(t) = \displaystyle\int_0^t \sin(t-s)\, f(s)\, ds$

10. $f(t) = (6te^{-3t} - e^{-3t} + 1)/9$

8.3   Application of Laplace Transforms to Ordinary Differential Equations

2. $y(t) = (5e^{2t} + 3e^{-2t})/4$

4. $y(t) = \begin{cases} 0, & 0 \le t < 1 \\ \dfrac{1}{9}(1 - \cos(3t - 3)), & t \ge 1 \end{cases}$

6. $y(t) = g_1(t) + g_2(t) + g_3(t)$ where

$$g_1(t) = \dfrac{e^{-t/2}}{\sqrt{3}}\left(3\cos\dfrac{\sqrt{3}}{2}t - \sin\dfrac{\sqrt{3}}{2}t\right)$$

$$g_2(t) = \begin{cases} 0, \ 0 \le t < 2 \\ -\dfrac{e^{-(t-2)/2}}{\sqrt{3}} 2 \sin \dfrac{\sqrt{3}}{2}(t-2), \ t \ge 2 \end{cases}$$

and

$$g_3(t) = \begin{cases} 0, \ 0 \le t < 1; \\ \dfrac{e^{-(t-1)/2}}{\sqrt{3}} 2 \sin \dfrac{\sqrt{3}}{2}(t-1), \ t \ge 1 \end{cases}$$

8.   a. $y_1(t) = (e^t + e^{-t})/2 = \cosh t$
   $y_2(t) = -(e^t - e^{-t})/2 = -\sinh t$
   b. $y_1(t) = -3t, \ y_2(t) = 3[(1+t)^2 - 1]/2$

Review Exercises for Chapter 8

3.   c. $\tilde{f}(z) = \dfrac{e^{-z}}{z^2+1}, \ \sigma(f) = 0.$

   d. $\tilde{f}(z) = \dfrac{1 - ze^{-z}}{z^2}$

   e. $\tilde{f}(z) = \log\left(\dfrac{z}{z-1}\right), \ \sigma = 1.$

4.   b. $f(t) = \begin{cases} 0, \ t \le 1 \\ \sin(t-1), \ t > 1 \end{cases}$

   c. $f(t) = \begin{cases} -te^{-t} + e^{-t}, \ t < 1 \\ -te^{-t} + e^{-t} + 1, \ t > 1 \end{cases}$

6.   a. $y(t) = (-15 + 23 \cos 2\sqrt{2}t)/8$
   b. $y(t) = 3(1 - e^{-t})$

   c. $y(t) = \begin{cases} 0, \ 0 \le t < 1 \\ 1 - \cos(t-1), \ t \ge 1 \end{cases}$

   d. $y(t) = 2te^{-t} + e^{-t}$

8.   If we substitute the expression for φ on page 419 into the boundary conditions that follow, we get four simultaneous equations involving $A, B, R,$ and T. After some algebraic manipulations, we get equation (10), as well as

$$R = \dfrac{T}{2ic_1c_2}(c_2^2 - c_1^2)e^{ia\omega/c_1}\sin(a\omega/c_2).$$

Together, these give $|R|^2 + |T|^2 = 1$.

13. $F(z) = \log\left(\dfrac{1-z}{z}\right)$

14. Exact evaluation gives

$$F(z) = \frac{2}{iz}\log\left(\frac{iz+1}{iz-1}\right).$$

In the integral form, $F(z)$ is clearly real when z is real. Convince yourself that the exact form for $F(z)$ is also real when z is.

16. c. Each term of the series has a logarithmic-type singularity at $j = \alpha$. But

$$F(j) = \frac{1}{1 - g\log(j-\alpha)}.$$

Thus, in addition to the logarithmic singularity at $j = \alpha$, $F(j)$ also has a simple pole singularity when the denominator vanishes, at $j = \alpha + e^{1/g}$.

Index